EXCELLENCE IN ELECTRICAL
ADDISON-WESLEY ▲ THE SIGN
AND COMPUTER ENGINEERING

Design and Analysis of Fault-Tolerant Digital Systems

Barry W. Johnson
University of Virginia, Charlottesville

Addison-Wesley Publishing Company

Reading, Massachusetts • Menlo Park, California • New York
Don Mills, Ontario • Wokingham, England • Amsterdam • Bonn
Sydney • Singapore • Tokyo • Madrid • San Juan

This book is in the ADDISON-WESLEY SERIES IN ELECTRICAL and COMPUTER ENGINEERING

Consulting Editor: Harold S. Stone

LIBRARY OF CONGRESS
Library of Congress Cataloging-in-Publication Data

Johnson, Barry W., 1957–
    Design and analysis of fault-tolerant digital systems/Barry W. Johnson.
    p. cm.—(Addison-Wesley series in electrical and computer engineering)
    Includes bibliographies and index.
    ISBN 0-201-07570-9
    1. Fault-tolerant computing. 2. System design. 3. System analysis.
I. Title. II. Series
QA76.9.F38J64 1989
621.39'2—dc19   88-3955
                                                                                                                                 CIP

Reprinted with corrections June, 1989

Copyright © 1989 by Addison-Wesley Publishing Company, Inc.

All rights reserved. No part of this book may be reproduced, stored in a retrieval system, or transmitted, in any form or by any means, electronic, mechanical, photocopying, recording, or otherwise, without the prior written permission of the publisher. Printed in the United States of America. Published simultaneously in Canada.

BCDEFGHIJ-HA-89

```
QA       Johnson, Barry W.
76.9
.F38  Design and analysis of
J64      fault-tolerant digital
1989     systems
621.392 J63d
```

Foreword *by Harold S. Stone*

On October 19, 1987, the world's financial markets suffered an upheaval of a magnitude reached only once before in this century. On that day, behind the scenes at the New York Stock Exchange, a fault-tolerant computer system processed transactions at a rate deemed unimaginable by the governors of the exchange. Six hundred million shares changed hands that day, and another six hundred million the following day. The system capacity was rated at only three hundred million transactions per day—a rate believed to be unreachable for at least the next five years.

While it is somewhat amazing that the computer system successfully processed the transactions, what is more amazing is that the system performed beyond its limits in spite of sporadic failures that occurred during the day. The worst outages were limited to processing halts of a few periods of a few minutes each. During the deluge of orders, as successive waves passed through the system, orders started backing up awaiting service. Although the large backlog resulted in unusually long processing delays, transaction throughput was virtuously continuous at the maximum sustainable output rate, thus enabling the exchange to remain open at a time when market liquidity was essential to maintain public confidence in the financial system.

Why did the computer system function as well as it did? It was no accident. That computer-system design is fault tolerant. Individual disks and processors in the system can fail, but the system has redundant processors and disks that automatically take the place of units that fail. The failures that day came at a rate much higher than normal. On the whole, the failures were not actually failures of hardware components, but were system failures of one type or another caused by the excessive number of transactions being processed that day. When a disk filled, for example, requests to store additional information on that disk were rejected. This is essentially the same response that is returned when a true hardware failure prevents a successful store operation from taking place. Thus, as various portions of the transaction system reached capacity, they rejected subsequent requests for service, and the rejection responses had much the same effect as hard failures. The apparent failure rate at the height of the transaction crunch was much higher than on a normal trading day. The massive processing load caused failures to occur precisely when the need for system availability was greatest.

Because of the inherent protection from failures in the design of the exchange's computer system, every failure point was backed up by another point. When a disk filled, transactions were routed to an alternate disk. When communications buffers filled, messages were directed to an alternate destination along an alternate route. This behavior cannot continue indefinitely, of course, because as a larger and larger fraction of system modules becomes inoperative, eventually some system module becomes the sole module of its type, and the system must shut down when this module fails. In a fault-tolerant computer system, failures force the system to reconfigure its activities. Reconfiguration pushes the system into a different operational state, possibly causing a monetary halt for some modules as well as some redoing of work that was lost because of the failure. The strategy for fault-tolerant operation of a system causes the operational state to be a stable state, in that when the system falls out of the operational state, it returns to an equivalent operational state almost immediately, unless failures have been sufficiently prevalent to prevent a return to operation.

The irony of events on October 19 is that two different systems were entwined that day—one the fault-tolerant computer system and the other the financial market system. The fault-tolerant computer system was designed for stability in the face of failure. The financial market demonstrated that it had an inherent instability. A slide in prices caused responses that day that further depressed prices and tended to exacerbate the fall rather than to correct it. On a normal trading day in an ideal market, a fall in price makes a stock more attractive to purchasers, who step in to buy the stock and thus create pressure for an upward move in price. This causes the stock price to stabilize at some level without experiencing wild gyrations in price. October 19 was not normal, however, so that downward price pressures were not compensated by upward price pressures, but instead produced responses that caused additional downward pressure. The inevitable result was a massive drop in the value of publicly held stock. So, the stable system, which was the computer system, continued to function through the day, even while the transactions reflected the inherent instability of the market system. Had the computer system been designed with instabilities equivalent to those of the market place, it would have ceased to operate early in the day. We can only speculate about the panic that might have been triggered if nervous investors had suddenly and unexpectedly been unable to liquidate securities.

From the experience of October 19, we realize that computers have become an indispensable part of everyday life. In critical application areas, such as transaction processing for a major financial market, of which the stock exchange is only one example, a computer outage during normal operation can wreak havoc. A mild result of such an outage may be massive inconvenience, but the more serious results may include loss of life and habitat. At risk on October 19 was the economic vitality of the United States and free world.

How do we go about building fault-tolerant computers that can run without stopping? Barry Johnson sets forth the basic principles of fault-tolerant computer design in this text. Many techniques described here were used in the design of the computers that ran without stopping on October 19.

The development of the fault-tolerant techniques has paralleled the development of computers, but the wide application of the techniques lagged the development of computers by about twenty years. Von Neumann contributed to the literature on how to build reliable computers from unreliable components while he was engaged in the design and construction of the first stored-program computer. The actual practice of building fault-tolerant computers was not extensive in the early days of computing. The high cost of hardware precluded the use of extra hardware for reliability except possibly in the most critical applications. Since computers were very new, they had not found their way into many applications where reliability was a critical issue. Nevertheless, rudimentary protection from failures, in the form of parity checks, was used in the first generation of commercial systems and has been evolving ever since.

In the 1960s, the space program provided a whole new set of applications that both demanded computer implementation and forced the computers to operate in a hostile environment where maintenance is impossible. In short, fault-tolerant computers are essential in space. Reliability techniques advanced rapidly during this period, and many successful spacecraft voyages through the solar system had on-board fault-tolerant computers controlling them. Manned space flights stimulated fault-tolerant computer designs of high sophistication because the lives of the astronauts depend on the nonstop functioning of the computer system. The computer systems developed for the manned flights were highly successful and demonstrated that computer systems can be built to function over extended periods of time without maintenance even in the presence of failures.

As a classic example of such behavior, during the first lunar landing an anxious world listened intently to Neil Armstrong's reports of computer alarms, unable to comprehend their significance. Mounting tension was followed by relief and joy when Armstrong's famous words "Tranquility base here" signaled a safe landing. The computers unexpectedly experienced a computational overload during the descent to Tranquility, but they continued to run without failure. The overload that day was quite analogous to the overload on October 19. Many of the principles put into practice in the Apollo flight found their way onto the floor of the New York Stock Exchange by 1987.

VLSI gave fault-tolerant techniques a tremendous boost. It greatly reduced the cost of redundancy and made replication of components a practical reality for all applications, not just the exotic ones. Meanwhile, with the drop in cost of computing power, computer technology became pervasive, finding applications in virtually all areas of science, business, and the arts. By the onset of the 1980s, computers had become critical components in the infrastructure of society. The stock exchange is only one of thousands of systems whose daily functioning is essential for maintaining the order of life.

One lesson of October 19 is that we cannot take the reliability of computers for granted. We must be able to design computer systems that function correctly in spite of failures. As every year brings greater reliance on computer systems, the need for fault tolerance becomes all the more important, while continual advances in device technology bring down the cost of achieving fault tolerance. This is the setting for Johnson's text.

Barry Johnson's treatment is quite thorough, carrying the reader through design techniques that describe what is possible to do, to evaluation techniques that enable the designer to determine how well the reliability goals have been achieved. Several case studies illustrate the principles as they have been put into practice. While these case studies make fascinating reading, the reader must understand that they represent the past and the present. The future will use the principles far more extensively than indicated in the case studies as new technology makes new practices feasible. But the underlying principles described in the text will remain unchanged.

To convey a better understanding of the impact of VLSI, Prof. Johnson concludes the text with material on VLSI design and testing. This technology is the enabling technology for fault tolerance, and the reader must be familiar with it to achieve the most innovative and effective fault-tolerant designs.

Today's reader who masters the material will not be the designer of a computer that guides the first lunar lander, nor of the machine that kept the transactions flowing during the crash of 1987. Today's reader, however, will be able to design machines that reach into every conceivable application, and if successful, the reader might well make a major contribution to the quality of life.

Harold S. Stone
Chappaqua, NY

Preface

Outline

The purpose of this book is to provide the reader an *introduction* to the design and analysis of fault-tolerant systems. The word *introduction* is key because it is assumed that the reader has not been exposed previously to the terminology and techniques used in the fault-tolerant computing field. The book is intended for a senior-level undergraduate course, a first-year graduate course, or as a self-study guide for those interested in learning the many facets of fault-tolerant system design. The book is suitable for a one semester course, and it assumes that the reader is familiar with combinational and sequential digital circuit design.

The key features of the book are as follows:

1. The book is introductory and can be of immediate use to individuals with no exposure to fault-tolerant computing. Consequently, the reader needs only an interest in fault tolerance and a background in basic digital circuit design to understand the material presented.
2. Each chapter is supplemented with problems that can be used as an aid to learning the material. The author has used each problem as a homework assignment or a test question in a class taught at the University of Virginia. The problems at the end of Chapter 5 are small projects that allow students to design and analyze a fault-tolerant system.
3. Each chapter concludes with a summary of the most important concepts and terms presented in the chapter. The reader can refer to the summary for a quick refresher on terminology or as a guide to the important items within the chapter.
4. Each chapter is augmented with suggested additional reading that can be useful, particularly in a graduate course, for more detailed study of

each topic. A typical graduate course will require students to read and possibly present selected articles from the additional reading sections at the end of each chapter.
5. Preliminary versions of the book have been used for the past three years in an introductory course on fault-tolerant computing taught at the University of Virginia. The use of the material in the classroom environment has stimulated better ways of presenting each topic, as well as uncovered and corrected errors in the presentation.
6. The book devotes considerable time to the design process. The various phases of a typical design process are outlined and illustrated with an example. Once again, the projects at the end of Chapter 5 provide the reader an opportunity to practice the actual design of fault-tolerant systems.
7. The book includes descriptions of 13 sample fault-tolerant systems that have been designed previously. The important problems and aspects of each design are summarized so that the reader has a feel for what others have done in the past and the historical developments in the fault-tolerant computing field.

The chapters of the book cover definitions and basic terminology (Chapters 1 and 2), fault tolerance techniques and concepts (Chapter 3), analysis procedures (Chapter 4), system design and illustrative fault-tolerant systems (Chapter 5), fault tolerance techniques in a VLSI design environment (Chapter 6), and test techniques and design for testability (Chapter 7). The material is intended for presentation in the classroom in the same order presented in the book.

Chapters 1 and 2 set the stage for the complete book by presenting historical background information and terminology that is used throughout each of the remaining chapters. The first two chapters are extremely important because they provide the background necessary to prepare the reader for studying the remainder of the book and the literature available in the fault-tolerant computing field. Definitions of such key terms as fault, error, failure, reliability, dependability, safety, and availability are presented and illustrated in the first two chapters. In addition, the various types of fault models available are described in Chapter 2. The author has attempted to use the commonly accepted definitions used by researchers and practitioners in the fault-tolerant computing field.

Chapter 3 presents a collection of techniques that can be used to design systems that are either fault tolerant or that possess the ability to detect their own faults. Various redundancy techniques including hardware, information, time, and software redundancy are presented and illustrated with simple examples. Techniques covered include, for example, majority voting, standby sparing, duplication with comparison, software-implemented self-test techniques, and error detecting and correcting codes. The intent of

Chapter 3 is to introduce the techniques; subsequent chapters provide extensive details on the use and evaluation of the various approaches.

Chapter 4 introduces various techniques for analyzing both redundant and nonredundant systems. The intent of Chapter 4 is to provide the reader with the approaches necessary to compare one or more design techniques using both quantitative and qualitative comparison methods. Reliability, availability, safety, and maintainability modeling is presented and illustrated with examples. Modeling approaches using both combinatorial and Markov models are presented. Perhaps one of the most interesting features of Chapter 4 is the inclusion of a detailed analysis example that demonstrates the process of comparing one or more design candidates for the purpose of selecting a preferred approach. The analysis example demonstrates the importance of safety in many applications in addition to the traditional metrics of reliability and availability.

Upon completion of the first four chapters, the reader should have the tools necessary to design fault-tolerant systems. The intent of Chapter 5 is to present and illustrate the various aspects of a typical design process. One unique feature of Chapter 5 is a discussion of several basic fault avoidance techniques that can be used in conjunction with fault tolerance techniques to achieve the desired attributes of a system. A second key feature of Chapter 5 is the emphasis on the importance of system analysis as an integral part of the design process. The process typically undertaken in the design of a fault-tolerant system is illustrated in Chapter 5 with the high-level design of an example flight control system for an aircraft. Chapter 5 also includes discussions of 13 fault-tolerant systems to illustrate the approaches taken by those that have actually developed fault-tolerant designs.

An important part of Chapter 5 is the projects included at the end of the chapter. Each project has been used by the author as a class project normally conducted over the last two or three weeks of a semester. The projects provide a mechanism for students to practice the process of designing a fault-tolerant system. The projects are appropriate for students to perform in small groups of two or three and can be completed to varying levels of detail. For example, in short projects students can perform only high-level designs, whereas for longer, more-detailed projects the students can complete the designs to the gate- or circuit-levels.

Chapter 6 focuses on the impact that VLSI technology has had on the design of fault-tolerant systems. Many techniques that were previously impractical are now feasible because of the decreased power consumption and increased chip density available through VLSI technology. Chapter 6 discusses the opportunities presented by VLSI, the problems presented, and techniques that can be employed on VLSI chips. Examples include complementary logic, totally self-checking logic, redundancy in array structures, and the use of redundancy to improve yield.

Finally, Chapter 7 presents test pattern generation and design for testability techniques. Testing is a crucial component of courses on fault-tolerant systems because of the impact that redundancy can have on testing. In addition, testing and design for testability must be considered throughout the design process to guarantee that a resulting design can be thoroughly tested. Chapter 7 covers such techniques as the D-algorithm test pattern generation technique, scan design as a means of design for testability, and testability analyzers.

It is intended that the individual completing this book will be thoroughly prepared to pursue more advanced studies in fault-tolerant computing, research in the field, or to practice the design of fault-tolerant systems. Also, it is intended that the book will be a valuable reference for people working in the field.

Acknowledgments

The successful completion of a project such as this book requires the coordinated efforts of many individuals. I would like to thank Paul M. Julich of Harris Corporaton who first provided me the opportunity to work on the design and analysis of fault-tolerant systems. My work with Paul formed the basis of much of the material contained within this book. I would also like to thank John Hadjilogiou of the Florida Institute of Technology (FIT) who allowed me to develop and teach a course on fault-tolerant computing within the Department of Electrical and Computer Engineering at FIT. Many of the methods of presenting the material were developed while working with John.

Edward A. Parrish, Jr., of Vanderbilt University and Robert J. Mattauch of the University of Virginia deserve special recognition for providing an excellent environment in which to teach, perform research, and write. Ed Parrish, who was my department chairman when the task of writing this book was first initiated, provided encouragement, support, and excellent facilities. Bob Mattauch, as my present department chairman, has been extremely supportive of this effort and has made the task of writing and working at the University of Virginia a pleasure.

I would like to give special thanks to James H. Aylor of the University of Virginia for his support. Jim has assisted me in numerous technical issues that have solidified the concepts presented in this book. In addition, Jim first stimulated my interest in fault-tolerant computing while I was a doctoral student studying under his direction. Finally, I would like to thank Jim for being a friend and supportive colleague during the difficult times associated with writing this book.

Thanks are also due to the many students that have used preliminary versions of this book during the past three years. Approximately 75 students have used the book in various stages and have provided numerous comments that I feel have extensively improved it. The true test of a book intended to be used in teaching is whether or not the students clearly understand the material. I feel strongly that the many constructive comments provided by my students have significantly improved this book, and for this I am very thankful.

I would also like to thank the many reviewers, especially Lee Higbie and Dhiraj K. Pradhan, that have provided excellent ideas and comments. Particular thanks is due to Harold S. Stone of IBM whose careful reading of the manuscript led to improvements in both the wording and the technical aspects of the text. In addition, Harold has provided numerous words of encouragement that have made the writing much easier.

Tom Robbins and his associates at Addison-Wesley have been excellent and deserve significant praise. I appreciate their patience and understanding when I missed my deadlines. I am also very thankful for the friendly attitudes shown toward me by everyone that I have contacted at Addison-Wesley. I cannot imagine a better publisher with which to work on a book.

I would especially like to thank my parents, Raymond and Clara Johnson for their unending support and encouragement. Without them I might never have reached a point where writing a book of this type was possible. I would also like to thank my wife's parents, Oadie and Ruth Rowland, for their continued support throughout the years that I have known them.

Finally and most importantly, I want to thank my wife, Susan, and my daughter, Ashby, who have stood by me throughout this endeavor. Without their patience, concern, and efforts, this project would not have been possible. At times during the past two years, they have carried not only their own responsibilities but many of mine as well. It is to Susan and Ashby that this book is dedicated; I hope that in some way it can repay them for all they have done for me.

Barry W. Johnson
Charlottesville, Virginia

Contents

1 Introduction   1

   1.1 Overview   1
   1.2 Origins of Fault-Tolerant Computing   3
   1.3 Goals of Fault Tolerance   4
       1.3.1 Reliability   4
       1.3.2 Availability   5
       1.3.3 Safety   6
       1.3.4 Performability   6
       1.3.5 Maintainability   7
       1.3.6 Testability   8
       1.3.7 Dependability   8
   1.4 Applications of Fault-Tolerant Computing   8
       1.4.1 Long-Life Applications   9
       1.4.2 Critical-Computation Applications   10
       1.4.3 Maintenance Postponement Applications   11
       1.4.4 High Availability Applications   13
   1.5 Fault Tolerance as a Design Objective   15
      Summary   16
      References   18
      Additional Reading   20

2 Fundamental Definitions   23

   2.1 Introduction   23
   2.2 Faults, Errors, and Failures   24
   2.3 Causes of Faults   28
   2.4 Characteristics of Faults   30

2.5	Fault Models		31
	2.5.1	The Logical Stuck-Fault Model	32
	2.5.2	Transistor Stuck-Fault Models	37
2.6	Error Models		37
2.7	Design Philosophies to Combat Faults		38
	Summary		40
	References		42
	Additional Reading		43
	Problems		45

3 Design Techniques to Achieve Fault Tolerance — 47

3.1	Introduction		47
3.2	Primary Design Issues		48
3.3	The Concept of Redundancy		48
3.4	Hardware Redundancy		51
	3.4.1	Passive Hardware Redundancy	51
		Triple Modular Redundancy	52
		N-Modular Redundancy	54
		Voting Techniques	54
	3.4.2	Active Hardware Redundancy	62
		Duplication with Comparison	63
		Standby Sparing	65
		Pair-and-a-Spare Technique	67
		Watchdog Timers	68
	3.4.3	Hybrid Hardware Redundancy	69
		N-Modular Redundancy with Spares	70
		Self-Purging Redundancy	71
		Sift-Out Modular Redundancy	75
		Triple-Duplex Architecture	78
	3.4.4	Summary of Hardware Redundancy	80
3.5	Information Redundancy		81
	3.5.1	Parity Codes	84
	3.5.2	m-of-n Codes	93
	2.5.3	Duplication Codes	95
	3.5.4	Checksums	98
	3.5.5	Cyclic Codes	102
	3.5.6	Arithmetic Codes	112
	3.5.7	Berger Codes	123
	3.5.8	Horizontal and Vertical Parity	125
	3.5.9	Hamming Error-Correcting Codes	127
	3.5.10	Error-Correcting Integrated Circuits	131

		3.5.11 Code Selection Issues	133
3.6	Time Redundancy		134
	3.6.1	Transient Fault Detection	135
	3.6.2	Permanent Fault Detection	136
	3.6.3	Recomputation for Error Correction	151
3.7	Software Redundancy		152
	3.7.1	Consistency Checks	153
	3.7.2	Capability Checks	154
	3.7.3	N-version Programming	154
	Summary		155
	References		159
	Additional Reading		160
	Problems		162

4 Evaluation Techniques 169

4.1	Introduction		169
4.2	Quantitative Evaluation Methods		170
	4.2.1	Failure Rate and the Reliability Function	170
	4.2.2	Failure Rate Calculation	175
	4.2.3	Mean Time to Failure	178
	4.2.4	Mean Time to Repair	180
	4.2.5	Mean Time Between Failure	180
	4.2.6	Fault Coverage	182
4.3	Reliability Modeling		185
	4.3.1	Combinatorial Models	185
		Series Systems	186
		Parallel Systems	189
	4.3.2	Fault Coverage and Its Impact on Reliability	193
	4.3.3	M-of-N Systems	197
	4.3.4	Markov Models	199
4.4	Safety Modeling		214
4.5	System Comparisons		216
4.6	Availability Models		219
4.7	Maintainability Models		223
4.8	Redundancy Ratios		226
4.9	Qualitative Methods		227
	4.9.1	Flexibility	228
	4.9.2	Technology Dependence	228
	4.9.3	Transparency to the User	228
	4.9.4	Testability	229
4.10	Tradeoff Analysis Example		229
	Summary		254

		References	256
		Additional Reading	257
		Problems	258

5 The Design of Practical Fault-Tolerant Systems — 263

- 5.1 Introduction — 263
- 5.2 The Design Process — 265
  - 5.2.1 Problem Definition — 266
  - 5.2.2 System Requirements — 267
  - 5.2.3 System Partitioning — 267
  - 5.2.4 Candidate Designs — 269
  - 5.2.5 High-Level Analysis — 270
  - 5.2.6 Hardware and Software Specifications — 271
  - 5.2.7 Hardware and Software Design and Analysis — 271
  - 5.2.8 Testing — 272
  - 5.2.9 System Integration and Test — 272
- 5.3 The Use of Fault Avoidance in the Design Process — 273
  - 5.3.1 Requirements Design Review — 274
  - 5.3.2 Conceptual Design Review — 275
  - 5.3.3 Specifications Design Review — 275
  - 5.3.4 Detailed Design Review — 276
  - 5.3.5 Final Review — 276
  - 5.3.6 Parts Selection — 276
  - 5.3.7 Design Rules — 277
  - 5.3.8 Documentation — 277
- 5.4 A Sample Design — 277
  - 5.4.1 Problem Definition and Initial Partitioning — 279
  - 5.4.2 Requirements Definition — 280
  - 5.4.3 System Partitioning — 282
  - 5.4.4 Candidate Designs — 284
  - 5.4.5 High-Level Analysis — 292
    - TTMR System Analysis — 292
    - TMR System Analysis — 295
    - TDTMR System Analysis — 297
    - 5MR System Analysis — 297
  - 5.4.6 Comparison of Approaches — 300
- 5.5 Sample Fault-Tolerant Systems — 304
  - 5.5.1 Long-life Applications — 304
    - Self-Testing and Repairing Computer — 305
    - Fault-Tolerant Spaceborne Computer — 310
    - Fault-Tolerant Building Block Computer — 315

		Space Shuttle	319
	5.5.2	Critical-Computation Applications	319
		Fault-Tolerant Multiprocessor	324
		Software Implemented Fault Tolerance	329
		August Systems CS-3001 Control Computer	332
		Multi-Microprocessor Flight Control System	333
		Agusta A129 Integrated Multiplex System	336
	5.5.3	High-Availability Applications	345
		The Tandem 16 NonStop System	346
		The Stratus/32 System	348
		Electronic Switching System	350
		The Synapse N+1 Architecture	353
Summary			355
References			356
Additional Reading			358
Projects			361

6 Fault-Tolerant Design of VLSI Circuits and Systems 375

6.1	Introduction		375
6.2	VLSI Technology		376
6.3	Failure Modes in VLSI Technology		378
	6.3.1	Metal Systems	379
	6.3.2	Diffusion	380
	6.3.3	Foreign Material	381
	6.3.4	Oxide	381
	6.3.5	Package and Bonding	382
	6.3.6	Mounting	382
	6.3.7	Misapplication	382
6.4	Distribution of Faults in VLSI Technology		383
6.5	Opportunities Presented by VLSI		385
6.6	Problems Presented by VLSI		387
	6.6.1	Common-mode Failures	387
	6.6.2	Increased Design Mistakes	389
	6.6.3	Increased Susceptibility to External Disturbances	390
6.7	Redundancy Techniques in a VLSI Design Environment		390
	6.7.1	Duplication with Complementary Logic	391
	6.7.2	Self-Checking Logic	394
	6.7.3	Totally Self-Checking Checkers for M-of-N Codes	402
	6.7.4	Reconfiguration Array Structures	404
		Fabrication-Time and Compile-Time Reconfiguration	410
		Real-Time Reconfiguration	421

Contents

6.7.5 Redundancy to Enhance Yield of VLSI Circuits	439
Summary	451
References	453
Additional Reading	454
Problems	457

7 Testing 463

7.1	Introduction	463
7.2	Fault-Testing	466
7.3	Test Pattern Generation	468
	7.3.1 Fault Tables	468
	7.3.2 Adaptive Experiments	477
	7.3.3 Boolean Differences	481
	7.3.4 Literal Propositions	488
	7.3.5 Path Sensitization	491
	7.3.6 The D-Algorithm	498
	7.3.7 Fault Simulation for Test Pattern Generation	510
	The Row Method	513
	The Column Method	513
7.4	Random Testing	516
7.5	Signature Analysis	517
7.6	Design for Testability	520
	7.6.1 Scanning as a Method of Design for Testability	524
	Level Sensitive Scan Design (LSSD)	525
	Scan Path	529
	Scan/Set Logic	532
	Random-Access Scan	535
	7.6.2 Sample Design Comparing LSSD and Scan Path	538
	7.6.3 Built-In Logic Block Observation	544
7.7	Testability Analysis	552
	7.7.1 Important Definitions	553
	7.7.2 Testability Analyzers	554
	SCOAP	555
	CAMELOT	559
	7.7.3 Comparison of Testability Analyzers	562
	7.7.4 Analysis of Circuits Containing Redundancy	564
	Summary	565
	References	568
	Additional Reading	570
	Problems	572

1

Introduction

1.1 Overview
1.2 Origins of Fault-Tolerant Computing
1.3 Goals of Fault Tolerance
1.4 Applications of Fault-Tolerant Computing
1.5 Fault Tolerance as a Design Objective
   Summary
   References
   Additional Reading

1.1 Overview

This textbook is devoted to the study of techniques for designing and analyzing fault-tolerant and easily-testable systems. A **fault-tolerant system** is one that can continue to correctly perform its specified tasks in the presence of hardware failures and software errors. For example, the effect of a software "bug" in a fault-tolerant system is overcome so that the system continues correct operation. Likewise, the failure of a hardware component in a fault-tolerant system does not inhibit that system's ability to correctly execute its design-specified functions. **Fault tolerance** is the attribute that enables a system to achieve fault-tolerant operation. Finally, the term **fault-tolerant computing** describes the process of performing calculations, such as those performed by a computer, in a fault-tolerant manner.

An **easily-testable system** is one whose ability to perform correctly can be verified in a simple and straightforward manner. The complexity of today's systems demands that special features be incorporated into the system to support testing. **Design for testability** is the process by which such features are included.

The concept of fault tolerance has become increasingly important during the past decade because of the increased use of computers in the vital aspects of almost everyone's life. Computers are no longer confined to use as powerful calculators where their incorrect performance can produce little more than frustration and lost time. Instead, computers are now integrated into commercial and military aircraft flight control systems ([Wensley et al. 1978], [Hopkins et al. 1978], and [Bosch and Kuehl 1977]), industrial controllers ([Ayache, Courtiat, and Diaz 1982] and [Wensley and Harclerode 1982]), space applications [Rennels 1978], and banking systems ([Katzman 1977], [Manual 1982], and [Herbert 1983]). In each application, erroneous computer performance can be devastating to financial records, environmental safety, national security, and even human life. In summary, fault tolerance has become more important simply because the functions of computers and other digital systems have become more crucial.

This book discusses the many aspects of designing and analyzing fault-tolerant systems. In addition, design for testability methods and techniques are presented. Specific topics covered include:

1. Fundamental terminology crucial to the understanding of fault tolerance and design for testability
2. Techniques for designing fault-tolerant systems
3. The use of fault tolerance to achieve design goals such as reliability
4. Measures of the quality of a fault-tolerant design
5. Practical examples of fault-tolerant systems
6. The impact of integrated circuit technology on the design of fault-tolerant systems
7. Techniques of design for testability

We begin our discussions in this chapter by examining some of the historical aspects of fault-tolerant computing. Also, we consider in more detail the design goals that might be satisfied through the use of fault tolerance, as well as the role of fault tolerance in the design process. Several key concepts presented in this chapter include:

1. Definitions of reliability, availability, maintainability, safety, dependability, performability, and testability
2. The distinction between fault tolerance and reliability
3. The applications in which fault tolerance is most frequently used
4. The role of fault tolerance in the design process

1.2 Origins of Fault-Tolerant Computing

Fault tolerance is certainly not a new field. The first digital computers made extensive use of error detection and fault tolerance techniques to overcome the low reliability of their basic components [Carter and Bouricius 1971]. Some of the early Bell Relay Computers (BRC), for example, had two central processing units [ERA 1950]; one unit would begin executing the next instruction when the other unit encountered an error. Later versions of the BRC used a retry mechanism to repeat an operation immediately after an error was detected. The IBM 650, UNIVAC, and the Whirlwind I computers [Weik 1955] incorporated parity to check the results of data transfers. The EDVAC computer [Carter and Bouricius 1971], designed in 1949, is generally considered to have been the first computer to completely duplicate the Arithmetic Logic Unit (ALU) and compare the results obtained by each unit; the processing continued as long as the two ALUs agreed.

The advent of the transistor, along with its increased reliability, led to a temporary decrease in the emphasis on fault-tolerant computing. For many designers, the major thrust was to increase computer performance and speed and to depend on the improved reliability of the transistor to guarantee correct computations. It was not until computers began performing much more critical tasks that fault tolerance again surfaced as a crucial issue. Perhaps the best examples are in the United States space program and in military applications. The increase in computational requirements in many of these applications mandated the use of digital computers, and the significant penalties for incorrect performance required that the computers perform their functions without error. As an example, the IBM Saturn V system [Kuehn 1969] used triplicated modules and parity checking to improve the fault tolerance capability of the system.

The first theoretical work in fault-tolerant computing is generally credited to John von Neumann [von Neumann 1956]. In 1952, von Neumann presented a series of lectures on the use of replicated logic modules to improve system reliability. von Neumann later developed an article entitled "Probabilistic Logics and the Synthesis of Reliable Organisms from Unreliable Components" [von Neumann 1956] in which he presented the concept of majority voting and analyzed the impact that such arrangements could have on the probability of a system producing erroneous results.

Since about 1970, the field of fault-tolerant computing has been rapidly developing. Several excellent journals such as *Computer*, *IEEE Micro*, the *Proceedings of the IEEE*, the *Journal of Design Automation and Fault-Tolerant Computing*, and the *IEEE Transactions on Computers* regularly present special issues that deal solely with fault-tolerant computing. In addition, the International Symposium on Fault-Tolerant Computing, commonly called the Fault-Tolerant Computing Symposium (FTCS), has been held each year, beginning as the Symposium of Fault-Tolerant Computing in 1971.

Despite the apparent progress and the prolific dissemination of information, the field of fault-tolerant computing is still relatively immature, particularly when it comes to the application of the technology. Much of this should change, however, partly because of the advent of Very Large Scale Integration (VLSI). VLSI technology has made the implementation of many fault tolerance techniques not only feasible, but in many cases, extremely practical and cost effective. VLSI, however, introduces new problems in the design of fault-tolerant systems that previously did not have to be addressed. Consequently, the advantages of VLSI do not come to us free of charge. Subsequent chapters of this book address the relationship between fault tolerance and VLSI in significant detail.

1.3 Goals of Fault Tolerance

It is natural to ask at this point why fault tolerance is so important and why it is the concern of so many designers. Fault tolerance is an attribute that is designed into a system to achieve some design goal. Just as a design must meet many functional and performance goals, it must satisfy numerous other requirements as well. The most prominent of the additional requirements are reliability, availability, safety, performability, dependability, maintainability, and testability. Fault tolerance is one system attribute capable of fulfilling such requirements. This chapter provides an overview of each requirement; Chapter 4 describes techniques that allow the evaluation of each attribute.

1.3.1 Reliability

The **reliability** $R(t)$ of a system is a function of time, defined as the conditional probability that the system will perform correctly throughout the interval $[t_0, t]$, given that the system was performing correctly at time t_0. In other words, the reliability is the probability that the system will operate correctly throughout a complete interval of time. The reliability is a conditional probability in that it depends on the system being operational at the beginning of the chosen time interval. The **unreliability** $Q(t)$ of a system is a function of time, defined as the conditional probability that a system will perform *incorrectly* during the interval $[t_0, t]$, given that the system was performing *correctly* at time t_0. The unreliability is often referred to as the *probability of failure*.

Reliability is most often used to characterize systems in which even momentary periods of incorrect performance are unacceptable, or in which it is impossible to repair the system. If repair is impossible, such as in many space applications, the time intervals being considered can be extremely long, perhaps as many as ten years. In other applications, such as aircraft flight control, the time intervals of concern can be no more than several hours, but the probability of working correctly throughout that interval can be 0.9999999 or higher. It is a common convention when reporting reliability numbers to use 0.9_i to represent the fraction that has i nines to the right of the decimal point. For example, 0.9999999 is written as 0.9_7.

It is important to understand the difference between fault tolerance and reliability. Fault tolerance is a technique that can improve reliability, but a fault-tolerant system does not necessarily have a high reliability. A system can be designed to tolerate any single hardware failure or software error that can occur, but the probability of such problems existing can be so high that the reliability is very low. Likewise, a highly-reliable system is not necessarily fault tolerant. A very simple system might be designed using extremely good components such that the probability of the hardware failing is very low, but if the hardware does fail, the system cannot continue its functions. In other words, the system can achieve a high reliability but not possess the attribute of fault tolerance.

In summary, fault tolerance can improve a system's reliability by keeping the system operational when hardware failures and software errors occur. For example, a computing system that has redundant processors can often be designed to continue the correct performance of its tasks, even when one or more of the processors becomes inoperable.

1.3.2 Availability

Availability is another design goal that we can achieve through the use of fault tolerance. **Availability** $A(t)$ is a function of time, defined as the probability that a system is operating correctly and is available to perform its functions at the instant of time t. Availability differs from reliability in that reliability depends on an *interval* of time, whereas availability is taken at an *instant* of time. A system can be highly available yet experience frequent periods of inoperability as long as the length of each period is extremely short. In other words, the availability of a system depends not only on how frequently it becomes inoperable but also on how quickly it can be repaired. The most common measure of availability is the expected fraction of time that a system is available to correctly perform its functions.

Availability is most often used as a design goal when the system's primary purpose is to provide its services as often as possible. Examples of high-availability applications include time-shared computing systems and certain transactions processing applications, such as airline reservation systems. The users of highly-available systems want those systems to possess a high probability of performing correctly at the instant they are requested to do so.

Fault tolerance offers numerous ways in which to improve the availability of a system. For example, the use of spare processors in a computing system can allow the functions of the system to be performed by a spare processor in the event that the primary processor becomes inoperable. In other words, the spare processor can perform the functions of the system while the primary processor is being repaired, thus keeping the system available for use.

1.3.3 Safety

One attribute that is often overlooked is the safety of a system. **Safety** $S(t)$ is the probability that a system will either perform its functions correctly or will discontinue its functions in a manner that does not disrupt the operation of other systems or compromise the safety of any people associated with the system. Safety is a measure of the *fail-safe* capability of a system; if the system does not operate correctly, you at least want the system to fail in a safe manner. For example, a pilot can safely fly an airplane, even if the autopilot fails, as long as the failure does not inhibit the aircraft's normal flight modes. Likewise, if a control valve for a chemical process fails, you often prefer that the valve fail in the closed position. Safety is the probability that these safe actions will result.

Safety and reliability differ because reliability is the probability that a system will perform its functions correctly, whereas safety is the probability that a system will either perform its functions correctly or will discontinue the functions in a manner that causes no harm. Certain techniques can be used to improve safety by turning a system off if a failure of some sort is detected. For example, in a nuclear power plant the reaction process should be stopped if some discrepancy is detected; this is the safe course of action.

1.3.4 Performability

In many cases, it is possible to design systems that can continue to perform correctly after the occurrence of hardware failures and software errors, but the *level* of performance is somehow diminished. For example, in today's era of multiprocessors, the failure of a single processor might not render the complete machine inoperable, but instead might simply decrease the speed of operation or decrease the amount of memory available to any one user. The multiprocessor, in this example, can still perform its tasks, but the relative quality of the performance has been decreased.

The **performability** $P(L, t)$ of a system is a function of time, defined as the probability that the system performance will be at, or above, some level

L at the instant of time t [Fortes and Raghavendra 1984]. If we relate performability to the multiprocessor example, the level of performance might simply be the number of processors available for computational use. Performability differs from reliability in that reliability is a measure of the likelihood that *all* of the functions are performed correctly, whereas performability is a measure of the likelihood that some subset of the functions is performed correctly.

Graceful degradation is an important feature that is closely related to performability. **Graceful degradation** is the ability of a system to automatically decrease its level of performance to compensate for hardware failures and software errors. For example, if an airplane's autopilot begins to perform incorrectly, the graceful degradation might consist of simply disabling the autopilot. The level of performance would be that coinciding with the loss of the autopilot, and the performability would be the probability of being at that level of performance at time t. Fault tolerance can provide graceful degradation and improve performability by eliminating failed hardware and software from a system, thereby allowing performance at some reduced level.

1.3.5 Maintainability

Almost every design has maintainability as a goal. **Maintainability** is a measure of the ease with which a system can be repaired, once it has failed. In more quantitative terms, maintainability $M(t)$ is the probability that a failed system will be restored to an operational state within a specified period of time t. The restoration process includes locating the problem, physically repairing the system, and bringing the system back to its operational condition. Maintainability is crucial in all systems, but it is particularly important when human lives, equipment, or the environment are placed in jeopardy while a system is repaired.

Many of the techniques that are so vital to the achievement of fault tolerance can be used to detect and locate problems in a system for the purpose of maintenance. Once the problem is located, maintenance personnel can then perform the necessary repairs. Automatic diagnostics can significantly improve the maintainability of a system because a majority of the time used to repair a system is often devoted to determining the source of the problem.

1.3.6 Testability

A **test** is a means by which the existence and quality of certain attributes within a system are determined. For example, if a computer is supposed to execute one million instructions per second, you would probably want to design a test to verify that the computer could indeed run at that particular rate. **Testability** is the *ability* to test for certain attributes within a system. Measures of testability allow us to assess the ease with which certain tests can be performed. As we will discover in subsequent chapters, certain tests can be automated and provided as an integral part of the system to improve the testability. Many of the techniques that are so vital to achieving fault tolerance can be used to detect and locate problems in a system for the purpose of improving testability. Testability is clearly related to maintainability because it is important to minimize the time required to identify and locate specific problems.

1.3.7 Dependability

The term **dependability** encompasses the concepts of reliability, availability, safety, maintainability, performability, and testability. Dependability is the quality of service that a particular system provides [Laprie 1985]. Reliability, availability, safety, maintainability, performability, and testability are measures used to quantify the dependability of a system.

1.4 Applications of Fault-Tolerant Computing

The use of fault-tolerant computing has spread into a number of fields for several reasons. First, a better understanding of fault tolerance techniques exists today. The field has grown from a handful of researchers twenty-five years ago to the point where several companies concentrate solely on fault-tolerant systems. Examples include Tandem Computers, Stratus Computers, and August Systems. Both Tandem and Stratus are competitors in the transactions processing industry, whereas August Systems concentrates on the development of reliable and fault-tolerant systems for industrial process control. Second, the advent of Large Scale Integration (LSI) and Very Large Scale Integration (VLSI) has made many fault tolerance techniques practical for the first time. Previously, designers could barely get a single computer to fit within their size, weight, and power consumption budgets. Fault-tolerant systems that used multiple computers were simply not practical in any but the most crucial applications. Finally, many systems that were previously mechanical are now electronic because of the tremendous capability and flexibility offered by electronics. Fault tolerance is now required because the new electronic systems are often less reliable.

Existing applications of fault-tolerant computing can be categorized into four primary areas: long-life applications, critical computations, maintenance postponement, and high availability. Each application presents differing design requirements and challenges.

1.4.1 Long-Life Applications

The most common examples of **long-life applications** are the unmanned space flight and satellites. The *Pioneer 10* spacecraft, for example, was launched on March 2, 1972, and became the first man-made object to pass beyond all known planets on June 13, 1983 [Lerner 1983]. During *Pioneer 10's* flight it has returned fascinating pictures of Jupiter and its moons, among other things. The information returned by *Pioneer 10* would have been severely restricted had not the electronics continued to function correctly throughout the time required for the spacecraft to reach its destination and perform its functions. The Mariner, Explorer, and Voyager missions are other examples of long-life space missions.

Satellites are also required to function correctly in space for extended periods of time. The cost of designing, building, and launching a satellite is much too high to allow electronic failures to render the satellite ineffective in space. Even though the space shuttle is now capable of retrieving satellites for repair, the cost of such repair is still extremely high, and many satellites are in orbits beyond the reach of the shuttle. Consequently, fault tolerance is required in satellite systems.

Typical requirements of a long-life application are to have a 0.95 probability of being operational at the end of a ten-year period. Unlike other applications, however, long-life systems can often allow extended outages as long as the system can eventually be made operational once again. For example, a one-week outage can be insignificant when you consider the five- or ten-year operational life of a satellite. In addition, long-life applications can frequently allow the system to be reconfigured manually by the operators. The Fault-Tolerant Spaceborne Computer (FTSC) [Stiffler 1976], the Self-Testing And Repairing (STAR) computer [Avizienis et al. 1971], and the Fault-Tolerant Building Block Computer (FTBBC) [Rennels 1980] are examples of systems designed for long-life applications.

As an example of a fault-tolerant computer system intended for long-life applications, Fig. 1.1 shows a general block diagram of the electronics found on board the *Voyager* spacecraft ([Jones 1979] and [Pradhan 1986]). The system consists of eight primary elements including the flight data system, the attitude control system, the command and control system, the radio system, a telemetry modulator, a command detector, a receiver, and a tape recorder. Two identical copies of each element are provided: one copy, called the *primary* copy, performs all the operations under normal circumstances and the second copy serves as a *backup*. The system is designed so

10   Introduction

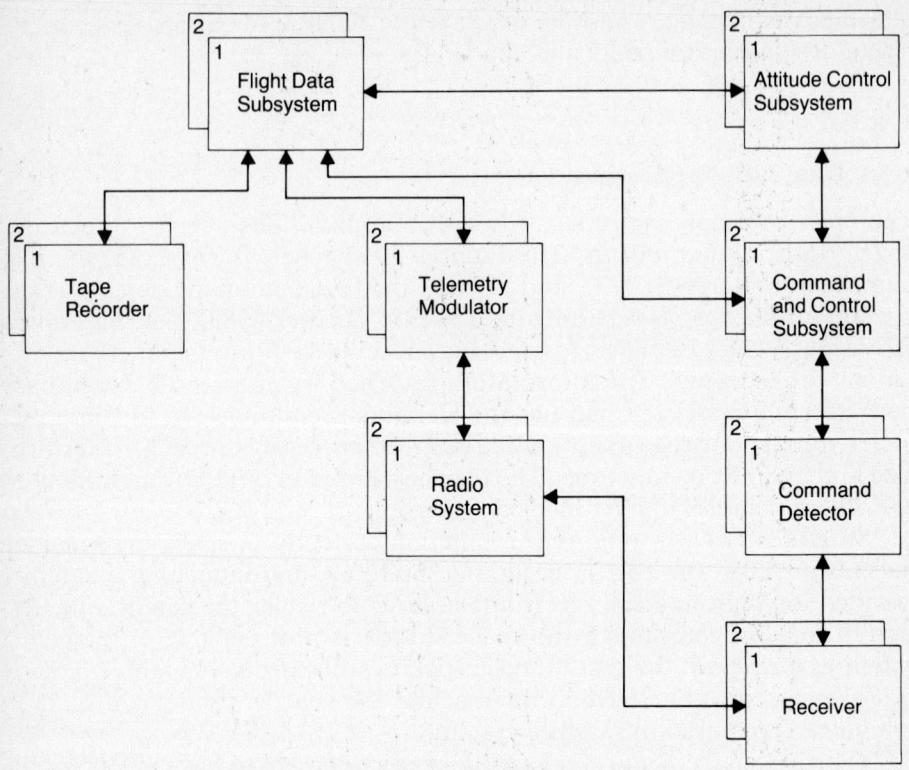

Fig. 1.1 The electronics on board the *Voyager* spacecraft achieve fault tolerance by using two identical copies of each major element.

that if the primary element fails, the backup element can be "switched in" to replace the primary and keep the system operational.

1.4.2 Critical-Computation Applications

Perhaps the most widely publicized applications of fault-tolerant computing are those in which the computations are critical to human safety, environmental cleanliness, or equipment protection. Examples include aircraft flight control systems, military systems, and certain types of industrial controllers. In **critical-computation applications,** the incorrect performance of the system will almost certainly yield devastating results. A typical requirement for a critical-computation application is to have a reliability of 0.9_7 at the end of a three-hour period. Requirements can vary, however, depending on the particular function that the system is performing.

The most publicly visible critical-computation application of fault-tolerant computing has been the space shuttle. A malfunction in the shuttle's flight control system during either ascent or descent can result in the loss of the shuttle. Consequently, extreme care has been taken to ensure that the system performs its tasks dependably. In fact, the shuttle can continue its mandatory flight control functions after as many as three computer failures [Sklaroff 1976].

Industrial control systems also perform critical computations. For example, chemical reactions may have to be precisely controlled to prevent explosions or other unwanted effects. The goal in almost all critical-computation applications is to prevent the electronics from being the *weak point* in the system; fault tolerance is a means of accomplishing this design goal.

As an example of a fault-tolerant system used in a critical-computation application, consider the architecture of the X-29 aircraft flight control system, as shown in Fig. 1.2 [Anderson 1983]. The forward swept wing technology that the X-29 uses to maximize aerodynamic benefits requires a stabilizing control system. If the control system fails to perform correctly, the airplane will not be flyable. A hybrid analog/digital fly-by-wire flight control system is used to provide closed-loop control of the aircraft. The term *fly-by-wire* simply means that there are no mechanical connections between the pilot's stick and the control surfaces (for example, the ailerons, elevators, and rudder). Instead, an electronic system samples the position of the pilot's stick, calculates the desired position of the control surfaces, and commands a motor to move the control surfaces. The connection between the pilot's stick and the control surfaces is completely electrical; consequently, the loss of the electronics implies the loss of the ability to fly the aircraft.

As shown in Fig. 1.2, the control system uses three identical computers performing the same operations. The results from each computer are examined, and the output from the system is formed via a majority vote of the three results. Consequently, a single computer performing incorrectly will be overruled by the two computers that are performing correctly. Each computer within the system consists of both a digital computer and an analog computer. The analog computer is used as a backup that can assume the functions of the system if the digital computer fails. The analog backup provides protection against software errors that could simultaneously affect all the digital computers.

1.4.3 Maintenance Postponement Applications

Maintenance postponement applications appear most frequently when maintenance operations are extremely costly, inconvenient, or difficult to

12 Introduction

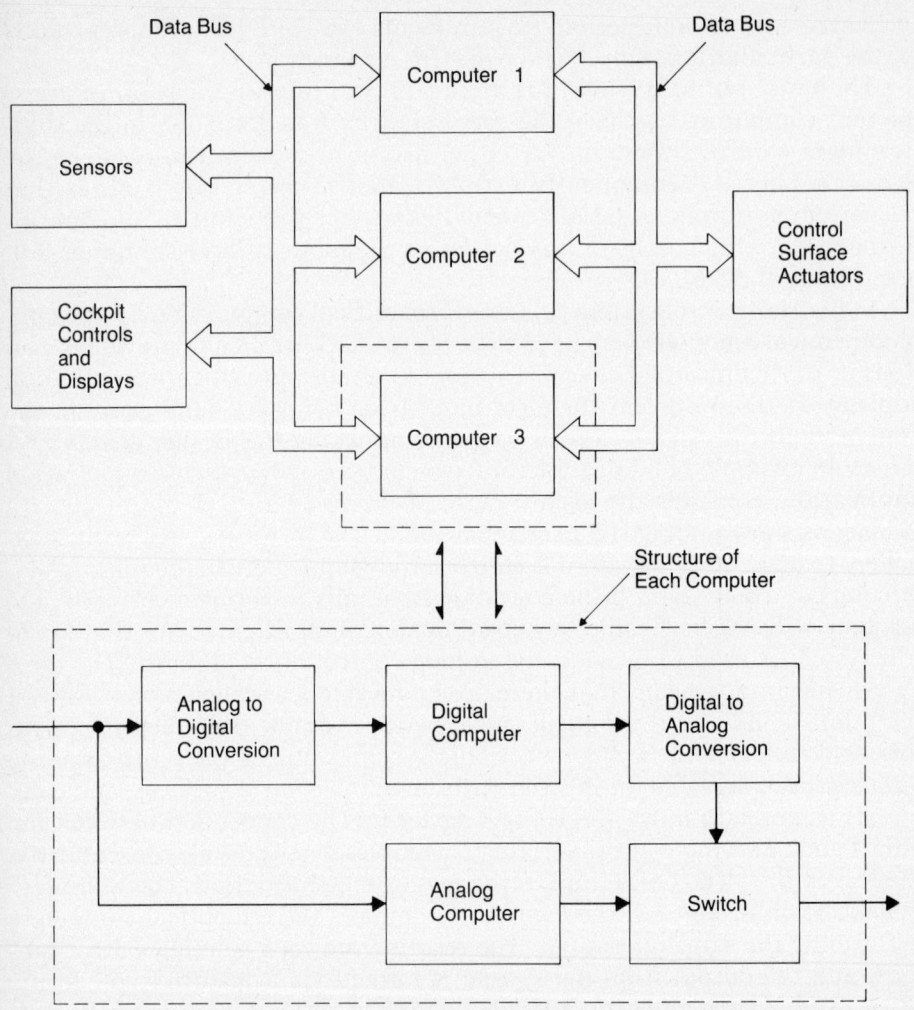

Fig. 1.2 The X-29 flight control system contains three complete computer systems. In addition, each computer system contains both an analog and a digital computer.

perform. Remote processing stations and certain space applications are good examples. In space, maintenance can be impossible to perform; at remote sites, the cost of unexpected maintenance can be prohibitive. The main goal is to use fault tolerance to allow maintenance to be postponed until a more convenient and cost-effective time. Maintenance personnel can visit a site monthly and perform any necessary repairs. Between maintenance visits, the system uses fault tolerance to continue to perform its tasks.

A telephone switching system [Toy 1978] is an example of a system that could require maintenance postponement. Many telephone switching systems are located in remote areas where it is necessary to provide telephone service, but it is costly to perform the maintenance and service operations. The primary objective is to design the system such that unscheduled maintenance can be avoided. Therefore, the telephone company can visit the facility periodically and repair the system or perform routine maintenance. Between maintenance visits, the system handles failures and service disruptions autonomously.

Figure 1.3 shows the block diagram of the 3B20D processor used in the Electronic Switching System (ESS) developed by AT&T Bell Laboratories [Serlin 1984]. Each element of this system is completely duplicated. One set of elements can be used to perform all the system functions, whereas the duplicate set serves as a backup in the event of a hardware failure. The duplicate set of elements can allow the system to remain functional while waiting for a repair to occur. Note in Fig. 1.3, for example, that either processor can access either storage disk, so the failure of a disk does not render the system inoperable.

1.4.4 High-Availability Applications

Availability is rapidly becoming a key parameter in many applications. Banking and other time-shared systems are good examples of **high-**

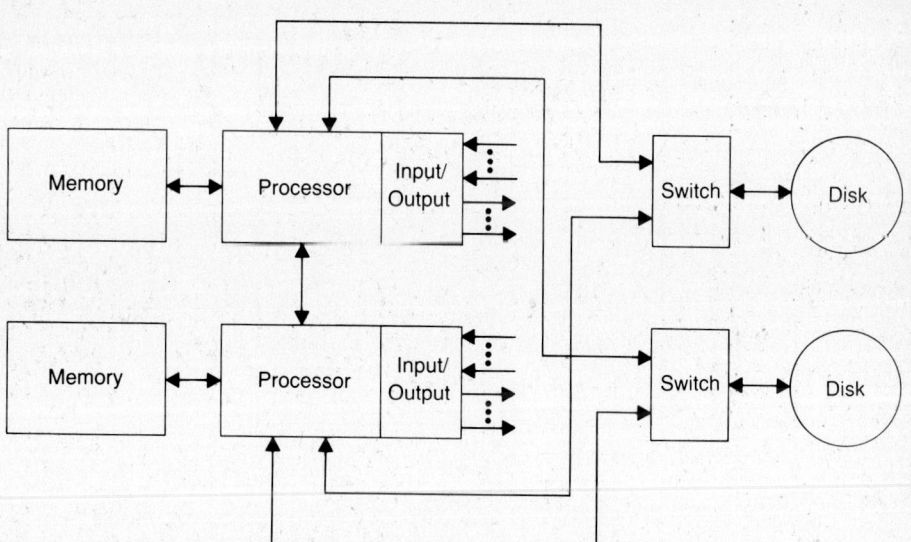

Fig. 1.3 Block diagram of the 3B20D processor used in the Bell Electronic Switching System. All critical components in this system are duplicated. (From [Serlin 1984] © 1984 IEEE)

14   Introduction

availability applications. Users of these systems want to have a high probability of receiving service when it is requested. The Tandem NonStop transaction processing system [Katzman 1977] is a good example of one designed for high availability. A major competitor of the Tandem computer is the Stratus system [Herbert 1983]. Both the Tandem and the Stratus computers are designed to achieve a high probability of being operational when their services are required.

Intel's 432 processor system [Siewiorek 1982] is an example design developed to support high availability in many general-purpose processing applications. Intel's 432 system employs a number of techniques that support fault-tolerant operation. For example, Fig. 1.4 shows the structure of the central processing unit (CPU) and illustrates how two CPUs can be operated as a pair. If CPU 1 in Fig. 1.4 is enabled, the system's outputs will come from CPU 1, and CPU 2 will check those outputs with its own. Similarly, CPU 2 could be enabled, and CPU 1 would serve as the checker.

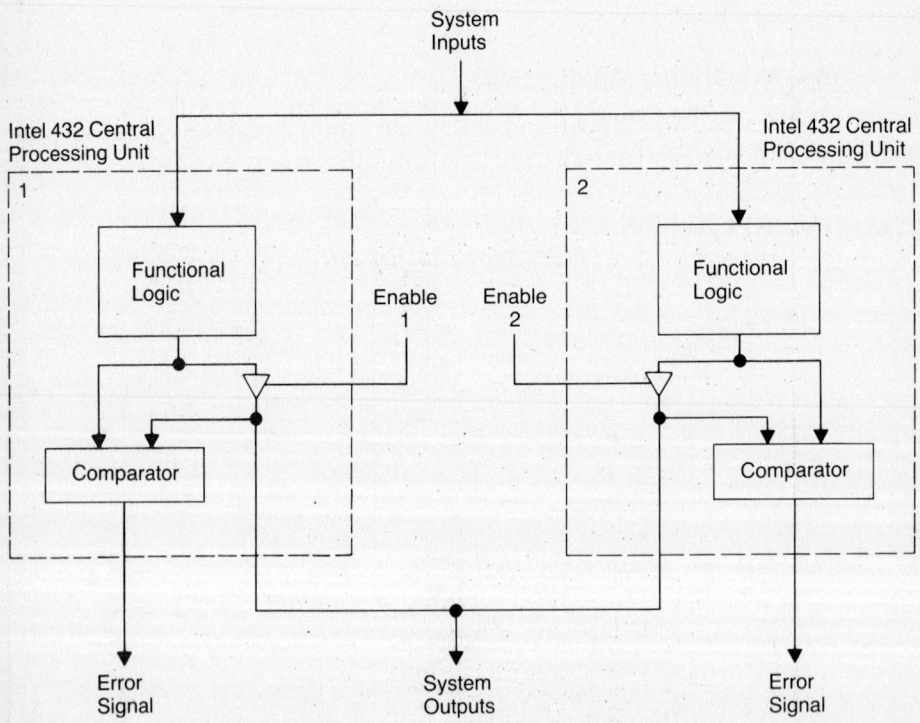

Fig. 1.4 The Intel 432 Central Processing Unit (CPU) supports operating two CPUs as a pair. The outputs of each CPU are compared. (From [Serlin 1984] © 1984 IEEE)

1.5 Fault Tolerance as a Design Objective

The design of a system is often conducted with many goals in mind. Certainly, we want the system to perform its intended function, but we also want the system to be dependable, cost effective, efficient, easy to test, and easy to repair. To accomplish our design goals, we begin with specific requirements. We may mandate that the cost and power not exceed certain values. Likewise, our design must meet specific reliability, availability, or maintainability goals. Fault tolerance plays an important role in attaining these goals, but it is not the only key part of the design process. Fault tolerance is a means of achieving our goals, but it must be coupled with other design techniques to be successful.

Figure 1.5 shows a top-level view of the design process. The system requirements, such as reliability, are achieved through two primary means: system design and system evaluation. The system design includes both fault avoidance and fault tolerance techniques. **Fault avoidance** techniques are performed to help prevent hardware failures and software errors. Examples include selecting high-quality components, enforcing design rules, and reviewing the designs periodically. Fault tolerance techniques, on the other hand, handle hardware failures and software errors when they occur. Examples include the use of redundant hardware, voting, and reconfiguration techniques.

The evaluation of a system is often overlooked as an integral part of the design process. Evaluation must be used in parallel with the design process, if the design is to be successful. System evaluation can uncover problems with a design early enough to allow corrections to be implemented. For example, if a design problem is discovered before the design is committed to hardware, the problem usually can be corrected easily. However, if problems remain in a system after it is built, correction can be impossible, or significant performance degradations may have to be accepted to allow the correction to be made. Numerous evaluation methods are available to analyze systems. Examples include Markov reliability models, system repair models, combinatorial reliability models, availability models, and maintainability models. In addition, evaluation techniques allow us to locate areas within a system that are prone to failure or where failure can be catastrophic. Each evaluation technique is crucial to a quality system design.

The primary purpose of this textbook is to study fault avoidance, fault tolerance, and system evaluation techniques that can be used in the design and analysis of fault-tolerant systems. This book presents the techniques that are available to achieve fault tolerance. More importantly, however, this book shows practical examples of how fault tolerance can be used to achieve design goals such as reliability, availability, and maintainability.

16 Introduction

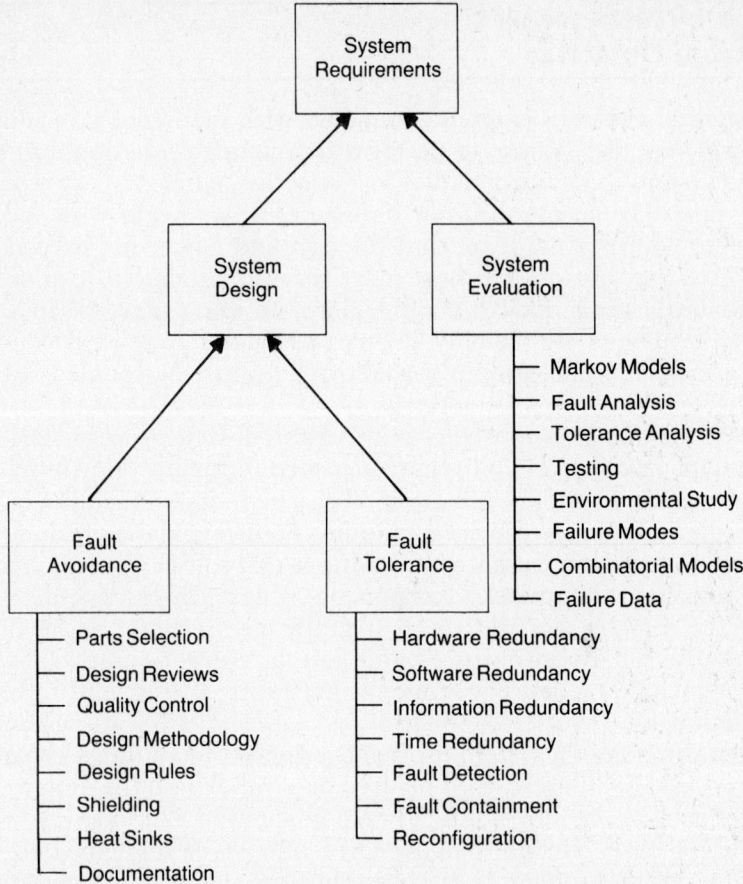

Fig. 1.5 A top-level view of the system design process illustrating the importance of fault avoidance, fault tolerance, and system evaluation.

Summary

This chapter has introduced the basic concept of fault tolerance and has illustrated the design goals often achieved via fault tolerance. The applications that require fault tolerance have been discussed, and examples from each application area have been presented. Perhaps the most important concept presented in this chapter is that fault tolerance is one aspect of a system design. To achieve design goals such as reliability or availability, a designer must use not only fault tolerance but fault avoidance and system evaluation as well.

The following list summarizes the important terminology and concepts that have been presented in this chapter.

Availability $A(t)$ — the probability that a system is operating correctly and is available to perform its functions at the instant of time t.

Critical-Computation Application — an application in which the incorrect performance of computations can create devastating results.

Dependability — the quality of service provided by a particular system.

Design for Testability — the process of including special features to make a system easily testable.

Easily-Testable System — a system whose ability to perform correctly can be verified in a simple and straight forward manner.

Fault Avoidance — the process of attempting to prevent hardware failures and software errors from occurring in a system.

Fault-Tolerant Computing — the process of performing calculations, such as those performed by a computer, in a fault-tolerant manner.

Fault-Tolerant System — a system that can continue the correct performance of its specified tasks in the presence of hardware failures and software errors.

Fault Tolerance — the quality or attribute that enables a system to behave in a fault-tolerant manner.

Graceful Degradation — the ability of a system to automatically decrease its level of performance to compensate for hardware failures and software errors.

High-Availability Application — an application in which availability is the crucial design requirement.

Long-Life Application — an application in which the longevity of operation is the crucial design requirement.

Maintainability, $M(t)$ — the probability that an inoperable system will be restored to an operational state within the time t.

Maintenance Postponement Application — an application in which it is desired to delay the process of repairing a system until the most convenient times.

Performability, $P(L,t)$ — the probability that a system is performing at or above some level of performance L at the instant of time t.

Reliability, $R(t)$ — the conditional probability that a system performs correctly throughout an interval of time $[t_0, t]$, given that the system was performing correctly at time t_0.

Safety, $S(t)$ — the probability that a system will either perform its functions correctly or will discontinue its functions in a well-defined, safe manner.

Test — a means by which the existence and quality of certain attributes within a system is determined.

Testability — the ability to test for certain attributes within a system.

Unreliability, $Q(t)$ — the conditional probability that a system will perform *incorrectly* during the interval of time $[t_0, t]$, given that the system was performing *correctly* at time t_0.

References

1. Anderson, D. "X-29 Forward swept wing flight control system," *Proceedings of the Joint AIAA-IEEE Fifth Digital Avionics Systems Conference*, Washington, D.C., December 1984, pp. 1–8.
2. Avizienis, A., Gilley, G.C., F.P. Mathur, D.A. Rennels, J.A. Rohr, and D.K. Rubin. "The STAR (Self-Testing And Repairing) computer: An investigation of the theory and practice of fault-tolerant computer design," *IEEE Transactions on Computers*, Vol. C-20, No. 11, November 1971, pp. 1312–1321.
3. Ayache, J., J. Courtiat, and M. Diaz. "REBUS: A fault tolerant distributed system for industrial real-time control," *IEEE Transactions on Computers*, Vol. C-31, No. 7, July 1982, pp. 637–647.
4. Bosch, J.A., and W.J. Kuehl. "Reconfigurable redundancy management for aircraft flight control," *Journal of Aircraft*, Vol. 14, No. 10, October 1977, pp. 966–971.
5. Carter, W.C., and W.G. Bouricius. "A survey of fault tolerant computer architecture and its evaluation," *Computer*, Vol. 4, No. 1, January 1971, pp. 9–16.
6. Engineering Research Association. *High Speed Computing Devices*, McGraw-Hill, 1950.
7. Fortes, J.A.B., and C.S. Raghavendra. "Dynamically reconfigurable fault-tolerant array processors," *Proceedings of the 14th International Conference on Fault-Tolerant Computing*, Kissimmee, Fla., June 20–22, 1984, pp. 386–392.
8. Herbert, E. "Computers: Minis and mainframes," *IEEE Spectrum*, Vol. 20, No. 1, January 1983, pp. 28–33.
9. Hopkins, A.L., T.B. Smith, and J.H. Lala. "FTMP: A highly reliable fault-tolerant multiprocessor for aircraft," *Proceedings of the IEEE*, Vol. 66, No. 10, October 1978, pp. 1221–1239.
10. Jones, C.P. "Automatic fault protection in the Voyager spacecraft," AIAA Paper No. 79–1919, American Institute of Aeronautics and Astronautics.
11. Katzman, J.A. "System architecture for nonstop computing," *Proceedings of the 14th Computer Society International Conference (Compcon)*, San Francisco, February 1977, pp. 77–80.

12. Kuehn, R.E. "Computer redundancy: Design, performance, and future," *IEEE Transactions on Reliability*, Vol. R-18, No. 1, February 1969, pp. 3–11.
13. Laprie, J.C. Dependable computing and fault tolerance: Concepts and terminology, *Proceedings of the 15th Annual International Symposium on Fault-Tolerant Computing*, June 19–21, 1985, Ann Arbor, Michigan, pp. 2–11.
14. Lerner, E.J. "Crossroads in space," *Spectrum*, Vol. 20, No. 9, September 1983, pp. 28–55.
15. Manual, T. "New architecture cuts redundancies in fail-safe processing," *Electronics*, Vol. 55, No. 17, August 25, 1982, pp. 45–46.
16. Pradhan, D.K. *Fault-Tolerant Computing Theory and Techniques*, Prentice-Hall, Englewood Cliffs, N.J., 1986.
17. Rennels, D.A. "Architectures for fault tolerant spacecraft computers," *Proceedings of the IEEE*, Vol. 66, No. 10, October 1978, pp. 1255–1268.
18. Rennels, D.A. "Distributed fault-tolerant computer systems," *IEEE Computer*, Vol. 13, No. 3, March 1980, pp. 55–64.
19. Serlin, O. "Fault-tolerant systems in commercial applications," *Computer*, Vol. 17, No. 8, August 1984, pp. 19–30.
20. Siewiorek, D.P., and R.S. Swarz. *The Theory and Practice of Reliable System Design*, Digital Press, Bedford, Mass., 1982.
21. Sklaroff, J.R. "Redundancy management technique for the space shuttle computers," *IBM Journal of Research and Development*, Vol. 20, No. 1, January 1976, pp. 20–28.
22. Stiffler, J.J. "Architectural design for near-100% fault coverage," *Proceedings of the International Symposium on Fault Tolerant Computing*, 1976, pp. 134–137.
23. Toy, W.N. "Fault-tolerant design of local ESS processor," *Proceedings of the IEEE*, Vol. 66, No. 10, October 1978, pp. 1126–1145.
24. von Neumann, J. "Probabilistic logics and the synthesis of reliable organisms from unreliable components," *Automata Studies, Annals of Mathematical Studies*, Princeton University Press, No. 34, pp. 43–98, 1956.
25. Weik, M.H. "A survey of domestic electronic digital computing systems," Report #971, Commerce Department, Ballistic Research Laboratories, Aberdeen Proving Grounds, Md., December 1955.
26. Wensley, J.H., L. Lamport, J. Goldberg, M.W. Green, K.N. Levitt, P.M. Melliar-Smith, R.E. Shostak, and C.B. Weinstock, "SIFT: Design and analysis of a fault tolerant computer for aircraft control," *Proceedings of the IEEE*, Vol. 66, No. 10, October 1978, pp. 1240–1255.
27. Wensley, J.H., and C.S. Harclerode. "Programmable control of a chemical reactor using a fault tolerant computer," *IEEE Transactions on Industrial Electronics*, Vol. IE-29, No. 4, November 1982, pp. 258–264.

Additional Reading

For the reader interested in learning more about the history and development of fault-tolerant computing, the following list of suggested references is provided. This list primarily includes tutorial and survey articles that give the reader a good feel for the development of the technology, terminology, and applications of fault-tolerant computing.

Avizienis, A. "Fault tolerance: The survival attribute of digital systems," *Proceedings of the IEEE*, Vol. 66, No. 10, October 1978, pp. 1109–1125.

Avizienis, A. "Fault tolerant computing: An overview," *Computer*, Vol. 4, No. 1, January 1971, pp. 5–8.

Avizienis, A. "Architecture of fault-tolerant computing systems," *Proceedings of the 1975 International Symposium on Fault-Tolerant Computing*, Paris, June 1975, pp. 3–16.

Baechler, D.O. "Aerospace computer characteristics and design trends," *Computer*, Vol. 4, No. 1, January/February 1971, pp. 45–57.

Cooper, A.E., and W.T. Chow. "Development of on-board space computer systems," *IBM Journal of Research and Development*, Vol. 20, No. 1, January 1976, pp. 5–19.

Deyst, J.J., Jr., J.V. Harrison, E. Gai, and K.C. Daly. "Fault detection, identification, and reconfiguration for spacecraft systems," *Journal of Astronautical Science*, Vol. 29, No. 2, April-June 1981, pp. 113–126.

Goldberg, J. "New problems in fault-tolerant computing," *Proceedings of the 1975 International Symposium on Fault-Tolerant Computing*, Paris, 1975, pp. 29–34.

Harris, R.L., and E.E. Jones. "Fault tolerance applications to future military system avionics," *Proceedings of the IEEE 1980 National Aerospace and Electronics Conference*, Dayton, Oh., May 1980.

Hecht, H. "Fault tolerant computers for spacecraft," *Journal of Spacecraft*, Vol. 14, No. 10, October 1977.

Hopkins, A.L., Jr. "Design foundations for survivable, integrated, on-board computation and control," *Proceedings of the 1977 Joint Automatic Control Conference*, San Francisco, Calif., June 22–24, 1977, pp. 232–237.

Hopkins, A.L., Jr. "Fault tolerant system design: Broad brush and fine print," *Computer*, Vol. 13, No. 3, March 1980, pp. 39–46.

Jennings, R. "Fault secure avionic system development," *Proceedings of the 1981 International Aerospace and Electronics Conference*, Vol. 1, Dayton, Oh., May 9–11, 1981.

Kime, C.R. "Fault tolerant computing: An introduction and a perspective," *IEEE Transactions on Computers*, Vol. C-24, No. 5, May 1975, pp. 457–460.

Koczela, L.J., and G.J. Burnett. "Advanced space missions and computer systems," *IEEE Transactions on Aerospace and Electronic Systems*, Vol. AES-4, No. 3, May 1968, pp. 456–467.

Nelson, V.P., and B.D. Carroll. *Tutorial: Fault-Tolerant Computing*, IEEE Computer Society Press, Washington, D.C., 1986.

Ramamoorthy, C.V. "Fault tolerant computing: An introduction and an overview," *IEEE Transactions on Computers*, Vol. C-20, No. 11, November 1971, pp. 1241–1244.

Rennels, D.A. "Fault-tolerant computing—Concepts and examples," *IEEE Transactions on Computers*, Vol. C-33, No. 12, December 1984, pp. 1116–1129.

Rennels, D.A. "Reconfigurable modular computer networks for spacecraft on-board processing," *Computer*, Vol. 11, No. 7, July 1978, pp. 49–59.

Short, R., and J. Goldberg. "A summary of Soviet activities in the design of fault tolerant digital machines," *Computer*, Vol. 11, No. 1, January/February 1971, pp. 28–33.

Short, R., and J. Goldberg. "A survey of Soviet activities in the design of fault tolerant digital machines," *IEEE Transactions on Computers*, Vol. C-20, No. 11, November 1971, pp. 1337–1352.

Siewiorek, D.P., D.E. Thomas, and D.L. Scharfetter. "The use of LSI modules in computer structures: Trends and limitations," *Computer*, Vol. 11, No. 7, July 1978, pp. 16–25.

Teschler, L. "Computers that won't fail," *Machine Design*, Vol. 50, No. 10, May 11, 1978, pp. 91–97.

2

Fundamental Definitions

2.1 Introduction
2.2 Faults, Errors, and Failures
2.3 Causes of Faults
2.4 Characteristics of Faults
2.5 Fault Models
2.6 Error Models
2.7 Design Philosophies to Combat Faults
    Summary
    References
    Additional Reading
    Problems

2.1 Introduction

Throughout the history of fault-tolerant computing, there has been substantial disagreement on the definitions of several key concepts. For example, the terms *fault*, *failure*, and *error* have often been used interchangeably in the literature. To many people, a failure has occurred when the time-shared computer they are using fails to respond to their requests or demands. To other people, a failure is a more specific physical defect within some electronic component. Some groups view the physical defects as faults instead of failures.

The purpose of this chapter is to introduce the basic terminology used in the fault-tolerant computing field. It is important to understand the causes of faults and the types of faults that can occur before considering the

23

techniques that are available to tolerate faults in a design. In addition, it is vital to have a clear understanding of how faults manifest themselves in digital systems. In other words, once a fault has occurred, how is it likely to propagate throughout the system and what are the probable impacts of that fault? Finally, we discuss the role of fault tolerance and fault avoidance in combating faults in digital systems. Perhaps the best available references on the fault tolerance terminology are [Avizienis 1982], [Laprie 1985], and [Johnson 1984].

2.2 Faults, Errors, and Failures

Three fundamental terms in fault-tolerant design are fault, error, and failure. There is a cause-and-effect relationship between faults, errors, and failures. Specifically, faults are the cause of errors, and errors are the cause of failures.

A **fault** is a physical defect, imperfection, or flaw that occurs within some hardware or software component. Essentially, the definition of a fault, as used in the fault tolerance community, agrees with the definition found in the dictionary. A fault is a blemish, weakness, or shortcoming of a particular hardware or software component. Examples of faults include shorts between electrical conductors, opens or breaks in conductors, or physical flaws or imperfections in semiconductor devices. Similarly, in software, an example of a fault is a program loop that when entered can never be exited.

An **error** is the manifestation of a fault. Specifically, an error is a deviation from accuracy or correctness. For example, suppose that a physical short results in a line within a circuit being permanently stuck at a logic 1. The physical short is a fault within the circuit. If some condition occurs that requires the line to transition to a logic 0, the value on the line will be in error. In other words, the correct value for the line will be logic 0, but the existence of the fault has caused the line to have an erroneous value. In other words, an error is the result of a fault.

Finally, if the error results in the system performing one of its functions incorrectly, a system failure has occurred. Essentially, a **failure** is the nonperformance of some action that is due or expected. Although it is often used interchangeably with the term **malfunction,** the term *failure* is rapidly becoming more commonly accepted. A failure is also the performance of some function in a subnormal quantity or quality. As an example, suppose that a line in a circuit is responsible for turning a valve on or off: a logic 1 turns the valve on and a logic 0 turns the valve off. If the line is stuck at logic 1, the valve is stuck on. As long as the user of the system wants the valve on, the system will be functioning correctly. However, when the user wants to turn the valve off, the system will experience a failure.

Figure 2.1 illustrates the cause-and-effect relationship between faults, errors, and failures. Faults result in errors, and errors can lead to system failures. One way to think of Fig. 2.1 is as a hierarchy. At the bottom of the hierarchy are faults. Errors are the effect of faults, and, finally, failures are the effect of errors.

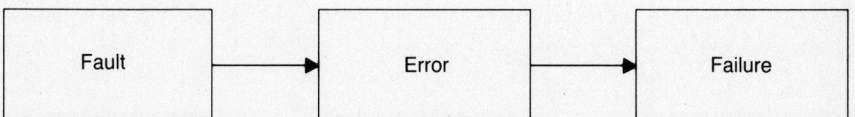

Fig. 2.1 Relationship between faults, errors, and failures. Failures are caused by errors, which are caused by faults.

The circuit shown in Fig. 2.2 further illustrates the distinction between faults and errors. The circuit of Fig. 2.2 is the logic diagram for a full-adder. The inputs A_1, B_1, and C_1 are the two bits of the operands and the carry bit, respectively. The truth table that shows the correct performance for this circuit is presented in Fig. 2.3. If a short occurs between line L and the power supply line resulting in line L becoming permanently fixed at a logic 1 value, a fault will have occurred. The fault is the actual short within the circuit.

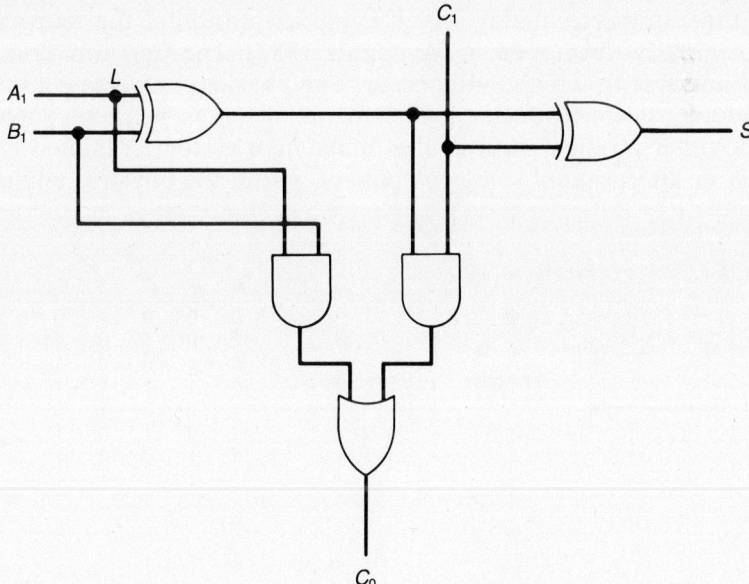

Fig. 2.2 Full-adder circuit to illustrate the difference between faults and errors.

26 Fundamental Definitions

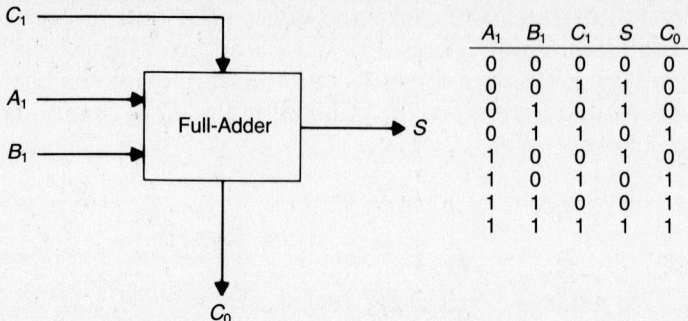

Fig. 2.3 Truth table for the fault-free full-adder circuit.

Figure 2.4 shows the truth table of the circuit that contains the physical short. By comparing the truth tables in Figs. 2.3 and 2.4, we can see that the circuit performs correctly for the input combinations 100, 101, 110, and 111, but not for 000, 001, 010, and 011. The physical short within the circuit is a fault. Whenever an input is applied that results in the circuit producing an incorrect output, an error has occurred. If the output of the circuit is controlling a relay and the relay is opened when it should be closed, a failure has occurred.

The concepts of faults, errors, and failures can be best presented by using a three-universe model, which is an adaptation of the four-universe model originally developed in [Avizienis 1982]. The first universe is the **physical universe** in which faults occur. The physical universe contains the semiconductor devices, mechanical elements, displays, printers, power supplies, and other physical entities that make up a system. A fault is a physical defect or alteration of some component within the physical universe.

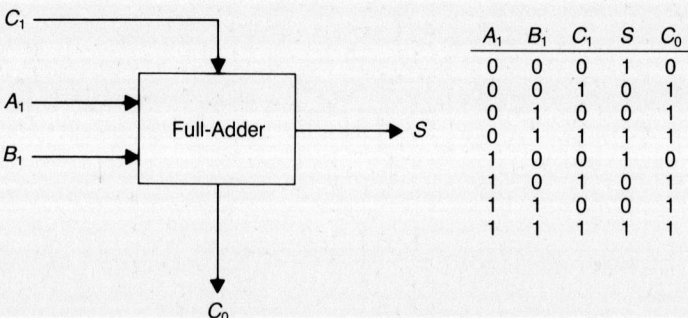

Fig. 2.4 Truth table for the full-adder circuit with an example fault.

The second universe is the **informational universe.** The informational universe is where the error occurs. Errors affect units of information such as data words within a computer or digital voice or image information. Terms that are applicable to the informational universe include parity errors, message errors, typographical errors, and bit errors. An error has occurred when some unit of information becomes incorrect.

The third universe is the **external universe** (or **user's universe**). The external universe is where the user of a system ultimately sees the effect of faults and errors. The external universe is where failures occur. The failure is any deviation that occurs from the desired or expected behavior of a system. For example, the user may expect a system to correctly print payroll checks. If, for some reason, a check is printed incorrectly, the user has witnessed a failure of the system, and the failure has been witnessed in the external universe.

Figure 2.5 illustrates the relationships implied in the three-universe model. In summary, faults are physical events that occur in the physical universe. Faults can result in errors in the informational universe, and errors can ultimately lead to failures that are witnessed in the external universe of the system.

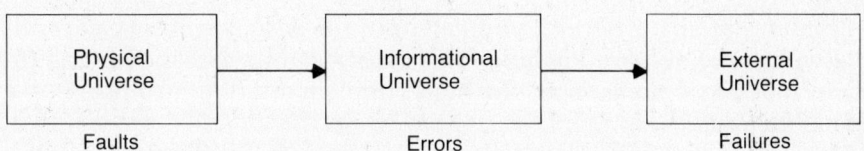

Fig. 2.5 Three-universe model representing the cause-and-effect relationship between faults, errors, and failures. Faults occur in the physical universe and cause errors to occur in the informational universe. Errors can result in failures that occur in the external universe.

The cause-and-effect relationship implied in the three-universe model leads to the definition of two important parameters: fault latency and error latency. **Fault latency** is the length of time between the occurrence of a fault and the appearance of an error due to that fault. A **latent fault** is one that is present in a system but has not yet produced an error. In other words, a latent fault has not yet produced any effect. **Error latency** is the length of time between the occurrence of an error and the appearance of the resulting failure. Based on the three-universe model, the total time between the occurrence of a physical fault and the appearance of a failure is the sum of the fault latency and the error latency.

2.3 Causes of Faults

Faults can be the result of a variety of things that occur within electronic components, external to the components, or during the component or system design process. It is very important to understand all the possible causes of faults. To understand the various causes of faults, we first examine a typical design process to identify areas where faults can occur.

A design process often begins with a somewhat vague statement of the problem. The designers generate the problem statement as their interpretation of the actual problem. From the problem statement, a high-level solution to the problem is outlined, and the development of specific algorithms and system architectures is begun. Hardware and software specifications are then developed from the initial algorithms and architectures. A complete hardware and software design and test procedure is then invoked.

Once the hardware and software are completed, the designs must pass through a complete integration and test procedure with the end result being an operational system that, designers hope, solves the original problem. The various parts of the design process are typically performed several times before the design is completed. For example, the preliminary design process can modify the problem statement, either by necessity or as the result of some design tradeoff.

Problems at any of several points within the design process can result in faults within the system. Possible **fault causes** can be associated with problems in four basic areas: specifications, implementation, components, and external factors.

The first cause of faults is the possibility of **specification mistakes.** These include incorrect algorithms, architectures, or hardware and software design specifications. For example, suppose that the designer of a digital circuit incorrectly specified the timing characteristics of some of the circuit's components. The design might perform correctly at times, but there could also be instances of incorrect performance.

The next cause of faults is **implementation mistakes.** Implementation, as defined here, is the process of transforming hardware and software specifications into the physical hardware and the actual software. The implementation can introduce faults due to poor design, poor component selection, poor construction, or software coding mistakes. For example, suppose that a printed circuit board is constructed such that adjacent lines of a circuit are shorted together. The components on the board can be performing correctly, but the board produces incorrect results because of a wiring mistake. A second crucial example of an implementation mistake is a software coding error. If the software is written incorrectly, a fault exists, and an error can result if the faulty software is executed at some point during the operation of the system.

The next cause of faults is **component defects.** Manufacturing imperfections, random device defects, and component wear-out are typical examples of component defects. Electronic components simply become defective sometimes. The defect can be the result of bonds breaking within the circuit or corrosion of the metal. Component defects are the most commonly considered cause of faults, but, as is evident from our discussions, component defects are only one of several causes of faults.

The final cause of faults is the **external disturbance,** for example, radiation, electromagnetic interference, battle damage, operator mistakes, and environmental extremes. If an electronic system is subjected to extreme temperature variations, the system can produce incorrect results. If the electronics in a military system are damaged during a battlefield encounter, a fault has been injected into the system by some external source. Also, electronic systems are usually very sensitive to electrostatic sources such as lightning or other weather-related effects. Finally, operator mistakes are considered to be external disturbances since the operator is external to the system's physical hardware and software. Clearly, an operator can provide incorrect commands to a system that can ultimately lead to system failures.

Figure 2.6 illustrates the general effects of faults in a system and follows the three-universe concepts discussed in the previous section. The four distinct causes—specification mistakes, implementation mistakes, component defects and external disturbances—result in hardware or software faults in the physical universe. The effect of a fault is to produce some error that is an unintentional deviation from correctness in the information of a system. The result of an error is a failure that occurs in the external universe.

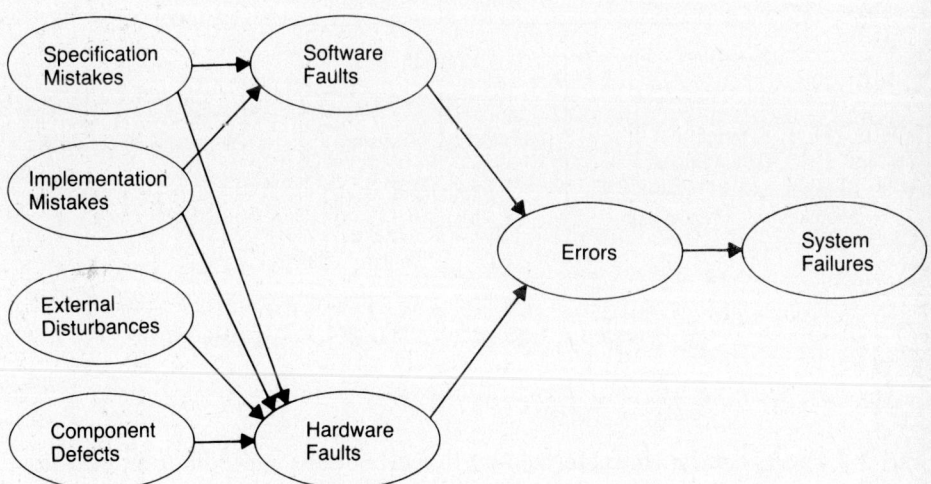

Fig. 2.6 Cause-and-effect relationship of faults, errors, and failures in a system.

2.4 Characteristics of Faults

We have spent quite some time developing the causes of faults and how faults, errors, and failures differ. To adequately describe faults, however, characteristics other than the cause are required. In addition to the cause, four major attributes are critical to the description of faults: nature, duration, extent, and value [Nelson and Carroll 1982]. Figure 2.7 illustrates each of the basic characteristics of faults.

The **fault nature** specifies the type of fault. For example, is the fault a hardware fault or a software fault? Also, is the fault in the analog or the digital circuitry? A power supply fault is an example of an analog hardware fault. A short circuit within a microprocessor is an example of a digital hardware fault. A loop within a subroutine that, when entered, can never be exited is an example of a software fault. In summary, the nature of a fault can be specified using the terms hardware, software, analog, and digital.

The **fault duration** specifies the length of time that a fault is active. First, there is the **permanent fault,** which remains in existence indefinitely if no corrective action is taken. Second, there is the **transient fault,** which can appear and disappear within a very short period of time. Third, there is the **intermittent fault,** which appears, disappears, and then reappears repeatedly. An example of a permanent fault is a logic line that is physically stuck at a logic 1. An example of a transient fault is one resulting from some external disturbance such as lightning. The lightning can temporarily dis-

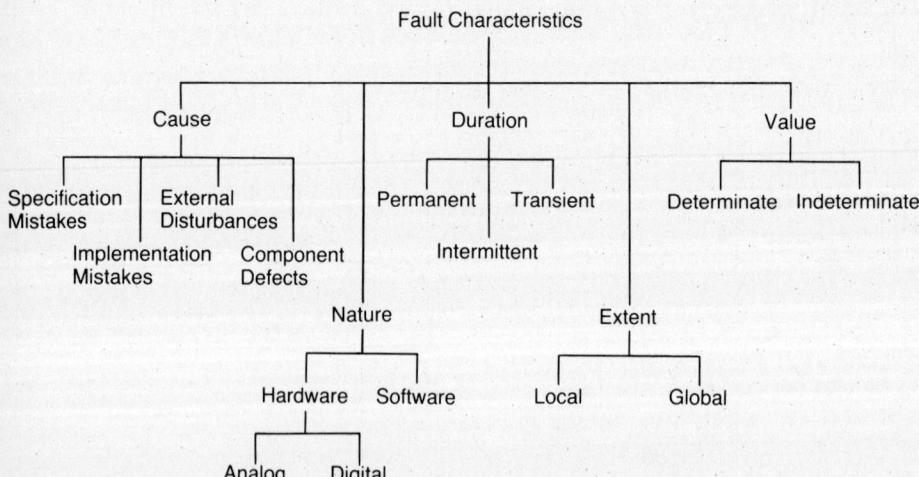

Fig. 2.7 Faults can be characterized by five attributes: cause, nature, duration, extent, and value.

rupt a system but do no permanent damage; as soon as the lightning disappears, the fault can also disappear. The effects of the external disturbance, that is, the errors, can take some time to disappear, but the actual fault may be present for only a very short time. An example of an intermittent fault is one resulting from a weak solder joint in a circuit. The solder joint can provide proper contact at certain times and improper contact at others.

The **fault extent** specifies whether the fault is localized to a given hardware or software module or whether it globally affects the hardware, the software, or both. A power supply fault is a good example of a fault that might globally affect a system. On the other hand, a fault in a memory that is used by only one processor might have a very localized impact on a system.

The **fault value** can be either determinate or indeterminate. A **determinate fault** is one whose status remains unchanged throughout time unless externally acted upon. For example, a fault that always results in a line being a logical 1 is a determinate fault. An **indeterminate fault** is one whose status at some time T may be different from its status at some increment of time greater than or less than T. For example, a number of hardware faults can produce state oscillations between a logical 1 and a logical 0. A good example is a fault that is sensitive to either the data or time.

2.5 Fault Models

In much of our work, it is necessary to assume that faults behave according to some fault model [Hayes 1985]. Although in practice faults can be transient in duration and indeterminate in value, it can be extremely difficult to analyze digital systems if we assume that faults have these characteristics. This is particularly true when we attempt to design test procedures for digital systems or to simulate faults within such systems. To make our problems more manageable, we need some way to restrict our attention to a subset of all faults that can occur.

Fault models allow us to specifically define the types of faults that will be considered and the behavior these faults will have. In addition, fault models allow us to represent the behavior of physical occurrences. We use models in all disciplines of engineering, and although we understand that models are not 100% accurate in all cases, we use them to make problems tractable and because their usage often introduces little error. The same is true of fault modeling.

Fault models attempt to represent the types of faults that can occur. In this section, we consider two primary fault models: the logical stuck-fault model that is used at the logic circuit level and the transistor stuck-fault model that is used at the transistor circuit level.

2.5.1 The Logical Stuck-Fault Model

The most common fault model is the **logical stuck-fault model** [Kohavi 1978], which has gained its popularity because of its effectiveness and simplicity. The logical stuck-fault model is sometimes referred to as the stuck-at-0 (s-a-0), stuck-at-1 (s-a-1) fault model or simply the stuck-fault model.

There are three basic assumptions of the stuck-fault model:

- a fault results in a module responding as if one of its inputs or outputs is physically stuck at a logic 1 or 0
- the basic functionality of the circuit is not altered by the fault
- the fault is permanent

The logic module can be a single gate or a collection of gates that implements some logic function.

The first assumption of the stuck-fault model is further illustrated in Fig. 2.8 by an AND gate having inputs A and B and output F. The input line A is assumed to remain free to take on either logic value (1 or 0), but the gate responds as if the line is physically stuck at either a logic 1 or logic 0. For example, a stuck-at-1 fault on line A results in the output of the AND gate in Fig. 2.8 being 1 whenever input B is 1, regardless of the actual value applied to line A. Line A is free, however, to assume any logic value. Likewise, a stuck-at-0 fault on line A causes the AND gate to always have an output of 0, regardless of the actual value applied to input A.

The second assumption of the stuck-fault model is that the basic functionality of the circuit is not affected by the fault. This assumption is often confusing, but it is very important in the stuck-fault model. It does not mean that the circuit continues to produce the correct results, but instead simply implies that the circuit produces the results expected of it, given the existence of the fault. For example, an AND gate that has a stuck-at-0 fault on one of its inputs should always produce a 0 at the output if the gate continues to behave as an AND gate. If, however, the fault transforms the AND

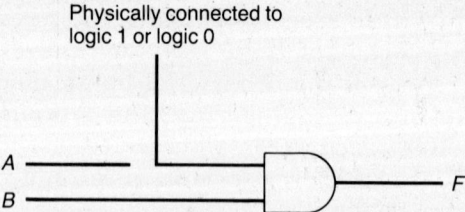

Fig. 2.8 Illustration of the basic concept of the logical stuck-fault model. The gate responds as if the input is permanently connected to a 1 or 0. The applied input, however, is free to assume any value.

2.5 ■ Fault Models

gate into an EXCLUSIVE-OR gate, the assumptions of the stuck-fault model are violated. Likewise, if a fault transforms a combinational circuit into a sequential circuit, the assumptions of the stuck-fault model are violated. This latter example is a significant problem in many of today's integrated circuit technologies.

The third assumption of the stuck-fault model is that the fault is permanent. Simply stated, the faulty module *always* performs as if a line is stuck at a specific logic value. This assumption simplifies the fault model by avoiding the difficulty of modeling intermittent or transient faults.

Most applications of the stuck-fault model restrict the number of faults that can occur at any one time. Typically, we assume that a circuit will never have more than one stuck fault. Although the single-fault assumption is not a specific property of the stuck-fault model, it is a very common assumption that is used in conjunction with the remaining stuck-fault assumptions to simplify the process of analyzing a circuit or generating test patterns. In a circuit that contains n lines, at most $2n$ unique, single, stuck faults can occur. Consequently, an upper bound is placed on the number of faults that must be examined. Even with the bound, it is easy to see that large circuits can have a prohibitively large number of potential faults.

The effectiveness of a fault model can often be quantified by a coverage parameter. A fault model is said to cover a fault if and only if the actual physical fault is accurately represented by the chosen fault model. Ideally, one hopes that a fault model covers 100% of all physical faults, but this is seldom the case.

The classic example of a case where the stuck-fault model does not cover a very specific and practical physical fault is found in the Complementary Metal Oxide Semiconductor (CMOS) implementation of a two-input NOR gate [Wadsack 1978]. The logic diagram and the transistor realization of the CMOS NOR gate are shown in Fig. 2.9. The circuit is a combination of two p-channel transistors in series and two n-channel transistors in parallel. Based on the values of the inputs, A and B, a path for current flow is established from either V_{DD} or V_{SS} to the output of the circuit. For example, if both A and B are at the logic 0 level, both p-channel transistors are conducting while both n-channel transistors are turned off. A path is established between V_{DD} and the output, thereby forcing the output to a logic 1 value. In a similar manner, if either or both inputs are 1, the corresponding p-channel transistors are turned off while one or both n-channel transistors are turned on; the result is that a path is established from the output to V_{SS}, thus forcing the output to a logic 0 value.

Several faults that behave as stuck faults can be easily recognized in the NOR circuit. For example, if input line A becomes physically shorted to

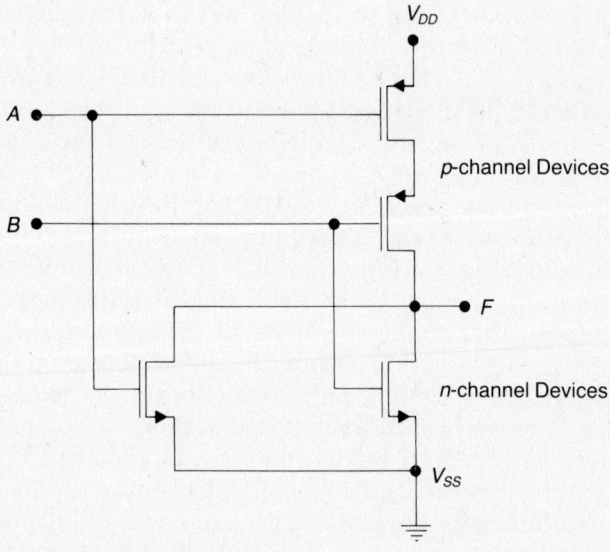

Fig. 2.9 Logic diagram and transistor implementation of the CMOS NOR gate. Inputs A and B determine if the output is pulled high (V_{DD}) or low (V_{SS}).

V_{DD}, the circuit behaves as if line A is stuck at 1; specifically, the output of the circuit will always be 0. Another example is when the drain and source of one of the n-channel devices become shorted together, causing the device to always have a logic 0 on the output. Consequently, the fault can be modeled as the output stuck at 0.

Several faults do not adhere to the stuck-fault model. One example is when a line within the circuit breaks. If a break in a line occurs, as shown in Fig. 2.10, and the input combination $AB = 10$ is presented to the circuit, a path does not exist from either V_{DD} or V_{SS} to the circuit output. In other words, neither the series network of p-channel transistors nor the parallel network of n-channel transistors is conducting. The output under such conditions will be pulled neither to logic 0 nor to logic 1, but, due to load capacitances present on the output, the output retains the value defined by the previous input. The length of time that the output remains at the previous value depends on the length of time required to discharge the load capacitance. This type of fault is often referred to as a **stuck-open fault**.

2.5 ■ Fault Models

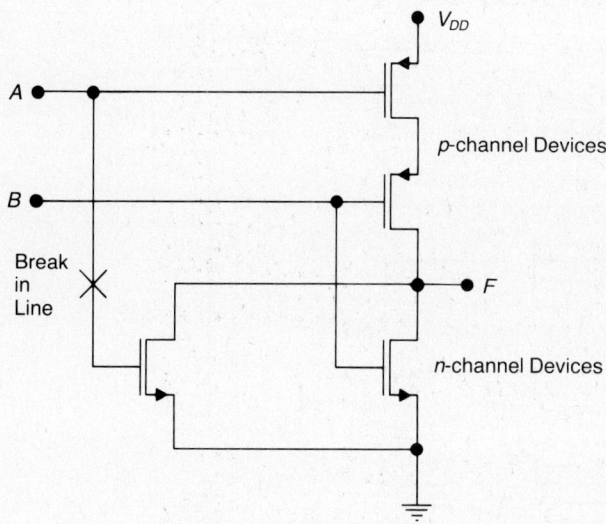

Fig. 2.10 A stuck-open fault exists when a break in a line occurs. Certain input patterns result in the output not being pulled either high or low. Instead the load capacitances on the output result in a previously defined output being retained.

When a fault results in the present output value depending on the previous output, the circuit has obtained a form of memory and is, therefore, no longer a combinational circuit. Instead, the circuit is a sequential circuit, and one of the basic assumptions of the stuck-fault model is violated. Therefore, the stuck-fault model cannot adequately model the stuck-open fault that can occur in the CMOS NOR gate.

Recognizing the limitations of the stuck-fault model, researchers have attempted to develop new and better fault models. For example, in [Wadsack 1978] logic circuit models of the various gates, such as the NOR gate, are developed to allow the effect of the stuck-open fault in CMOS circuits to be simulated. However, the fault model requires that additional gates be added to the description of the circuit, thereby increasing the complexity.

As an example, Fig. 2.11 shows the modified representation of a CMOS NOR gate. In Fig. 2.11, the D-type flip flop is used to model the memory that is introduced when the stuck-open fault occurs. The flip flop is a level-triggered device such that when the clock line is 1, the output of the flip flop simply follows the D input. When the clock transitions from a 1 to a 0, however, the value on the D input is latched, and the flip flop output retains that value until the clock returns to 1. Note that under all fault-free operating conditions, the clock remains at a logic 1, and the output of the flip flop is just the output of the NOR gate G_1. Traditional stuck-at-1 and stuck-at-0

36   **Fundamental Definitions**

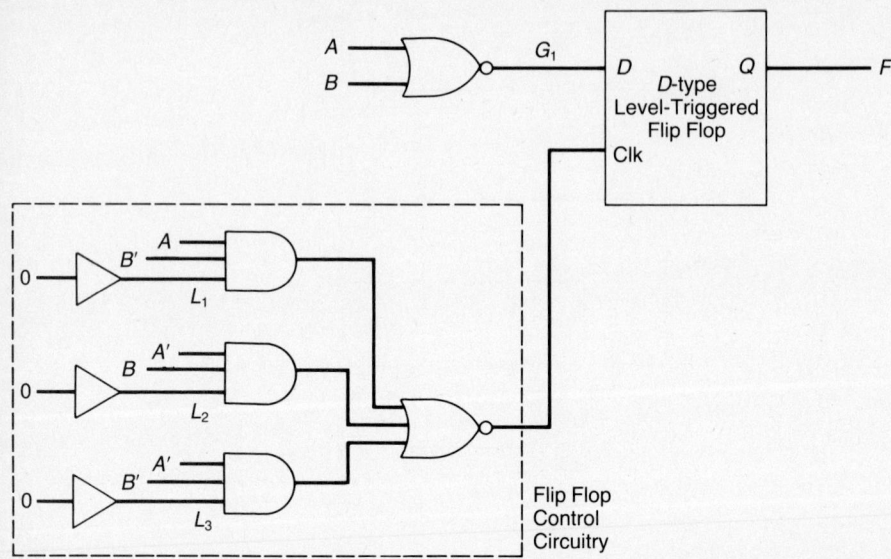

Fig. 2.11 Logical model of a NOR gate to account for the stuck-open fault. The D-type flip flop is used to simulate the memory effect produced by the stuck-open fault.

faults can be injected by simply placing them on the inputs A and B, and on the output G_1 of the NOR gate.

Stuck-open faults can be modeled by placing a stuck-at-1 fault on either line L_1, L_2, or L_3. L_1 stuck-at-1 represents a stuck-open fault affecting the n channel device driven by input A. L_2 stuck-at-1 represents a stuck-open fault affecting the n-channel device driven by input B. Finally, L_3 stuck-at-1 represents a stuck-open fault affecting either p-channel device. For example, if L_1 is stuck-at-1, the clock line on the flip flop will be logic 0 whenever the input combination $AB = 10$ occurs. In other words, the n-channel device driven by input A should be on, but it is not because of the stuck-open fault. Consequently, the output of the gate is simply the output specified by the previous input combination. Any other input combination results in the gate performing as expected.

The difficulty with the modeling scheme developed in [Wadsack 1978] is the additional complexity that must be added to a circuit to allow for the modeling of stuck-open faults. For each gate, we must add four additional gates and a flip flop to allow the effects of stuck-open faults to be adequately modeled. Consequently, the complexity of the circuit, as far as simulation is concerned, is substantially increased.

2.5.2 Transistor Stuck-Fault Models

In some cases, such as the stuck-open fault, the logical stuck-fault model is inaccurate. One way to make the fault model more accurate is to construct the model at the transistor level as opposed to the gate level. Such models are often referred to as switch-level models because the transistor is essentially performing a switching function. Here, however, we use the terminology **transistor stuck-fault model** to describe the techniques employed.

In the transistor stuck-fault model, faults are represented as either a transistor stuck on or a transistor stuck off. For example, in the CMOS NOR gate of Fig. 2.9, the faults are assumed to result in transistors being permanently on or permanently off. The advantage of the transistor stuck-fault model is that the stuck-open fault described previously can be represented as a transistor permanently stuck off. For example, in Fig. 2.9, if the n-channel transistor driven by input A is permanently off, the response of the circuit to the input combination $AB = 10$ will be that the output is forced neither to V_{DD} nor to V_{SS}. Consequently, the output retains its previous value due to the load capacitances present on the output.

The clear disadvantage of the transistor stuck-fault model is the additional complexity. For example, a circuit containing 100 NOR gates contains 400 transistors, so the number of elements that must be represented and manipulated has quadrupled. Other gates such as EXCLUSIVE-ORs require even more transistors. In circuits containing many gates, the impact on complexity of using the transistor stuck-fault model can be overwhelming.

2.6 Error Models

Another technique that has been proposed for the modeling of the effects of faults is to model the effect in the informational universe. In other words, the *error* is modeled rather than the fault [Patel and Fung 1982]. The underlying assumption of this approach is that regardless of the fault, the effect of that fault in the informational universe is to change a logic value at some time and in some result produced by the system. For combinational circuits, this can be viewed as a modification of the circuit's truth table.

The truth table modification that results can vary as a function of time, but the truth table will be either correct or modified in some way. In other words, we may not know what form the fault takes, but we can look at the response of the circuit and determine whether or not the results are correct. We use this type of model in our daily lives when we compare pieces of information. For example, when we balance our checkbooks, we detect the existence of an error by the deviation between the bank's balance and our balance. Once we have detected the error, we must then examine the details of the checkbook and the statement to locate the problem. However, if the bank's balance agrees with our own, we can be reasonably confident that both balances are correct.

Error models are useful in the design and verification of many self-testing schemes, but they are not normally used as a means of test pattern generation, as the stuck-fault model is. Also, the error model implies that some time can elapse before we recognize that a fault exists.

Various other forms of fault models have been developed in the past for modeling faults in digital systems [Hayes 1985]. For the purposes of this text, however, the logical stuck-fault model is used during our discussions of test pattern generation. The interested reader is referred to the additional reading for descriptions of other available fault models.

2.7 Design Philosophies to Combat Faults

There are three primary techniques for attempting to improve or maintain a system's normal performance in an environment where faults are of concern: fault avoidance, fault masking, and fault tolerance. Figure 2.12 illustrates the barriers that are constructed by each of the available techniques.

Fault avoidance is any technique that is used to prevent faults in the first place. Fault avoidance can include such things as design reviews, component screening, testing, and other quality control methods. If a design review, for example, is conducted appropriately, many of the specification mistakes that might otherwise result in faults can be eliminated. Also, a system can often be shielded to prevent external disturbances such as lightning or radiation from causing faults in the system. Shielding is a form of fault avoidance. Finally, if a design is effectively tested, many of the faults that might be in a system after manufacture can be detected and eliminated before the system is placed into operation.

Fault masking is any process that prevents faults in a system from introducing errors into the informational structure of that system. Error correcting memories, for example, correct a memory's data before a system uses the data. Thus the system never experiences the impact of the fault within the memory. Error correction in a memory is a form of fault masking. Another example of fault masking is majority voting. If a committee of three people makes a decision by voting yes or no on a proposition, any two votes that agree determine the decision of the committee. The ultimate decision of the committee represents the wishes of a majority of the committee members and masks the desires of a member that happens to disagree with the majority. Similar techniques can be applied to digital systems such that two modules can mask the effect of a faulty module.

2.7 ■ Design Philosophies to Combat Faults

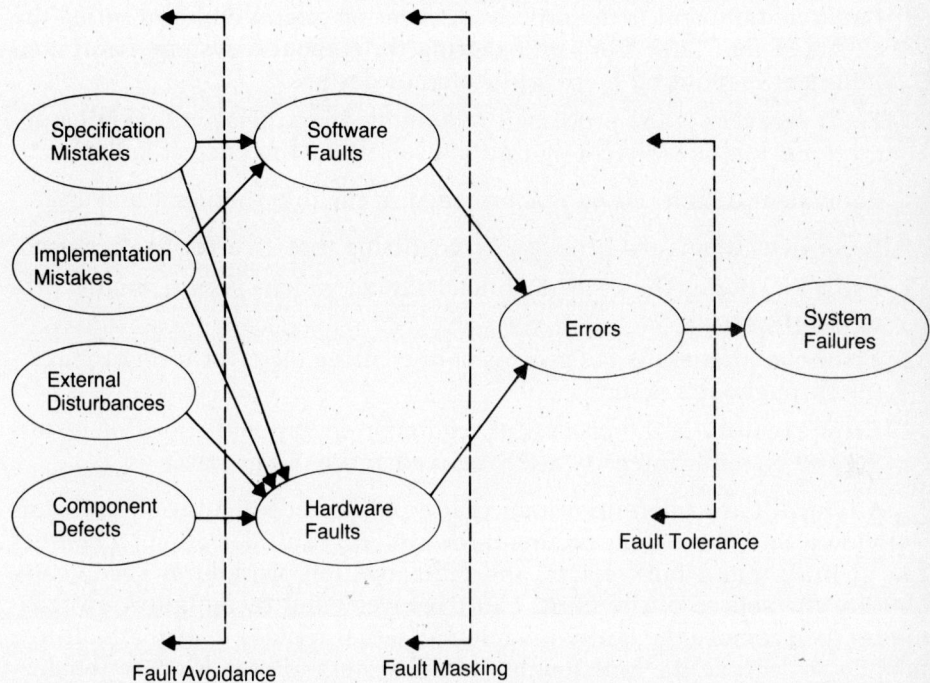

Fig. 2.12 Barriers constructed by design techniques of fault avoidance, fault tolerance, and fault masking.

Fault tolerance is the ability of a system to continue to perform its tasks after the occurrence of faults. The ultimate goal of fault tolerance is to prevent system failures from ever occurring. Since failures are directly caused by errors, the terms fault tolerance and error tolerance are often used interchangeably. For our purposes, however, we use the term fault tolerance.

Fault tolerance can be achieved by many techniques. Certainly, fault masking is one approach to tolerating faults that have occurred. Another approach is to detect and locate the fault that has occurred and reconfigure the system to remove the faulty component. **Reconfiguration** is the process of eliminating a faulty entity from a system and restoring the system to some operational condition or state. If the reconfiguration technique is used, the designer must be concerned with the following processes:

1. **Fault detection** is the process of recognizing that a fault has occurred. Fault detection is often required before any recovery procedure can be implemented.
2. **Fault location** is the process of determining where a fault has occurred so that an appropriate recovery can be implemented.

3. **Fault containment** is the process of isolating a fault and preventing the effects of that fault from propagating throughout a system. Fault containment is required in all fault-tolerant designs.
4. **Fault recovery** is the process of remaining operational or regaining operational status via reconfiguration even in the presence of faults.

Equivalent definitions can be provided in the informational universe:

1. **Error detection** is the process of recognizing that an error has occurred.
2. **Error location** is the process of determining which specific module produced the error.
3. **Error containment** is the process of preventing the error from propagating throughout a system.
4. **Error recovery** is the process of regaining operational status or restoring the system's integrity after the occurrence of an error.

A typical fault-tolerant system that employs reconfiguration performs as follows. Suppose a fault occurs in the system. Fault detection techniques can identify that a fault exists, and fault location procedures specifically identify the source of the fault. Fault recovery and reconfiguration techniques then remove the faulty module and either replace it with a fault-free module or degrade the functionality of the system to keep it operational.

Summary

This chapter has presented the definitions of several terms that are crucial to an understanding of fault-tolerant system design. In addition, we have discussed the basic causes and characteristics of faults in digital systems. Each fundamental term and concept is summarized in the following list.

Component Defects—physical imperfections or flaws in an electronic component.

Determinate Fault—a fault whose status remains unchanged throughout time.

Error—the occurrence of an incorrect value in some unit of information within a system.

Error Containment—the process of preventing an error from propagating throughout a system.

Error Detection—the process of recognizing that an error has occurred.

Error Latency—the length of time between the occurrence of an error and the appearance of a system failure.

Error Location—the process of determining the source of an error.

Error Recovery—the process of regaining operational status and restoring a system's integrity after the occurrence of an error.

External Disturbances—an action external to the system that produces a hardware or software fault.

External Universe—the domain where the user of a system ultimately sees the effects of faults and errors. The external universe is where failures occur.

Failure—a deviation in the expected performance of a system.

Fault—a physical defect, imperfection, or flaw that occurs in hardware or software.

Fault Avoidance—a technique that attempts to prevent the occurrence of faults.

Fault Cause—one of four basic items that can result in faults: specification mistakes, implementation mistakes, component defects, and external disturbances.

Fault Containment—the process of confining the effects of a fault to a limited locality.

Fault Detection—the process of recognizing that a fault has occurred.

Fault Duration—the length of time that a fault is active in a system.

Fault Extent—the characteristic that specifies whether a fault is localized to a given module or globally affects the system.

Fault Latency—the length of time between the occurrence of a fault and the appearance of an error.

Fault Location—the process of determining where a fault has occurred.

Fault Masking—the process of preventing faults from introducing errors.

Fault Nature—a characteristic of a fault that describes its type. The fault type can be hardware, software, digital, or analog.

Fault Recovery—the process of maintaining or regaining operational status after a fault has occurred.

Fault Tolerance—the ability to continue the correct performance of functions in the presence of faults.

Fault Value—the characteristic that specifies whether a fault is determinate or indeterminate.

Implementation Mistakes—incorrect actions occurring during the transformation of hardware and software specifications into physical hardware and actual software.

Indeterminate Fault—a fault whose status changes from time to time.

Informational Universe—the domain that contains units of information and is the location of errors.

Intermittent Fault—a fault that appears, disappears, and then reappears within a system.

Latent Fault—a fault that is present within a system but that has not yet produced an error.

Logical Stuck-Fault Model—a representation that assumes all faults will appear as lines in the logic diagram being physically stuck at a logic 1 or logic 0 value.

Malfunction—the incorrect performance of some system function. A malfunction is equivalent to a failure.

Permanent Fault—a fault that remains in a system indefinitely.

Physical Universe—a domain that contains physical entities and is the location of faults.

Reconfiguration—the process of eliminating a faulty entity from a system and restoring the system to some operational state.

Specification Mistakes—incorrect algorithms, architectures, or hardware/software design specifications that lead to either hardware or software faults.

Stuck-open Fault—a physical fault resulting in the output of a gate depending on the present input and the previous output.

Transient Fault—a fault that appears and then disappears a short time later.

Transistor Stuck-Fault Model—a representation that assumes all faults will appear as transistors in the circuit diagram being physically stuck on or stuck off.

User's Universe—equivalent to the external universe. The domain where the user sees the effects of faults and errors.

References

1. Avizienis, A. "The four-universe information system model for the study of fault tolerance," *Proceedings of the 12th Annual International Symposium on Fault-Tolerant Computing*, Santa Monica, California, June 22-24, 1982, pp. 6–13.
2. Hayes, J.P. "Fault modeling," *IEEE Design and Test*, Vol. 2, No. 2, April 1985, pp. 88–95.
3. Johnson, B.W. "Fault-tolerant microprocessor-based systems," *IEEE Micro*, Vol. 4, No. 6, December 1984, pp. 6–21.
4. Kohavi, Z. *Switching and Finite Automata Theory*, McGraw-Hill, New York, 1978.
5. Laprie, J-C. "Dependable computing and fault tolerance: Concepts and terminology," *Proceedings of the 15th Annual International Symposium on Fault-Tolerant Computing*, Ann Arbor, Mich., June 19-21, 1985, pp. 2–11.

6. Nelson, V.P., and B.D. Carroll. "Fault-tolerant computing (A tutorial)," presented at the AIAA Fault Tolerant Computing Workshop, November 8-10, 1982, Fort Worth, Tex.
7. Patel, J.H., and L.Y. Fung. "Concurrent error detection in ALUs by recomputing with shifted operands," *IEEE Transactions on Computers*, Vol. C-31, No. 7, July 1982, pp. 589–595.
8. Wadsack, R.L. "Fault modeling and logic simulation of CMOS and MOS integrated circuits," *The Bell System Technical Journal*, Vol. 57, No. 5, May-June 1978, pp. 1449–1475.

Additional Reading

The following list of references is provided for the reader who is interested in pursuing the topics of this chapter in more detail.

Anderson, T., and P.A. Lee. *Fault Tolerance Principles and Practices*, Prentice-Hall International, London, 1981.

Anderson, T., and P.A. Lee. "Fault tolerance terminology proposals," *Proceedings of the 12th Annual International Symposium on Fault-Tolerant Computing*, Santa Monica, Calif., June 22-24, 1982, pp. 29–33.

Avizienis, A. "Fault tolerance: The survival attribute of digital systems," *Proceedings of the IEEE*, Vol. 66, No. 10, October 1978, pp. 1109–1125.

Banerjee, P., and J.A. Abraham. "Characterization and testing of physical failures in MOS logic circuits," *IEEE Design and Test*, Vol. 1, No. 4, August 1984, pp. 76–86.

Beh, C.C., K.H. Arya, C.E. Radke, and K.E. Torku. "Do stuck fault models reflect manufacturing defects?" *Proceedings of the 1982 International Test Conference*, October 1982, pp. 35–42.

Bryant, R.E. "A switch-level model and simulator for MOS digital systems," *IEEE Transactions on Computers*, Vol. C-33, No. 2, February 1984, pp. 160–177.

Bryant, R.E., and M.D. Schuster. "Fault simulation of MOS digital circuits," *VLSI Design*, Vol. 4, No. 10, October 1983, pp. 24–30.

Carter, W.C. "A time for reflection," *Proceedings of the 12th Annual International Symposium on Fault-Tolerant Computing*, Santa Monica, Calif., June 22–24, 1982, p. 41.

Case, G.R. "Analysis of actual fault mechanisms in CMOS logic gates," *Proceedings of the 13th Design Automation Conference*, June 1976, pp. 265–270.

Chandramouli, R. "On testing stuck-open faults," *Proceedings of the Fault Tolerant Computing Symposium*, 1983, pp. 258–265.

Courtois, B. "Failure mechanisms, fault hypothesis, and analytical testing on LSI NMOS (HMOS) circuits," *VLSI 81*, University of Edinburgh, August 1981, Academic Press, 1981.

El-ziq, Y.M. "Classifying, testing, and eliminating VLSI MOS failures," *VLSI Design*, Vol. 4, No. 9, September 1983, pp. 30–35.

El-ziq, Y.M. "Automatic test generation for stuck-open faults in CMOS VLSI," *Proceedings of the 18th Design Automation Conference*, June 1981, pp. 347–352.

Galiay, J., Y. Crouzet, and M. Verniault. "Physical versus logical fault models in MOS LSI circuits: Impact on their testability," *IEEE Transactions on Computers*, Vol. C-29, No. 6, June 1980, pp. 527–531.

Goldberg, J. "A time for integration," *Proceedings of the 12th Annual International Symposium on Fault-Tolerant Computing*, Santa Monica, Calif., June 22–24, 1982, p. 42.

Gupta, A.K., and J.R. Armstrong. "Functional fault modeling and simulation for VLSI devices", Department of Electrical Engineering, Virginia Polytechnic Institute and State University, Blacksburg, Va., 1985.

Hayes, J.P. "Modeling faults in digital logic circuits," in *Rational Fault Analysis*, R. Saeks and S.R. Liberty, ed., Marcel-Dekker, New York, 1977, pp. 78–95.

Hayes, J.P. "Fault modeling for digital MOS integrated circuits," *IEEE Transactions on Computer-Aided Design*, Vol. CAD-3, No. 3, July 1984, pp. 200–207.

Hayes, J.P. "An introduction to switch-level modeling," *IEEE Design and Test of Computers*, Vol. 4, No. 4, August 1987, pp. 18–25.

Kopetz, H. "The failure-fault (FF) model," *Proceedings of the 12th Annual International Symposium on Fault-Tolerant Computing*, Santa Monica, California, June 22-24, 1982, pp. 14–17.

Lala, P.K. *Fault Tolerant and Fault Testable Hardware Design*, Prentice-Hall International, London, England, 1985.

Lee, P.A., and D.E. Morgan. "Fundamental concepts of fault-tolerant computing—progress report," *Proceedings of the 12th Annual International Symposium on Fault-Tolerant Computing*, Santa Monica, California, June 22-24, 1982, pp. 34–38.

Mangir, T.E. "Sources of failure and yield improvement for VLSI," *Proceedings of the IEEE*, Vol. 72, No. 6, June 1984, pp. 690–708.

Pradhan, D.K. *Fault-Tolerant Computing—Theory and Techniques*, Volumes I and II, Prentice-Hall, Englewood Cliffs, N.J., 1986.

Reddy, S.M., M.K. Reddy, and V.D. Agrawal. "Robust tests for stuck-open faults in CMOS combinational logic circuits," *Proceedings of the 14th International Fault Tolerant Computing Symposium*, June 1984, pp. 44–49.

Rennels, D.A. "Fault-tolerant computing—concepts and examples," *IEEE Transactions on Computers*, Vol. C-33, No. 12, December 1984, pp. 1116–1129.

Robinson, A.S. "A user oriented perspective of fault-tolerant systems models and terminologies," *Proceedings of the 12th Annual International Symposium on Fault-Tolerant Computing*, Santa Monica, Calif., June 22-24, 1982, pp. 22–28.

Siewiorek, D.P., and R.S. Swarz. *The Theory and Practice of Reliable System Design*, Digital Press, Bedford, Mass., 1982.

Timoc, C., M. Buehler, T. Griswold, C. Pina, F. Stott, and L. Hess. "Logical models of physical failures," *Proceedings of the International Test Conference*, 1983, pp. 546–553.

Problems

2.1. Devise an original example to illustrate the differences between faults, errors, and failures. As you illustrate these concepts, relate them to the three-universe model.

2.2. Some systems are designed for reliability, whereas others are designed to be highly available. Explain the difference between these two concepts and give an original example of an application requiring high reliability and one that needs high availability. How would you expect the four different causes of faults to complicate the designs of high reliability applications as compared to high availability applications?

2.3. Faults can be characterized by five major attributes. Give original examples of faults that illustrate each of these attributes.

2.4. Develop a design methodology for digital systems that incorporates fault avoidance at each of the major steps. In other words, examine the design process and determine where fault avoidance techniques such as designs reviews should be incorporated. Pretend you are making recommendations to your employer on how fault avoidance can be used to improve the quality of your company's products.

2.5. Fault masking is an attractive technique for use in systems that cannot allow even momentary erroneous results to be generated. However, fault masking does have several serious limitations. In your opinion, what are the disadvantages of using a fault masking approach?

2.6. Develop a glossary of the key terms used in the fault-tolerant computing field. Keep the glossary available to use as a reference and study guide.

2.7. Explain the difference between the stuck-at-1, stuck-at-0 fault model and the transistor stuck-fault model. Use the transistor diagram of a CMOS NAND gate to illustrate a physical fault that is covered by the stuck-at-1, stuck-at-0 fault model but is not covered by the transistor stuck-fault model. Also, determine a physical fault that is covered by the transistor stuck-fault model but is not covered by the stuck-at-1, stuck-at-0 fault model.

2.8. The stuck-at-0, stuck-at-1 fault model was developed when transistor transistor logic (TTL) was extremely popular. Contrast the differences between TTL and CMOS, and develop an explanation of whether the stuck-at-0, stuck-at-1 fault model is more appropriate for TTL than CMOS. If the model is more appropriate, provide a detailed discussion of why.

3

Design Techniques to Achieve Fault Tolerance

3.1 Introduction
3.2 Primary Design Issues
3.3 The Concept of Redundancy
3.4 Hardware Redundancy
3.5 Information Redundancy
3.6 Time Redundancy
3.7 Software Redundancy
Summary
References
Additional Reading
Problems

3.1 Introduction

The two preceding chapters covered the fundamental definitions and concepts in fault-tolerant computing. In this chapter, we examine the techniques that are available to achieve fault tolerance. In particular, we begin by identifying the fundamental design issues, defining the concept of redundancy, and examining the forms that redundancy can take. The overall purpose of this chapter is to provide the reader with the fundamental techniques that are used in the design of fault-tolerant systems. Chapter 4 then examines evaluation techniques that allow us to compare two or more approaches and select the best approach for a particular application. Chapter 5 provides examples of systems that use the various techniques presented in this chapter.

3.2 Primary Design Issues

The development of a fault-tolerant system requires the consideration of many design issues. Among these are fault detection, fault containment, fault location, fault recovery, and fault masking. Each of these concepts was defined in Chapter 2, but here we consider the role of each of these functions in the design of fault-tolerant systems.

A system that employs fault masking achieves fault tolerance by "hiding" faults that occur. Such systems do not require that the fault be detected before it can be tolerated, but it is required that the fault be contained. In other words, we want to make the *effect* attribute of all faults local; we do not want one fault to globally affect the performance of a system. Fault containment techniques prevent the effects of faults from spreading throughout a system. Fault masking is often an effective method of achieving fault containment.

Systems that do not use fault masking require fault detection, fault location, and fault recovery to achieve fault tolerance. Fault detection is the cornerstone of many fault-tolerant systems. The ability to detect faults is essential to the fault location and fault recovery processes. Fault location is required after fault detection to identify exactly which component is faulty such that a fault recovery procedure can be implemented. Typically, fault recovery involves some form of reconfiguration that is usually accomplished by disabling, either physically or logically, a faulty component and enabling, again either physically or logically, a replacement component.

This chapter introduces techniques for achieving fault detection, fault location, fault containment, fault masking, fault recovery, and fault tolerance, each of which requires the use of some form of redundancy. Before discussing the specific forms of redundancy, a fundamental definition of redundancy is presented and illustrated.

3.3 The Concept of Redundancy

In the early days of fault-tolerant designs, the concept of redundancy was almost always considered as physical hardware redundancy. The most common technique used to achieve some form of fault tolerance was the physical replication of boxes or hardware components within a system. We now have a better understanding of redundancy and the various forms of redundancy that can occur [Johnson 1984].

Redundancy is simply the addition of information, resources, or time beyond what is needed for normal system operation. The redundancy can take one of several forms:

1. **Hardware redundancy** is the addition of extra hardware, usually for the purpose of either detecting or tolerating faults.
2. **Software redundancy** is the addition of extra software, beyond what is needed to perform a given function, to detect and possibly tolerate faults.
3. **Information redundancy** is the addition of extra information beyond that required to implement a given function; for example, error detecting codes use a form of information redundancy.
4. **Time redundancy** uses additional time to perform the functions of a system such that fault detection and often fault tolerance can be achieved.

To illustrate the concept of redundancy, consider a simple digital filtering application. The hardware that might be used in the digital filter is shown in Fig. 3.1 and includes an analog-to-digital converter, a small microprocessor, and a digital-to-analog converter. All the hardware shown in Fig. 3.1 is required to implement the function of the digital filter, so there is no hardware redundancy in this system. Likewise, suppose that the software that executes in the microprocessor performs no functions other than the minimum required to implement the filter. Consequently, the digital filter contains no redundancy in the software. In a similar manner, we can see that the simple digital filter contains no redundancies of any form as long as the system is designed strictly to implement the basic requirements of the digital filter.

Suppose now that we want to enhance the design of the digital filter to allow some primitive forms of fault detection. One simple and familiar approach is to add several lines of software to perform a validity check on the results that are being generated. For example, we may know that the values generated by the digital filter should never overflow the range that can be

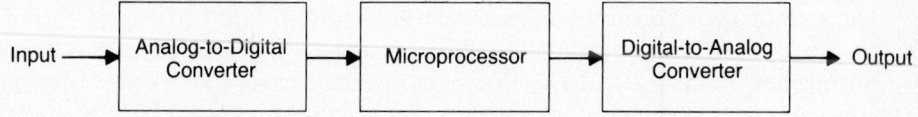

Fig. 3.1 A simple digital system used to implement a digital filter. The system is designed such that no redundancy is present.

represented by an 8-bit digital word. Consequently, we can add several lines of software to verify that we never have an overflow from the 8-bit range of representation. The extra software that has been added is redundant in that it is not needed to perform the basic functions of the digital filter, but it was added to give an additional capability. If adding the extra software requires adding more memory to store the program, hardware redundancy has been introduced to the system because the extra memory was not mandatory for the basic digital filtering operation.

Now suppose that we are concerned about transient faults upsetting the digital filtering system. In an attempt to detect transients, we can perform each calculation of the filter twice at separate times. If the two results differ, some transient fault may have upset one of the two calculations. The required computations of the filtering application now take longer, so we have introduced time redundancy into the system.

Finally, suppose that we modify the output of the digital filter such that each word is appended with a single bit. The extra bit assumes a value of 1 if the unappended output of the filter has an even number of 1s and a value of 0 if the unappended output has an odd number of 1s; this is the concept of odd parity. The added bit has additional information that can be used for the purpose of detecting faults; specifically, the extra bit indicates whether the total number of bits in the information is even or odd. For example, if one bit in the resulting appended output becomes corrupted, the number of 1s in the information changes, and the erroneous condition can be detected. Simple parity schemes such as this are one form of information redundancy. The extra information is not needed to perform the basic task of digital filtering, but it can provide some additional capability such as error detection.

The addition of redundancy never comes cheaply. The extra software in the preceding example would require extra time to develop as well as extra expense. Likewise, the extra memory necessary to store the extra lines of code would increase the filter's cost, weight, power consumption, and size. If the time redundancy approach is employed, we might be forced to use a faster, more expensive processor to allow the performance of all computations twice and still meet the sampling rate requirements of the filtering application. Finally, the information redundancy used in the filtering application requires either additional software or hardware to generate and store the extra bit of information and to interpret what that information means.

The use of redundancy can provide additional capabilities within a system. In fact, if fault tolerance or fault detection is required, some form of redundancy is also required. But, redundancy can have a very important impact on a system's performance, size, weight, power consumption, and reliability. We must understand the types of redundancy techniques available and the methods that can be used to evaluate the impact of the redundancy.

3.4 Hardware Redundancy

The physical replication of hardware is perhaps the most common form of redundancy used in digital systems today. As semiconductor components have become smaller and less expensive, the concept of hardware redundancy has become more common and more practical. The costs of replicating hardware within a system are decreasing simply because hardware costs are decreasing.

There are three basic forms of hardware redundancy [Johnson 1984]: passive, active, and hybrid. **Passive** techniques use the concept of fault masking to hide the occurrence of faults and prevent the faults from resulting in errors. Passive approaches are designed to achieve fault tolerance without requiring any action on the part of the system or an operator. Passive techniques, in their most basic form, mask faults rather than detect them.

The **active** approach, which is sometimes called the dynamic method, achieves fault tolerance by detecting the existence of faults and performing some action to remove the faulty hardware from the system. In other words, active techniques require that the system be reconfigured to tolerate faults. Active hardware redundancy uses fault detection, fault location, and fault recovery in an attempt to achieve fault tolerance.

Hybrid techniques combine the attractive features of both the passive and active approaches. Fault masking is used in hybrid systems to prevent erroneous results from being generated. Fault detection, fault location, and fault recovery are also used in the hybrid approaches to improve fault tolerance by removing faulty hardware and replacing it with spares. A spare element within a system is one that is extra; the spare element is not needed until another element becomes faulty, at which time the spare is used to replace the faulty element. Providing spares is one form of providing redundancy in a system. Hybrid methods are most often used in the critical-computation applications where fault masking is required to prevent momentary errors, and high reliability must be achieved. Hybrid hardware redundancy is usually a very expensive form of redundancy to implement.

3.4.1 Passive Hardware Redundancy

Passive hardware redundancy relies on voting mechanisms to mask the occurrence of faults. Most passive approaches are developed around the concept of majority voting. As previously mentioned, the passive approaches achieve fault tolerance without the need for fault detection or system reconfiguration; the passive designs inherently tolerate the faults.

Triple Modular Redundancy

The most common form of passive hardware redundancy is called **triple modular redundancy** (TMR). The basic concept of TMR (illustrated in Fig. 3.2) is to triplicate the hardware and perform a majority vote to determine the output of the system. If one of the modules becomes faulty, the two remaining fault-free modules mask the results of the faulty module when the majority vote is performed. In typical applications, the replicated modules are processors, memories, or any hardware entity. In addition, TMR can be applied to software where three different versions of programs that perform the same function are used to protect against software faults in any one of the three. For the moment, however, we will focus on hardware techniques.

The primary difficulty with TMR is the voter; if the voter fails, the complete system fails. In other words, the reliability of the simplest form of TMR can be no better than the reliability of the voter. Any single component within a system whose failure leads to a failure of the system is called a **single point of failure.** Several techniques can be used to overcome the effects of voter failure. One approach is to triplicate the voters and provide three independent outputs, as shown in Fig. 3.3. In Fig. 3.3, the three functional modules each receive identical inputs and perform identical functions using those inputs. The results generated by the three modules are voted on to produce three results. Each result is correct as long as no more than one module, or input, is faulty. Several stages of TMR can be interconnected using this approach, as shown in Fig. 3.4. If a voter fails in one stage

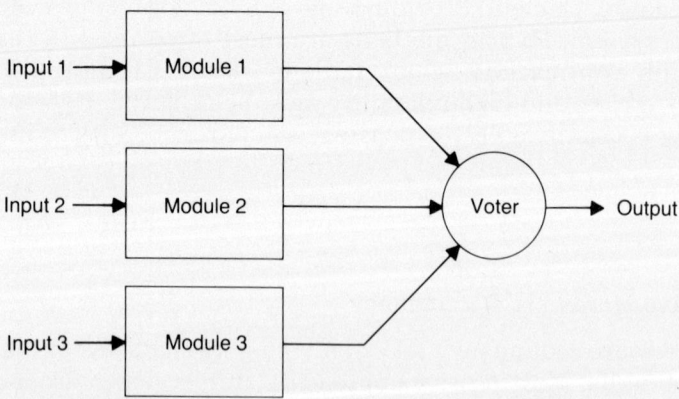

Fig. 3.2 Triple modular redundancy (TMR) uses three identical modules, performing identical operations, with a majority voter determining the output.

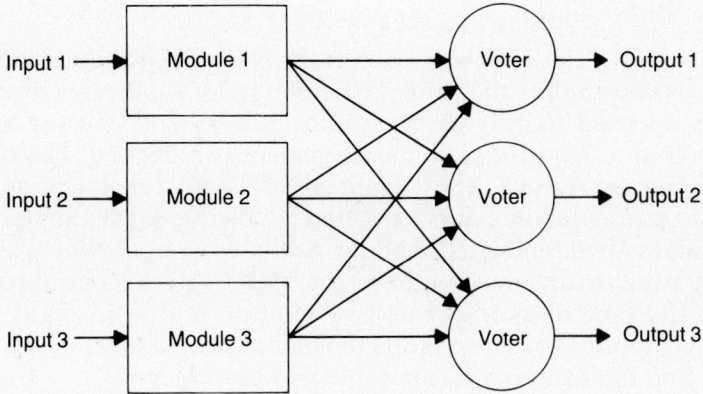

Fig. 3.3 Triple modular redundancy with triplicated voters can be used to overcome susceptibility to voter failure. The voter is no longer a single point of failure in the system.

of the system, the subsequent stage sees the failure as one input becoming corrupted. Voting at the output of the stage that gets the erroneous input corrects the erroneous result. A TMR system with triplicated voters is commonly called a **restoring organ** because the configuration produces three correct outputs even if one input is faulty. In essence, the TMR with triplicated voters restores the error-free signal.

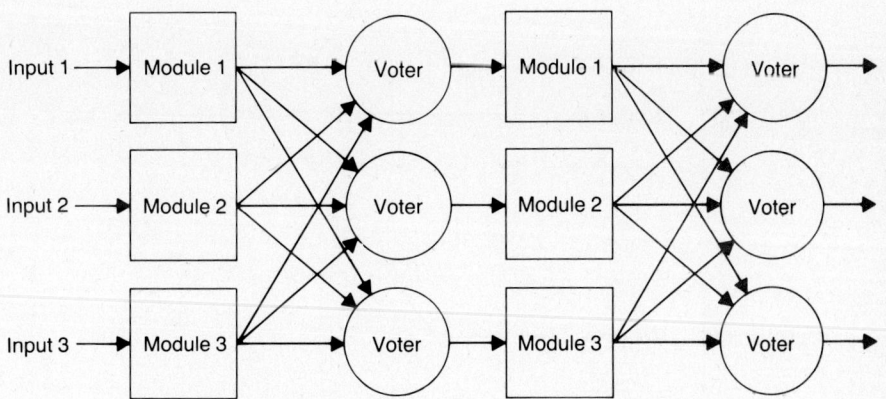

Fig. 3.4 In multiple-stage TMR systems, voting occurs between each stage so that errors are corrected before being passed to a subsequent module.

N-Modular Redundancy

A generalization of the TMR approach is the **N-modular redundancy (NMR)** technique. NMR applies the same principle as TMR but uses N of a given module as opposed to only three. In most cases, N is selected as an odd number so that a majority voting arrangement can be used. The concept of NMR is shown in Fig. 3.5. The advantage of using N modules rather than three is that more module faults can often be tolerated. For example, a 5MR system contains five replicated modules and a voter as shown in Fig. 3.6. A majority voting arrangement allows the 5MR system to produce correct results in the face of as many as two module faults. In many critical-computation applications, two faults must be tolerated to allow the required reliability and fault tolerance capabilities to be achieved.

The primary tradeoff in NMR is the fault tolerance achieved versus the hardware required. Clearly, practical applications must limit the amount of redundancy that can be employed. Power, weight, cost, and size limitations very often determine the value of N in an NMR system.

Voting Techniques

Voting within NMR systems can occur at several points. For example, an industrial controller can sample the temperature of a chemical process

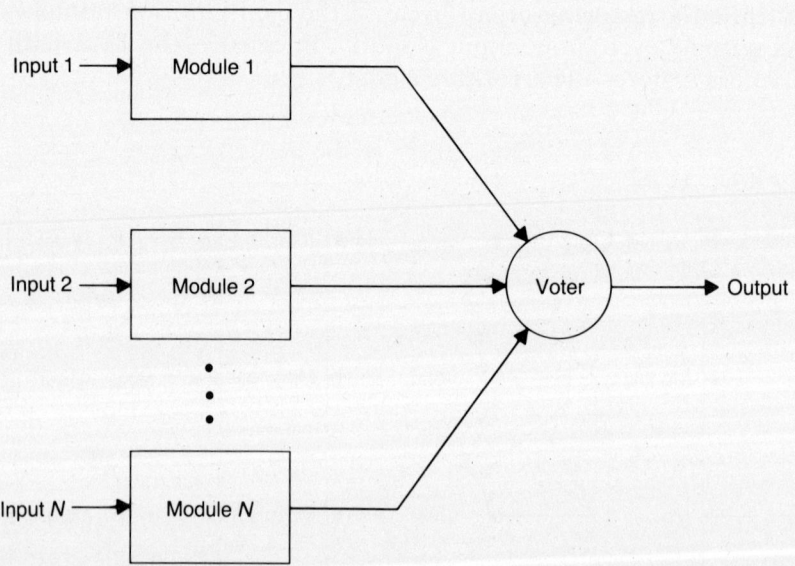

Fig. 3.5 N-modular redundancy (NMR) is a generalization of TMR with N identical modules performing identical operations. If N is odd, majority voting can be used to produce an output.

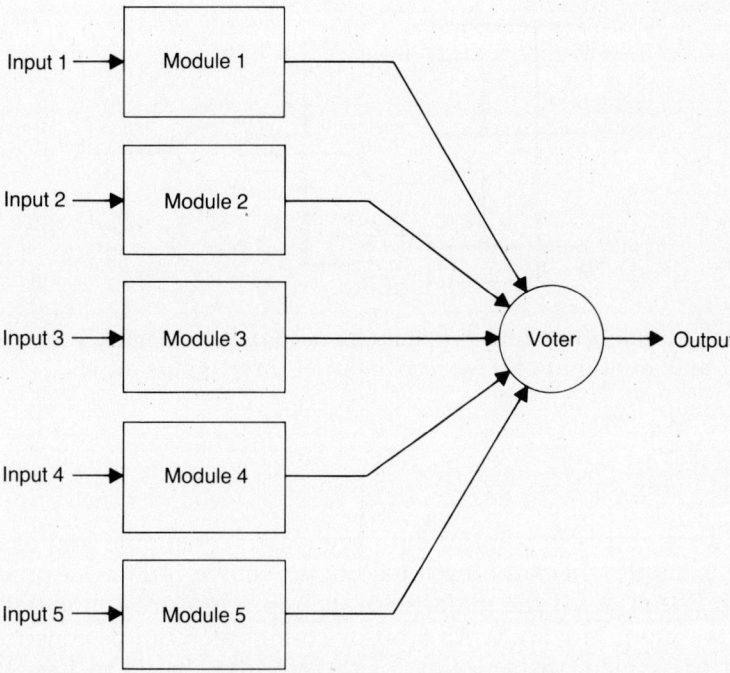

Fig. 3.6 5MR is an example of NMR with five identical modules. Majority voting allows the failure of two modules to be tolerated.

from three independent sensors, perform a vote to determine which of the three sensor values to use, calculate the amount of heat or cooling to provide to the process (the calculations being performed by three or more separate modules), and then vote on the calculations to determine a result. The voting can be performed on both analog and digital data. The alternative, in this example, might be to sample the temperature from three independent sensors, perform the calculations, and then provide a single vote on the final result. The primary difference between the two approaches is fault containment. If voting is not performed on the temperature values from the sensors, the effect of a sensor fault is allowed to propagate beyond the sensors and into the primary calculations. Voting at the sensors, however, masks, and contains, the effects of a sensor fault. Providing several levels of voting, however, does require additional redundancy, and the benefits of fault containment must be weighed against the cost of the extra redundancy.

Voting not only involves a number of design tradeoffs, it also poses several procedural problems. The first is deciding whether a hardware voter will be used or whether the voting process will be implemented in software.

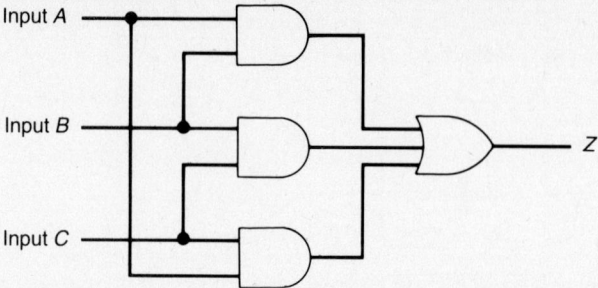

Fig. 3.7 A 1-bit majority voter produces an output of 1 when two out of three inputs are 1 and an output of 0 when two out of three inputs are 0.

Hardware voters for digital data are relatively simple and easy to design. For example, the combinational circuit shown in Fig. 3.7 produces an output bit Z that is 1 if the majority of the input bits are 1, and 0 if the majority of the input bits are 0. An 8-bit or 16-bit voter can be constructed using eight or sixteen, respectively, of the circuits shown in Fig. 3.7. Each circuit operates independently on the appropriate bits of each word of digital data. The time required to perform the vote using hardware is simply the propagation delay through the digital logic circuit.

In most practical applications, timing is critical in the voting procedure. If values arrive at the voter at slightly different times, incorrect results can be generated temporarily. In many applications, an incorrect result cannot be allowed for even a very small period of time. To overcome timing problems, flip flops can be used at the inputs of the voter to synchronize the voting process. An example is shown in Fig. 3.8 where master-slave flip flops are used to solve timing problems at the inputs of a 1-bit voter. Master-slave flip flops are shown here because of their extensive use in traditional digital systems design. Each D flip flop shown in Fig. 3.8 is positive-edge triggered. In other words, each D flip flop stores the value present on the D input at the time when the clock goes from 0 to 1.

The timing diagram in Fig. 3.8 shows a two-phase clock that drives the master-slave flip flops. The inputs to the voter are stored in the master flip flop on the positive edge of the Phase 1 clock pulse. On the positive edge of the Phase 2 clock, the data is stored in the slave flip flop, which then applies the data to the combinational circuit that actually performs the voting. The output of the combinational circuit is stored in the output master flip flop on the positive edge of the Phase 1 clock, and the voter output Z becomes valid after the rising edge of the Phase 2 clock. An 8-, 16-, 32-, or, in general,

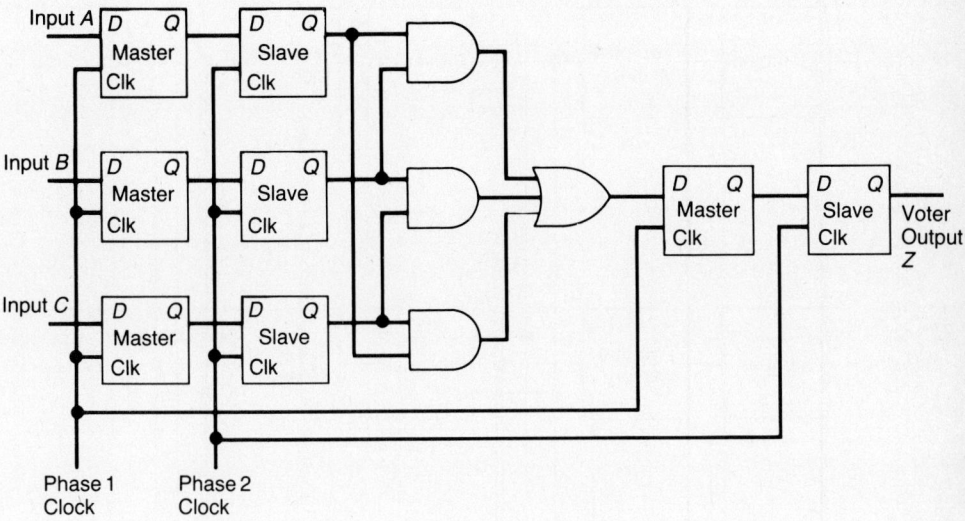
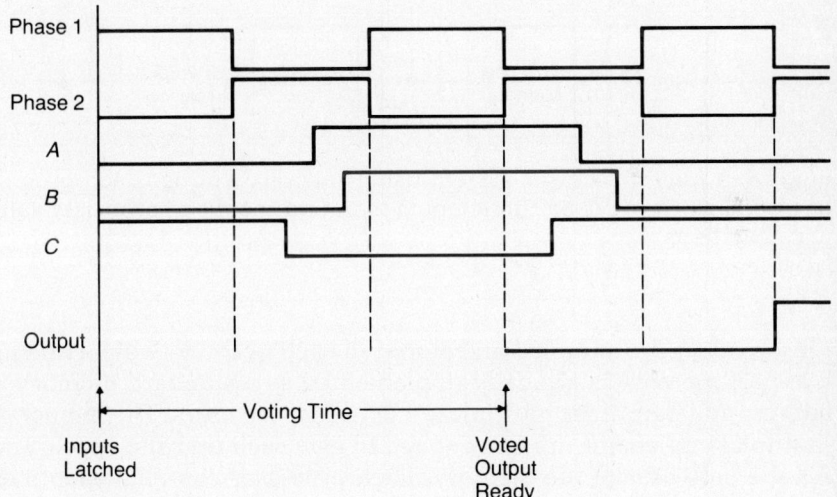

Fig. 3.8 A synchronized majority voter uses latches on the inputs and outputs to synchronize the arrival of inputs and availability of the output.

an n-bit voter can be constructed using an appropriate number of the circuits of Fig. 3.8 operating in parallel.

If voting is performed using software, a mechanism must be available to provide the software routine with the data on which to vote. An example of a microprocessor system that uses software voting is shown in Fig. 3.9. The

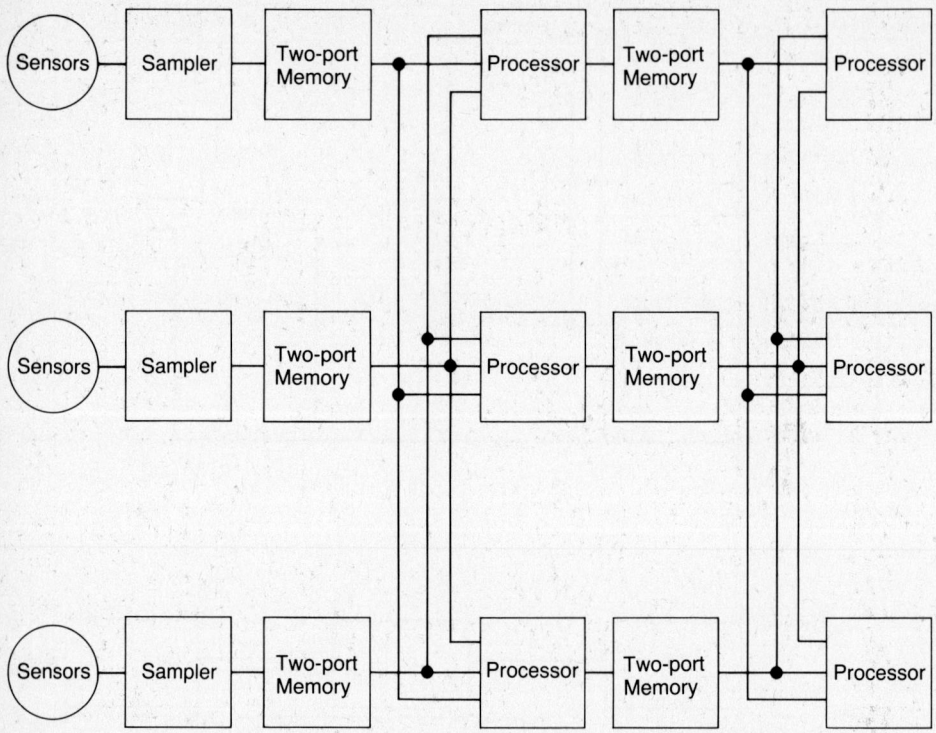

Fig. 3.9 A microprocessor system using software voting. Each processor performs a majority vote on three inputs to determine the appropriate value to use in calculations.

sensor values are sampled and stored in each of three two-port memories. A two-port memory is typically implemented as a standard memory with the address and data lines multiplexed between two users. The memory and the multiplexing scheme are fast enough to give each user the appearance of being the only user of the memory. Each processor can read each memory to obtain all the sensor values. Once each sensor value is read from memory, each processor votes independently on the data to determine which of the three results to use in its calculations. The voting program can be as simple as a sequence of three comparisons, with the outcome of the vote being the value that agrees with at least one of the other two. After completing a set of computations, each processor stores the results in a second set of two-port memories, as shown in Fig. 3.9, so that other processors can use the results. This concept is a software implementation of the triplicated voting scheme presented in Fig. 3.3. Each voter is software implemented and distributed among three processors such that the failure of any one processor does not disable the voting mechanism.

The primary tradeoff between hardware voting and software voting is speed versus hardware. In a hardware voter, the actual delay between the application of the inputs and the availability of the output can be made very small, depending on the propagation delays of the various electronic components used to construct the circuit. The disadvantage of the hardware voter is the number of logic elements that must be used; for example, a 16-bit voter using the circuit of Fig. 3.8 requires 64 logic gates and 64 master-slave flip flops. Each master-slave flip flop is actually two D flip flops, so a total of 128 D flip flops must be provided. Also, one hardware voter is required for each "voted" output that must be provided. For example, if TMR with triplicated voters is employed, three distinct hardware voters are required. The impact of the hardware required for the voter is to increase the system's power consumption, weight, and size.

By taking advantage of a processor's computational capabilities, a software voter performs the voting process with a minimum amount of additional hardware. Also, by simply modifying the software, the software voter can modify the manner in which the voting is performed. The disadvantage of the software voter is that the voting can require more time to perform simply because the processor cannot execute instructions and process data as rapidly as a dedicated hardware voter.

The decision to use hardware or software voting typically depends on:

1. The availability of a processor to perform the voting
2. The speed at which voting must be performed
3. The criticality of space, power, and weight limitations
4. The number of different voters that must be provided
5. The flexibility required of the voter with respect to future changes in the system.

A second major problem with the practical application of voting is that the three results in a TMR system, for example, may not completely agree, even in a fault-free environment. The sensors that are used in many control systems can seldom be manufactured such that their values agree exactly. Also, an analog-to-digital converter can produce quantities that disagree in the least significant bits, even if the exact signal is passed through the same converter multiple times. When values that disagree slightly are processed, the disagreement can propagate into larger discrepancies. For example, suppose that two signals x_1 and x_2 are supposed to have the same value, which is A; however, x_1 has a value of A and x_2 has a value of $A + \Delta$. Comparing x_1 and x_2 clearly produces a difference of Δ. However, if a multiplication of each signal by some constant C is performed, the difference is amplified if C is greater than 1. Specifically, $Cx_1 = CA$ and $Cx_2 = C(A + \Delta)$. The difference between the resulting products is now $C\Delta$ In other words, small differences in inputs can produce large differences in outputs that can significantly affect the voting process. Consequently, a majority voter may find that no two results agree exactly in a TMR system, even though the system may be functioning perfectly.

One approach that alleviates the problem of disagreeing results is called the **mid-value select** technique. Basically, the mid-value select approach chooses a value from the three available in a TMR system by selecting the value that lies between the remaining two. As an example, consider Fig. 3.10. Of three available signals, if two signals are the result of fault-free computations and the third is the result of a faulty calculation, one of the fault-free results should lie between the other fault-free result and the faulty result. The mid-value select technique can be applied to any system that uses an odd number of modules such that one signal must lie in the middle of the others.

Another approach that is often used when quantities never exactly agree is to ignore the least significant bits of the information. In other words, a majority vote is performed but only on the k most significant bits of the data. The assumption is that acceptable disagreements will occur only in the least-significant bits of the data. Disagreements that affect the most sig-

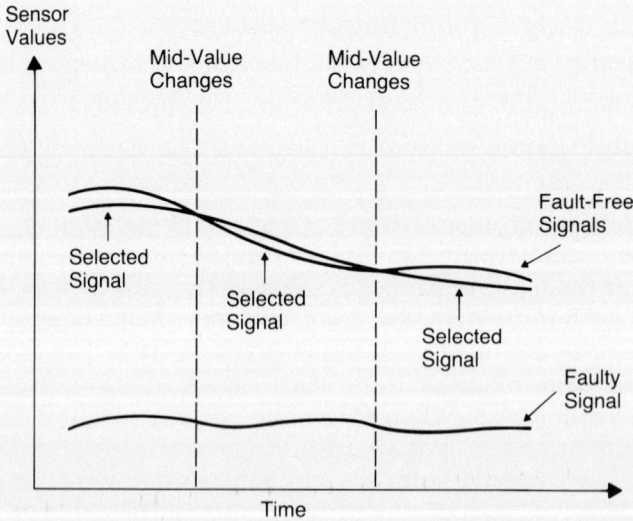

Fig. 3.10 The mid-value select technique chooses the signal that has a value in between the remaining signals. The total number of signals must be odd.

nificant bits of the information are not acceptable and must be corrected. The number of bits that are ignored depends on the application and is a function of the accuracy of the components being used.

The major difficulty with most techniques that use some form of voting is that a single result must ultimately be produced, thus creating a potential point where one failure can cause a system failure. Clearly, single points of failure are to be avoided if a system is to be truly fault-tolerant. The need for a single result is apparent in many applications. For example, banking systems must display one balance for each checking account, not three. Even though the bank's computers may vote internally on some of the results, one result must ultimately be created. The same is true in critical-computation applications such as aircraft flight control. Most aircraft, even military aircraft, do not have redundancy of the actuators, or motors, that physically move the control surfaces. Consequently, a single control signal must be provided.

Several approaches have been used successfully to create single results from redundant computations. One approach is the **flux-summing** technique, which is illustrated in Fig. 3.11. In Fig. 3.11, a TMR system is employed to control the armature current of a small motor. The flux-summing approach uses the inherent properties of closed-loop control systems to compensate for faults. The flux-summer, in this example, is a transformer that has three primary windings and a single secondary winding. The current produced in the secondary winding is proportional to the sum of the individual currents in the three primary windings. Under fault-free circum-

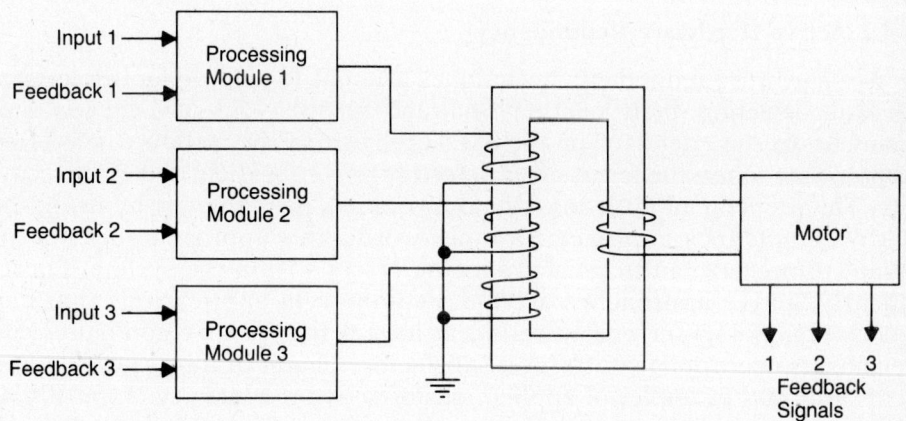

Fig. 3.11 Flux-summing uses the inherent properties of closed-loop control systems to tolerate faults. If one module becomes faulty, the remaining modules automatically compensate.

stances, each module provides approximately one-third of the total current produced in the secondary winding.

If a module fails, several scenarios can result. First, the faulty module may stop providing current to the transformer. In this case, the motor loses approximately one-third of the current necessary to maintain the present shaft position, or shaft velocity, depending on what quantity is being controlled. The remaining two modules sense, via the feedback path, that the motor is deviating from the desired position or velocity. In other words, the error signal produced by each module as part of the closed-loop control process increases. The result is that the two fault-free modules will change the current they are providing to offset the loss of current from the faulty module.

In a second failure scenario, the faulty module may provide a maximum current to the flux-summer, regardless of the input signal values. The inherent feedback of the system once again compensates for the condition by modifying the currents produced by the remaining fault-free modules. To understand the response of the flux-summer, visualize one fault-free module as providing a current of equal magnitude but of opposite polarity to the faulty module such that the effect of the faulty module is canceled. The remaining fault-free module is then capable of controlling the system.

Note that the flux-summing approach is not a voting process, but it has the same effect of masking faults. The flux-summer can be used in the basic TMR approach or in the more general NMR technique. The primary limitation is the number of coils that can be physically mounted on an iron core. The flux-summers can be designed in a very reliable manner and are extremely insensitive to external disturbances of various types.

3.4.2 Active Hardware Redundancy

Active hardware redundancy techniques attempt to achieve fault tolerance by fault detection, fault location, and fault recovery. Because the faults of many designs are detected on the basis of the *errors* they produce, it is often appropriate to use the terms error detection, error location, and error recovery. The property of fault masking, however, is not achieved by using the active redundancy approach. In other words, this approach does not attempt to prevent faults from producing errors within the system. Consequently, active approaches are most common in applications that can tolerate temporary, erroneous results as long as the system reconfigures and regains its operational status in a satisfactory length of time. Satellite systems are good examples of applications of active redundancy. Typically, it is not catastrophic if satellites have infrequent, temporary failures. In fact, it is usually preferable to have temporary failures than to accommodate the high degree of redundancy necessary to achieve fault masking.

Duplication with Comparison

The first example of active redundancy is the simple **duplication with comparison** scheme shown in Fig. 3.12. The basic concept of duplication with comparison is to develop two identical pieces of hardware, have them perform the same computations in parallel, and compare the results of those computations. In the event of a disagreement, an error message is generated. In its most basic form, the duplication concept can only detect the existence of faults, not tolerate them, because there is no method for determining which of the two modules is faulty. However, duplication with comparison can be used as the fundamental fault detection technique in an active redundancy approach.

The duplication with comparison method of active redundancy poses several potential problems. First, if the modules both receive the same input, a failure of the input device or in the lines over which the input signals must be transmitted will cause both modules to produce the same, erroneous results. Second, the comparator may not be able to perform an exact comparison, depending on the application area. Although duplicated telephone switching processors may always exactly agree if they are fault free, the processors in a digital control application may never exactly agree. Finally, faults in the comparator can cause an error indication when no error exists, or, worse yet, the comparator can fail such that eventual faults in the duplicated modules are never detected.

A technique that can be used in a duplicated microprocessor system to overcome some of the problems just mentioned is illustrated in Fig. 3.13. The basic concept is to implement the comparison process in software

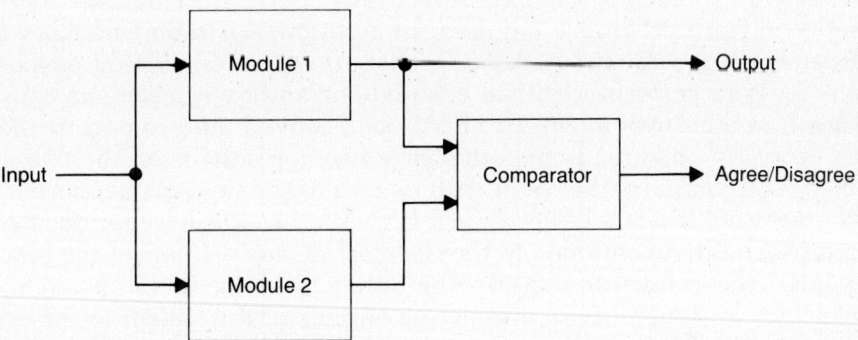

Fig. 3.12 Duplication with comparison uses two identical modules performing the same operations and compares their results. Fault detection is provided but not fault tolerance.

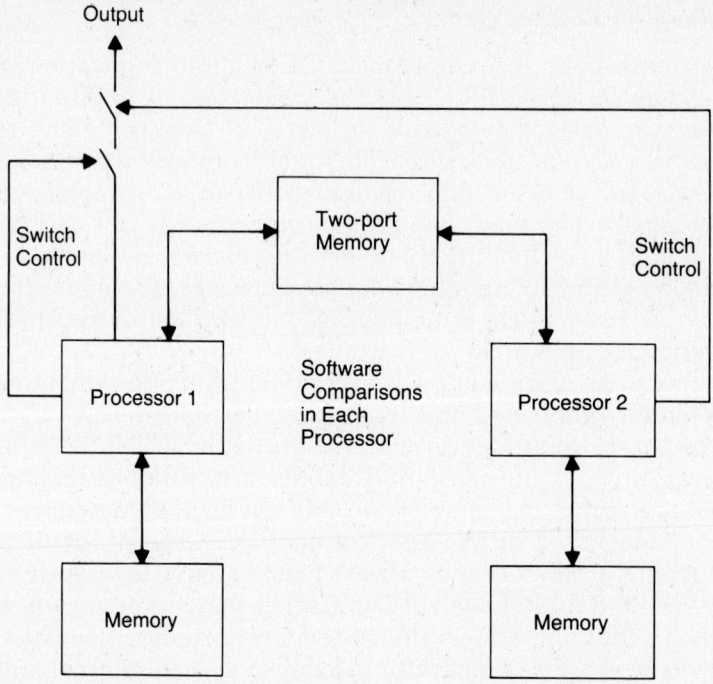

Fig. 3.13 The necessary comparisons in duplication with comparison can be implemented in software. Both processors must agree that results match before an output is generated.

that executes in each of the microprocessors. Each processor has its own memory to store programs and data. In addition, a two-port memory can transfer results from one processor to the other for comparison purposes. The processors perform identical calculations and each place one copy of the result in their own memory and a second copy in the two-port memory. Each processor then reads the other processor's results from the two-port memory and compares the result with its own. If the two-port memory fails, both processors detect a disagreement between the data stored in their own memories and that contained in the two-port memory. If one of the processors fails, the condition can also be detected using this approach. To provide for the capability to disable the outputs of the system in the event that a disagreement occurs, two sets of switches are provided. The switches can be implemented in either the digital or the analog domain, depending on the application. If either processor detects a disagreement with the other, the processor detecting the problem opens its switch and disables the outputs. Both processors must agree that their computations are identical before the system is allowed to produce an output.

The comparisons between the two processors can be performed in one of several ways. The first, and most straightforward technique, is to simply compare each digital word bit by bit. In hardware, a bit-by-bit comparison can be performed using two-input EXCLUSIVE-OR gates. The two bits to be compared are provided as inputs to the EXCLUSIVE-OR gate, and the output of the gate is 1 if the two bits disagree and 0 if they agree. n EXCLUSIVE-OR gates can be operated in parallel to achieve the desired comparison of two n-bit words. If the output of any one EXCLUSIVE-OR gate is 1, the two words do not exactly agree. In software, the comparisons are easily implemented using a COMPARE instruction that is commonly found in the instruction sets of almost all microprocessors.

Bit-by-bit comparisons have the same problems as found in voting circuits. Specifically, many applications require comparing digital words that do not agree exactly, even though the system is fault free. In such cases, it is common to compare only the most significant bits of words. For example, in the comparison of two 16-bit words, the two least significant bits might be ignored because their impact on the system is negligible. The feasibility of using such a comparison technique clearly depends on the application. In banking systems, for example, bank balances calculated on duplicate processors must agree exactly. In many real-time control applications, however, minor differences are expected and have little, if any, impact. For example, an analog signal ranging in voltage from 0 to +5 volts and generated via an 8-bit digital-to-analog converter changes by no more than approximately 20 millivolts when the least significant bit changes. In fact, the analog signal changes by no more than approximately 80 millivolts when the two least significant bits are changed.

Standby Sparing

A second form of active hardware redundancy is called the **standby sparing** (or standby replacement) technique and is illustrated in Fig. 3.14. In standby sparing, one module is operational and one or more modules serve as standbys, or spares. Various fault detection or error detection schemes are used to determine when a module has become faulty, and fault location is used to determine exactly which module, if any, is faulty. If a fault is detected and located, the faulty module is removed from operation and replaced with a spare. The reconfiguration operation in standby sparing can be viewed conceptually as a switch whose output is selected from one, and only one, of the modules providing inputs to the switch. The switch examines error reports from the error detection circuitry associated with each module to decide which module's output to use. If all modules are provid-

66   Design Techniques to Achieve Fault Tolerance

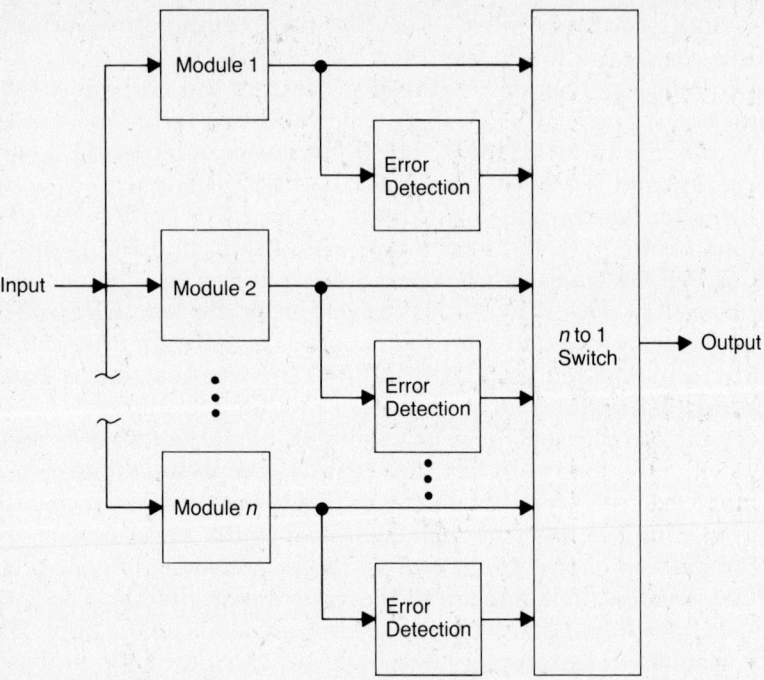

Fig. 3.14 In standby sparing, one of n modules is used to provide the system's output, and the remaining $n - 1$ modules serve as spares. Error detection techniques identify faulty modules so that a fault-free module is always selected to provide the system's output.

ing error-free results, the selection can be made using a fixed priority. Any module that provides erroneous results is eliminated from consideration.

Standby sparing can bring a system back to full operational capability after the occurrence of a fault, but it requires that a momentary disruption in performance occur while the reconfiguration is performed. If the disruption in processing must be minimized, **hot standby sparing** can be used. In the hot standby sparing technique, the spares operate in synchrony with the on line modules and are prepared to take over at any time. In contrast to hot standby sparing is **cold standby sparing** where the spares are unpowered until needed to replace a faulty module. The disadvantage of the cold standby sparing approach is the time required to apply power to a module and perform initialization prior to bringing the module into active service. The advantage of cold standby sparing is that spares do not consume power until needed to replace a faulty module. A satellite application where power consumption is extremely critical is an example where cold standby sparing may be desirable, or required. A process control system that controls a chemical reaction is an example where the reconfiguration time needs to be minimized, and cold standby sparing is undesirable, or unusable.

A key advantage of standby sparing is that a system containing n identical modules can often provide fault tolerance capabilities with significantly fewer than n redundant modules. For example, consider a multiprocessor containing n processing modules. If each processor is identical and any processor can perform the functions of the other, a single spare processor can protect the system against the failure of any one of the original n processors. In general, k spare processors can protect the system against the failure of any k of the original n processors. The percentage of redundant processors in this case is simply $(k/n) \times 100$.

A key component of the standby sparing approach is the fault detection or error detection scheme used to identify the faulty module. Throughout Chapter 3 we identify several approaches that can be employed, the first of which is the pair-and-a-spare technique.

Pair-and-a-Spare Technique

The **pair-and-a-spare** technique, shown in Fig. 3.15, combines the features present in both standby sparing and duplication with comparison. In essence, the pair-and-a-spare approach uses standby sparing; however, two modules are operated in parallel at all times and their results are compared to provide the error detection capability required in the standby sparing approach. The error signal from the comparison is used to initiate the reconfiguration process that removes faulty modules and replaces them with spares.

The reconfiguration process can be viewed conceptually as a switch that accepts the modules' outputs and error reports and provides the comparator with the outputs of two modules, one of which forms the output of the system. As long as the two selected outputs agree, the spares are not used. When a miscompare occurs, however, the switch uses the error reports from the modules to first identify the faulty module and then select a replacement module. In other words, the switch uses the error information from the comparator and the individual modules to maintain two fault-free modules operating in a duplication with comparison arrangement.

A variation on the pair-and-a-spare technique is to *always* operate modules in pairs. During the design, modules are permanently paired together, and when one module fails, neither module in the pair is used. In other words, modules are always operated and discarded in pairs so that the specific identification of which module is faulty is never required, only the identification of a faulty pair is necessary. Faulty pairs are easily identified based on the outcome of the comparison process.

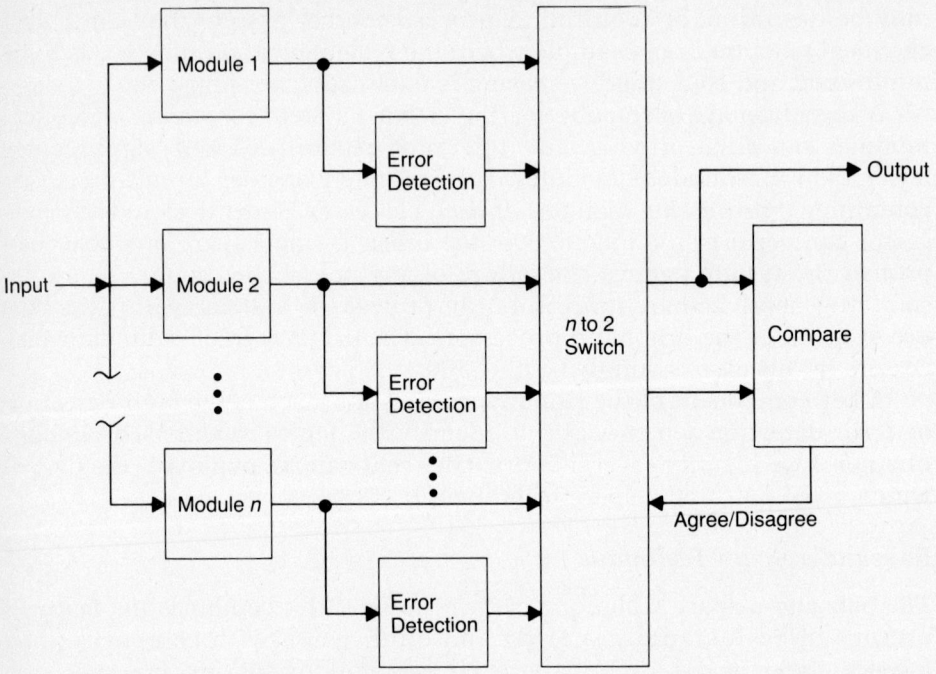

Fig. 3.15 The pair-and-a-spare technique combines duplication with comparison and standby sparing. Two modules are always online and compared, and any spare can replace either of the online modules.

Watchdog Timers

One form of active hardware redundancy that is extremely useful for detecting faults in a system is the **watchdog timer**. A watchdog timer is an active form of hardware redundancy because some action is required on the part of the system to indicate a fault-free status. The concept of a watchdog timer is that the lack of an action is indicative of a fault.

A watchdog timer is a timer that must be reset on a repetitive basis. The failure of the system to perform the reset function results in the system being reset or turned off to prevent a system failure from occurring. The fundamental assumption is that the system is fault free if it possesses the capability to repetitively perform a function such as setting a timer. For example, in standby sparing, each module that successfully and repetitively resets a timer would be interpreted as having fully functional capabilities. A module that cannot reset the timer is removed from consideration as a viable module for use within the system.

The frequency at which the timer must be reset is a function of the system. In an aircraft control system, for example, it may be imperative to detect faults within 100 milliseconds, or less, of their occurrence. Consequently, the watchdog timer must be reset at intervals of less than 100 milliseconds to allow the timer to detect a fault before any catastrophic effects of the fault occur. In banking systems, however, the faults may only need to be detected within one second, for example. Therefore, the timer is set to expire at 1-second intervals.

The watchdog timer provides good fault detection capability for certain types of faults. For example, if a processor simply ceases its functions, the watchdog timer can detect the problem. Also, if a processor becomes overloaded and requires unusual amounts of time to perform its functions, the watchdog timer detects the existence of such conditions. Watchdog timers are particularly useful for detecting a lack of response. As an analogy, if you fail to receive any mail for three or four days, you do not know whether there has not been any mail for you or if the mail has simply not run for some reason. If, however, your postman always left a blank piece of paper in your box, regardless of whether or not you had mail, and you removed the paper each day, you could determine that the postman had indeed come. The concept of the watchdog timer is identical to this simple example.

A watchdog timer can be used to detect faults in both the hardware and the software of a system. In many applications, software routines must execute in prespecified lengths of time. In digital control systems, for example, the routines execute repetitively at specific intervals. If a routine suddenly begins requiring more than the expected time to execute, a fault may have appeared in the software; for example, the fault may be an infinite loop. A timer can be used to detect the condition of a software routine requiring more than the specified length of time. The timer is set to a value that corresponds to the expected execution time at the beginning of the execution period. If the timer expires before the routine completes its tasks, the timer generates an interrupt for the processor. If the routine completes its functions before the timer expires, the timer is reset.

3.4.3 Hybrid Hardware Redundancy

The fundamental concept of hybrid hardware redundancy is to combine the attractive features of both the active and the passive approaches. Fault masking is used to prevent the system from producing erroneous results; and fault detection, fault location, and fault recovery are used to reconfigure the system in the event of a fault. Hybrid redundancy is usually very expensive in terms of the amount of hardware required to implement a system. Consequently, hybrid redundancy is most often used in applications that require extremely high integrity of the computations.

N-Modular Redundancy with Spares

Although there are several approaches to hybrid redundancy, most are based on the concept of **N-modular redundancy (NMR) with spares**. The idea of NMR with spares is to provide a basic core of N modules arranged in a voting, or a form of voting, configuration. In addition, spares are provided to replace failed units in the NMR core. The benefit of NMR with spares is that a voting configuration can be restored after a fault has occurred. For example, a design that uses TMR with one spare will mask the first module fault that occurs. If the faulty module is then replaced with the spare unit, the second module fault can also be masked, thus providing tolerance of two module faults. For a passive approach to tolerate two module faults, five modules must be configured in a fault masking arrangement. The hybrid approach can accomplish the same results using only four modules and some fault detection, location, and recovery techniques.

The NMR with spares technique is illustrated in Fig. 3.16. The system remains in the basic NMR configuration until the disagreement detector determines that a faulty unit exists. One approach to fault detection is to compare the output of the voter with the individual outputs of the modules. A

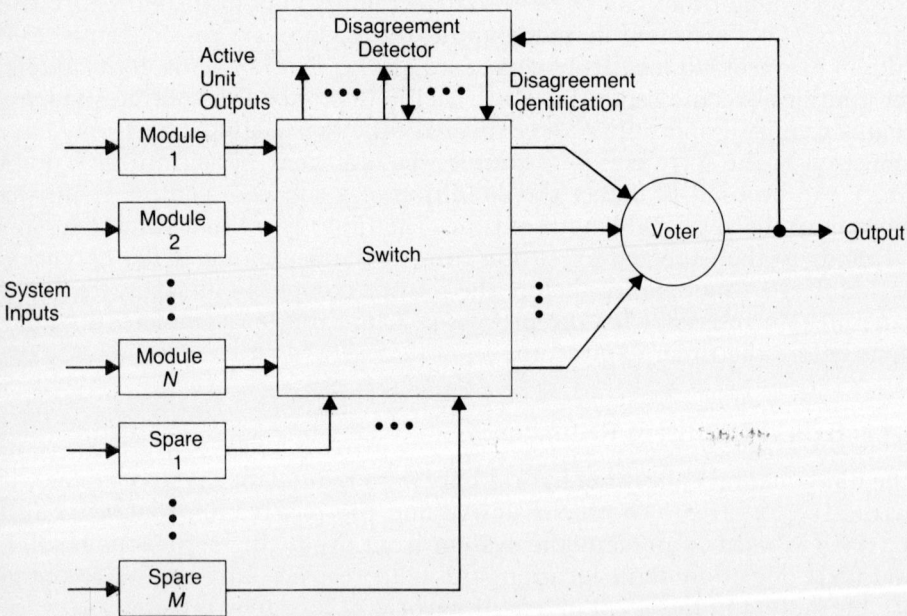

Fig. 3.16 N-modular redundancy with spares combines NMR and standby sparing. The voted output is used to identify faulty modules, which are then replaced with spares.

module that disagrees with the majority is labeled as faulty and removed from the NMR core. A spare unit is then switched in to replace the faulty module. The reliability of the basic NMR system is maintained as long as the pool of spares is not exhausted. Voting always occurs among the active participants in the NMR core, masking faults and ensuring continuous, error-free computations.

Self-Purging Redundancy

A second approach to hybrid redundancy is called **self-purging redundancy** [Losq 1976]. The basic concept of self-purging redundancy is similar to that of the NMR with spares approach. The major difference is that all units are actively participating in the system in the self-purging technique, whereas some units function as spares in the NMR approach and may not be an active part of the system until a fault occurs. The overall concept of self-purging redundancy is illustrated in Fig. 3.17. Each of the N identical modules is designed with the capability to remove itself from the system in the event that its output disagrees with the voted output of the system.

There are three basic features of the self-purging redundancy concept. First, N identical modules are obtained. Each module is capable of performing the functions required of the system. Second, a set of N switches is developed, and one switch is associated with each of the N modules. The function of the switch is to remove, or purge, its associated module from the system in the event that the module fails. Third, a voter is developed to produce the system output and provide masking of any faults that occur. The voter used in the self-purging technique is a threshold gate. Before proceeding to discuss the self-purging approach, it should be beneficial to review the basic concept of a threshold gate [Kohavi 1978].

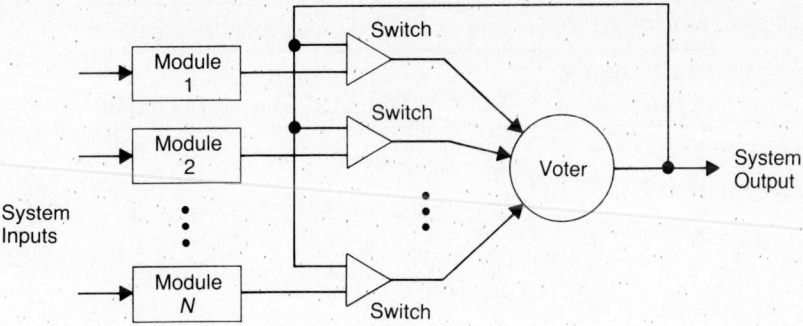

Fig. 3.17 Self-purging redundancy uses the system output to remove modules whose output disagrees with the system output. (From [Losq, 1976] © 1976 IEEE)

A binary threshold gate has n binary inputs, $x_1, x_2, \ldots, x_n$, and one binary output z. Each input is weighted according to the parameters $w_1, w_2, \ldots, w_n$. The output z, of the threshold gate is determined by comparing the weighted sum of the inputs to some prespecified threshold T. In other words, the output z, is given by

$$z = \begin{cases} 1, & \text{if } \sum_{i=1}^{n} w_i x_i \geq T \\ 0, & \text{if } \sum_{i=1}^{n} w_i x_i < T \end{cases}$$

The addition and multiplication functions are the conventional decimal operations. The summation $\sum_{i=1}^{n} w_i x_i$ is typically called the weighted sum, and T is the threshold of the gate.

The use of a threshold gate is illustrated in Table 3.1 where three input variables are used, and the threshold is set as 2. Each input is weighted by a factor of one. Any time the weighted sum of the inputs is 2 or greater, meaning at least two of the inputs have a value of 1, the output assumes the value of 1. If the weighted sum is less than 2, meaning at least two of the inputs have a value of 0, the output assumes a value of 0. In this special case, the threshold gate has the same characteristics as a majority voter.

One approach to using a threshold gate in a fault-tolerant design is to force to zero the weight of all modules that have been identified as faulty. Consequently, faulty modules do not contribute to the output of the system. For example, four modules using a threshold gate with a threshold of 2 can produce the correct output after two modules have failed as long as the contribution of the failed modules to the threshold gate is made zero. In a similar manner, the threshold can be varied to allow fewer modules to control the system. For example, if the threshold gate has three inputs, two

TABLE 3.1 Operation of a three-input threshold gate

Inputs				
X_1	X_2	X_3	Weighted sum	Output
0	0	0	0	0
0	0	1	1	0
0	1	0	1	0
0	1	1	2	1
1	0	0	1	0
1	0	1	2	1
1	1	0	2	1
1	1	1	3	1

Threshold = 2

of which have been assigned zero weights because they are faulty, the threshold can be reduced to 1 to allow the remaining fault-free module to control the system.

Perhaps the most difficult problem with threshold gates is the fact that they are analog elements; usually constructed from operational amplifiers, resistor and transistor circuits, or magnetic cores that perform the summation and multiplication operators. Consequently, threshold elements are often not practical for use in fault-tolerant digital systems. This is one reason that the self-purging technique is not commonly used.

In addition to the threshold gate, the switch is the next most basic element of the self-purging approach. The purpose of the switch is to force a module's contribution to the threshold gate to be 0 if that module produces an output that disagrees with the output of the threshold gate. In other words, the switch sets the weight associated with a given module to zero if the output of that module disagrees with the output of the system. This is accomplished through the use of a flip flop with inputs J and K and outputs Q and $\overline{Q}$ that disables the output of a module if that module produces an erroneous output. The basic structure of the switch is shown in Fig. 3.18. Typically, the flip flop has an initialization mechanism that allows a disabled module to once again contribute to the summation process, in the event that the module is restored to a fault-free state. While excluded from the

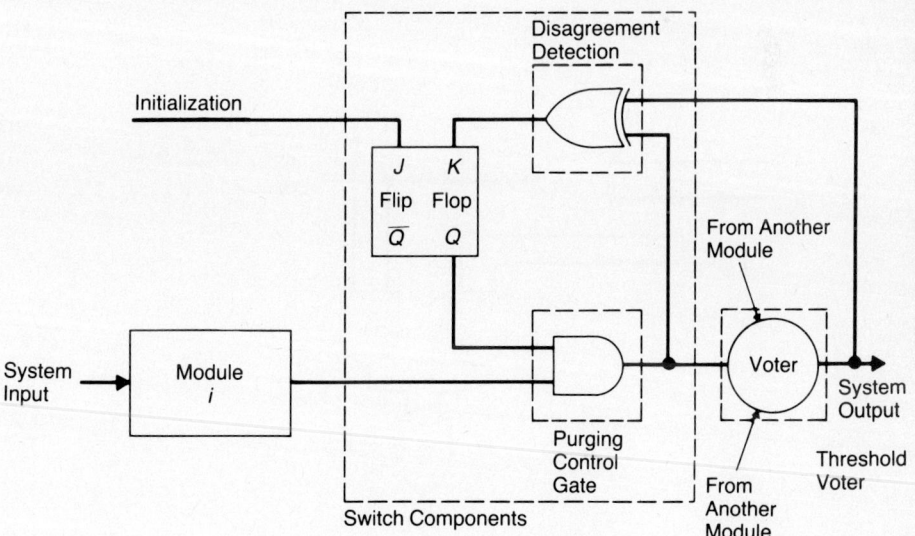

Fig. 3.18 Basic structure of the switch in a self-purging system. The switch removes the module from the voting process if the system output disagrees with the module output. (From [Losq, 1976] © 1976 IEEE)

summation, modules can be removed from the system without affecting the results generated by the system. This is an extremely attractive feature of the self-purging approach that facilitates maintenance and repair; maintenance personnel can disable individual modules and replace them without interrupting the service provided by the system.

As an example of the use of self-purging redundancy, consider the design of a 1-bit full-adder. Three full-adders are used to provide triple redundancy. Each 1-bit full-adder must produce both a sum bit and a carry bit. The logic diagram for the full-adder is shown in Fig. 3.19. Two threshold voters are required in this application; one voter for the sum bit and one for the carry bit. The switching mechanism must be modified to accommodate the multiple outputs. Two approaches can be taken. The first approach is to completely remove a module from both threshold voters in the event that either the carry bit or the sum bit is found to be in error. The problem with this concept is that the module may be perfectly capable of producing the correct carry bit even though the sum bit circuitry is faulty. If the complete module is removed, the redundancy available in both the sum circuitry and the carry circuitry has been reduced when only one may have been faulty. The second approach is to consider the sum and carry circuitry as completely separate. The major problem with this approach is the additional hardware required; a flip flop must be provided for each module's sum output and a separate flip flop provided for each module's carry output. This doubles the number of flip flops required compared to the first approach.

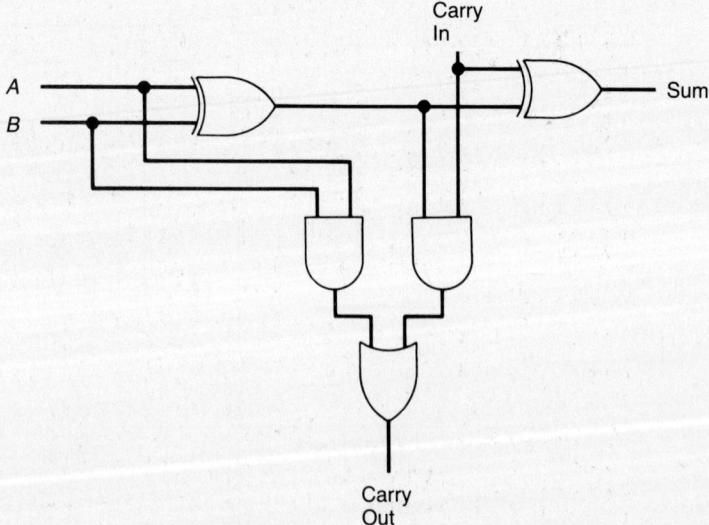

Fig. 3.19 Logic diagram of the full-adder circuit.

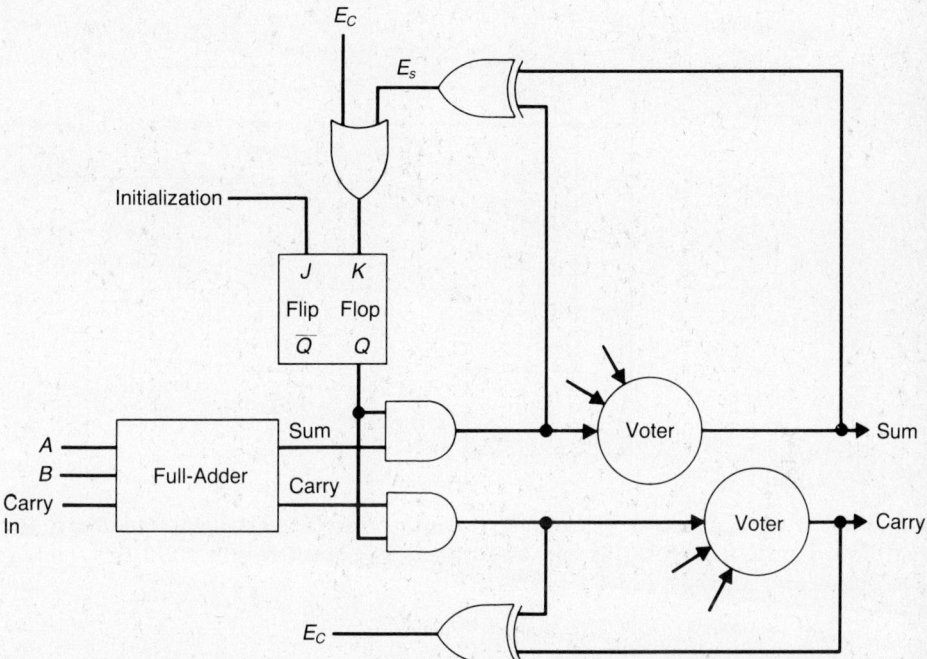

Fig. 3.20 Switching mechanism for implementation of self-purging full adder—both the sum and carry bits are considered. Only one of three switching mechanisms is shown.

The switching mechanism using the first approach is shown in Fig. 3.20. Only one of the three full-adder modules is shown. The remaining two full-adders have identical switching circuitry associated with them. Note that a single flip flop can purge both the contribution of the sum bit and the carry bit by a module. Also note that each module's carry and sum bits are compared to those produced by the voter to create the carry error signal E_c and the sum error signal E_s. The voters in this case are threshold gates.

Sift-Out Modular Redundancy

Another hybrid redundancy method is called **sift-out modular redundancy** [deSousa and Mathur 1978]. Sift-out modular redundancy also uses N identical modules that are configured into a system using special circuits called comparators, detectors, and collectors. The basic structure of the sift-out modular redundancy technique is illustrated in Fig. 3.21.

The function of the comparator is to compare each module's output with the remaining modules' outputs. Thus, the comparator produces one signal for each comparison that can be performed. For example, if five mod-

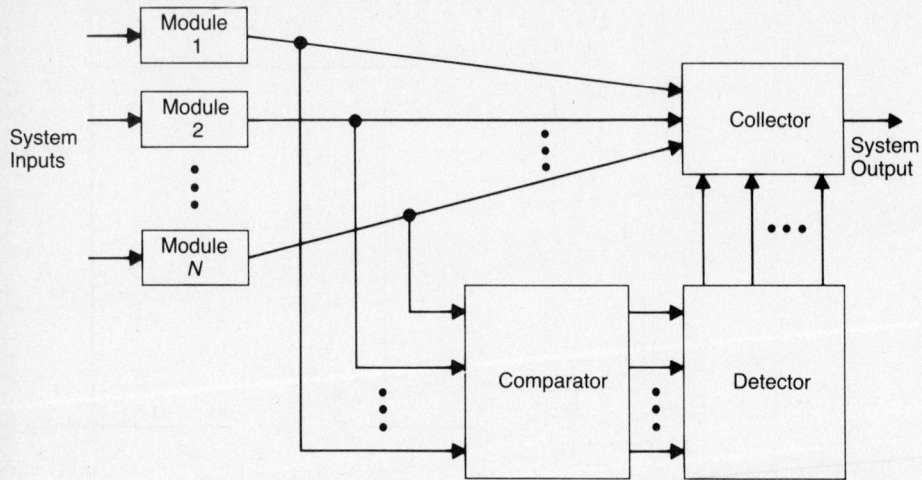

Fig. 3.21 Sift-out modular redundancy uses a centralized collector to create the system output. All modules are compared to detect faulty modules. (From [De Sousa and Mather, 1978] © 1978 IEEE)

ules are used, ten comparisons are performed; if four modules are used, six comparisons are made, and if three modules exist, three comparisons are made. Each signal generated by the comparator is 1 if the two units being compared disagree; and 0, otherwise. The logic diagram for a comparator for three modules is shown in Fig. 3.22.

The function of the detector is to determine which disagreements are reported by the comparator and to disable a unit that disagrees with a majority of the remaining modules. The detector produces one signal for each

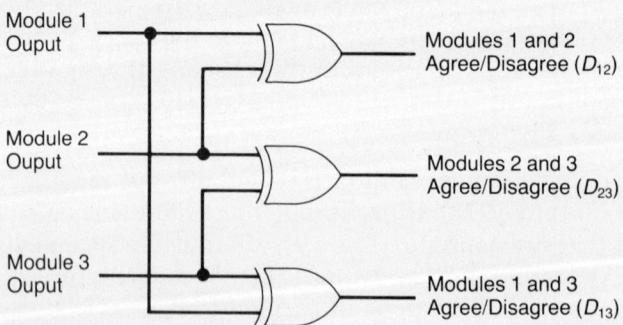

Fig. 3.22 The comparator compares the output of each module with all others to produce disagreement signals (D_{ij}).

module. The value of that signal is 1 if the module disagrees with a majority of the remaining modules, and it is 0 if the module agrees with a majority of the remaining modules. In a manner similar to that of self-purging redundancy, the detector uses flip flops to force modules that are identified as faulty to remain identified as faulty until an initialization is generated. An initialization signal allows modules to be placed back into the system, in a functional sense, once it is decided proper to do so. This feature allows the system to recover from transient faults and facilitates maintenance and repair. The logic diagram of a detector for a three-module system is shown in Fig. 3.23.

The last major component of the sift-out modular redundancy approach is the collector. The function of the collector is to produce the system's output, given the outputs of the individual modules and the signals from the detector that indicate which modules are faulty. A module that is properly identified as faulty is not allowed to influence the output of the system. The logical structure of the collector is simple, as is illustrated in the three-channel collector of Fig. 3.24.

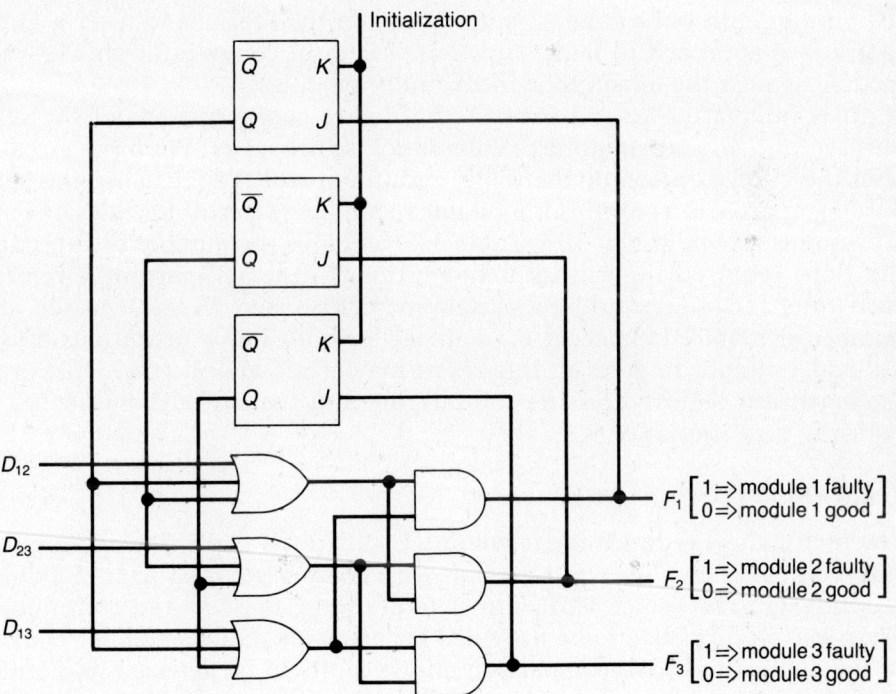

Fig. 3.23 The detector uses the disagreement signals to identify modules as faulty or fault-free.

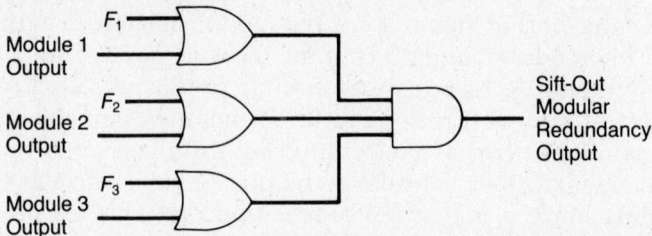

Fig. 3.24 The collector combines module outputs to produce the system output. Contributions from modules identified as faulty are ignored.

The self-purging redundancy concept and the sift-out modular redundancy technique are similar in many respects. Both approaches depend on removing modules from the functional operation of the system to achieve high reliability. The major difference in the two methods is the technique used to identify a module as faulty. The self-purging method compares each module's output with the output of the system, in a way that allows modules to determine independently if they are good or faulty; thus, the name self-purging. Sift-out modular redundancy, on the other hand, uses a more centralized approach to fault detection. A central comparator checks each module against the others to identify faulty modules.

It is interesting to compare the hardware configurations of the self-purging and sift-out modular redundancy approaches. We have already seen the configuration of the triply redundant full-adder using the self-purging approach. The sift-out modular redundancy circuit for the same application is shown in Fig. 3.25. Table 3.2 compares the number of gates and flip flops required in each implementation. In the self-purging circuitry each voter is assumed to be a simple majority voter. As you can see, the number of required elements is identical for both implementations of the full-adder circuit. In general, this result may or may not be true. However, the hardware required for the self-purging and the sift-out approaches is typically very similar.

Triple-Duplex Architecture

The final hybrid redundancy technique is called the **triple-duplex architecture** because it combines duplication with comparison and triple modular redundancy. The use of TMR allows faults to be masked and continuous, error-free performance to be provided for up to one faulty module. The use of the duplication with comparison allows faults to be detected and faulty modules removed from the TMR voting process. The triple-duplex architecture is typically used in control applications where the flux-summing arrangement can be employed as the voting mechanism.

3.4 ■ Hardware Redundancy

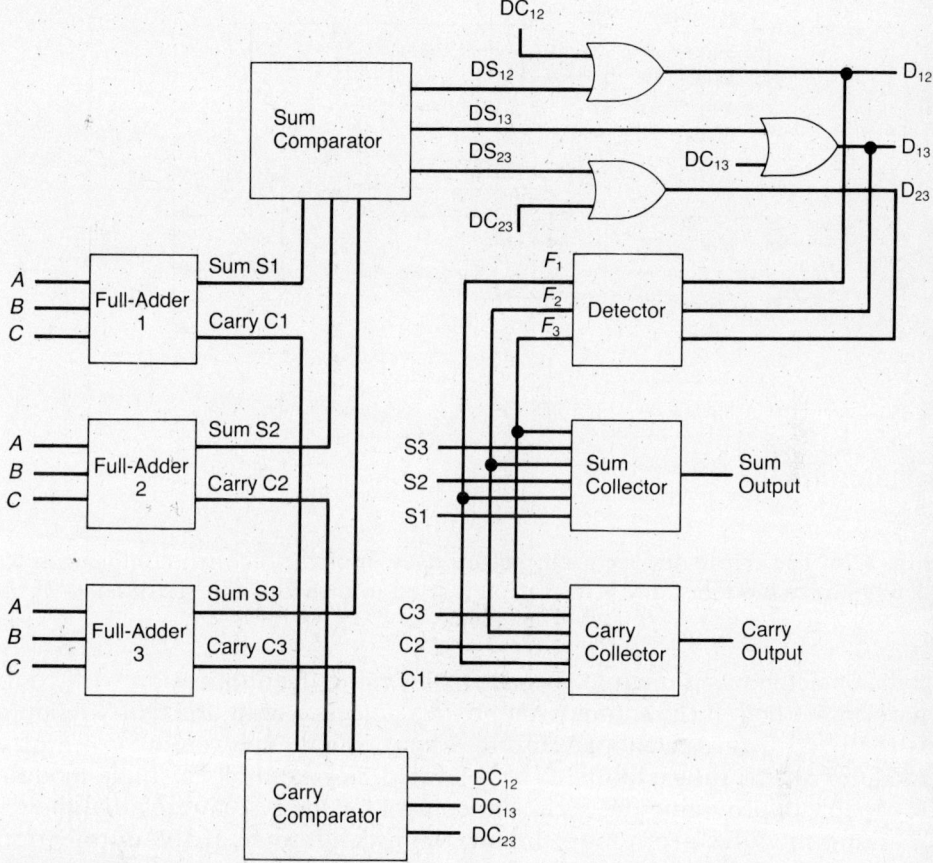

Fig. 3.25 Implementation of a full-adder using the sift-out modular redundancy technique.

The basic structure of the triple-duplex architecture using the flux-summing technique is shown in Fig. 3.26. The flux-summing process is capable of tolerating any single fault that occurs in the system. To allow faults

TABLE 3.2 Comparison of self-purging and sift-out full-adder implementations

Redundancy technique	Number of gates	Number of flip flops
Self-purging	38	3
Sift-out	38	3

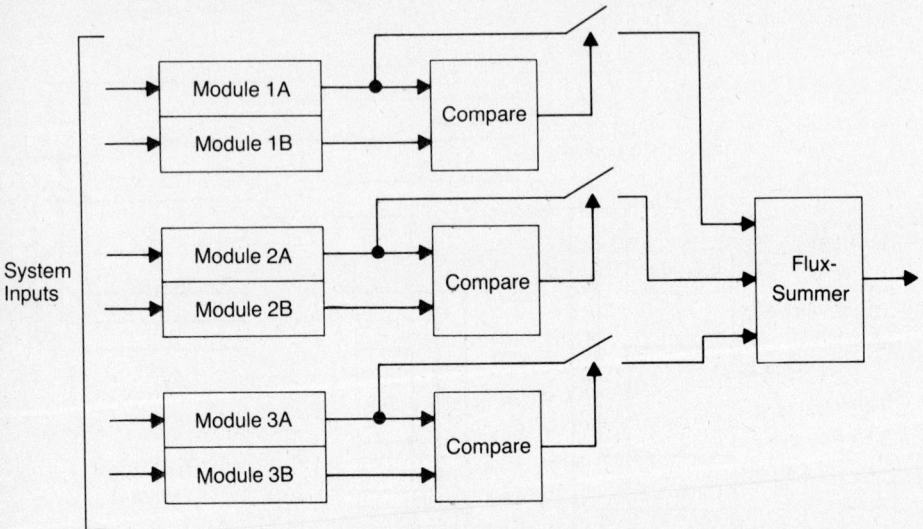

Fig. 3.26 The triple-duplex architecture uses duplication with comparison to detect faulty modules, and triplication is used to provide fault masking.

to be detected, each module is constructed using the duplication with comparison method. If the comparison process detects a fault, the faulty module is removed from the flux-summing arrangement. The removal of faulty modules allows future faults to be tolerated. For example, a single module is capable of providing the necessary control for the system provided the remaining modules are removed from the flux-summer. If the duplication with comparison can detect the faults, the triple-duplex architecture with flux summing can tolerate up to two module faults.

The output of the comparison process can be used to disconnect a module from the flux-summer, as shown in Fig. 3.26. If the comparison detects a fault, the appropriate switch can be opened to remove the affected module from the flux-summer. The remaining modules are then capable of providing the necessary current to the flux-summer. In practical applications, when a module is removed from the flux-summing arrangement, that particular input to the flux-summer is automatically connected to a known, fixed value such as a ground reference.

3.4.4 Summary of Hardware Redundancy

The three primary hardware redundancy techniques—passive, active, and hybrid—each have advantages and disadvantages that are important in different applications. The key differences are as follows:

1. Passive techniques rely strictly on fault masking.
2. Active techniques do not use fault masking but instead employ detection, location, and recovery techniques (reconfiguration).
3. Hybrid approaches employ both fault masking and reconfiguration.

The choice of a hardware approach depends heavily on the application. Critical-computation applications usually mandate some form of either passive or hybrid redundancy because momentary, erroneous results are not acceptable in such systems. The highest reliability is usually achieved using the hybrid techniques. In long-life and high-availability applications, active approaches are often used because it is typically acceptable to have temporary, erroneous outputs; the important thing is that the system can be restored quickly to an operational state using reconfiguration techniques.

The cost, in terms of hardware, of the redundancy techniques increases as we go from active to passive and finally to hybrid. Active techniques typically use less hardware, but they have the disadvantage of potentially producing momentary, erroneous outputs. Passive techniques provide fault masking, but they use substantial investments in hardware. Finally, hybrid approaches provide the advantage of fault masking, but they require enough hardware to use voting *and* they require hardware for spares. Hybrid approaches are typically the most costly in terms of hardware and are used when the highest levels of reliability are required.

3.5 Information Redundancy

Information redundancy is the addition of redundant information to data to allow fault detection, fault masking, or possibly fault tolerance. Good examples of information redundancy are error detecting and error correcting codes, formed by the addition of redundant information to data words, or by the mapping of data words into new representations containing redundant information. Before beginning the discussions of various codes, we will define several basic terms that will appear throughout the textbook [Tang and Chien 1969].

In general, a **code** is a means of representing information, or data, using a well-defined set of rules. For example, a telephone in an office complex can be designed to produce one long ring when the incoming call originated within the complex and two short rings when the incoming call originated outside the complex. The number of rings is a code that allows the recipient of the call to identify, at least partially, the origin of the call.

A **code word** is a collection of symbols, often called digits if the symbols are numbers, used to represent a particular piece of data based on a specified code. A **binary code** is one in which the symbols forming each code word consist of only the digits 0 and 1. For example, a binary coded decimal

(BCD) code defines a 4-bit code word for each decimal digit. The BCD code, for example, is clearly a binary code. A code word is said to be *valid* if the code word adheres to all the rules that define the code; otherwise, the code word is said to be *invalid*.

The **encoding process** is the process of determining the corresponding code word for a particular data item. In other words, the encoding process takes an original data item and represents it as a code word using the rules of the code. For example, given the decimal digit 9, the encoding process determines the BCD code word to be 1001. The **decoding process** is the process of recovering the original data from the code word. In other words, the decoding process takes a code word and determines the data that it represents. For example, the decoding process transforms the BCD code word 0011 into the decimal digit 3.

Of primary interest in this chapter are binary codes. In many binary code words, a single error in one of the binary digits causes the resulting code word to no longer be correct, but, at the same time, the code word is valid. Consequently, the user of the information has no means of determining the correctness of the information. For example, if the BCD code word 0011 is corrupted by complementing the least significant bit to yield 0010, the user of the code word has no way of knowing that the corresponding decimal digit should be 3 instead of 2.

It is possible, however, to create a binary code for which the valid code words are a subset of the total number of possible combinations of 1s and 0s. If the code words are formed correctly, errors introduced into a code word force it to lie in the range of illegal, or invalid, code words, and the error can be detected. This is the basic concept of the error *detecting* codes. The basic concept of the error *correcting* code is that the code word is structured such that it is possible to determine the correct code word from the corrupted, or erroneous, code word.

An **error detecting code** is a specific representation allowing errors introduced into a code word to be detected. For example, suppose that a code word is created and then transferred from one point to another within a system. If the code word is constructed according to an error detecting code, errors that are introduced during the transfer process can be detected at the receiving point.

The basic concept of the error detecting codes is very simple. If a piece of data contains n bits, there are 2^n possible combinations of 1s and 0s for that data. If the code words are structured such that only some of those 2^n combinations are considered to be valid, the occurrence of one of the invalid combinations can be used to signal the existence of an error. For example, the binary coded decimal (BCD) representation encodes each of the ten decimal digits as a 4-bit binary number. Because 4 bits are sufficient to represent 16 unique combinations, several of the combinations are not used in the BCD code. If one of the unused combinations occurs, an error has been generated, and a relatively simple circuit could be designed to detect that error. The BCD code cannot detect all errors, however. If the BCD representation 0000 were to have the least significant bit corrupted yielding the quantity 0001, the error could never be detected because 0001 is a valid BCD representation. The key to an error detecting code, therefore, is to structure the code words such that some maximum number of errors is guaranteed to result in invalid representations.

In many applications, it is not sufficient to simply detect errors. Often one would like to correct the error before its effects can impact the system. Real-time correction enables the operation of the system to continue in an uninterrupted manner—a mandatory attribute of systems that perform critical computations. A number of techniques are available to allow coding schemes to correct errors. Such codes are called **error correcting codes**. Typically, the code is described by the number of bit errors that can be corrected. For example, a code that can correct single-bit errors is called a *single-error* correcting code. A code that can correct 2-bit errors is called a *double-error* correcting code, and so on.

In general, an error correcting code is a specific representation allowing errors introduced into a code word to be corrected. Once again, suppose that a code word is created and then transferred from one point in a system to another. If the code word is constructed according to an error correcting code, errors that are introduced during the transfer process can be corrected at the receiving point. For example, error *detecting* codes can be used to initiate reconfiguration in an active redundancy scheme, whereas error *correcting* codes can be used to provide fault masking in passive redundancy.

A fundamental concept in the characterization of both error detecting and error correcting codes is the **Hamming distance**. The Hamming distance between any two binary words is the number of bit positions in which the two words differ. For example, the binary words 0000 and 0001 differ in only one position, and therefore have a Hamming distance of 1. The binary words 0000 and 0101, however, differ in two positions; consequently, their Hamming distance is 2. Clearly, if two words have a Hamming distance of 1, it is possible to change one word into the other simply by modifying one bit in one of the words. If, however, two words differ in two bit positions, it is impossible to transform one word into the other by changing one bit in one of the words.

The Hamming distance gives insight into the requirements of error detecting codes and error correcting codes. We define the **code distance** as the minimum Hamming distance between any two valid code words. If a code has a distance of two, any single-bit error introduced into a code word results in the erroneous word being an invalid code word because all valid code words differ in at least two bit positions. If a code has a distance of three, any single-bit error or any double-bit error results in the erroneous word being an invalid code word because all valid code words differ in at least three positions. However, a code distance of three allows any single-bit error to be corrected, if it is desired to do so, because the erroneous word with a single-bit error is a Hamming distance of 1 from the correct code word and a Hamming distance of 2 from all others. Consequently, the correct code word can be identified from the corrupted code word.

In general, a code can correct up to c bit errors and detect up to d additional bit errors if and only if

$$2c + d + 1 \leq H_d$$

where H_d is the Hamming distance of the code [Nelson and Carroll 1986]. For example, a code with a Hamming distance of 2 cannot provide any error correction, but it can detect single-bit errors. Similarly, a code with a Hamming distance of 3 can correct single-bit errors or detect double-bit errors.

A second fundamental concept of codes is separability. A **separable code** is one in which the original information is appended with new information to form the code word, thus allowing the decoding process to consist of simply removing the unwanted information and keeping the original data. In other words, the original data is obtained from the code word by stripping away extra bits, called the code bits or check bits, and retaining only those associated with the original information. A **nonseparable code** does not possess the property of separability, and, consequently, requires more complicated decoding procedures.

3.5.1 Parity Codes

Perhaps the simplest form of a code is the **parity code.** The basic concept of parity is very straightforward, but there are variations on the fundamental idea. Single-bit parity codes require the addition of an extra bit to a binary word such that the resulting code word has either an even number of 1s or an odd number of 1s. If the extra bit results in the total number of 1s in the code word being odd, the code is referred to as *odd* parity. If the resulting number of 1s in the code word is even, the code is called *even* parity. If a code word with odd parity experiences an error in one of its bits, the parity becomes even. Likewise, if a code word with even parity encounters a single-bit error, the parity becomes odd. Consequently, a single-bit error can be detected by checking the number of 1s in the code words. The single-bit parity code (either odd or even) has a Hamming distance of 2, therefore allowing any single-bit error to be detected but not corrected. The single-bit parity codes (both odd and even) for BCD words are listed in Table 3.3. It is important to note that the parity code is a separable code.

The most common application of parity is in the memories of computer systems. Before being written to memory, a data word is encoded to achieve the correct parity. The encoding is the appending of a bit to force the result-

TABLE 3.3 Odd and even parity codes for BCD data

Decimal digit	BCD	BCD odd parity	BCD even parity
0	0000	0000 1	0000 0
1	0001	0001 0	0001 1
2	0010	0010 0	0010 1
3	0011	0011 1	0011 0
4	0100	0100 0	0100 1
5	0101	0101 1	0101 0
6	0110	0110 1	0110 0
7	0111	0111 0	0111 1
8	1000	1000 0	1000 1
9	1001	1001 1	1001 0
		↑ Parity bit	↑ Parity bit

ing word to have the appropriate number of 1s. When the data word is subsequently read from memory, the parity must be checked to verify that it has not changed as a result of a fault within the memory. If an error is detected, the user of the memory is notified via an error signal that a potential problem exists. Additional hardware is required to handle the extra information (the extra bit appended to each word). For example, the memory must contain one extra bit per word to store the extra information, and the hardware must be designed to create and check the parity bit. As can be seen, the information redundancy concept often requires hardware redundancy as well.

The organization of a memory with parity coding is shown in Fig. 3.27. The data entering the memory also passes into a parity generator that per-

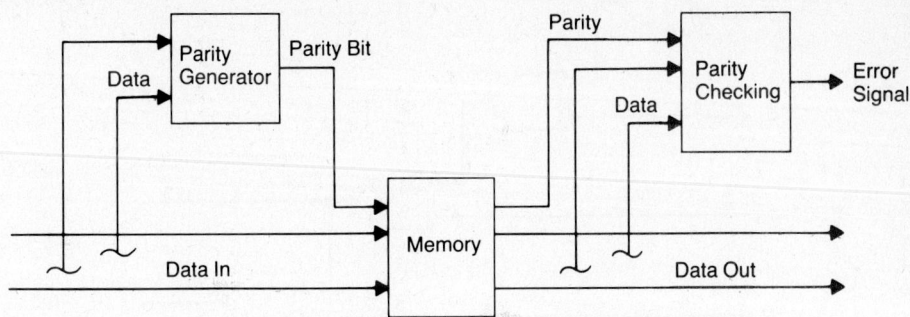

Fig. 3.27 Organization of a memory that uses single-bit parity. The parity bit is generated when data is written to memory and checked when data is read.

forms the encoding process. The generated parity bit and the original data are then stored in memory as a complete code word. When the code word is read from memory, the data portion of the code word passes through a parity checker that regenerates the parity bit and compares it to the original parity bit stored in memory as part of the code word. A disagreement between the original parity bit and the regenerated parity bit causes an error signal to be generated.

Circuits to generate and check parity bits are relatively simple. Each data bit must be examined to determine the total number of 1s present, such that the value of the parity bit can be specified. Recall that the EXCLUSIVE-OR operation, when performed on several bits, produces a 1 if the group of bits contains an odd number of 1s, and a 0 if the group contains an even number of 1s. This is exactly the function required to generate and check parity. Note that the parity generation process and the parity checking process are the same function. Therefore, a circuit that can check parity can also generate parity. If we limit ourselves to use two-input, EXCLUSIVE-OR gates, a circuit that generates and checks even parity for 4-bit data words is shown in Fig. 3.28. The generated parity line is 1 if the data bits d_0 through d_3, contain an odd number of 1s; otherwise, the generated parity bit is 0. When the parity is checked (for example, when the data is read from memory), parity is first regenerated for the data bits, and the regenerated parity bit is compared to the parity bit P that was stored in memory. If a disagreement between the two bits occurs, the error signal becomes 1.

The single-bit parity code has a minimum Hamming distance of 2. This is easily seen by examining the combinations of 1s and 0s available when the code is constructed. Suppose that the original data word consists of n

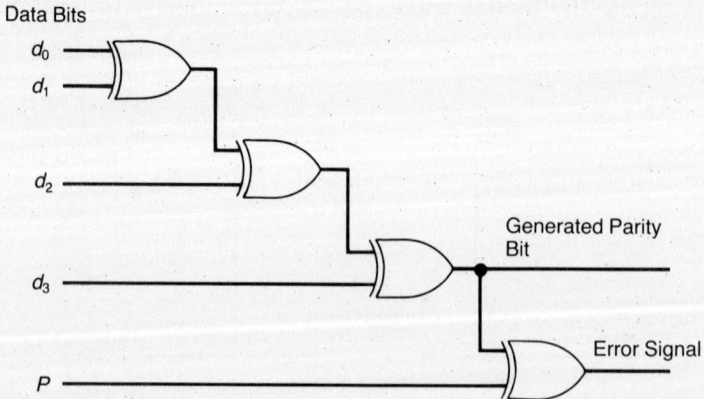

Fig. 3.28 A 4-bit parity generation and checking circuit for even parity.

bits, such that 2^n possible combinations are available. The addition of a parity bit yields a code word with $n + 1$ bits, or 2^{n+1} different combinations. In other words, the addition of the parity bit results in twice as many combinations being available. Of the 2^{n+1} combinations, $2^{n+1}/2$ have an odd number of 1s and $2^{n+1}/2$ have an even number of 1s. For an odd parity code, only those combinations that have an odd number of 1s are valid code words; the others are invalid. Likewise, for an even parity code, only those combinations that have an even number of 1s are valid code words. Changing any single bit in a code word changes the parity of the code word. For example, changing a 1 to a 0 decreases the number of 1s by one, and changing a 0 to a 1 increases the number of 1s by one. In either case, a word with odd parity is mapped into a word with even parity by a 1-bit error. Likewise, a word with even parity is mapped into a word with odd parity by a 1-bit error. To map an odd parity code word into another odd parity code word requires changing a minimum of two bits. Similarly, changing an even parity code word into another even parity code word requires changing a minimum of two bits. Therefore, the distance of the single-bit parity code is two. Any single-bit error can be detected by the single-bit parity code.

One of the biggest problems with single-bit parity codes is their inability to guarantee the detection of some very common multiple-bit errors. For example, a memory can be constructed from individual chips that each contain several bits; 4 bits is a very common number. If a chip in the memory becomes faulty, the simple parity code may be unable to detect the resulting error because multiple bits are affected. The basic parity scheme can be modified to provide additional error detection capability. There are five basic parity approaches:

1. Bit-per-word parity
2. Bit-per-byte parity
3. Bit-per-chip parity
4. Bit-per-multiple-chips parity
5. Interlaced parity

The basic concept of each approach is illustrated in Fig. 3.29.

The basic idea of **bit-per-word parity** is to append one parity bit to each word. The primary disadvantage of the bit-per-word approach is that certain errors can go undetected. For example, if a word, including the parity bit, becomes all 1s because of a complete failure of a bus or a set of data buffers, the odd parity method only detects the condition if the total number of bits, including the parity bit, is even. Likewise, even parity only detects this problem if the total number of bits is odd. In a similar manner, the condition of all bits becoming 0 can never be detected by the even bit-per-word parity method because 0 is considered to be an even number of 1s. Odd bit-per-word parity always detects the condition of all bits being 0.

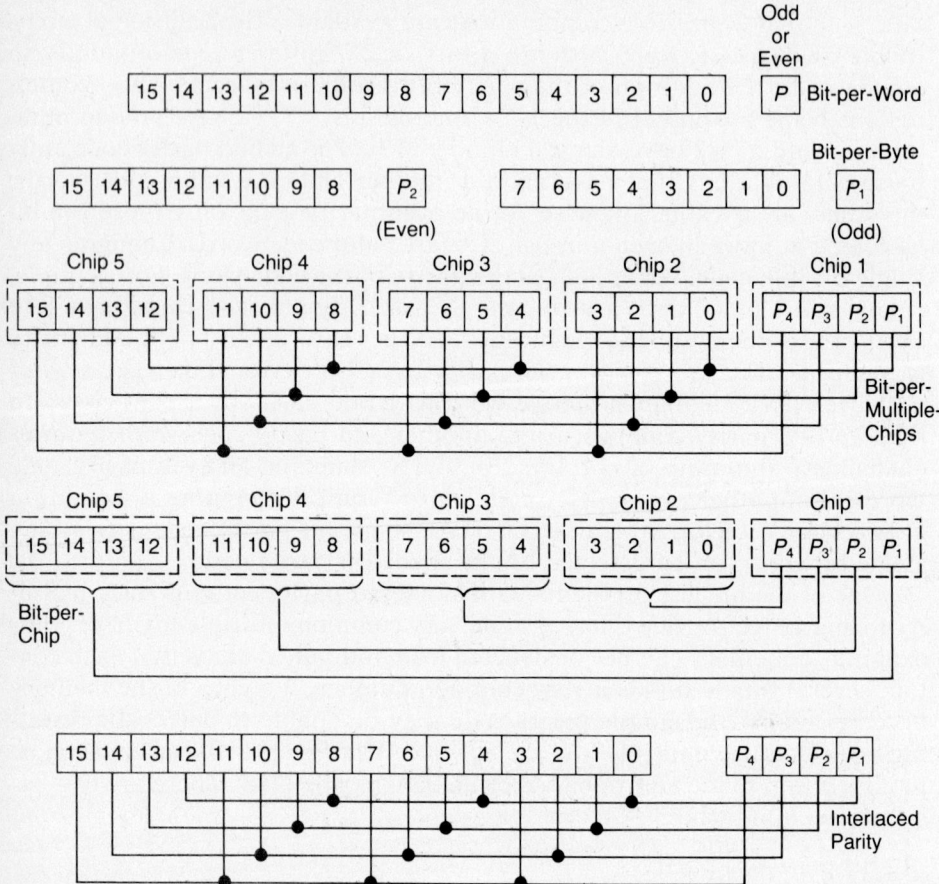

Fig. 3.29 The five basic forms of the parity code include: bit-per-word, bit-per-byte, bit-per-multiple-chips, bit-per-chip, and interlaced parity.

An alternate approach is the **bit-per-byte parity** technique. Here, two parity bits are used on two separate portions of the original data. The technique is called bit-per-byte, but the parity groups can be any number of bits; not just the 8 bits normally associated with the term byte. To gain the full advantages of the approach, however, the number of information bits associated with each parity bit should be even. Also, the parity of one group should be even and the parity of the other group should be odd. The primary advantage of this approach is the ability to detect both the *all 1s* and *all 0s* conditions. If the complete code word becomes all 1s, the even parity bit is erroneous. If the complete code word becomes all 0s, the odd parity bit is erroneous. In both cases, the erroneous conditions are detected.

3.5 ■ Information Redundancy

The bit-per-byte technique also provides additional protection against multiple-bit errors; for example, 2-bit errors will always be detected as long as one bit is in the even parity group and the other is in the odd parity group.

The fundamental disadvantage of both the bit-per-word and bit-per-byte parity approaches is the ineffective detection capability for multiple-bit errors. Many memories are organized using memory chips that contain either 4 bits, 8 bits, or more, of memory. Several of these chips are then used in parallel to form the complete number of bits of each word in the memory. If one chip fails (this is called the whole-chip failure mode), several bits of each word of memory can be affected. Therefore, the single-bit error assumption is often ineffective.

One approach that is useful for detecting the failure of a complete chip is the **bit-per-multiple-chips parity** method, as illustrated in Fig. 3.29. The basic concept is to have one bit from each chip of the memory associated with a single parity bit. Sufficient parity bits are provided to allow each data bit within a chip to be associated with a distinct parity bit. In Fig. 3.29, for example, parity bit P_1 establishes the parity of a group of bits including bits 0, 4, 8, and 12. Note that each parity group includes one, and only one, bit from each chip. If one chip fails, all the parity groups are affected, but no more than one bit in each parity group is corrupted, so the parity code detects the error.

At first it might appear that the bit-per-multiple-chips approach is relatively expensive because at least one chip of the memory is necessary to store the parity bits. If each chip is 4 bits wide, a memory capable of storing 16-bit words requires four of these chips. If bit-per-word parity is used, an extra chip must be added to store just that one parity bit. Typically, the same type of chip is used to store the parity bit and to store the data, so a 4-bit chip is used to store a single bit in the bit-per-word approach. The bit-per-multiple-chips approach simply takes advantage of the extra bits that probably are present in the design anyway. Consequently, the cost of using the bit-per-multiple-chips approach in this simple example is minimal.

One disadvantage of the bit-per-multiple-chip parity approach is that the failure of a complete chip is detected, but it is not located. The failure of any one chip causes all parity groups to be in error, so the cause of the problem cannot be identified. One procedure that overcomes this problem is the **bit-per-chip parity** organization. Here, each parity bit is associated with one chip of the memory, as illustrated in Fig. 3.29. Specifically, parity bit P_1 establishes correct parity for the group containing data bits 12, 13, 14, and 15. If a single bit becomes erroneous, the existence of the error is detected, and the chip that contains the erroneous bit is identified. This is extremely valuable from a maintenance standpoint; not only does the system have the capability to warn of the occurrence of a problem, but the system can direct maintenance personnel to the source of the problem. The primary disadvantage of the bit-per-chip parity method is the susceptibility to the whole-chip failure mode. Because the basic parity code can detect only single errors, the multiple error condition associated with the failure of a complete chip can go undetected.

An alternate organization of the parity code is called **interlaced parity**. Interlaced parity is very similar in form to the bit-per-multiple-chips approach with one key difference. In interlaced parity, the parity groups are formed without regard to the memory's physical organization. This is in contrast to the bit-per-multiple-chip organization, which is intimately tied to the physical structure of the memory. In interlaced parity, the information bits are divided into equal-sized groups, and one parity bit is associated with each group. The bits of each group are then positioned such that no two adjacent bits are from the same parity group. This is accomplished by placing the first bit from group 1 in the least significant bit position, the first bit from group 2 in the next most significant position, the first bit from group 3 in the next position, and so on. Once the first bits of each group are placed, the remaining bits are added to the word in a similar manner. The basic idea is illustrated in Fig. 3.29.

The interlaced parity method is most often used when errors in adjacent bits are of major concern. Because no two adjacent bits are in the same parity group, errors in any two adjacent bits are detected. A good example is a parallel bus; in many buses, two adjacent bits can become shorted together. The interlaced organization of parity detects errors due to this type of fault.

In the approaches we have considered so far, each bit was contained in one and only one parity group. In the **overlapping parity** approach, however, parity groups are formed with each bit appearing in more than one parity group. The primary advantage of overlapping parity is that errors can be located in addition to being detected. Once the erroneous bit is located, it can be corrected by a simple complementation, if desired. Overlapping parity is the basic concept of some of the Hamming error correcting codes.

Figure 3.30 illustrates the basic idea of overlapping parity when applied to 4 bits of information. Three parity groups are required to uniquely identify each erroneous bit in the 4 bits of information. The concept of overlapping parity is to place each bit in a unique combination of the parity groups. For example, referring to Fig. 3.30, bit 3 appears in each parity group, whereas bits 0, 1, and 2 appear in different combinations of two groups. If any one bit becomes erroneous, the impact is unique, as is illustrated in Fig. 3.30. For example, if bit 3 is in error, all the parity groups are affected, but if bit 1 is erroneous, the parity groups associated with P_0 and P_2 are affected and P_1 is unaffected. As shown in Fig. 3.30, each possible single-bit error produces a unique impact on the parity of the three groups.

The overlapping parity approach can be transformed into an error correction scheme by using several comparators (EXCLUSIVE-OR gates) and a

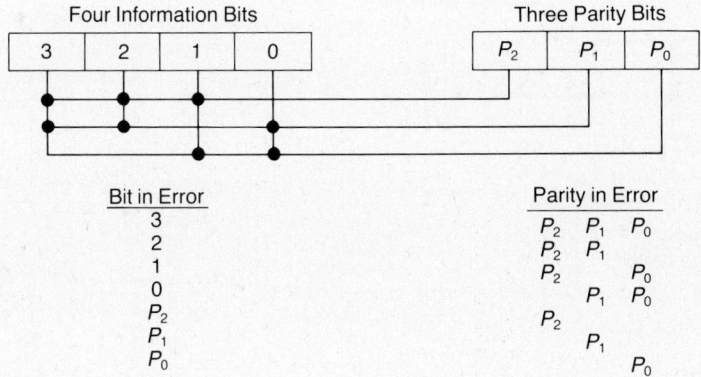

Fig. 3.30 Overlapping parity assigns each bit to multiple parity groups.

decoder in conjunction with the parity checking circuits. In addition, the correction process is performed by complementing the appropriate bit using an EXCLUSIVE-OR gate as a programmable inverter. The concept is illustrated in Fig. 3.31 for 4 bits of information and three parity groups. When the 4 bits of information are written to memory, the three parity bits are generated and stored with the original four bits of information as a single code word. When the code word is subsequently read from memory, the parity bits are regenerated using the parity generation circuits. The regenerated parity bits P_{r0}, P_{r1}, and P_{r2}, are compared to the parity bits that were stored with the original information in memory. The results of the comparisons are fed to a 3-8 decoder that produces a logic 1 on one of its outputs and a logic 0 on all the others, based on the values of its three inputs. For example, the decoder is wired such that all 0s on its inputs forces a 1 to occur on the "no error" line. If the input to the decoder is all 1s, the output labeled "correct bit 3" is set to 1 indicating that bit 3 is the erroneous bit that needs correcting. The correction is performed using a simple collection of EXCLUSIVE-OR gates. If the correction line on one of the EXCLUSIVE-OR gates is 1, the bit is complemented, and therefore corrected. If the correction line is 0, the associated bit is passed through the EXCLUSIVE-OR gate uncomplemented, and therefore unchanged.

The penalty for using overlapping parity on 4 bits of information is high; 3 parity bits are required to detect and locate errors for the 4 bits of information, a redundancy of 75%. As the number of information bits increases, however, the number of parity bits required becomes a smaller percentage of the number of actual information bits. The required relationship between the number of information bits and the number of required parity bits can be determined in a fairly simple manner. Let m be the number of

92   Design Techniques to Achieve Fault Tolerance

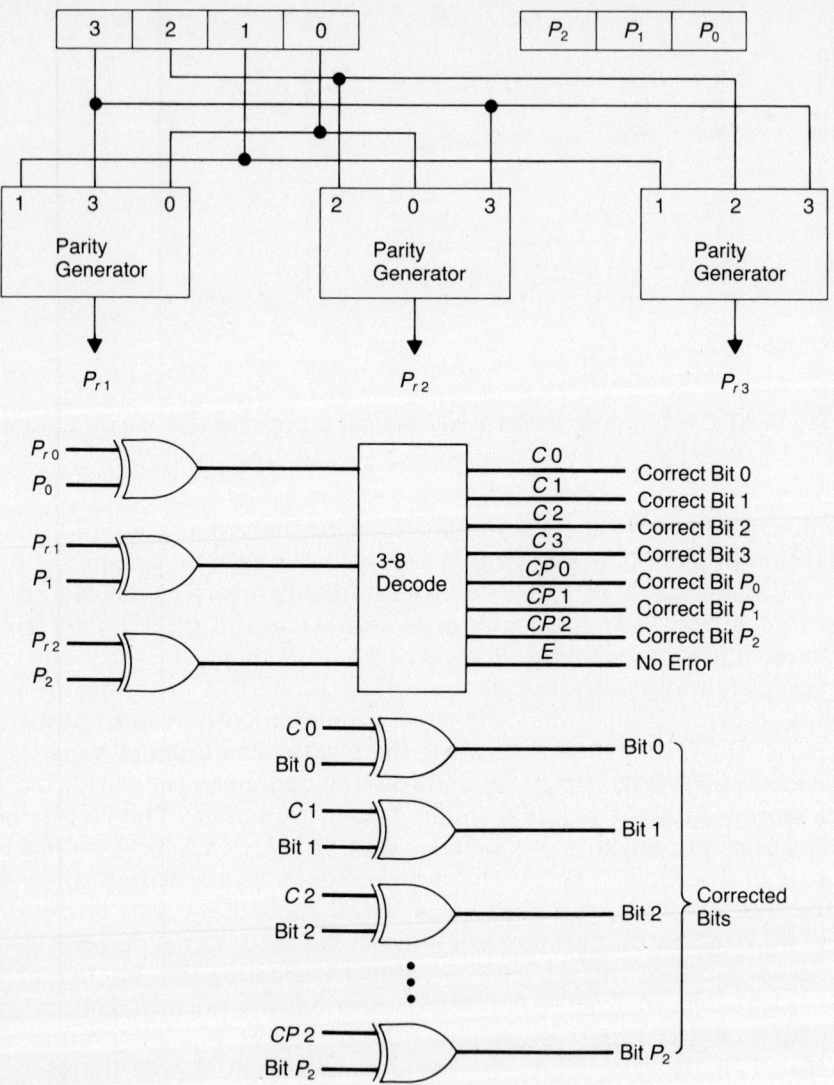

Fig. 3.31 Error correction using overlapped parity.

information bits to be protected using an overlapping parity approach, and let k be the number of parity bits required to protect those m information bits. Each bit error that can occur must produce a unique result when the parity is checked. With k parity bits, there are 2^k unique outcomes of the parity checking process. With k parity bits and m information bits, there are $m + k$ different, single errors that can occur. So, we know that 2^k must be at least as large as $m + k$. Also, we must have a unique result of the parity checks when there is no error, so the total number of unique combinations must be at least as large as $m + k + 1$. Therefore, the relationship between k and m must be

$$2^k \geq m + k + 1$$

Table 3.4 shows the minimum number of parity bits required, as a function of the number of information bits, for the overlapping parity approach. Note that the percentage of redundancy decreases substantially as the number of information bits increases.

3.5.2 *m*-of-*n* Codes

The basic concept of the ***m*-of-*n* code** is to define code words that are n bits in length and contain exactly m 1s. As a result, any single-bit error forces the resulting erroneous word to have either $m + 1$ or $m - 1$ 1s, and the error can be detected. The primary advantage of the m-of-n code is the conceptual simplicity; it is very easy to visualize the error detection process. The major disadvantage, however, is that the encoding, decoding, and the detection processes are often extremely difficult to perform, despite their conceptual simplicity.

The easiest way to construct an m-of-n code is to take the original i bits of information and append i bits. The appended bits are chosen such that the resulting $2i$-bit code words each have exactly i 1s, therefore producing an i-of-$2i$ code. For example, the 3-of-6 code words constructed from 3 bits of information are shown in Table 3.5. The obvious disadvantage of using the i-of-$2i$ code is that twice as many bits are required to represent the in-

TABLE 3.4 Minimum number of parity bits for overlapping parity code

Number of information bits	Number of parity bits	Redundancy percentage
2	3	150.0%
4	3	75.0
6	4	66.7
8	4	50.0
10	4	40.0
12	5	41.7
16	5	31.25
24	5	20.8
32	6	18.75
64	7	10.9

TABLE 3.5 3-of-6 code for representing three bits of information

Original information	3-of-6 code	
000	000	111
001	001	110
010	010	101
011	011	100
100	100	011
101	101	010
110	110	001
111	111	000
	Original information	Appended information

formation; consequently, the redundancy of the code is 100%. The advantage of creating an i-of-$2i$ code is that both the encoding and the decoding processes are simple because the code is separable. The encoding procedure can be performed by examining the pattern of 1s in the information to be encoded and looking up in a table the desired bits to append, based on the pattern of 1s in the original information. The decoding can be performed by simply removing the appended bits from the code word and retaining the original information.

It is easy to see that m-of-n codes have a distance of two. Any single-bit error in an m-of-n code word changes the number of 1s to either $m + 1$ or $m - 1$, depending on whether the error changed a 0 to a 1 or a 1 to a 0. A second bit-error, however, can change the number of 1s back to m. For example, if one bit is changed from 0 to 1 and a second bit is changed from 1 to 0, the number of 1s in the code word remains unchanged. Consequently, the error goes undetected. If the errors are all unidirectional, meaning that all errors are either a change of a 1 to a 0 or a change of a 0 to a 1, but not combinations of the two changes, the m-of-n code detects the multiple errors. Consequently, the m-of-n code provides detection of all single errors and all multiple, unidirectional errors.

The m-of-n codes can often be constructed more efficiently, but the separable nature of the code is usually lost. For example, the 2-of-5 code for BCD data is shown in Table 3.6. The code words each have 5 bits, and the code is capable of detecting any single-bit error that occurs. The 2-of-5 code of Table 3.6 possesses the same error detection capability as the simple parity code and uses the same number of extra bits. The difficulty is that the information is not readily available from the code word. The code words must be converted back to the original information words by performing a

TABLE 3.6 Nonseparable 2-of-5 code for BCD data

Decimal digit	BCD data	2-of-5 code
0	0000	00011
1	0001	11000
2	0010	10100
3	0011	01100
4	0100	10010
5	0101	01010
6	0110	00110
7	0111	10001
8	1000	01001
9	1001	00101

decoding process. The decoding can be easily performed by look-up tables, but the process is much more complicated than needed for the separable m-of-n codes.

3.5.3 Duplication Codes

Duplication codes are based on the concept of completely duplicating the original information to form the code word. The primary advantage of the duplication code is simplicity. The major disadvantage is clearly the number of extra bits that must be provided to allow the code to be constructed.

Duplication codes are found in many applications, including memory systems and some communication systems. The encoding process for the duplication code consists of simply appending the original i bits of information to itself to form a code word of length $2i$ bits. If a single-bit error occurs, the two halves of the code word disagree, and the error can be detected. In communication systems, the duplication concept is often applied by transmitting all information twice; if both copies agree, the information is assumed to be correct. The penalty paid in the communications application is a decrease in the information rate of the system because $2i$ bits must be transmitted to obtain i bits of information.

A variation on the basic duplication code is to complement the duplicated portion of the code word. The use of **complemented duplication** is particularly advantageous when the original information and its duplicate must be processed by the same hardware. For example, consider the memory system shown in Fig. 3.32. Each word of the memory is i bits. As part of the encoding process, each i-bit word of information is followed sequentially in memory by its complemented duplicate. Suppose that one bit slice of the memory becomes faulty such that every bit stored in that slice assumes a value of 1, regardless of the desired value. Each word of informa-

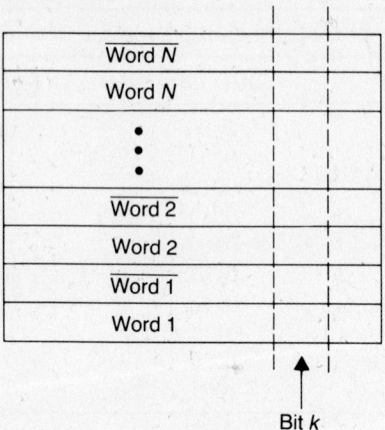

Word i = ith word of original data
$\overline{\text{Word } i}$ = complement of ith word of original data

Fig. 3.32 Example of complemented duplication in a memory containing N words of data.

tion and its complement now has one bit position in which the bits are not complements but are exactly the same. Because the complemented duplication is used, the error can be detected. If the simple duplication method had been employed, the error would not have been detected.

The duplicated complement approach is also effective in communication systems where the same physical media is used by both the original information and its duplicate. As illustrated in Fig. 3.33, a faulty line in the

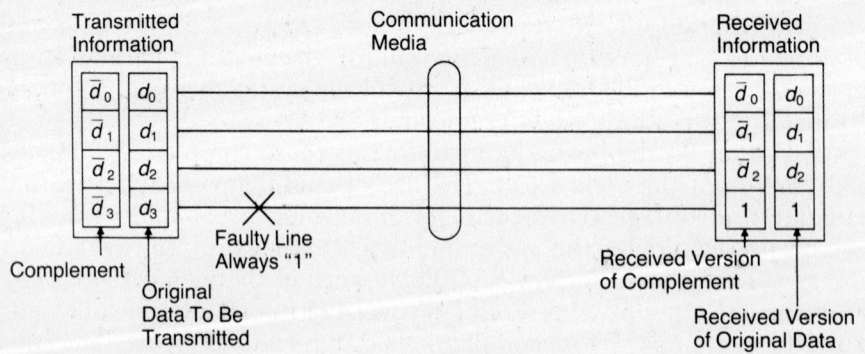

Fig. 3.33 Example of complemented duplication for error detection in a communication system.

communication media causes one bit to be the same in both the original information and the duplicate. If the original information and its duplicate are supposed to be complements, the error is detected; otherwise, the error goes undetected.

A final variation on the duplication code is the **swap and compare** method. The basic concept of swap and compare is to maintain two copies of the original information, but to swap the upper and lower halves of the second copy. Figure 3.34 illustrates the idea when applied to a memory system. A single bit slice that is faulty affects the upper half of one copy of the information and the lower half of the other copy. By comparing the appropriate halves, the error can be detected.

The primary advantage of all variations of the duplication codes is the simplicity associated with generating the code words and the ease of obtaining the original information from the code word. The advantage is usually offset, however, by the cost of completely duplicating the original information. Also, it is usually very time consuming to implement the duplication codes. In memory applications, each word must be written and read twice. In communication applications, each word must be transmitted twice. In both cases, the time required to perform the operation is doubled. So, the duplication code often requires not only the 100% redundancy in information, but typically a 100% redundancy in hardware and time as well.

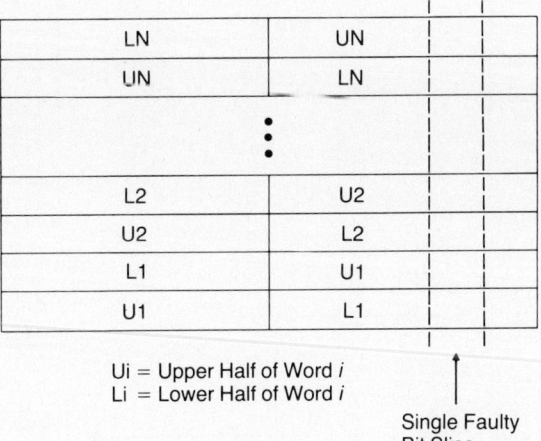

Fig. 3.34 Swap and compare technique applied to a memory system containing N words of data.

3.5.4 Checksums

The checksum is another form of separable code that is most applicable when blocks of data are to be transferred from one point to another. For example, checksums are used frequently in data transfers between mass storage devices—such as disks—and a computer, and packet-switched networks. The **checksum** is a quantity of information that is added to the block of data to help detect errors. Four primary types of checksums can be used: single-precision, double-precision, Honeywell, and the residue checksum.

The basic concept of the checksum is illustrated in Fig. 3.35. When the original data is created, an additional piece of information, called the checksum, is appended to the block of data. The checksum is then regenerated when the data is received at the destination or, in some applications, when the data is read from memory. The regenerated checksum and the original checksum are compared to determine if an error has occurred in the data, the checksum generation, checksum regeneration, or the checksum comparison.

The checksum is basically the sum of the original data. The difference between the various forms of the checksum is the way in which the summa-

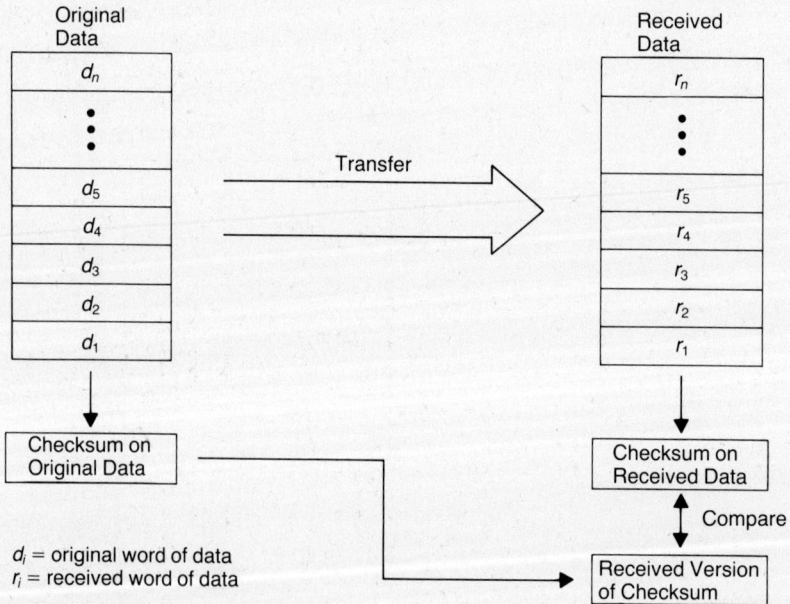

Fig. 3.35 In checksum coding, the sum of the original data words is appended to the block of data.

3.5 ■ Information Redundancy

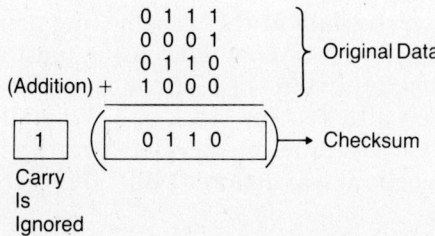

Fig. 3.36 A single-precision checksum is formed by adding the data words and ignoring any overflow.

tion is generated. The simplest form of the checksum is the **single-precision checksum**. The single-precision checksum is formed by performing the binary addition of the data that is to be protected by the checksum and ignoring any overflow that occurs. For example, if each data word is n bits, the checksum is n bits as well. If the true binary sum of the data exceeds $2^n - 1$, an overflow has occurred; in the single-precision checksum the overflow is ignored. In other words, the single-precision checksum is formed by adding the n-bit data in a modulo-2^n fashion. Figure 3.36 shows an example of the single-precision checksum for a block of four words, each of which is four bits.

The primary difficulty with the single-precision checksum is that information, and, as a result, the ability to detect errors are lost in the ignored overflow. As an example, consider the use of the single-precision checksum in the communications system shown in Fig. 3.37. The single-precision

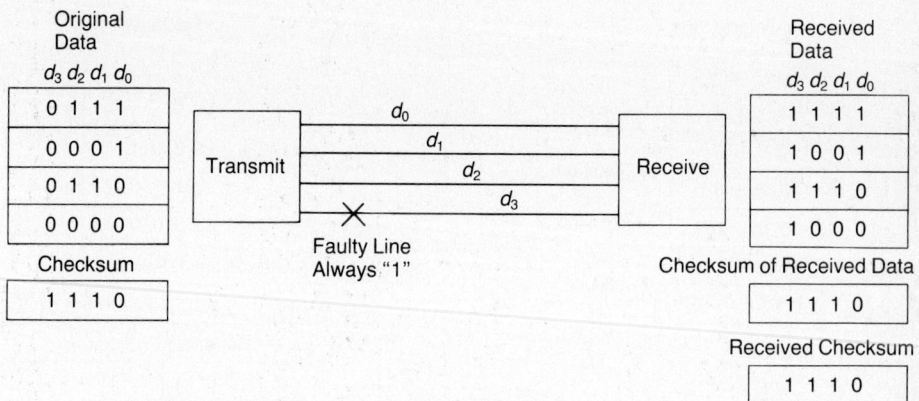

Fig. 3.37 The single-precision checksum is unable to detect certain types of errors. The received checksum and the checksum of the received data are equal, so no error is detected.

checksum is formed on the data at the transmitting point and is sent to the receiver along with the four words of data. If the most significant data line becomes logically stuck at a value of 1, the most significant bit of both the data and the checksum also become stuck at 1. When the checksum is generated on the received data, the regenerated and the received checksums agree because the overflow was ignored. But, the received information is certainly not correct.

One technique that is often used to overcome the limitations of the single-precision checksum is to compute the checksum in double precision, thus the name **double-precision checksum.** The basic concept of the double-precision checksum is to compute a $2n$-bit checksum for a block of n-bit words using modulo-2^{2n} arithmetic. Overflow is still a concern, however, but it is now overflow from a $2n$-bit sum as opposed to an n-bit sum. As an example, consider the problem illustrated in Fig. 3.37. We saw that the single-precision checksum failed to detect the error that occurred in the example. The double-precision checksum, as shown in Fig. 3.38, detects the stuck-at-1 fault. The checksum that is generated at the transmission point is sent as two 4-bit quantities. At the receiving point, each word has its most significant bit stuck at the logic 1 value. Also, the checksum has the most significant bit of each of its 4-bit halves stuck at 1. The regenerated checksum now disagrees with the received checksum, and the error is detected.

A third form of the checksum is called the **Honeywell checksum.** The basic idea of the Honeywell checksum is to concatenate consecutive words to form a collection of double-length words. For example, if there are k n-bit words, a set of $k/2$ $2n$-bit words is formed. The double-length words are

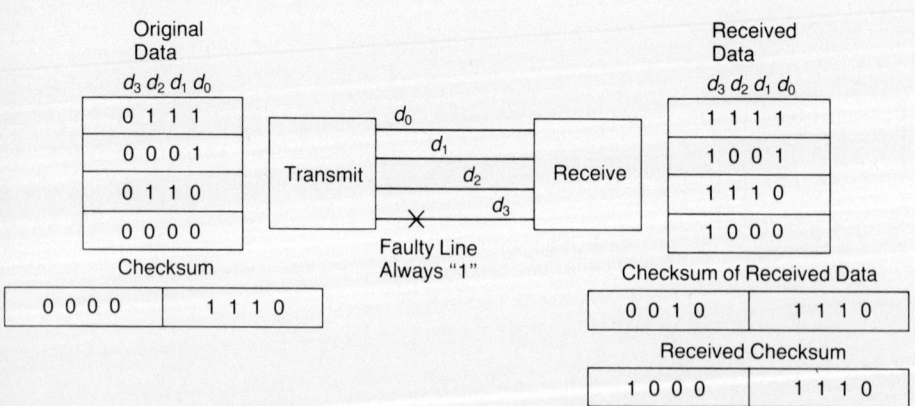

Fig. 3.38 A double-precision checksum is formed by adding the data using double-precision arithmetic. The received checksum and the checksum of the received data are not equal, so the error is detected.

3.5 ■ Information Redundancy

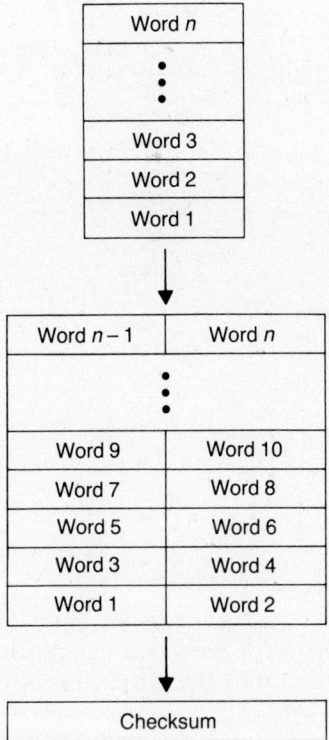

Fig. 3.39 In the Honeywell checksum, adjacent data words are concatenated prior to forming the checksum.

structured as shown in Fig. 3.39, and a checksum is formed over the newly structured data. The primary advantage of the Honeywell checksum is that a bit error that appears in the same bit position of all words will affect at least two bit positions of the checksum. For example, if a complete column of the original data is erroneous, the modified data structure has two erroneous columns.

An example of the Honeywell checksum is shown in Fig. 3.40. At the transmitting point, the new data structure is formed and the checksum is computed. If line d_3 is faulty, the regenerated checksum at the receiving point will differ from the checksum that is transmitted.

The final form of the checksum is the **residue checksum**. The concept of the residue checksum is the same as the single-precision checksum except that the carry bit out of the most significant bit position is not ignored but is added back to the checksum in an end-around carry fashion, as illus-

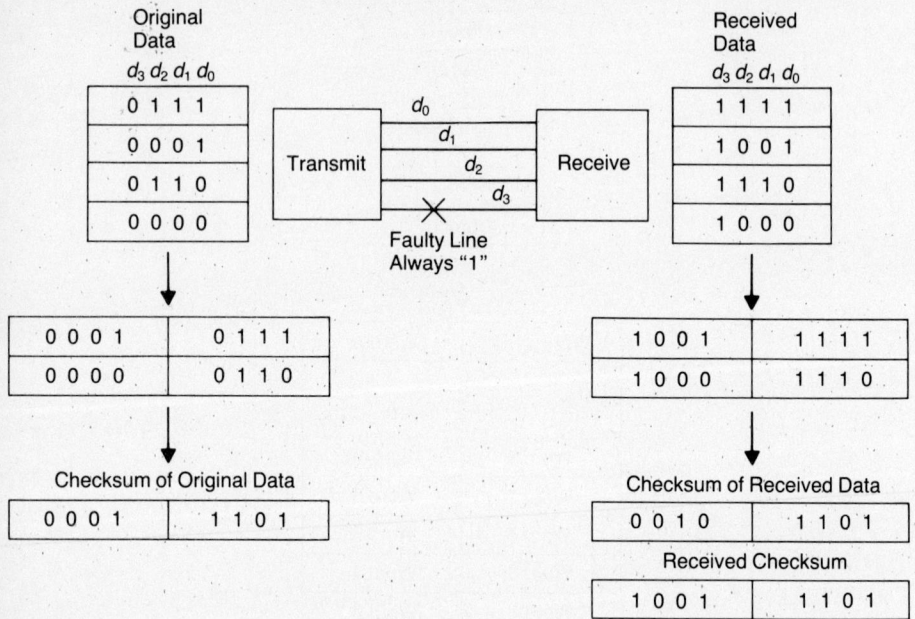

Fig. 3.40 Illustration of the error detection capability of the Honeywell checksum. The received checksum and the checksum of the received data are not equal, so the error is detected.

trated in Fig. 3.41. Using the same example as before (where the single-precision checksum was unable to detect the error), Fig. 3.42 shows that the residue checksum does indeed allow the error to be detected. The checksum regenerated at the receiving point differs from the checksum generated at the transmission point and transmitted to the receiver.

Note that checksums can detect errors but not locate them. If the checksum generated at the receiving point differs from the checksum generated at the transmission point, an error is indicated, but there is not enough information available to determine where the error has occurred. The complete block of data over which the checksum was formed must be corrected.

3.5.5 Cyclic Codes

The fundamental feature of **cyclic codes** is that any end-around shift of a code word will produce another code word [Lin 1983]. In other words, the cyclic code is invariant to the end-around shift operation. Cyclic codes are frequently applied to sequential-access devices such as tapes, bubble memories, and disks. In addition, cyclic codes are extremely popular for use in data links. One reason that cyclic codes are attractive is because the en-

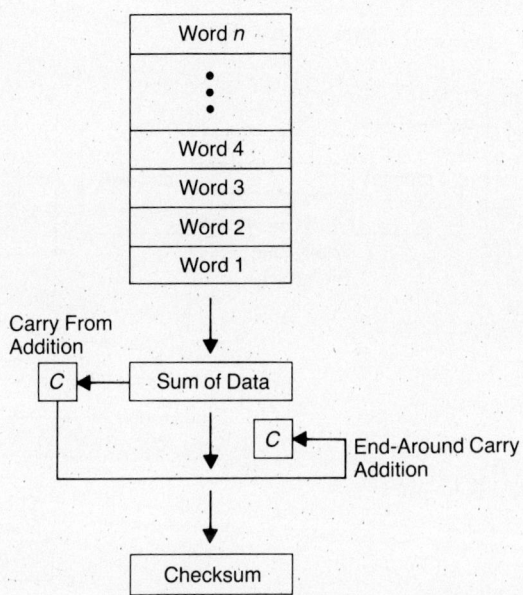

Fig. 3.41 The residue checksum is formed using end-around carry addition so that information in the carry bit is not lost.

coding operation can be implemented using simple shift registers with feedback connections.

A cyclic code is characterized by its generator polynomial $G(X)$, which is a polynomial of degree $n - k$ or greater, where n is the number of bits contained in the complete code word produced by $G(X)$, and k is the number of bits in the original information to be encoded. For binary cyclic codes, the coefficients of the generator polynomial are all either 0 or 1. The integers n and k specify the characteristics of the cyclic code. A cyclic code with a generator polynomial of degree $(n - k)$ is called an (n, k) cyclic code. Such codes possess the property of being able to detect all single errors and all multiple, adjacent errors affecting fewer than $(n - k)$ bits [Lin and Costello 1983]. The error detection property of cyclic codes is particularly important in communications applications where burst errors can occur. A **burst error** is the result of a transient fault and usually introduces a number of adjacent errors into a given data item. For example, a word that is transmitted serially can have several adjacent bits corrupted by a single disturbance; one would hope that the coding scheme could detect such errors. (n, k) cyclic codes can detect adjacent errors as long as the number of adjacent bits affected does not exceed $(n - k)$.

Cyclic codes depend on the representation of data by polynomials. We are very much accustomed to these types of representations because of the

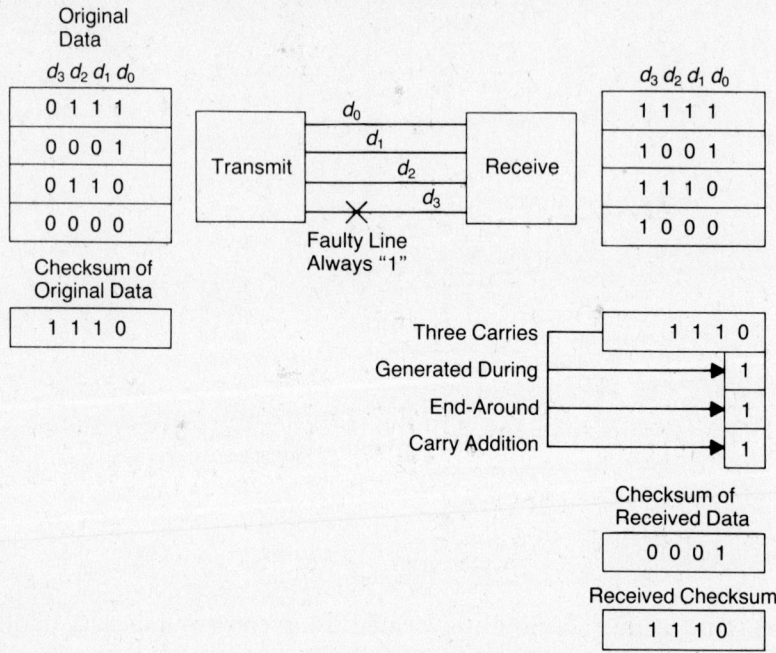

Fig. 3.42 Illustration of the error detection capability of the residue checksum. The checksum of the received data and the received checksum are not equal, so the error is detected.

number systems in which we work. For example, the decimal equivalent of the binary number 1011 can be represented as the polynomial $1 * r^3 + 0 * r^2 + 1 * r^1 + 1 * r^0$, where r is the base of the number system, in this case, $r = 2$, and $*$ is decimal multiplication. The coefficients of the polynomial represent the digits of the number.

The properties of cyclic codes are generated by representing the code words as coefficients of a polynomial. For example, suppose we have the code word $v = (v_0, v_1, \ldots, v_{n-1})$. This code word corresponds to the polynomial $V(X)$, where

$$V(X) = v_0 + v_1 X + v_2 X^2 + \cdots + v_{n-1} X^{n-1}$$

Each n-bit code word is represented by a polynomial of degree $(n - 1)$ or less. If the coefficient v_{n-1} is 0, the degree of the polynomial will be less than $n - 1$; but if the coefficient v_{n-1} is 1, the polynomial will have a degree of $(n - 1)$. The polynomial $V(X)$ is called the **code polynomial** of the code word v.

The code polynomials for a nonseparable cyclic code are generated by multiplying a polynomial, representing the data to be encoded, by another polynomial known as the **generator polynomial.** The generator polynomial determines the characteristics of the cyclic code. Any additions required during the multiplication of the two polynomials are performed using modulo-2 addition. For example, suppose that we have a generator polynomial, $G(X) = 1 + X + X^3$ and we wish to encode the binary data (1101). The data (1101) can be represented by the data polynomial $D(X) = 1 + X + X^3$. The code polynomial is generated by multiplying the data polynomial and the generator polynomial. Specifically, the code polynomial is generated as $V(X) = D(X) * G(X) = (1 + X + X^3)(1 + X + X^3) = 1 + X^2 + X^6$. In more exact terms, the code polynomial is given by $V(X) = 1 + 0*X + 1*X^2 + 0*X^3 + 0*X^4 + 0*X^5 + 1*X^6$ and the code word v consists of the coefficients of that code polynomial. In other words, v = (1010001).

Suppose we look at another example to clarify the basic concepts. Once again, consider the generator polynomial to be $G(X) = 1 + X + X^3$. Now, suppose that we wish to encode the data (1111), which is represented by the data polynomial $D(X) = 1 + X + X^2 + X^3$. To determine the code polynomial $V(X)$ we multiply the data polynomial $D(X)$ by the generator polynomial $G(X)$. The resulting code polynomial is $1 + 0*X + 0*X^2 + 1*X^3 + 0*X^4 + 1*X^5 + 1*X^6$. Note that the coefficients are added in a modulo-2 fashion, which allows for some very simple implementations of the cyclic codes. The resulting code word in this example is v = (1001011).

The nonseparable cyclic codes generated for four bits of information data using the generator polynomial, $G(X) = 1 + X + X^3$, are shown in Table 3.7. Note that the cyclic code shown in Table 3.7 is a code of distance 3; thus, any 2-bit errors can be detected. The distance can be easily determined by comparing all code words and seeing that all possible pairs of code words differ in at least three bit positions. The "cost" of the error detection capability is reflected in the three extra bits required to represent the information.

Perhaps the most interesting aspect of the nonseparable cyclic codes is the manner in which they can be generated. Recall that the code polynomial is generated by multiplying the data polynomial by the generator polynomial and adding the coefficients in a modulo-2 fashion. If we consider the blocks labeled X as multipliers by the factor X, and the addition elements as modulo-2 adders, the circuit shown in Fig. 3.43 performs the multiplication of two polynomials. For example, if $D(X) = 1$, the output $V(X)$ of the circuit will be $1 + X^2 + X^3$. Likewise, if $D(X) = 1 + X + X^3$, the output will be given by

$$V(X) = 1 + X + X^3 + [X(1 + X + X^3) + (1 + X + X^3)]X^2$$

or

$$V(X) = 1 + X + X^2 + X^3 + X^4 + X^5 + X^6$$

TABLE 3.7 Cyclic code words for 4-bit information words.

Information (d_0, d_1, d_2, d_3)	Code $(v_0, v_1, v_2, v_3, v_4, v_5, v_6)$
0000	0000000
0001	0001101
0010	0011010
0011	0010111
0100	0110100
0101	0111001
0110	0101110
0111	0100011
1000	1101000
1001	1100101
1010	1110010
1011	1111111
1100	1011100
1101	1010001
1110	1000110
1111	1001011

Data polynomial = $d_0 + d_1 x + d_2 x^2 + d_3 x^3$
Generator polynomial = $1 + x + x^3$
Code polynomial = $v_0 + v_1 x + v_2 x^2 + v_3 x^3 + v_4 x^4 + v_5 x^5 + v_6 x^6$

Therefore, if the generator polynomial is $G(X) = 1 + X^2 + X^3$, the circuit shown in Fig. 3.43 will generate the code polynomial by multiplying the data polynomial by the generator polynomial. For example, if the data to be

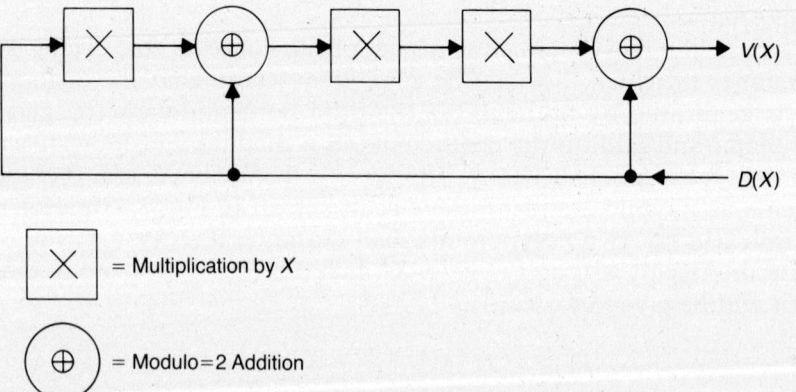

☒ = Multiplication by X

⊕ = Modulo=2 Addition

Fig. 3.43 Example circuit for generating a cyclic code word by multiplying an incoming data polynomial $D(X)$ by the generator polynomial.

encoded is (1101), the data polynomial will be $D(X) = 1 + X + X^3$. The resulting code polynomial yields the code word (1111111).

The generation circuits for cyclic codes can be implemented in digital hardware using storage elements and EXCLUSIVE-OR gates. The function of the storage elements is to implement a time delay. The circuit shown in Fig. 3.44 creates the code associated with the generator polynomial $G(X) = 1 + X + X^3$. Initially, the storage elements, or registers, are loaded with all 0s. The data to be encoded appears on the line $D(X)$ serially. The bits of the code word will appear on the line $V(X)$ serially. The occurrence of a clock pulse causes the registers to be loaded with the values on their inputs. Table 3.8 illustrates the operation of the circuit during the encoding of the data (1101).

Having examined the process of generating cyclic codes, we now consider the decoding procedure. The version of the cyclic code that has been presented thus far is not a separable code, so the decoding process involves more than simply picking certain bits from the code word. The structure of the cyclic code, however, makes the decoding process relatively easy.

Suppose that we wish to determine if the code word $(r_0, r_1, r_2, \ldots, r_{n-1})$ is valid. We know that this code word can be represented by the code polynomial $R(X) = r_0 + r_1 X + r_2 X^2 + \cdots + r_{n-1} X^{n-1}$. We also know that the correct code polynomial was generated by multiplying the original data polynomial by the generator polynomial. In other words, if $R(X)$ is a valid code polynomial, it was generated as $R(X) = D(X)G(X)$, where $G(X)$ is the generator polynomial and $D(X)$ is the original data polynomial. If we write

$$R(X) = D(X)G(X) + S(X)$$

then the quantity $S(X)$ should be zero if the polynomial $R(X)$ is a valid code polynomial. In other words $R(X)$ should be an exact multiple of the generator polynomial. One way to determine if $R(X)$ is indeed an exact multiple of the generator polynomial is to divide the polynomial $R(X)$ by the generator

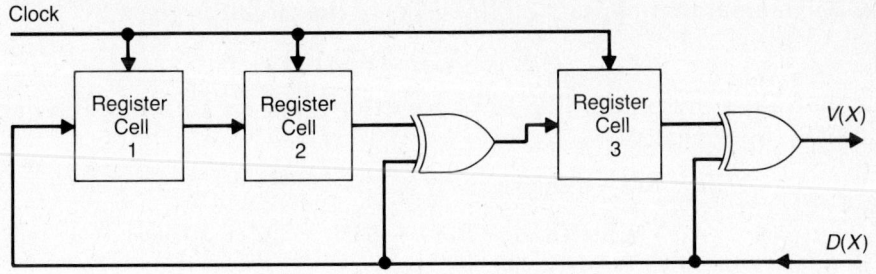

Fig. 3.44 Circuit for generating cyclic code words for the generator polynomial $G(X) = 1 + X + X^3$.

TABLE 3.8 The encoding process for the circuit of Fig. 3.44

Clock period	Register values			$D(x)$	$V(x)$
	1	2	3		
0	0	0	0		
				1	1
1	1	0	1		
				1	0
2	1	1	1		
				0	1
3	0	1	1		
				1	0
4	1	0	0		
				0	0
5	0	1	0		
				0	0
6	0	0	1		
				0	1
7	0	0	0		

polynomial $G(X)$ and see if the remainder of the division is zero. If the remainder is zero, the polynomial is an exact multiple of the generator polynomial and is a valid code polynomial. The quantity $S(X)$ is called the **syndrome polynomial.**

The process of division may at first seem complicated and difficult to implement. It turns out to be quite simple, however, when feedback circuits similar to the cyclic code generators are used. The circuit shown in Fig. 3.45, for example, is capable of dividing a polynomial by the polynomial $1 + X + X^3$. Once again, the blocks labeled as X perform multiplication by the factor X. The adders in the circuit of Fig. 3.45 are modulo-2 adders. The polynomial that appears on line $B(X)$ of the circuit is given by

$$B(X) = (X^3 + X)D(X)$$

But the values present on line $D(X)$ are determined by both line $B(X)$ and line $V(X)$. Specifically,

$$V(X) + B(X) = D(X)$$

$$V(X) = D(X) - B(X) = D(X) - (X^3 + X)D(X)$$

Because the functions of addition and subtraction are the same in the modulo-2 system, we obtain

3.5 ■ Information Redundancy

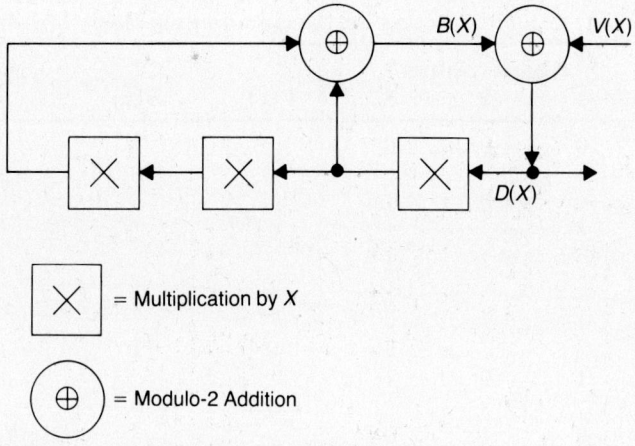

Fig. 3.45 A division circuit for use in decoding cyclic code words.

$$V(X) = (X^3 + X + 1)D(X)$$

$$D(X) = \frac{V(X)}{(X^3 + X + 1)}$$

The circuit of Fig. 3.45 divides the polynomial $V(X)$ by the polynomial $(X^3 + X + 1)$.

If the multiplicative elements of the circuit of Fig. 3.45 are replaced with storage elements and the modulo-2 adders are constructed using EXCLUSIVE-OR gates, the circuit of Fig. 3.46 results. If the storage elements, or registers, are initialized to 0, the polynomial represented by the data stream appearing on line $V(X)$ is divided by $(X^3 + X + 1)$, and the result appears as a data stream on line $D(X)$. Once the division has been completed, the registers contain the value of the syndrome, or remainder, of the division process. If the syndrome is zero, a valid code word has been received. Otherwise, the word received was an invalid code word. Table 3.9 illustrates the decoding process for the code word (1010001).

As an example, consider the results generated by the circuit of Fig. 3.46 when the code word is erroneous. Suppose that the code word (1010001) was the intended code word, but an error resulted in the received code word being (1011001). In other words, a single error has resulted in the 0 in the center bit position becoming a 1. Table 3.10 details the functions of the circuit when the received data stream on line V is (1011001). As can be seen, the syndrome that results is nonzero, clearly indicating an invalid code word; the received code word is not evenly divisible by the generator polynomial.

TABLE 3.9 The decoding process for the circuit of Fig. 3.46

Clock period	Register values 1	2	3	$V(x)$	$B(x)$	$D(x)$
0	0	0	0	1	0	1
1	0	0	1	0	1	1
2	0	1	1	1	1	0
3	1	1	0	0	1	1
4	1	0	1	0	0	0
5	0	1	0	0	0	0
6	1	0	0	1	1	0
7	0	0	0	↑ Code word		↑ Original information

Syndrome (Register cells 1, 2, 3)

The primary disadvantage of the cyclic codes discussed thus far is that they are not separable. It is possible, however, to generate a separable, cyclic code [Nelson and Carroll 1986]. To generate an (n,k) code, the original data polynomial $D(X)$ is first multiplied by X^{n-k}, and the result is divided by the generator polynomial $G(X)$ to obtain a remainder of $R(X)$. The code polynomial is then computed as $V(X) = R(X) + X^{n-k}D(X)$, and $V(X)$ is an exact multiple of the generator polynomial $G(X)$. Note that the multipli-

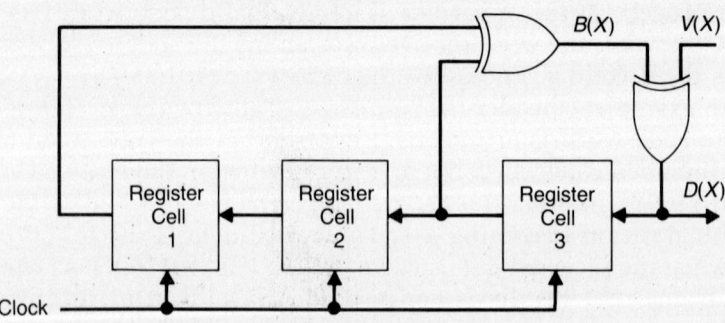

Fig. 3.46 Decoding circuit for the cyclic code with generator polynomial, $G(X) = 1 + X + X^3$.

TABLE 3.10 The decoding process with erroneous information

Clock period	Register values 1	2	3	V(x)	B(x)	D(x)
0	0	0	0			
				1	0	1
1	0	0	1			
				0	1	1
2	0	1	1			
				1	1	0
3	1	1	0			
				1	1	0
4	1	0	0			
				0	1	1
5	0	0	1			
				0	1	1
6	0	1	1			
				1	1	0
7	1	1	0	↑ Received word		

Nonzero syndrome (columns 1, 2, 3)

cation by X^{n-k} can be performed by simply shifting the coefficients of the data polynomial. Also note that the addition of the remainder polynomial is equivalent to simply appending the remainder to the polynomial $X^{n-k}D(X)$.

The validity of the encoding process just described is relatively obvious. Suppose, however, that we have an arbitrary data polynomial $D(X)$ and a generator polynomial $G(X)$. The code polynomial $V(X)$, is given by

$$V(X) = X^{n-k}D(X) + R(X)$$

where $R(X)$ is the remainder obtained when $X^{n-k}D(X)$ is divided by $G(X)$. In other words,

$$\frac{X^{n-k}D(X)}{G(X)} = Q(X) + \frac{R(X)}{G(X)}$$

where $Q(X)$ is the quotient computed in the division process.

Multiplying both sides of the previous equation by $G(X)$ yields

$$X^{n-k}D(X) = G(X)Q(X) + R(X)$$
$$X^{n-k}D(X) - R(X) = G(X)Q(X)$$

Recall, however, that all addition and subtraction operations are performed using modulo-2 arithmetic, so addition and subtraction are identical. Consequently,

$$X^{n-k}D(X) - R(X) = X^{n-k}D(X) + R(X) = G(X)Q(X) = V(X)$$

Therefore, the code polynomial $V(X)$, formed as

$$V(X) = X^{n-k}D(X) + R(X)$$

is an exact multiple of the generator polynomial $G(X)$.

As an example of the construction of the separable (7,4) cyclic code consider the data (1001) and the generator polynomial $G(X) = 1 + X + X^3$. The data polynomial corresponding to (1001) is $D(X) = 1 + X^3$. Therefore, $X^{n-k}D(X) = X^3D(X) = X^3+X^6$. Dividing $X^3 + X^6$ by $G(X)$ yields a remainder of $R(X) = X^2 + X$. Adding the remainder polynomial to $X^3 + X^6$ results in $V(X) = X + X^2 + X^3 + X^6$. So, the code word is given by (0111001). Note that the last four bits of the code word are the coefficients of the original data polynomial, and the first three bits are the coefficients of the remainder polynomial.

3.5.6 Arithmetic Codes

Arithmetic codes are very useful when it is desired to check arithmetic operations such as addition, multiplication, and division [Avizienis 1971]. The basic concept is the same as all coding techniques. The data presented to the arithmetic operation is encoded before the operations are performed. After completing the arithmetic operations, the resulting code words are checked to make sure that they are valid. If they are not, an error condition exists.

An arithmetic code must be invariant to a set of arithmetic operations. An arithmetic code A has the property that $A(b * c) = A(b) * A(c)$, where b and c are operands, $*$ is some arithmetic operation, and $A(b)$ and $A(c)$ are the arithmetic code words for the operands b and c, respectively. Stated in words, the performance of the arithmetic operation on two arithmetic code words will produce the arithmetic code word of the result of the arithmetic operation. To completely define an arithmetic code, the method of encoding and the arithmetic operations for which the code is invariant must be specified. Examples of arithmetic codes are the AN codes, residue codes, inverse-residue codes, and the residue number system.

AN Codes

The simplest arithmetic code is the AN code which is formed by multiplying each data word N by some constant A. The AN codes are invariant to addition and subtraction but not multiplication and division. If N_1 and N_2 are two operands to be encoded, the resulting code words are AN_1 and AN_2, respectively. If the two code words are added, the sum is $A(N_1 + N_2)$, which is the code word of the correct sum. The operations performed under an AN code can be checked by determining if the results are evenly divisible by the constant A. If they are not, an error condition exists.

The magnitude of the constant A determines both the number of extra bits required to represent the code words and the error detection capability provided. The selection of the constant A is critical to the effectiveness and efficiency of the resulting code. First, for binary codes, the constant must not be a power of two. To see the reason for this limitation, suppose that we encode the binary number $(a_{n-1}a_{n-2} \cdots a_2a_1a_0)$ by multiplying by the constant $A = 2^a$. Multiplication by 2^a is equivalent to a left arithmetic shift of the original binary word, so the resulting code word is $(a_{n-1}a_{n-2} \cdots a_2a_1a_000 \cdots 0)$, where a 0s have been appended to the original binary number. The decimal representation of the code word is given by

$$a_{n-1}2^{a+n-1} + \cdots + a_22^{a+2} + a_12^{a+1} + a_02^a + 02^{a-1} + \cdots + 02^1 + 02^0$$

which is clearly, evenly divisible by 2^a. It is also easy to see, however, that changing just one coefficient still yields a result that is evenly divisible by 2^a. For example, if the coefficient of the 2^a term changes from 0 to 1, the result remains evenly divisible by 2^a. Thus, an AN code that has $A = 2^a$ is not capable of detecting single-bit errors.

An example of a valid AN code is the $3N$ code, where all words are encoded by multiplying them by 3. If the original data words are n bits long, the code words for the $3N$ code require $n + 2$ bits. Table 3.11 shows the $3N$ code for 4-bit information. An example of the effectiveness of the $3N$ code is shown in Fig. 3.47 where a binary adder is protected by the $3N$ code. Under fault-free circumstances, the result of adding the two code words (010010) and (000011) results in the valid code word (010101), which is evenly divisible by 3. If, however, line S_1 is stuck at the logic 1 value, the result of the addition of (010010) and (000011) is (010111), which is not evenly divisible by 3. Therefore, the error can be detected by checking each resulting code word to determine if it is evenly divisible by 3.

The encoding of operands in the $3N$ code can be performed by a simple addition if we recognize that we can multiply any number by 3 by adding the original number to a value that is twice that number. In other words, we form $3N$ by adding N and $2N$. The quantity $2N$ is easily created by shifting the binary number left by one place. The numbers N and $2N$ can then be added. Figure 3.48 illustrates the use of an $(n+1)$-bit adder to produce the $(n + 2)$-bit code word for an n-bit operand. An $(n + 1)$-bit adder can be used if the carry-bit out of the adder is used as the most significant bit of the resulting sum.

TABLE 3.11 Resulting $3N$ code words for 4-bit information words

Original information	$3N$ code word
0000	000000
0001	000011
0010	000110
0011	001001
0100	001100
0101	001111
0110	010010
0111	010101
1000	011000
1001	011011
1010	011110
1011	100001
1100	100100
1101	100111
1110	101010
1111	101101

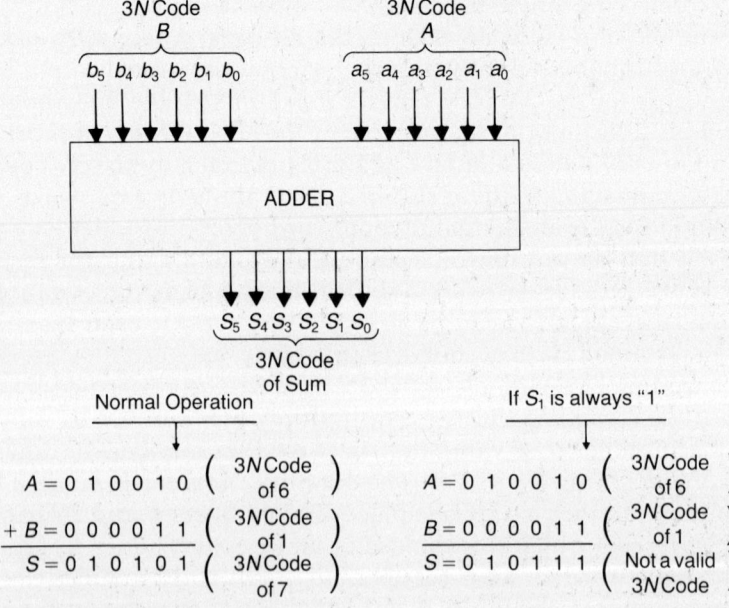

Fig. 3.47 Illustration of the error detection capabilities of the $3N$ arithmetic code. The presence of the fault results in the sum being an invalid $3N$ code.

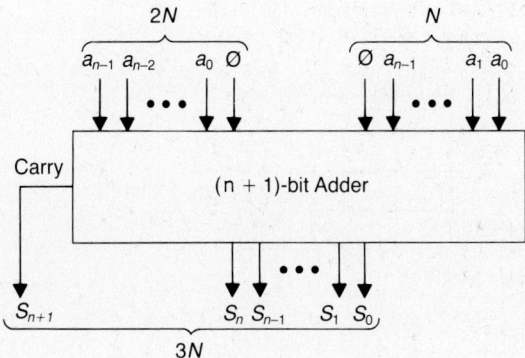

Fig. 3.48 Illustration of the use of an $(n + 1)$-bit adder to create $3N$ code words.

The $3N$ code can be checked by using a simple combinational circuit. For example, suppose that we wish to design an error checker for the $3N$ code of 2-bit data; in this case, the code words are 4 bits long. The Karnaugh map of the combinational circuit that performs the checking operation is shown in Fig. 3.49. The circuit produces a value of 1 when the inputs to the circuit represent a valid $3N$ code word; otherwise, the output of the circuit is 0. The combinational circuit can be implemented using a multiplexer, as shown in Fig. 3.49.

Residue Codes

The next class of arithmetic codes to be studied are the residue codes. A residue code is a separable arithmetic code created by appending the residue of a number to that number. In other words, the code word is constructed as $D\,|\,R$, where D is the original data and R is the residue of that data. The encoding operation consists of determining the residue and appending it to the original data. The decoding process involves simply removing the residue, thus leaving the original data word.

The residue of a number is simply the remainder generated when the number is divided by an integer. For example, suppose we have an integer N and we divide N by another integer m. N may be written as an integer multiple of m as

$$N = Im + r$$

$$\frac{N}{m} = I + \frac{r}{m}$$

where r is the remainder, sometimes called the residue, and I is the quotient. The quantity m is called the check base, or the modulus. For example,

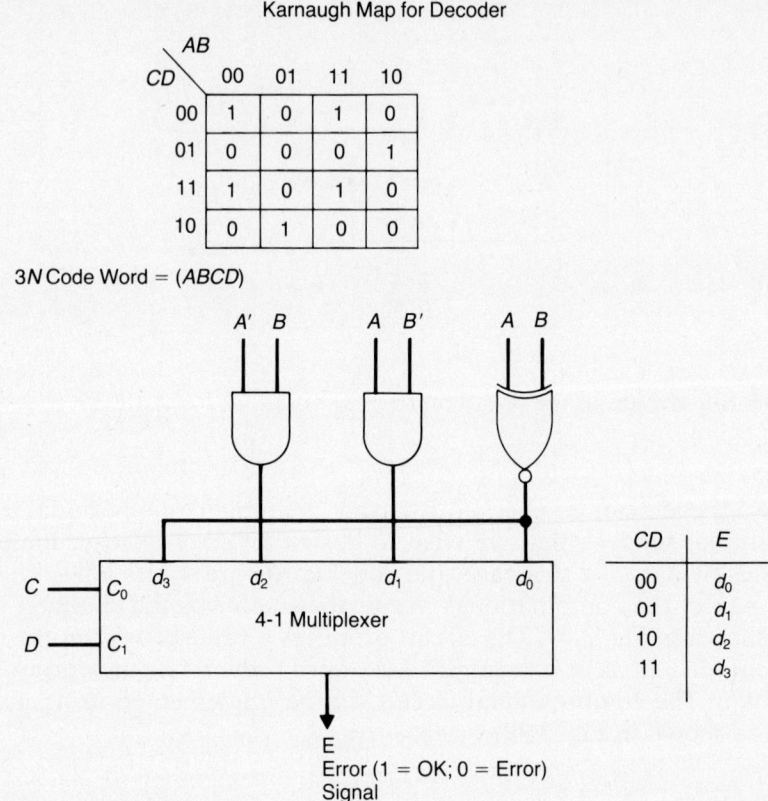

Fig. 3.49 A simple error detection circuit for the 3N code can be constructed using combinational logic.

if $N = 14$ and $m = 3$, the quotient I will be 4 and the residue will be 2. We often write this as

$$14 = 2 \text{ modulo}(3)$$

Separable residue codes, as mentioned previously, are formed by appending the residue of a data word to that data word. The number of extra bits required to represent the code word depends on the particular modulus selected. The residue will never be larger than the modulus; in fact, $0 \leq r < m$. For example, if the original data is n bits and the modulus is 3, the code word will require $n + 2$ bits. Table 3.12 illustrates the residue code that results when 4-bit data is encoded using a modulus of 3.

The primary advantages of the residue codes are that they are invariant to the operation of addition, and the residues can be handled separately

TABLE 3.12 Residue code words for 4-bit information words using a modulus of three

Information	Residue	Code word
0000	0	0000 00
0001	1	0001 01
0010	2	0010 10
0011	0	0011 00
0100	1	0100 01
0101	2	0101 10
0110	0	0110 00
0111	1	0111 01
1000	2	1000 10
1001	0	1001 00
1010	1	1010 01
1011	2	1011 10
1100	0	1100 00
1101	1	1101 01
1110	2	1110 10
1111	0	1111 00

from the data during the addition process. The structure of an adder that uses the separable residue code for error detection is shown in Fig. 3.50. The two data words D_1 and D_2 are added to form a sum word S. The residues r_1 and r_2 of D_1 and D_2, respectively, are also added using a modulo-m adder, where m is the modulus used to encode D_1 and D_2. If the operations are performed correctly, the modulo-m addition of r_1 and r_2 yields the residue r_s, of the sum S. A separate circuit is then used to calculate the residue of S. If the calculated residue r_c, differs from r_s, an error has occurred in one part of the process. For example, errors can be detected that occur in the generation of S, r_s, or r_c.

If the modulus for the residue code is selected in a special manner, a low-cost residue code results. Specifically, low-cost residue codes have a modulus of $m = 2^b - 1$, where b is some integer greater than or equal to 2. The number of extra bits required in a low-cost residue code is equal to b. For example, the residue code shown in Table 3.12 for a modulus of 3 is a low-cost residue code.

The main advantage of the low-cost residue code is the ease with which the encoding process can be performed. Recall that we must determine a remainder to encode information using a residue code; therefore, a division is necessary. The low-cost residue codes, however, allow the division to be recast as an addition process. The information bits to be encoded, are first di-

118   Design Techniques to Achieve Fault Tolerance

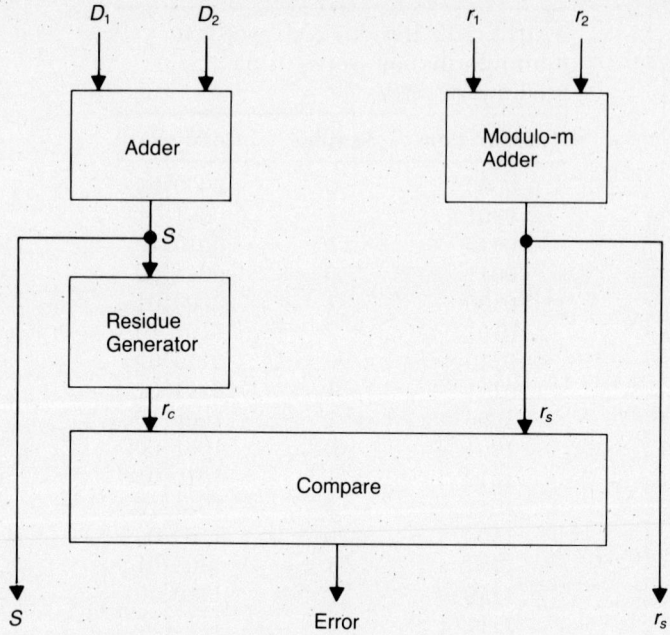

Fig. 3.50 The structure of an adder designed using the separable residue code.

vided into groups, each group containing b bits. The groups are then added in a modulo-$(2^b - 1)$ fashion to form the residue of the information bits. For example, Fig. 3.51 shows the procedure for determining the residue for the information bits (10100111) using a modulus of 3. Eight-bit information

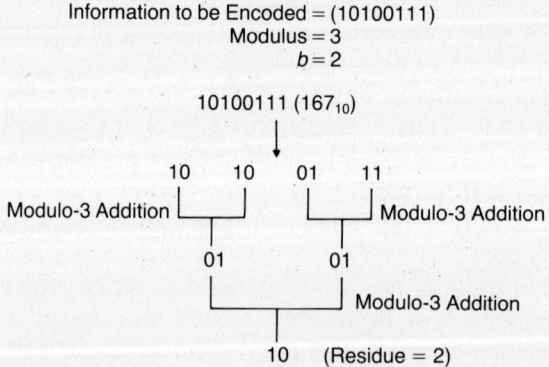

Fig. 3.51 The residue calculation for a low-cost residue code can be performed using successive additions.

words can be encoded using three, 2-bit, modulo-3 adders, as shown in Fig. 3.52, when the modulus is chosen as three.

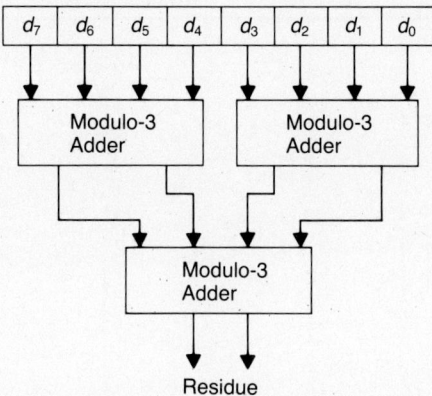

Fig. 3.52 The residue generation for 8 bits of data can be performed using three modulo-3 adders.

Inverse-Residue Codes

A modification of the separable residue code is the separable **inverse-residue code.** The inverse-residue code is formed in a manner similar to that of the residue code by appending information to the original data. Rather than append the residue, the inverse residue is calculated and appended. The inverse residue Q is calculated for a data word N as $m - r$, where m is the modulus and r is the residue of N. The code word for N then becomes $N|Q$. Table 3.13, for example, shows the inverse-residue code for 4-bit data using a modulus of 3.

The inverse-residue codes have been found to have better fault detection capability for repeated-use faults [Avizienis 1971]. A **repeated-use fault** is one that is encountered multiple times before the code is checked because the hardware is used multiple times before the code is checked. For example, if repeated addition is used to perform multiplication and the adder has a fault of some type, a repeated-use fault occurs. Repeated-use faults are particularly difficult to detect because subsequent effects of the fault can cancel the previous effects of the fault, thus rendering the fault undetectable.

Residue Number System

The final arithmetic code to consider is the **residue number system** (RNS) [Taylor 1984]. The residue number system has certain advantages in the speed at which arithmetic operations can be performed. In the RNS, num-

TABLE 3.13 Inverse-residue code words for 4-bit information words using a modulus of three

Information	Residue	Inverse residue	Code
0000	0	3	0000 11
0001	1	2	0001 10
0010	2	1	0010 01
0011	0	3	0011 11
0100	1	2	0100 10
0101	2	1	0101 01
0110	0	3	0110 11
0111	1	2	0111 10
1000	2	1	1000 01
1001	0	3	1001 11
1010	1	2	1010 10
1011	2	1	1011 01
1100	0	3	1100 11
1101	1	2	1101 10
1110	2	1	1110 01
1111	0	3	1111 11

bers are represented by a set of residues. The RNS does not produce a separable code, as did the residue and the inverse-residue codes.

To represent a number in the RNS, we first define a set of relatively prime moduli. The concept of relatively prime implies that the largest number that divides evenly into any two moduli is 1. The moduli set is given as

$$P = \text{moduli set} = [p_1, p_2, \ldots, p_L]$$

The range of numbers that can be represented using this moduli set is $0 \leq N \leq M$, where N is the number to be represented and M is the product of all the moduli. The representation of a number N in the RNS is determined by computing the residue of N for each of the moduli p_i. In other words N is represented as the collection of residues

$$N = (x_1, x_2, \ldots, x_L)$$

where x_i is the residue of N calculated using the modulus p_i. For example, suppose we wish to represent the integer 32 using the moduli set [3,4,5]. If we divide 32 by 3, we obtain a remainder of 2. If we divide 32 by 4, we obtain a remainder of 0. If we divide 32 by 5, the remainder is 2. Consequently, the representation of 32, using the moduli set [3,4,5] is given by (2,0,2).

One of the biggest advantages of the RNS is the fact that it is a carry-free number system. Therefore, arithmetic operations such as addition can

3.5 ■ Information Redundancy

be performed on the individual digits of numbers independent of the remaining digits of the number. In contrast to the decimal and the binary number systems where carry information from the previous digits must be known before the result for the present digit can be calculated, the RNS allows the calculation of all digits of the result of an operation in parallel. As an example, suppose we wish to add the two numbers 32 and 14. If the moduli set is [3,4,5], the RNS representation of 32 is (2,0,2) and the representation of 14 is (2,2,4). The sum of 32 and 14 is 46, and the RNS representation of 46 is (1,2,1), also using the moduli set [3,4,5]. The RNS representation of 46 can be obtained by adding the RNS representations of 32 and 14 using the appropriate modulus to perform the addition of each digit. This process is illustrated in Fig. 3.53. Stated mathematically, the sum of two RNS numbers $(x_1, x_2, \cdots, x_L)$ and $(y_1, y_2, \cdots, y_L)$ is given by $(z_1, z_2, \cdots, z_L)$, where each z_i is computed as $z_i = (x_i + y_i)$ modulo-p_i.

To better see the possibility of speed improvements in an RNS addition compared to a binary addition, consider Fig. 3.54 where the ripple-carry addition of two 6-bit binary numbers is compared to the addition of the same two numbers when RNS representations and arithmetic are used. Note that modulo adders can be used to add the corresponding residues of the RNS representations. The RNS addition allows the residues to be added in parallel, and each residue has a smaller number of bits than the 6 bits obtained in the binary representations. Also, RNS adders can be designed

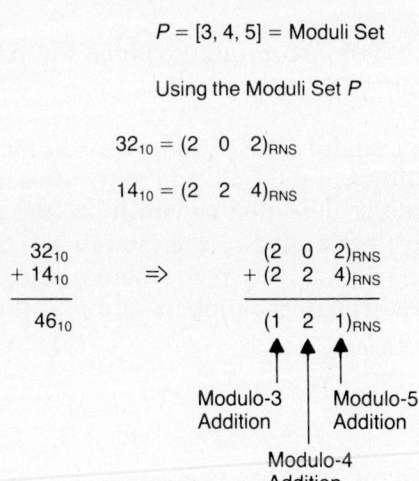

Fig. 3.53 In residue arithmetic, each modulus is added independently to produce the result.

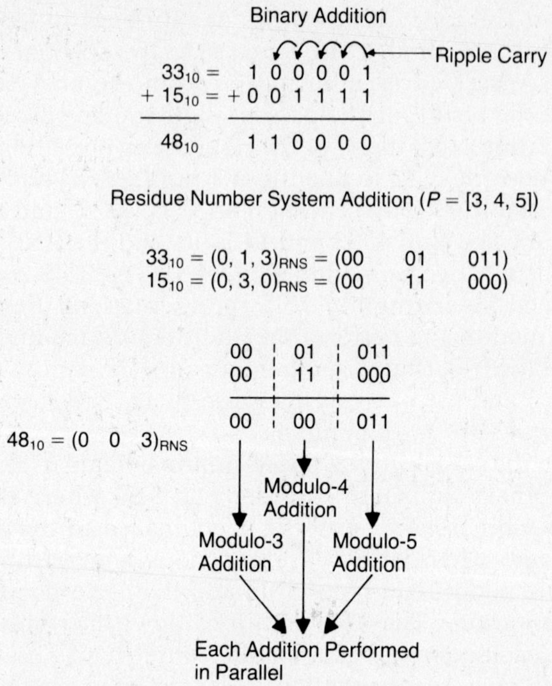

Design Techniques to Achieve Fault Tolerance

Binary Addition

$$33_{10} = 1\ 0\ 0\ 0\ 0\ 1$$
$$+15_{10} = +0\ 0\ 1\ 1\ 1\ 1$$
$$\overline{48_{10} 1\ 1\ 0\ 0\ 0\ 0}$$

Ripple Carry

Residue Number System Addition ($P = [3, 4, 5]$)

$$33_{10} = (0, 1, 3)_{RNS} = (00 \quad 01 \quad 011)$$
$$15_{10} = (0, 3, 0)_{RNS} = (00 \quad 11 \quad 000)$$

$$\begin{array}{c|c|c} 00 & 01 & 011 \\ 00 & 11 & 000 \\ \hline 00 & 00 & 011 \end{array}$$

$48_{10} = (0\ \ 0\ \ 3)_{RNS}$

Modulo-4 Addition

Modulo-3 Addition Modulo-5 Addition

Each Addition Performed in Parallel

Fig. 3.54 The residue number system offers the advantage of speed over binary addition because the residues are added in parallel.

using techniques such as look-up tables. Since the residues are typically small, look-up tables are practical and allow for very-high-speed arithmetic operations.

Although speed is certainly the primary reason for using the RNS, the error detection capabilities of the code add to its attractiveness in many applications. To obtain error detection capability in the residue number system, redundant moduli are added to the moduli set. Suppose we selected the moduli set $[p_1, p_2, \cdots, p_L, p_{L+1} \cdots, p_M]$, where p_1 through p_L are sufficient to represent the desired range of numbers and p_{L+1} through p_M are redundant. Define M and T as

$$M = p_1 p_2 \cdots p_L$$
$$T = p_{L+1} p_{L+2} \cdots p_M$$

The nonredundant moduli represent the range of numbers from 0 through M. The addition of the redundant moduli allows the representable range of numbers to be expanded to 0 through MT. A code can now be structured such that valid code words lie in the range from 0 through M, and invalid

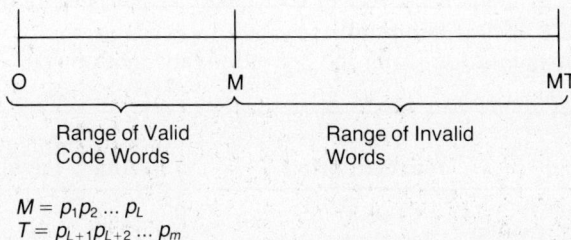

Fig. 3.55 Error detection capability is created in a residue representation because of the use of redundant moduli.

code words lie in the range from $M + 1$ through MT. This concept is illustrated in the simple number line shown in Fig. 3.55. As an example, consider Table 3.14. Table 3.14a shows the RNS representations of the numbers from 0 to 7 using the nonredundant moduli set [3,4]. This representation clearly has no error detection capability. A single-bit error can map a valid code word, for example, (1010), into another valid code word, in this example, (0010). If redundant moduli are used, however, error detection capability can be provided. Table 3.14b shows the RNS representations obtained when the moduli set [3,4,5] is used. The modulus 5 is redundant in this example. As can be seen, the RNS representation using the redundant moduli of this example forms a distance-2 code that allows the detection of any single-bit errors.

The residue number system offers advantages of speed and error detection in many applications, however, its use is not widespread. The primary difficulty with using the RNS is the conversion between a representation such as binary or decimal and the RNS. Human beings have difficulty interpreting numbers in the RNS, consequently applications involving human/computer interfaces must provide conversions between the RNS and a representation such as decimal. Because such conversions are extremely difficult, the RNS has not found widespread applicability.

3.5.7 Berger Codes

A very simple form of coding is the **Berger code** [Lala 1985]. Berger codes are formed by appending a special set of bits, called the check bits, to each word of information. Therefore, the Berger code is a separable code. The check bits are created based on the number of 1s in the original information. A Berger code of length n has I information bits and k check bits, where $k = \lceil \log_2(I + 1) \rceil$ and $n = I + k$. A code word is formed by first creating a binary number that corresponds to the number of 1s in the original I

TABLE 3.14 Residue number system representations
(a) Representation using the nonredundant moduli set
$P = [3, 4]$

Number	RNS representation		RNS (binary form)		
0	0	0	00	00	
1	1	1	01	01	
2	2	2	10	10	← Single-bit
3	0	3	00	11	error can
4	1	0	01	00	map one
5	2	1	10	01	into the
6	0	2	00	10	← other.
7	1	3	01	11	

(b) Representation using the redundant moduli set
$P = [3, 4, 5]$

Number	RNS representation			RNS (binary form)		
0	0	0	0	00	00	000
1	1	1	1	01	01	001
2	2	2	2	10	10	010
3	0	3	3	00	11	011
4	1	0	4	01	00	100
5	2	1	0	10	01	000
6	0	2	1	00	10	001
7	1	3	2	01	11	010

bits of information. The resulting binary number is then complemented and appended to the I information bits to form the $(I + k)$-bit code word. For example, suppose that the information to be encoded is (0111010), such that $I = 7$. The value of k is then $k = \lceil \log_2(7 + 1) \rceil = 3$. The number of 1s in this word of information is four, and the 3-bit binary representation of four is (100). The complement of (100) is (011), so the resulting code word is (0111010011), which is simply the original information with 011 appended.

If the number of information bits is small, the redundancy of a Berger code is high. Table 3.15, for example, shows the number of check bits required as a function of the number of information bits. As can be seen, the redundancy is 50% or greater if the number of information bits is eight or less. However, as the number of information bits increases, the efficiency of the code improves substantially.

If the number of information bits is related to the number of check bits by the relationship

$$I = 2^k - 1$$

TABLE 3.15 Number of required check bits in a Berger code

Number of information bits	Number of check bits	Percentage redundancy
4	3	75.00%
8	4	50.00
16	5	31.25
32	6	18.75
64	7	10.94

the resulting code is called a maximal length Berger code. For example, the code constructed for $I = 7$ and $k = 3$ is a maximal length Berger code.

The primary advantages of the Berger codes are that they are separable and they detect all multiple, unidirectional errors. For the error detection capability it provides, the Berger codes use the fewest number of check bits of the available separable codes [Lala 1985]. The resulting Berger codes for $I = 4$ and $k = 3$ are illustrated in Table 3.16.

3.5.8 Horizontal and Vertical Parity

The use of both horizontal and vertical parity is a very simple extension of the basic parity scheme. Suppose that we have a memory consisting of

TABLE 3.16 Berger code words for 4-bit information words

Original information	Berger code
0000	0000 111
0001	0001 110
0010	0010 110
0011	0011 101
0100	0100 110
0101	0101 101
0110	0110 101
0111	0111 100
1000	1000 110
1001	1001 101
1010	1010 101
1011	1011 100
1100	1100 101
1101	1101 100
1110	1110 100
1111	1111 011

three 4-bit words, as shown in Fig. 3.56. By forming an odd parity bit, for example, for both the columns and the rows, we can detect and locate an error because the error affects the parity in both a column and a row. The erroneous bit is at the intersection of the row and column that has erroneous parity. Once the location of the erroneous bit is known, the error can be corrected by simply complementing the affected bit. If the bit is a 0 and it is incorrect, the correct value is a 1. Likewise, if the bit is a 1 and it is incorrect, the correct value is a 0.

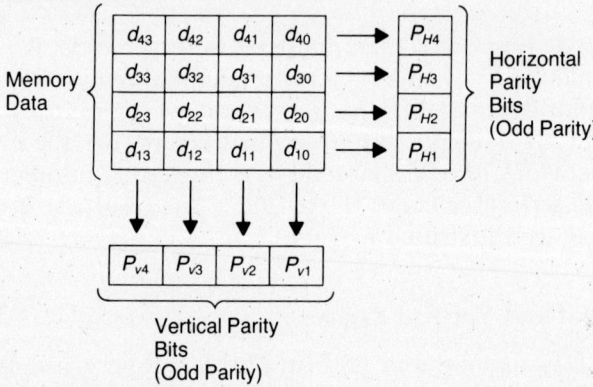

Fig. 3.56 Vertical and horizontal parity uses a parity bit for each row and each column. If bit d_{31}, for example, becomes in error, both P_{H3} and P_{V2} will be erroneous. All other parity bits will be correct.

Horizontal and vertical parity is a very useful technique for correcting errors in groups of data words. Suppose, for example, that a block of n k-bit words is to be transmitted from one point within a system to another. Each word can be extended to include a single parity bit, so that each word is $k + 1$ bits. Also, the n words can be EXCLUSIVE-OR'ed together on a bit-by-bit basis to create one $(k + 1)$-bit word that is the parity word for the block of data. At the receiving point, the parity bit for each word and the parity word for the group are recomputed, and the location of any single error is determined and corrected.

Horizontal and vertical parity, however, cannot help correct *multiple* errors within the block of data words. The existence of multiple columns and rows with erroneous parity prevents successful identification of the erroneous bit. For example, if rows i and j and columns k and m have erroneous parity, the errors can be in bit positions (i,k), (i,m), (j,k), or (j,m). Consequently, correction is not possible, but the detection of a problem is still accomplished.

3.5.9 Hamming Error-Correcting Codes

Possibly the most common extension of the fundamental parity approach is the **Hamming error-correcting code** [Hamming 1950]. Many memory designs incorporate error correction for several reasons. First, Hamming error correction is relatively inexpensive; typically, the Hamming codes require anywhere from 10% to 40% redundancy. Second, the Hamming codes are efficient in terms of the time required to perform the correction process; the encoding and the decoding processes inject relatively small time delays. Third, the error correction circuit is readily available on inexpensive chips. Finally, the memory can contribute as much as 60% to 70% of the faults in a system. In addition, transient faults are becoming much more prevalent as memory chips become denser. The combination of permanent and transient faults in memories makes the use of error correction very attractive.

The Hamming codes are best thought of as overlapping parity. The Hamming single error-correcting code uses c parity check bits to protect k bits of information. The relationship between the values of c and k is

$$2^c \geq c + k + 1$$

The total length of the code word is $n = c + k$. As we saw in the overlapping parity approach, the c check bits provide one unique combination for each possible information bit that can be erroneous, one combination for each parity check bit that can be erroneous, and one combination for the error-free case.

The Hamming code is formed by partitioning the information bits into parity groups and specifying a parity bit for each group. The ability to locate which bit is erroneous is obtained by overlapping the groups of bits. In other words, a given information bit will appear in more than one group in such a way that if the bit is erroneous, the parity bits that are in error will identify the erroneous bit. For example, suppose that there are four information bits (d_3, d_2, d_1, d_0) and, as a result, three parity check bits (c_1, c_2, c_3). The bits are partitioned into groups as (d_3, d_1, d_0, c_1), (d_3, d_2, d_0, c_2), and (d_3, d_2, d_1, c_3). Each check bit is specified to set the parity, either even or odd, of its respective group. Now, if bit d_0, for example, is erroneous, both c_1 and c_2 are incorrect. However, c_3 is correct because the value of d_0 has no impact on the value of c_3. Table 3.17 shows the check bits that are affected for each possible erroneous bit when four bits of information and three check bits are used. Note that the effect of each erroneous bit is unique; a unique combination of parity check bits is produced in each case, so the erroneous bit can be located.

The basic process involved in the Hamming codes is no different from that of other codes. First, the original data is encoded by generating a set, call it C_g, of parity check bits. When the information is to be checked for

TABLE 3.17 Check bits affected by single data bit errors

Erroneous bit	Check bits affected
d_0	c_1, c_2
d_1	c_1, c_3
d_2	c_2, c_3
d_3	c_1, c_2, c_3
c_1	c_1
c_2	c_2
c_3	c_3

correctness, the encoding process is repeated and a set, call it C_c, of parity check bits is regenerated. If C_g and C_c agree, the information is correct. If, however, C_g and C_c disagree, the information is incorrect and must be corrected. To aid in the correction, we define the syndrome S as the result obtained by forming the EXCLUSIVE-OR of C_g and C_c. The syndrome is a binary word that is 1 in each bit position in which C_g and C_c disagree. A syndrome that is all 0s indicates correct information. Using the example of four information bits and three check bits, Table 3.18 illustrates the syndromes that result for each possible erroneous bit when the group partitions of Table 3.17 are used.

The syndrome can be used to point directly to the erroneous bit if the bits are arranged appropriately. Suppose that we number the bit positions from right to left with 1 representing the rightmost position and n representing the leftmost position in an n-bit word. Each bit is then placed in the bit position corresponding to the syndrome that will result when that bit becomes erroneous. For example, the syndromes shown in Table 3.18 suggest the bit ordering of Fig. 3.57. The value of the syndrome now locates specifically the bit that is in error.

TABLE 3.18 Resulting syndromes for each possible single bit error

Erroneous bit	Syndromes
d_0	110
d_1	101
d_2	011
d_3	111
c_1	100
c_2	010
c_3	001

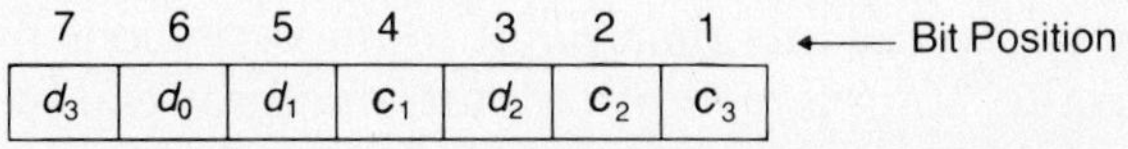

Fig. 3.57 Required ordering of bits to allow syndrome to identify specific erroneous bit.

The structure of a Hamming single-error–correction unit for four bits of information is shown in Fig. 3.58. When the information and the check bits are received, the parity check bits are regenerated and the syndrome bits

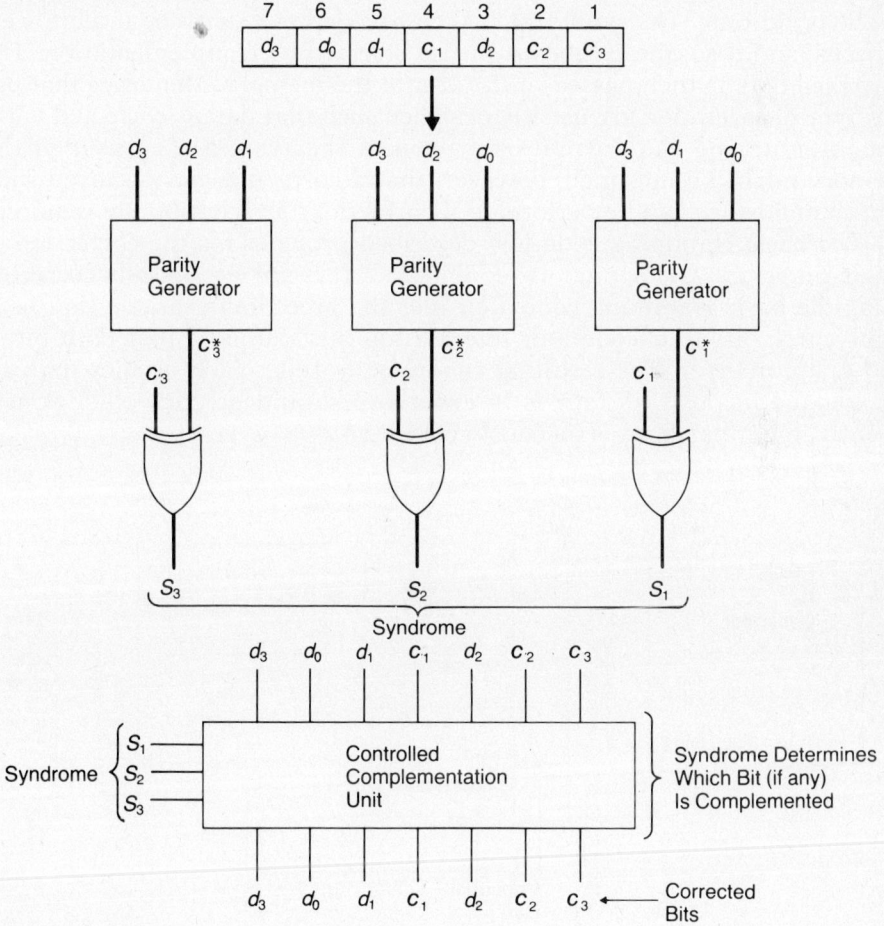

Fig. 3.58 Hamming single error correction unit for four information bits and three check bits.

are created. The syndrome bits then point to the erroneous bit, if one exists. By using a controlled complementation unit, the error can be corrected. The syndrome can specify which output, if any, of a multiplexer is set to the logic 1 state. If none of the multiplexer outputs is set to 1, the original information is passed through the EXCLUSIVE-OR gates in an uncomplemented fashion. If a bit is erroneous, the syndrome value results in the appropriate bit being complemented, thus the correction process has been implemented.

The structure of a memory that uses the Hamming single-error–correcting code is shown in Fig. 3.59. When data is written to memory, the check bits are generated and stored in memory along with the original information. Upon reading the information and the check bits from memory, the check bits are regenerated and compared to the stored check bits to generate the syndrome. The syndrome is then decoded to determine if a bit is erroneous, and if so, the erroneous bit is corrected by complementation. The corrected data is then passed to the user of the memory. Memories that use this type of correction are usually designed such that data is corrected without interrupting the normal operation of the system. The user of the memory might be informed, however, that a correction has occurred such that maintenance can be performed if corrections are continually required.

The basic Hamming code just described provides for the correction of single-bit errors. Unfortunately, double-bit errors are erroneously corrected using the basic Hamming code. Consider the preceding example that used three check bits to encode four information bits. Suppose that both bit d_0 and c_2 are in error. The resulting syndrome is 100, which implies that the bit located in position four is in error and should be corrected. Consequently, bit c_1 has been erroneously corrected.

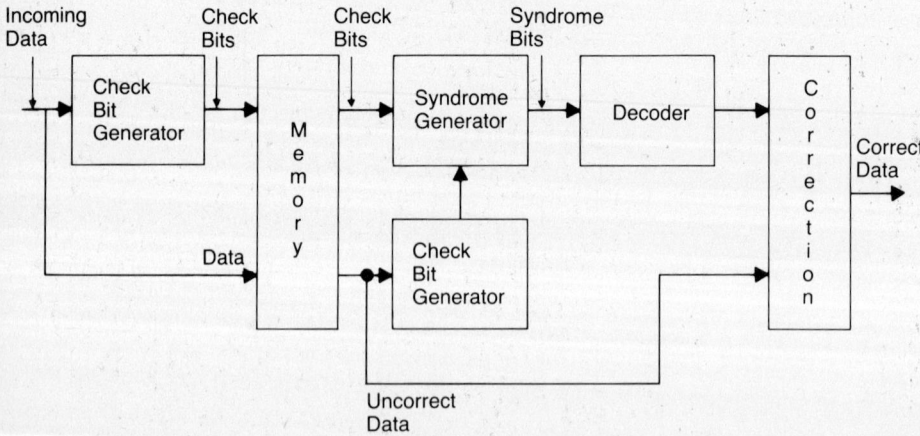

Fig. 3.59 Basic structure of a memory using Hamming single error correcting code.

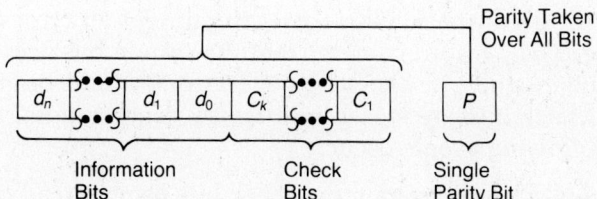

Fig. 3.60 Hamming code modification to achieve double error detection and single error correction.

To overcome the problem of erroneous correction and provide a code that can correct single-bit errors and identify double-bit errors, the basic Hamming code is modified. The resulting code is called the *modified* Hamming code. The modification consists of simply adding one additional parity check bit that checks parity over the entire Hamming code word, as shown in Fig. 3.60. If a single bit is in error, the additional parity bit indicates that the overall parity is incorrect. The syndrome then points to the correct bit that is erroneous. If a double-bit error occurs, the additional parity bit indicates that the overall parity is correct because a single parity check cannot detect a double-bit error. But, the syndrome is nonzero because the remaining parity checks indicate an error. Therefore, the double error can be detected, and an erroneous correction prevented.

To summarize, if the overall parity is incorrect and the syndrome is not 0, a single-bit error is corrected. If the overall parity is correct and the syndrome is not 0, a double-bit error is identified and no correction occurs. If the overall parity is correct and the syndrome is 0, the data is assumed to be correct.

3.5.10 Error-Correcting Integrated Circuits

Several commercial integrated circuits (ICs) are available to detect, locate, and correct errors. Examples include the Intel 8206, Motorola MC68540, Advanced Micro Devices AM2960 and AMZ8160, National Semiconductor DP8400, and Fujitsu MB1412A. The ICs are designed to generate the check bits, generate the syndrome, decode the syndrome, and perform the data correction. Typical units are organized to support 16-bit data but can usually be expanded to provide error detection and correction for larger word sizes. For example, the Intel 8206 can support up to 80-bit words in a memory [Intel 1985]. The ICs almost always use the modified Hamming code to detect double errors and correct single errors.

A block diagram of the fundamental structure of an error-correcting IC is shown in Fig. 3.61. A word coming from the system bus goes directly into the memory and into a write check bit generator. Both the original data and the generated check bits are stored in memory. Upon reading a word from memory, the following events occur:

1. The data is used to regenerate the check bits using a read check bit generator.
2. A syndrome is created by comparing the original check bits with the regenerated check bits.

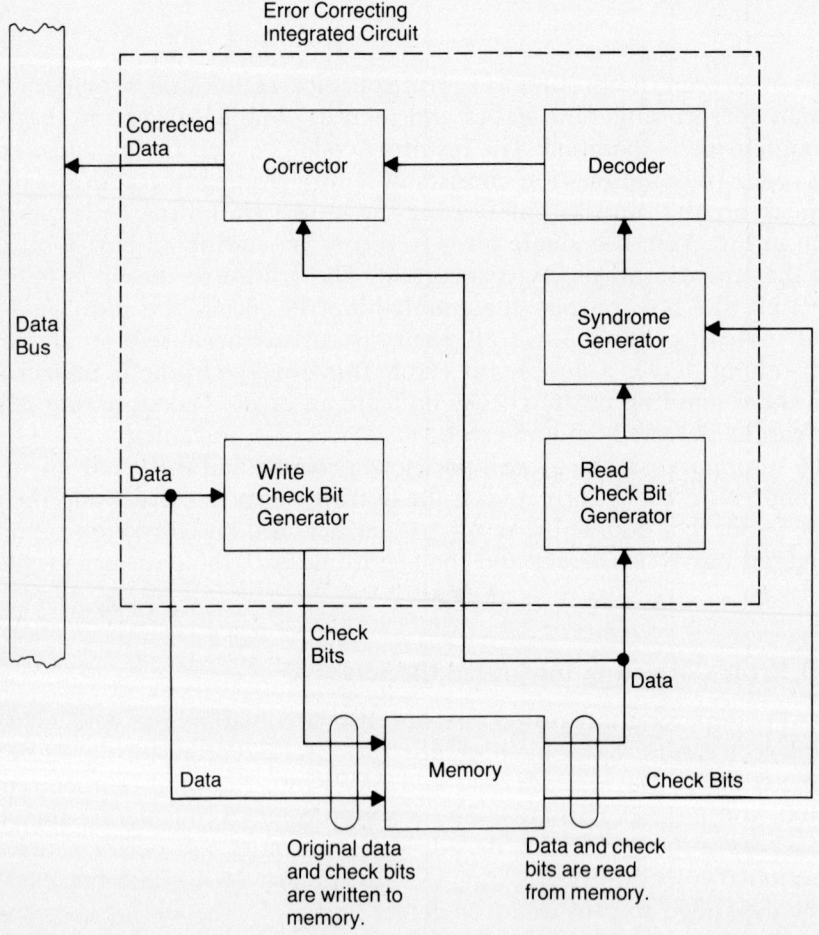

Fig. 3.61 Typical organization of a commercially available error correction integrated circuit.

3. The syndrome is used to identify any erroneous bits.
4. The data is passed through a correction unit.
5. The corrected data is placed on the system bus.

The device typically provides an error indication so that the user can record the occurrence of the error for maintenance purposes. Often, the number of error occurrences is used to indicate a permanent fault in the memory.

The additional time required during a memory write operation is simply the time necessary to generate the check bits. When data is read from memory, however, the check bits must be regenerated, the syndrome calculated and decoded, and the correction completed. Once the syndrome is generated and decoded the existence or nonexistence of an error is known. The final step of correction is necessary, however, before the data is used.

The Intel 8206, for example, can perform the detection process in a 16-bit system in a maximum of 52 nanoseconds [Intel 1985]. The combination of the detection and the correction requires no more than 67 nanoseconds. Designed using Intel's HMOS technology, the 8206 is available in 68-pin leadless packages and operates from a single 5-volt power supply.

3.5.11 Code Selection Issues

Several tradeoffs must be performed prior to the selection of a particular coding technique. As previously mentioned, codes are considered a form of information redundancy because the fundamental concept is to modify the original data in such a manner that sufficient information is provided to allow error detection and/or correction to be performed. The information redundancy requires, however, that other forms of redundancy exist as well. For example, the encoding and decoding processes require time, so redundancy of time is produced. Likewise, the additional bits found in the code word require additional storage and processing circuitry, so redundancy of hardware is produced. The amount of redundancy required in a code is a good measure of the cost of the code. On the other hand, the effectiveness of the code is usually measured in terms of the number of bit errors that can be detected or corrected. The key to the design process is to select a code that fulfills the desired error detection and/or correction capability while maintaining costs at an acceptable level.

A major decision in the selection of a code is whether or not the code needs to be separable. Separable codes are easier to use because the decoding process consists of simply ignoring the check bits that have been added. As a result, the information is available immediately, and, in many cases, a system can begin to use the information in parallel with the checking of the information. For example, a memory that uses single-bit parity can provide the data to a processor concurrent with checking the parity of that data. Nonseparable codes, however, require that the decoding process be performed before the data is available for use. In some applications, encoding and decoding can be performed in parallel with transmission and reception to minimize the impact of the encoding and decoding times. For example, encoding and decoding of cyclic codes can be performed bit by bit as the information is transmitted and received in a serial fashion. Consequently, cyclic codes find wide applicability in systems in which data is transmitted serially.

A second major decision in the selection of a code is whether error detection, error correction, or both are required. In applications in which temporary, erroneous results are acceptable—for example, in active redundancy systems that require reconfiguration—error detection is normally sufficient. However, if fault masking is mandatory to prevent erroneous data from propagating throughout a system, error correction is needed. If single-error correction, as opposed to single-error detection, is needed, a code with a larger distance is mandatory, and the redundancy required increases.

A third major decision in the selection of a code is the number of bit errors that need to be detected or corrected. As we have seen, error detection and correction capability is directly related to the distance of the code and, as a result, the necessary redundancy. The application dictates, to a large degree, the necessary distance of the code. For example, in serial communication systems, a single fault can easily affect several consecutive bits in a code word; so the ability to handle multiple-bit errors is important. In many memory designs, however, single-bit errors are common, and single-error detection or single-error correction codes are typically employed.

3.6 Time Redundancy

The fundamental problem with the forms of redundancy discussed thus far is the penalty paid in extra hardware for the implementation of the various techniques. Both hardware redundancy and information redundancy can require large amounts of extra hardware for their implementation. In an effort to decrease the hardware required to achieve fault detection or fault tolerance, time redundancy has recently received much attention. Time redundancy methods attempt to reduce the amount of extra hardware at the expense of using additional time. In many applications, the time is of much less importance than the hardware because hardware is a physical entity that impacts weight, size, power consumption, and cost. Time, on the other hand, may be readily available in some applications.

The selection of a particular type of redundancy is very dependent upon the application. For example, some systems can better stand additional hardware than additional time; others can tolerate additional time much more easily than additional hardware. The selection in each case must be made by examining the requirements of the application and the available techniques that can meet such requirements.

3.6.1 Transient Fault Detection

The basic concept of time redundancy is the repetition of computations in ways that allow faults to be detected. Time redundancy can function in a system in several ways, but the most basic form of time redundancy is illustrated in Fig. 3.62. The fundamental concept is to perform the same computation two or more times and compare the results to determine if a discrepancy exists. If an error is detected, the computations can be performed again to see if the disagreement remains or disappears. Such approaches are often good for detecting errors resulting from transient faults, but they cannot protect against errors resulting from permanent faults.

A second time redundancy approach makes use of other detection techniques that can be built into a system. Suppose, for example, that an error-detecting code provides the primary means of error detection within a

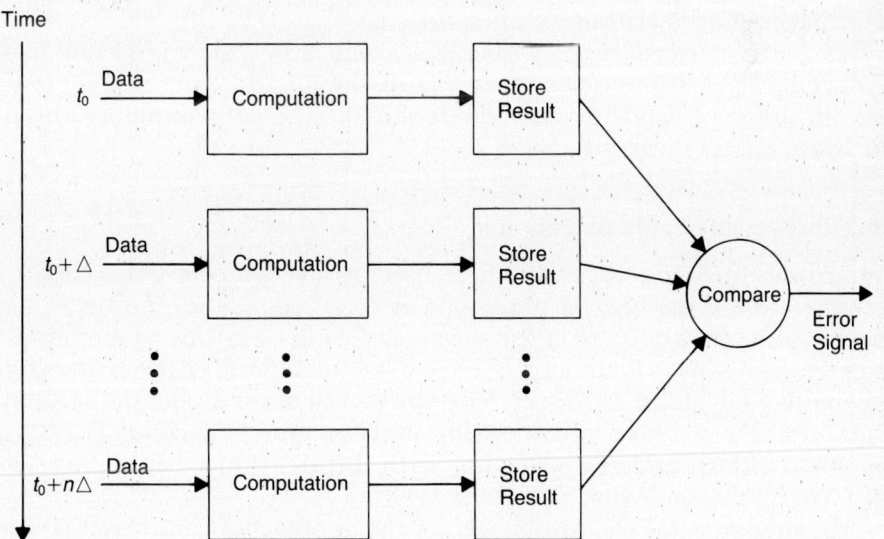

Fig. 3.62 In time redundancy, computations are repeated at different points in time and then compared.

system. When the coding scheme detects an error, either of two conditions exists:

1. A permanent fault has produced the error, and the correct course of action is to shut down the faulty portions of the system.
2. A transient fault has produced the error, in which case the hardware is still usable, and it would be a waste of resources to shut down the system.

Time redundancy can often be employed to distinguish between the permanent and the transient faults. The system can perform the computations one or more times after the detection of the first error; if the error condition clears, the fault that caused the error can be assumed to have been transient. If, however, the problem continues to be detected, the fault is most likely permanent, and the faulty parts of the system must be removed from operation.

The main problem with many time redundancy techniques is assuring that the system has the same data to manipulate each time it redundantly performs a computation. If a transient fault has occurred, a system's data may be completely corrupted, making it difficult to repeat a given computation.

Another example of time redundancy actually combines both information and time redundancy. The concept requires that the same computations be performed multiple times using different coding schemes in each case. For example, suppose that a processor uses an *AN* arithmetic code to detect errors in the arithmetic operations performed by that processor. The processor can perform the functions first using a $3N$ code and second using a $5N$ code. The assumption is that a fault should affect the two computations in different ways because the data that is being manipulated by the hardware differs in the two cases.

3.6.2 Permanent Fault Detection

In the past, time redundancy has been used primarily to detect transients in systems. One of the biggest potentials of time redundancy, however, now appears to be the ability to detect permanent faults while using a minimum of extra hardware. Four approaches are considered; alternating logic [Reynolds and Metze 1978], recomputing with shifted operands (RESO) [Patel and Fung 1982], recomputing with swapped operands (RESWO) [Johnson 1984], and recomputing with duplication with comparison (REDWC) [Johnson, Aylor, and Hana 1988].

The fundamental concept of each of the approaches considered is illustrated in Fig. 3.63. During the first computation or transmission, the operands are used as presented and the results are stored in a register. Prior

3.6 ■ Time Redundancy

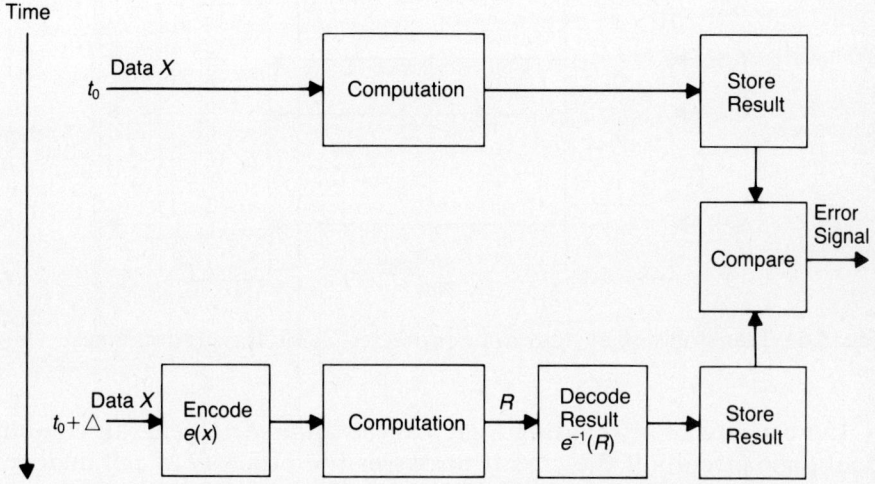

Fig. 3.63 Permanent faults can be detected using time redundancy by modifying the way in which computations are performed the second time.

to the second computation or transmission, the operands are encoded in some fashion using the encoding function e. After the operations have been performed on the encoded data, the results are then decoded and compared to those obtained during the first operation. The selection of the encoding function is made so as to allow faults in the hardware to be detected. The alternating logic approach uses the complementation operator as the encoding function. RESO uses an arithmetic shift as the encoding function, RESWO uses a swapping function to encode the operands and REDWC is a variation of RESWO. Each approach has both advantages and disadvantages.

Alternating Logic

The **alternating logic** concept has been applied to the transmission of digital data over wire media and the detection of faults in digital circuits. Suppose that we wish to detect errors in data that is transmitted over a parallel bus, as shown in Fig. 3.64, using the time redundancy approach. At time t_0 we transmit the original data, and at time $t_0 + \delta$ we transmit the complement of the data. As shown in Fig. 3.65, if a line of the bus is stuck at either a 1 or a 0, the two versions of the information that are received will not be complements of each other. Therefore, the fault can be detected. In general, if a sequence of information is transmitted using this approach, each bit line should alternate between a logic 1 and a logic 0; provided the transmission is error free. Thus, the reason for the name *alternating logic*.

138   Design Techniques to Achieve Fault Tolerance

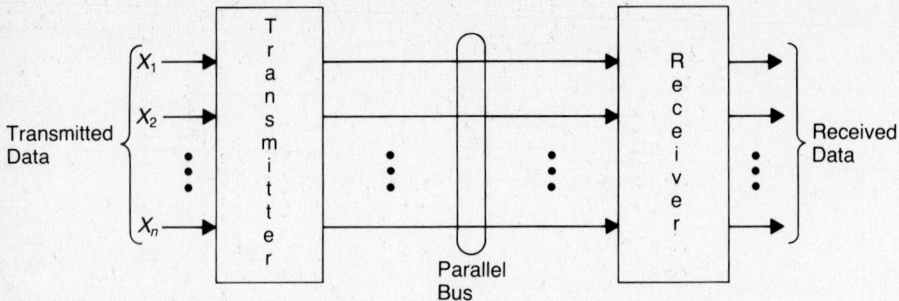

Fig. 3.64 Transmission system to be protected using time redundancy.

The concept of alternating logic can be applied to general, combinational logic circuits if the circuit possesses the property of self-duality. A combinational circuit is said to be self dual if and only if $f(X) = \overline{f(\overline{X})}$, where f is the output of the circuit and X is the input vector for the circuit [Kohavi 1978]. Stated verbally, a combinational circuit is self-dual if the output for the input vector X is the complement of the output when the input vector $\overline{X}$ is applied. For example, the full-adder circuit, shown in Fig. 3.66 with its truth table, is a self-dual circuit. For a self-dual circuit, the application of an input X followed by the input $\overline{X}$, produces outputs that alternate between 1 and 0. The key to the detection of faults using the alternating logic ap-

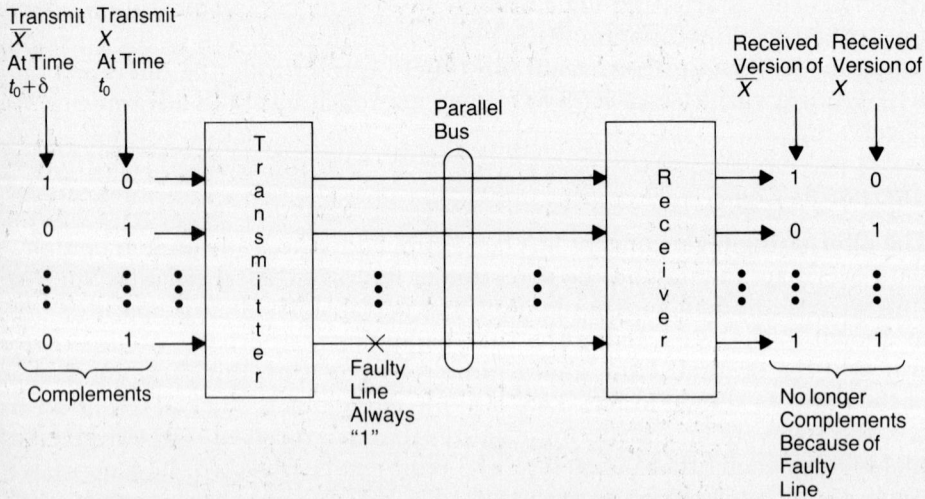

Fig. 3.65 Illustration of alternating logic time redundancy—the second transmission is the complement of the first.

proach is determining that at least one input combination exists for which the fault does not result in alternating outputs.

A key advantage of the alternating logic approach is that any combinational circuit with n input variables can be transformed into a self-dual circuit with no more than $n + 1$ input variables. To see this, first define the dual of a function. The dual f_d of an n-variable function f is given by [Kohavi 1978]

$$f_d = \overline{f}(\overline{x}_1, \overline{x}_2, \cdots, \overline{x}_n)$$

In other words, the dual of the function f is obtained by first complementing f and then replacing each variable with the complement of the variable. The function f_{sd} given by

$$f_{sd} = x_{n+1}f + \overline{x}_{n+1}f_d$$

is then a self-dual function because when x_{n+1} is 1, the value of f_{sd} is f, and when x_{n+1} is 0, the value of f_{sd} is f_d. Thus, x_{n+1} is a control line that determines which of two functions appears on the output line. So, defining $f_{sd}(x_1, x_2, \cdots, x_n, x_{n+1})$ to be a function of $n + 1$ variables, complementary inputs produce complementary outputs.

The use of alternating logic detects a set of faults if for every fault within the set there is at least one input combination that produces nonalternating outputs. For example, Table 3.19 shows the truth table that results when one stuck-at-1 or stuck-at-0 fault is present in the full adder circuit of Fig. 3.66. The notation A_0 and A_1 is used to represent the case where line A is stuck at logic 0 and logic 1, respectively. Similar notation is used for all other lines. As can be seen, each stuck-type fault results in at least one set of nonalternating outputs being produced for complementary inputs at either the carry or the sum output.

Note, however, that faults may not be immediately detected using alternating logic. For example, suppose that the full-adder contains a stuck-at-0 fault on line D. The sum output is not affected by this fault, so the carry output is depended upon for the detection of the fault. The carry output, however, will have alternating outputs for the complementary input combinations (000) and (111) as well as (001) and (110). In other words, the fault is not detected until one of the remaining four input combinations is applied to the circuit. Depending on the application, the time elapsed before the detection of the fault can be significant.

Recomputing with Shifted Operands

Another form of time redundancy is called **recomputing with shifted operands (RESO)** [Patel and Fung 1982]. RESO was developed as a method to provide concurrent error detection in arithmetic logic units (ALUs). RESO uses the basic time redundancy method that was shown in Fig. 3.63,

TABLE 3.19 Truth table for single faults in the full-adder circuit of Fig. 3.66

(a) Sum output truth table

			Fault-free sum	Sum when specified fault is present														
A	B	C_1	S	A_0	B_0	P_0	C_{10}	D_0	E_0	C_{00}	S_0	A_1	B_1	P_1	C_{11}	D_1	E_1	C_{01}	S_1
0	0	0	0	0	0	0	0	0	0	0	0	1	1	1	1	0	0	0	1
0	0	1	1	1	1	1	0	1	1	1	0	0	0	0	1	1	1	1	1
0	1	0	1	1	0	0	1	1	1	1	0	0	1	1	0	1	1	1	1
0	1	1	0	0	1	1	1	0	0	0	0	1	0	0	0	0	0	0	1
1	0	0	1	0	1	0	1	1	1	1	0	1	0	1	0	1	1	1	1
1	0	1	0	1	0	1	1	0	0	0	0	0	1	0	0	0	0	0	1
1	1	0	0	1	1	0	0	0	0	0	0	0	0	1	1	0	0	0	1
1	1	1	1	0	0	1	0	1	1	1	0	1	1	0	1	1	1	1	1

(b) Carry output truth table

			Fault-free carry	Carry when specified fault is present														
A	B	C_1	C_0	A_0	B_0	P_0	C_{10}	D_0	E_0	C_{00}	S_0	A_1	B_1	P_1	C_{11}	D_1	E_1	C_{01}	S_1
0	0	0	0	0	0	0	0	0	0	0	0	0	0	0	1	1	1	0
0	0	1	0	0	0	0	0	0	0	0	1	1	1	0	1	1	1	0
0	1	0	0	0	0	0	0	0	0	0	1	0	0	1	1	1	1	0
0	1	1	1	1	0	0	0	0	1	0	1	1	1	1	1	1	1	1
1	0	0	0	0	0	0	0	0	0	0	0	1	0	1	1	1	1	0
1	0	1	1	0	1	0	0	0	1	0	1	1	1	1	1	1	1	1
1	1	0	1	0	0	1	1	1	0	0	1	1	1	1	1	1	1	1
1	1	1	1	1	1	1	1	1	0	0	1	1	1	1	1	1	1	1

and the encoding function is selected as the left shift operation with the decoding function being the right shift operation. In many cases, the shift operations can be either arithmetic or logical shifts. The RESO technique was derived assuming bit-sliced organizations of the hardware.

In logical operations, it is relatively easy to understand the error detection capability of the RESO technique. Suppose that bit slice i of a circuit is faulty and produces an erroneous value for the function's output at that bit slice. During the first computation with the operands not shifted, the ith output of the circuit is erroneous. When the operands are shifted by one bit, the faulty bit slice then operates on, and corrupts, the $(i - 1)$th bit. When the result is shifted back to the right, the two results—the first with unshifted operands and the second with shifted operands—are either both

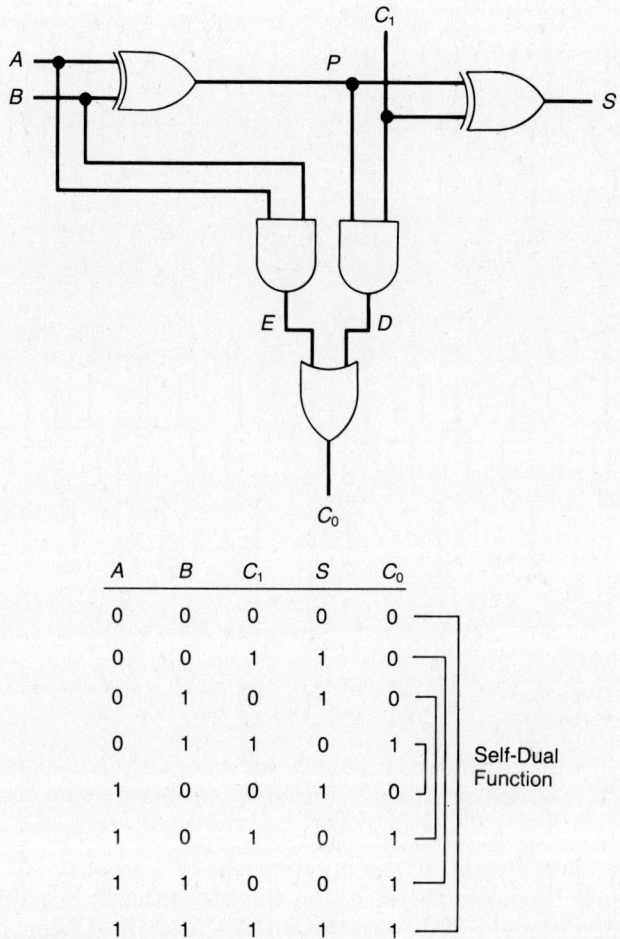

Fig. 3.66 The full-adder is a self-dual circuit—complementary inputs produce complementary outputs.

correct or they disagree in either (or both) the ith or the $(i-1)$th bits. As an example, consider the logical AND operation performed on two 8-bit operands, as illustrated in Fig. 3.67. If only one bit slice is faulty and that faulty slice has no impact on any other bit slices, a single left shift detects the errors that occur in logical operations.

For a bit-sliced, ripple-carry adder, a 2-bit arithmetic shift is required to guarantee the detection of errors that can occur [Patel and Fung 1982]. Once again suppose that bit slice i is faulty. In a ripple-carry adder, a faulty bit slice can have one of three effects: the sum bit out of the bit slice can be

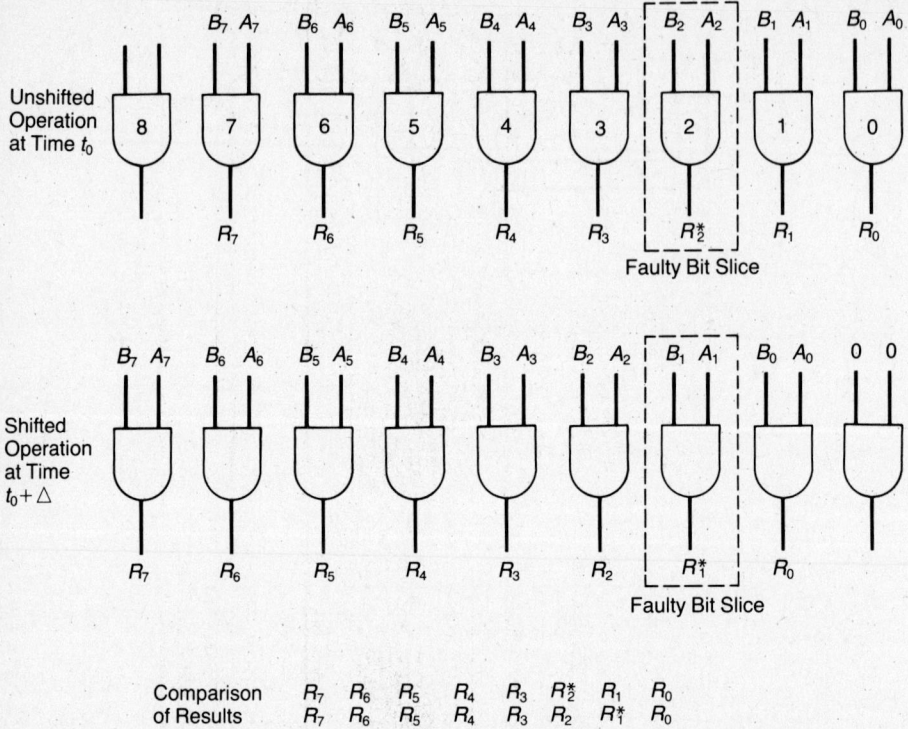

Fig. 3.67 Illustration of time redundancy for a logical operation. The fault affects different bits of the result each time and can therefore be detected.

erroneous, the carry bit out of the bit slice can be erroneous, or both may be in error. Table 3.20 shows the effect on the sum of each possible error.

In summary, the result generated for the unshifted operands if bit i is faulty is incorrect by one of $[0, \pm 2^i, \pm 2^{i+1}, \pm 3 \cdot 2^i]$.

TABLE 3.20 Errors and their effect on sum

Error	Effect on Sum
Sum is 0	-2^i
Sum is 1	$+2^i$
Carry is 0	-2^{i+1}
Carry is 1	$+2^{i+1}$
Sum is 0, carry is 0	$-(2^{i+1} + 2^i) = -3 \cdot 2^i$
Sum is 0, carry is 1	$2^{i+1} - 2^i = +2^i$
Sum is 1, carry is 0	$-(-2^i + 2^{i+1}) = -2^i$
Sum is 1, carry is 1	$2^{i+1} + 2^i = +3 \cdot 2^i$

3.6 ■ Time Redundancy

When the operands are shifted to the left by two bits, a similar analysis can show that the result will be incorrect by one of $[0, \pm 2^{i-2}, \pm 2^{i-1}, \pm 3 \cdot 2^{i-2}]$. As can be seen, the results of the two computations cannot agree unless both are correct. Therefore, the error will be detected.

The structure of an ALU that uses the RESO technique is shown in Fig. 3.68. The additional hardware required for the technique is the three shifters, the storage register to hold the results of the first computation, and

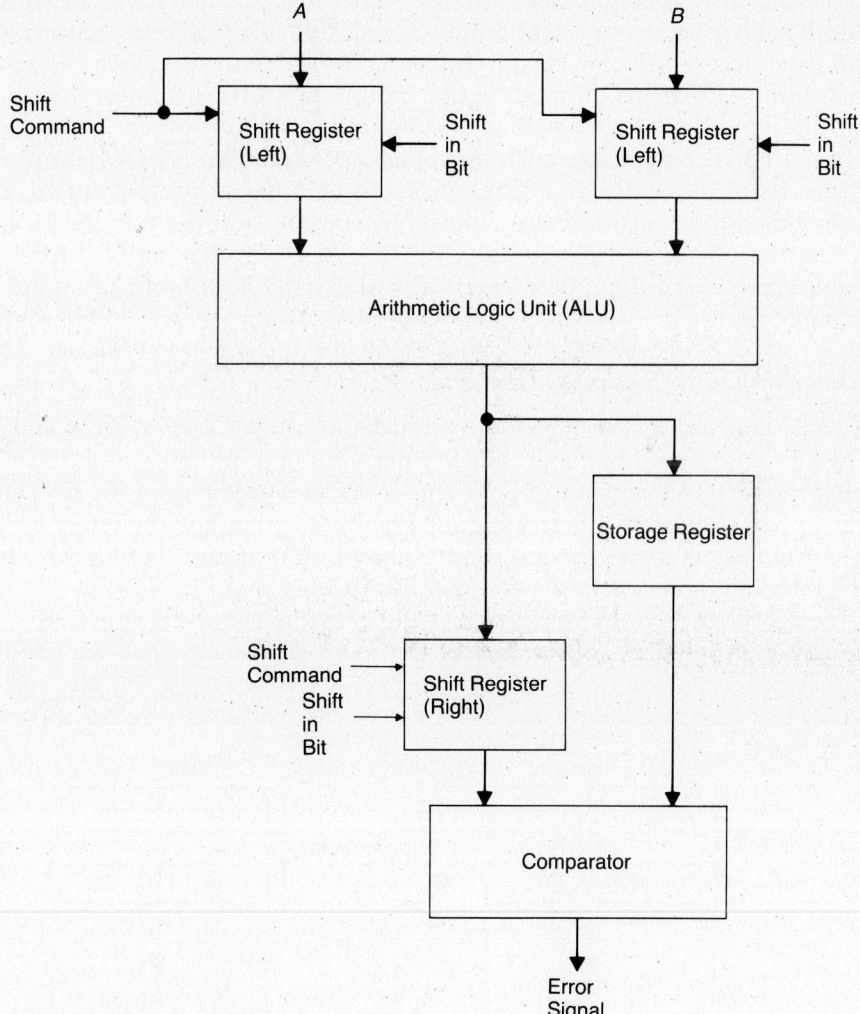

Fig. 3.68 ALU structure using RESO.

the comparator. Also, the ALU must be extended by 2 bits to allow a two-bit, arithmetic shift to be performed.

The primary problems with the RESO approach are the additional hardware required, the lack of coverage provided for faults in the shifters, and the requirement that the comparator be totally self-checking such that faults in the comparator do not render the approach ineffective.

The extra bits required for the ALU in the RESO approach can be eliminated if a cyclic shift is used instead of an arithmetic shift. But, a cyclic shift of one or two bits requires that the carry bits between bit slices be directed either to the neighboring bit slice when the operands are not shifted or another bit slice when the operands are shifted. The circuitry required to handle the carry bits can become more complex than the extra bit slices added to the ALU to accommodate the arithmetic shifts. A compromise is to swap the upper and lower halves of the operands as proposed in the technique called **recomputing with swapped operands (RESWO)** [Hana and Johnson 1986]. Several carry bits must still be appropriately handled, but the affected carries are the carry-in, the carry-out, and the auxiliary carry between the upper and lower halves of the ALU. Traditionally, ALUs are often designed such that these carry bits are easily accessible and controllable anyway.

Recomputing with Swapped Operands

The basic concept of the RESWO technique is shown in Fig. 3.69. During the first computation, the operands are manipulated in their standard form. During the second computation, the upper and lower halves of the operands are swapped such that a faulty bit slice operates on either the lower or upper half of the operands during the first computation and the opposite half of the operands during the second computation.

The RESWO technique can detect the existence of any single faulty bit slice in a ripple-carry adder. Using the same approach that we used to

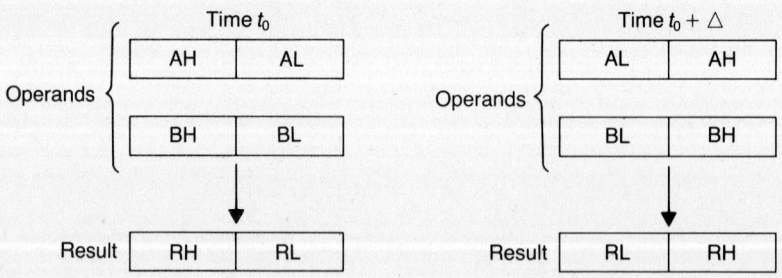

Fig. 3.69 In RESWO, operand halves are swapped prior to repeating the operation.

verify the fault detection capability of the RESO technique, consider the following two-step procedure. Suppose the words to be added are n bits each, with the lower half containing bits 0 through r and the upper half containing bits $r + 1$ through $n - 1$. If bit slice i is faulty, the sum and carry bits from slice i can impact the sum by $\pm 2^i$ and $\pm 2^{i+1}$, respectively. If both the sum and the carry of bit slice i are erroneous, the resulting sum may be off by $\pm 2^i \pm 2^{i+1}$. During the first computation, with the operand halves not swapped, the faulty bit slice i can cause the sum to be in error by one of $[0, \pm 2^i, \pm 2^{i+1}, \pm 2^i \pm 2^{i+1}]$. If $i \leq r$, the result of the recomputation with swapped operand halves is in error by one of $[0, \pm 2^{i+r+1}, \pm 2^{i+r+2}, \pm 2^{i+r+1} \pm 2^{i+r+2}]$. For $i > r$, the result of the recomputation with the operand halves swapped is in error by one of $[0, \pm 2^{i-r-1}, \pm 2^{i-r}, \pm 2^{i-r-1} \pm 2^{i-r}]$. In both cases, the results of the two computations either are not in error or they disagree, in which case the error can be detected.

For all logic operations, it is fairly easy to see the effectiveness of the swapped operand approach. During the computations with the operand halves not swapped, a faulty bit slice i affects bit i of the result. Once the operand halves are swapped, a faulty bit slice i affects bit $i \pm (r + 1)$, depending on whether the faulty bit slice is in the upper or lower half of the logical unit. In either case, the error is detected.

The structure of an ALU designed to accommodate the RESWO technique is shown in Fig. 3.70. C_{in} is the carry-in for the addition process and C_{out} is the carry-out of the addition process. Multiplexers are positioned appropriately to switch the carry bits when the operand halves are swapped. During the recomputation with swapped operand halves, the multiplexers direct C_{in} to the upper half of the adder and C_{out} is taken from the carry-out of the lower half of the adder. The carry-in for the lower half of the adder is then selected as the carry-out of the upper half of the adder. The multiplexers are very simple circuits that do not add significant circuitry to the overall ALU. For example, Fig. 3.71 shows the circuit required for the 2-to-1 multiplexer.

Recomputing with Duplication with Comparison

An alternative method that takes advantage of both time redundancy and hardware redundancy concepts is called **recomputing with duplication with comparison (REDWC)** [Johnson, Aylor, and Hana 1988]. The method with which error detection is accomplished resembles that of duplication with comparison. Time redundancy is then used to complete the calculation and obtain the final result. For example, a 32-bit addition operation can be realized by using each of two 16-bit adders twice. During each calculation, the error detection is accomplished by comparing the results from using the two adders. REDWC is similar in many respects to a method described in [Toy 1982].

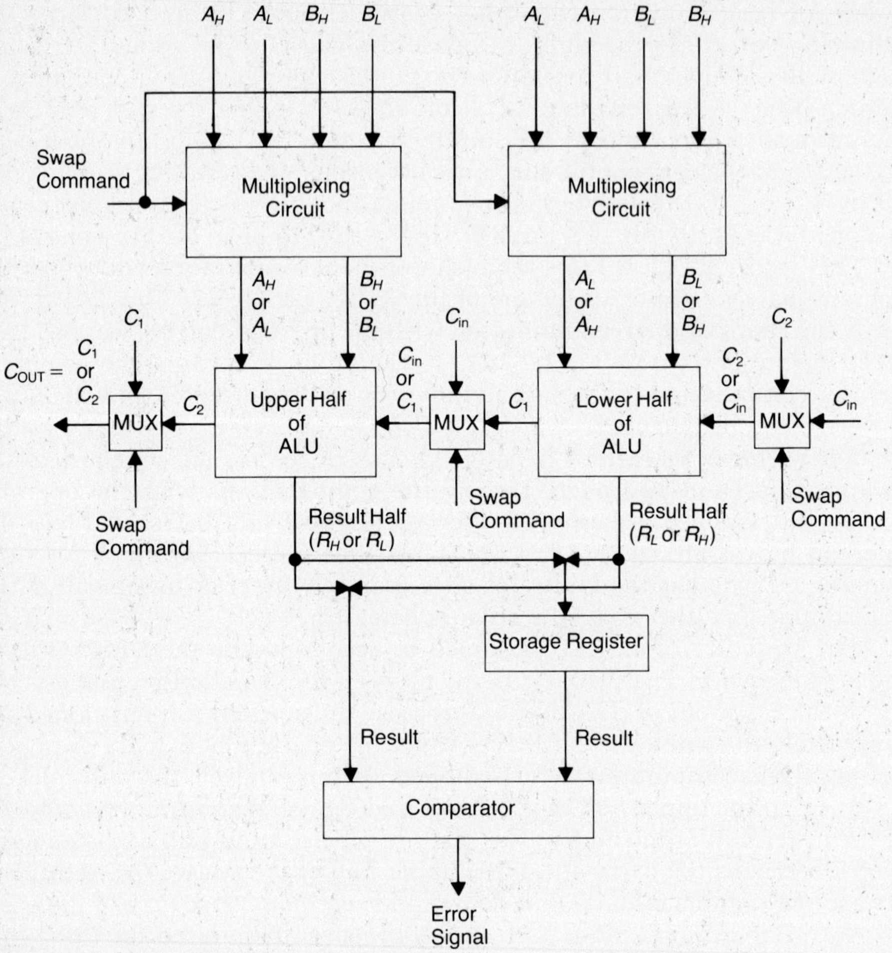

146 Design Techniques to Achieve Fault Tolerance

Fig. 3.70 ALU structure using RESWO.

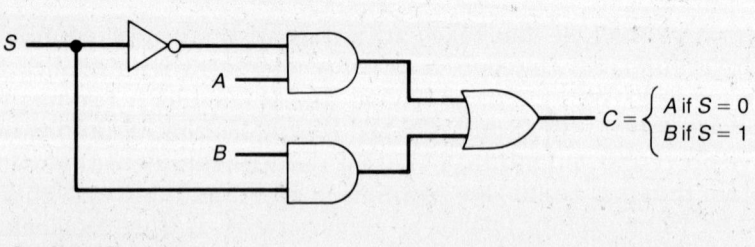

S = Swap Command
$\left.\begin{array}{l}A\\B\end{array}\right\}$ Inputs

Fig. 3.71 2-to-1 multiplexer to perform single-bit swapping.

To best describe the basis of REDWC, the operation of an n-bit full-adder is shown as an example. The n-bit adder, as well as the operands, is divided into lower and upper halves with each half consisting of $n/2$ bits. To perform an addition, two calculations are executed. In the first calculation, the lower halves of the operands are added in both halves of the adder. In other words, the adder performs the addition of the lower halves of the operands twice but in parallel on the upper and lower halves of the adder, as seen in Fig. 3.72. The results are then compared and one result is stored to represent the lower half of the final output. The adder hardware is then used a second time to perform a second calculation. The second calculation starts by selecting the upper halves of the operands with the output carry of the first computation to be added in both halves of the adder. The results of the second calculation are then compared and are available as the upper half of the final output.

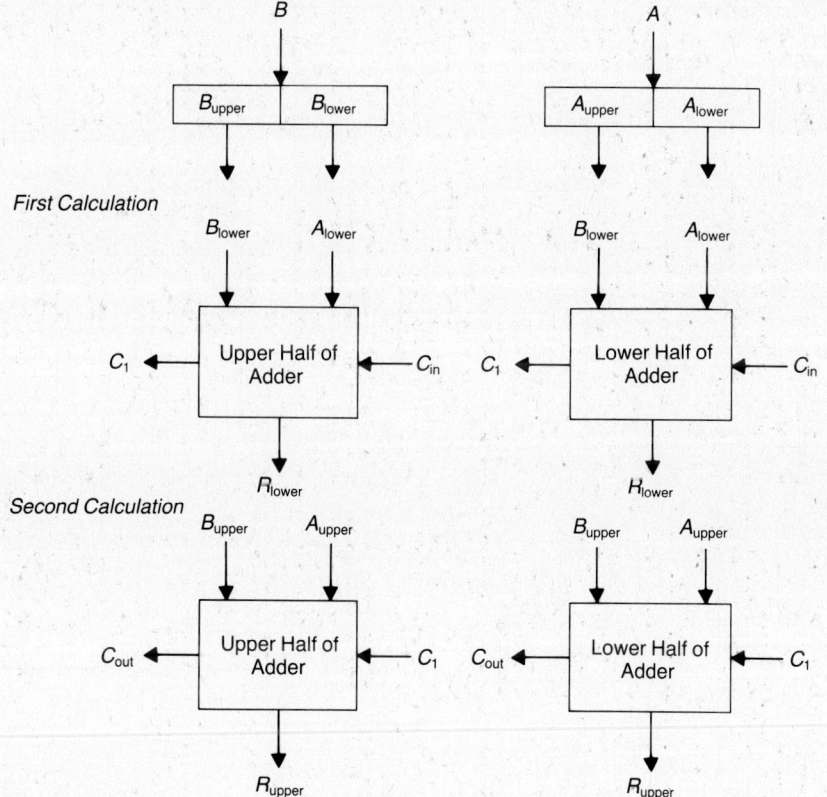

Fig. 3.72 REDWC uses both the upper and lower halves of the adder to perform identical calculations. The adder hardware is used twice to complete the final result.

148 Design Techniques to Achieve Fault Tolerance

Selection of the appropriate operand half is performed using multiplexers (MUXs). Also, the carries at the boundaries of the adder are handled using one MUX. Figure 3.73 shows a block diagram of a 32-bit adder that uses the REDWC technique for error detection.

The checking for errors is performed during each calculation by comparing the outputs of the upper and lower halves of the adder. Although error detection is accomplished by a simple comparison between the results

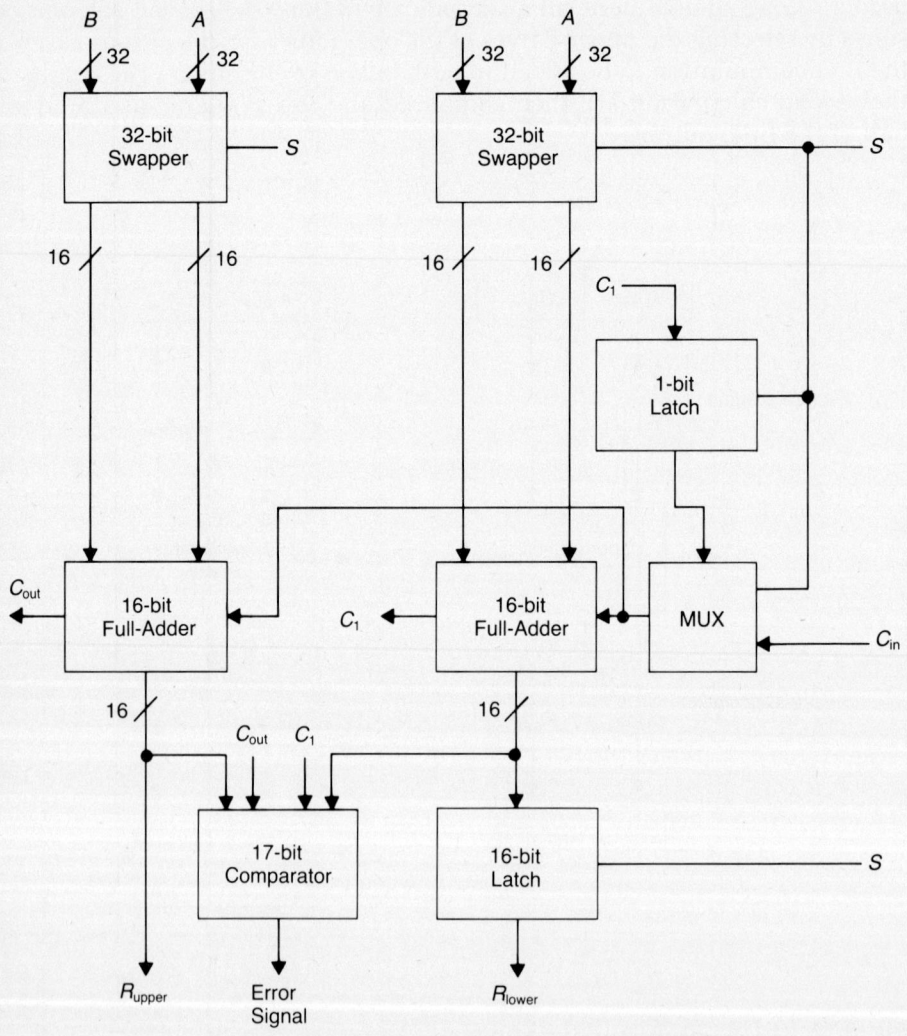

Fig. 3.73 A 32-bit adder designed using the REDWC technique.

of the two halves, the same hardware is used twice to obtain the final results. Therefore, the concept of duplication with comparison is employed as the error detection mechanism, and the concept of time redundancy is applied to complete the calculation and obtain the final output.

Since the error detection mechanism of REDWC is identical to that of duplication with comparison, both techniques possess the same ability to detect faults; that is, the ability to detect all single faults confined to one half of the adder in arithmetic and logic operations, as long as both halves do not become faulty in a similar manner and at the same time. REDWC is also capable of detecting errors in lookahead-carry adders provided that the lookahead-carry operation does not overlap the boundary between the two halves of the adder. In other words, the carry between the two halves must ripple from the lower to the upper half of the adder, but within each half of the adder, lookahead-carry operations can be performed.

The 32-bit adder, shown in Fig. 3.73, illustrates the structure required to implement REDWC. The adder consists of two "swappers," the adder itself, a multiplexer (MUX), a storage register, and a comparator. Each swapper consists of 32 2-1 MUXs that select the appropriate operand half. A control signal S is provided to manipulate the MUXs and, consequently, control the operation of the adder. When S is low, the lower halves are selected to be added in both halves of the adder; and when S is high, the upper halves are selected. Each half of the adder consists of four 4-bit lookahead-carry adder cells. Each cell possesses a lookahead-carry capability, but the carry is rippled between each cell, as shown in Fig. 3.74. The

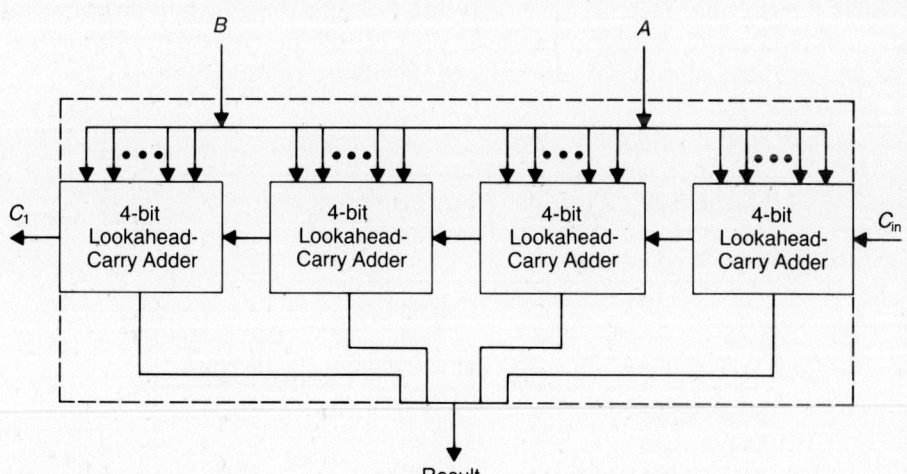

Fig. 3.74 A 16-bit adder constructed from four 4-bit lookahead carry units — the carry bit ripples between lookahead units.

TABLE 3.21 Best-case and worst-case cell counts for three error detection techniques compared to the nonredundant case

Technique	Number of cells		Average number of cells	Percentage increase in average number of cells
	Best case	Worst case		
Nonredundant	253.33	320	286.67	0%
Duplication with comparison	542.33	685	613.67	114
RESO	665.00	876	770.50	169
REDWC	446.67	582	514.34	79

output carry of the lower half (C1) is stored in a 1-bit latch to be used as the input carry for the second calculation. The 2-1 MUX is used to appropriately handle the carry at the adder boundaries.

Table 3.21 shows the best-case and the worst-case cell count, as well as the average cell count for a gate array implementation of adders designed using duplication with comparison (DWC), RESO, REDWC, and also an adder designed without redundancy [Johnson, Aylor, and Hana 1988]. The best-case cell count corresponds to optimal wire routing conditions in the VLSI implementation, whereas the worst-case cell count corresponds to the worst possible routing conditions. The average number of cells is used as a means of comparison to eliminate routing considerations. Note that the REDWC technique uses significantly fewer cells than either duplication with comparison or RESO. Also note that all of the time redundancy approaches require significant amounts of hardware redundancy.

TABLE 3.22 Calculation times for adders with concurrent error detection compared to the nonredundant case

Technique	Calculation time (nanoseconds)	Percentage increase
Nonredundant	37.5	0.00%
Duplication with comparison	48.3	28.80
RESO	83.5	122.67
REDWC	52.6	40.27

An equally important criterion for comparison is the resulting system throughput. Table 3.22 summarizes simulation results for the nonredundant, RESO, DWC, and REDWC adders [Johnson, Aylor, and Hana 1988]. The table shows the amount of time (in nanoseconds) required to complete a calculation (including error detection), as well as the percentage increase in calculation time over that of the nonredundant adder. The calculation time represents the maximum time between the application of the input carry to the adder and the availability of the error signal on the output of the comparator. The comparators used for the various adders have the same 10.8 nanoseconds of propagation delay. The result of the addition is actually available 10.8 nanoseconds earlier than the times shown in Table 3.22, however, the error signal is not available until approximately 10.8 nanoseconds after the addition is completed.

From Table 3.22, one can see that the penalty, in time, paid by using REDWC for error detection is considerably less than that of RESO and comparable to that of DWC. Although time redundancy is part of the REDWC technique, the extra time required to complete the calculation is only 40.27% over that of the nonredundant operation.

Time redundancy techniques form an important class of options for designing fault-tolerant systems. Just as all redundancy approaches, however, time redundancy cannot be used in all applications because of the additional time that must be employed. If time is available, however, time redundancy techniques do provide an opportunity to minimize the amount of additional hardware required.

3.6.3 Recomputation for Error Correction

The time redundancy approach can also provide for error correction if the computations are repeated three or more times. Consider the logical AND operation that was illustrated in Fig. 3.67. Suppose the operation is performed three times: first, without shifting the operands; second, with a 1-bit arithmetic shift of the operands; and third, with a 2-bit arithmetic shift of the operands. The results generated using the shifted operands are then shifted the appropriate number of bits to the right to properly position the bits of the results. Because each of the three operations used operands that were displaced from each other by at least one bit position, a different bit in each result will be affected by the faulty bit slice. If the bits in each position are then compared, the results due to the faulty bit slice can be corrected by performing a majority vote on the three results obtained for each bit position. This process is illustrated in Fig. 3.75.

Unfortunately, the approach just described does not work for arithmetic operations because the adjacent bits are not independent. A single, faulty bit slice can affect more than one bit of the result.

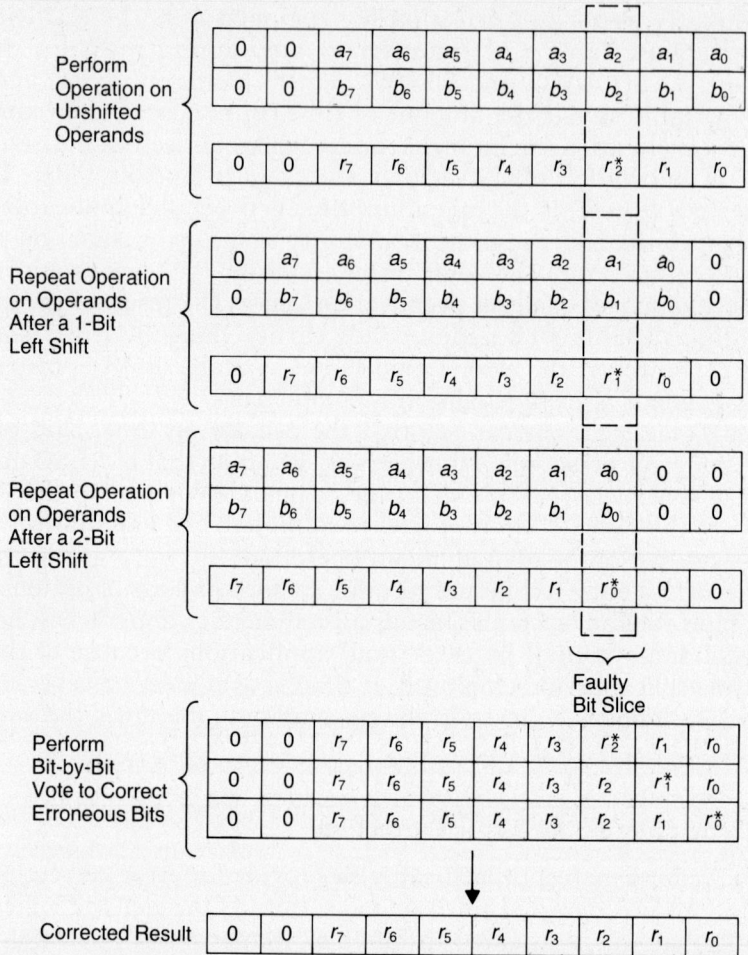

Fig. 3.75 Example of error correction using time redundancy.

3.7 Software Redundancy

In applications that use computers, many fault detection and fault tolerance techniques can be implemented in software. The redundant hardware necessary to implement the capabilities can be minimal, whereas the redundant software can be substantial. Redundant software can occur in many forms; you do not have to replicate complete programs to have redundant software. Software redundancy can appear as several extra lines of code used to check the magnitude of a signal or as a small routine used to periodically test a memory by writing and reading specific locations. In this section, we consider three major software redundancy techniques: consistency checks, capability checks, and software replication methods [Chen and Avizienis 1978].

3.7.1 Consistency Checks

A **consistency check** uses *a priori* knowledge about the characteristics of information to verify the correctness of that information. For example, in some applications, it is known in advance that a digital quantity should never exceed a certain magnitude. If the signal exceeds that magnitude, an error of some sort is present. A consistency check can often be implemented easily in hardware, but it is most likely to appear in a system's software. For example, a processing system can sample and store many sensor readings in a typical control application. Each sensor reading can be checked to verify that it lies within an acceptable range of values. As another example, the amount of cash requested by a patron at a bank's teller machine should never exceed the maximum withdrawal allowed. Likewise, the address generated by a computer should never lie outside the address range of the available memory.

An example of consistency checking that can be performed in hardware is the detection of invalid instruction codes in computers. Many computers use n-bit quantities to represent 2^k possible instruction codes where $2^k < 2^n$. In other words, there are $2^n - 2^k$ instruction codes that are illegal. Each instruction code can be checked to verify that it is not one of the illegal codes. If an illegal code occurs, the processor can be halted to prevent an erroneous operation from occurring. This technique is particularly useful in detecting a "run-away" processor that is erroneously interpreting data as instructions.

Another form of consistency checking that can prove valuable in many control applications is to compare the measured performance of the system with some predicted performance. This technique is particularly useful in control applications where some dynamic system is under control. The dynamic system can be modeled and the predicted performance obtained from a software implementation of the model. The actual performance of the system can then be measured and compared with the model-predicted performance. Any significant deviations of the measured performance from the predicted performance can indicate a fault. The difficulty with this approach is twofold. First, the model must be accurate if good results are to be obtained. Second, it is difficult to establish the level of deviation that will be allowed before an error is signaled. In some applications, the nonlinearity of a system can result in a linearized model deviating substantially from the actual performance under certain input conditions.

Another form of consistency check that is very useful in systems where data is transferred over buses is the word count overflow check. The data is often transferred in packets, with each packet possessing a specific number of data words. At the beginning of each packet can be a header that states the number of words in that packet. If the receiving device detects a disagreement between the actual number of words received and the number specified in the header, an error can be indicated. The process of counting the words and comparing that count with the received count can be implemented in software in many applications; thus, the redundancy occurs in software.

3.7.2 Capability Checks

Capability checks are performed to verify that a system possesses the capability expected. For example, you would like to know whether or not you have your complete memory available or if all the processors in your multiprocessor system are working properly. As another example, you might want to know if the ALU in your processor is working properly.

Several forms of capability checks exist. The first is a simple memory test. A processor can simply write specific patterns to certain memory locations and read those locations to verify that the data was stored and retrieved properly. In many cases, it is not necessary to write and read a large number of locations to achieve reasonably good fault coverage. The memory test can be a supplement to parity as protection against faults in the memory.

Another form of capability check is a set of ALU tests. Periodically, a processor can execute specific instructions on specific data and compare the results to known results stored in a read-only memory (ROM). This form of capability check can verify both the ALU and the memory in which the known results are stored. The instructions that are executed can consist of adds, multiplies, logical operations, and data transfers.

Another form of capability check consists of verifying that all processors in a multiple processor system are capable of communicating with each other. This can consist of periodically passing specific information from one processor to another. For example, each processor may be required to set a specific bit in a shared memory to indicate their capability to communicate with that memory, and, as a result, communicate with other processors through that memory.

3.7.3 *N*-Version Programming

The software redundancy techniques that we have considered thus far have been those that use extra, or redundant, software to detect faults that can occur in hardware. We have not yet considered approaches for detecting, or possibly tolerating, faults that can occur in the software of a system. Software faults are unusual entities. Software does not break as hardware does, but instead software faults are the result of incorrect software designs or coding mistakes. Therefore, any technique that detects faults in software must detect design flaws. A simple duplication and comparison procedure will not detect software faults if the duplicated software modules are identical, because the design mistakes will appear in both modules.

The concept of **N-version programming** was developed to allow certain design flaws in software modules to be detected [Chen and Avizienis 1978]. The basic concept of N-version programming is to design and code the software module N times and to compare the N results produced by these modules. Each of the N modules is designed and coded by a separate group of programmers. Each group designs the software from the same set of specifications such that each of the N modules performs the same function. However, it is hoped that by performing the N designs independently, the same mistakes will not be made by the different groups. Therefore, when a fault occurs, the fault either does not occur in all modules or it occurs differently in each module, so that the results generated by the modules will differ.

Certainly, the importance of software fault tolerance is easy to see. If we design a microprocessor-based system to be fault tolerant using a TMR configuration, the hardware redundancy is of little use if a single software fault can disable each of the redundant processors. The N-version programming technique states that each of the three processor's software should be designed and coded independently such that a common fault is less likely to occur.

The primary difficulties with the N-version approach are twofold. First, software designers and coders can tend to make similar mistakes. Therefore, we are not guaranteed that two completely independent versions of a program will not have identical faults. Second, the N versions of a program are still developed from a common specification, so the N-version approach will not allow the detection of specification mistakes.

To overcome many of the problems associated with N-version programming, software designers employ rigid design rules and methods to attempt to prevent faults from occurring. This approach we know as fault avoidance, and it is very important in the design of reliable software. If the software is designed correctly in the first place, fault tolerance techniques for the software will not be necessary.

Summary

This chapter has presented the basic techniques for achieving fault tolerance; hardware redundancy, information redundancy, time redundancy, and software redundancy. In most cases, the various techniques have been illustrated with examples of their actual use. In many applications, a combination of the techniques must be used to meet the requirements of reliability and fault tolerance. Subsequent chapters illustrate the design and evaluation of fault-tolerant systems using the techniques presented in this chapter. The following list summarizes the important concepts introduced in this chapter.

Active Hardware Redundancy—techniques that achieve fault tolerance through the use of fault detection and reconfiguration.

Alternating Logic—a time redundancy technique in which a calculation is repeated after complementing the original data.

AN Code—an arithmetic code in which code words are created by multiplying the original data N by a constant A.

Arithmetic Code—a code in which code words are invariant to arithmetic operations.

Berger Code—a separable code formed by concatenating the original data and the complement of the binary word representing the number of 1s in the original data.

Binary Code—a code in which code words contain only symbols that are either 0 or 1.

Bit-per-Byte Parity—a parity code in which one parity bit is assigned to each of two groups of $n/2$ bits in an n-bit word.

Bit-per-Chip Parity—a parity code in which one parity bit is assigned to each chip in a memory design.

Bit-per-Multiple-Chips Parity—a parity code in which each parity group has one, and only one, bit from each chip.

Bit-per-Word Parity—a parity code in which one parity bit is assigned to each n-bit word.

Burst Error—a condition in which consecutive bits of a code word are corrupted by a single fault.

Capability Check—a verification of an expected capability or feature.

Checksum—the sum of a list of data items.

Code—a specific set of rules for representing information.

Code Distance—the minimum Hamming distance between any two valid code words within a given code.

Code Polynomial—a polynomial of degree $n - 1$ whose coefficients are the n bits of a code word.

Code Word—a collection of symbols, or digits, use to represent information according to the rules of a given code.

Cold Standby Sparing—the use of unpowered standby spares.

Complemented Duplication — a code in which data is duplicated but in a complemented form.

Consistency Check — a check for consistency between actual results and expected results.

Cyclic Code — a code in which code words are invariant to end-around shifts.

Decoding Process — the process of determining a data word from a code word.

Double-Precision Checksum — a code in which the checksum is formed using double-precision arithmetic.

Duplication Codes — a code in which data is completely and exactly duplicated.

Duplication with Comparison — the use of two identical modules performing identical computations and the comparison of their results.

Encoding Process — the process of determining a code word from an original data word.

Error Detecting Code — a code that allows errors to be detected.

Error Correcting Code — a code that allows errors to be corrected.

Flux-Summing — an alternative to majority voting that uses the inherent properties of feedback systems to compensate for faults in replicated modules.

Generator Polynomial — a polynomial $G(X)$ used to create a cyclic code word $V(X)$ by forming $V(X) = D(X)G(X)$, where $D(X)$ is the original data polynomial.

Hamming Distance — the number of bit positions in which two binary code words differ.

Hamming Error-Correcting Codes — a class of parity codes that allows error correction through the use of overlapping parity groups.

Hardware Redundancy — the addition of redundant hardware to a system.

Honeywell Checksum — a code in which the checksum is calculated after concatenating consecutive words to form double words.

Hot Standby Sparing — the use of powered standby spares.

Hybrid Hardware Redundancy — techniques that use a combination of fault masking and reconfiguration.

Information Redundancy — the addition of redundant information to a data item.

Interlaced Parity — a parity code in which adjacent bits are assigned to different parity groups.

Inverse Residue Code — a code in which code words are created by appending the inverse residue to the original data.

m-of-n Code — a code in which each n-bit code word has exactly m 1s.

Mid-Value Select — a form of voting that selects the middle value from amongst an odd number of signals.

N Modular Redundancy — a form of passive redundancy that uses N modules and majority voting.

N Modular Redundancy with Spares — a hybrid technique that combines N modular redundancy with standby sparing.

Nonseparable Code — a code in which the original data cannot be separated from the check bits.

N-version Progamming — a technique that compares the results from N separate programs, each of which performs the same operations.

Overlapped Parity — a parity code in which each data bit belongs to more than one parity group.

Pair-and-a-Spare Technique — the combination of duplication with comparison and standby sparing.

Parity Code — a code in which all code words have an even (odd) number of 1s. If the number of 1s is even, the code is called even parity. If the number of 1s is odd, the code is called odd parity.

Passive Hardware Redundancy — techniques that achieve fault tolerance through the use of fault masking.

Recomputing with Duplication with Comparison (REDWC) — a redundancy technique that combines time redundancy and duplication with comparison.

Recomputing with Shifted Operands (RESO) — a time redundancy technique in which a calculation is repeated after shifting the operands.

Recomputing with Swapped Operands (RESWO) — a time redundancy technique in which a calculation is repeated after swapping the upper and lower halves of each operand.

Redundancy — the addition of information, resources, or time beyond what is needed for normal system operation.

Repeated Use Faults — a fault within a hardware or software unit used multiple times to create a single result.

Residue Checksum — a code in which the checksum is calculated using end-around carry addition.

Residue Code — a code in which code words are created by appending the residue to the original data.

Residue Number System — a system in which numbers are represented using multiple residues.

Restoring Organ — a unit that creates a correct output even though one of its inputs is incorrect.

Self-Purging Redundancy — a hybrid technique where faulty modules remove themselves from the voting arrangement.

Separable Code — a code formed by appending check bits to the original data.

Sift-Out Modular Redundancy — a hybrid technique that uses a centralized process to remove faulty units from a voting arrangement.

Single point of failure — a single component or element whose failure results in the failure of the system.

Single-Precision Checksum — a code in which the checksum is formed using single-precision arithmetic.

Software Redundancy — the addition of redundant software to a system.

Standby Sparing — the use of spare units that replace faulty ones during a reconfiguration process.

Swap-and-Compare Code — a code formed by duplicating the data and swapping the upper and lower halves of the duplicated data.

Syndrome Polynomial — the remainder polynomial found by dividing a cyclic code polynomial by the generator polynomial.

Time Redundancy — the use of redundant time in a system's operations.

Triple-Duplex Architecture — triple modular redundancy where each triplicated unit uses duplication with comparison.

Triple Modular Redundancy (TMR) — a form of passive redundancy that triplicates modules and uses majority voting.

Watchdog Timer — a timer that if not set at a specific frequency is used to indicate faulty behavior.

References

1. Avizienis, A. "Arithmetic error codes: Cost and effectiveness studies for application in digital system design," *IEEE Transactions on Computers*, Vol. C-20, No. 11, November 1971, pp. 1322–1331.
2. Chen, L., and A. Avizienis. "N-version programming: A fault tolerant approach to reliability of software operation," *Proceedings of the International Symposium on Fault Tolerant Computing*, 1978, pp. 3–9.
3. de Sousa, P.T., and F.P. Mathur. "Sift-out modular redundancy," *IEEE Transac-*

tions on Computers, Vol. C-27, No. 7, July 1978, pp. 624–627.

4. Hamming, R.W. "Error detecting and error correcting codes," *Bell System Technical Journal*, Vol. 26, No. 2, April 1950, pp. 147–160.

5. Hana, H.H., and B.W. Johnson. "Concurrent error detection in VLSI circuits using time redundancy," *Proceedings of SOUTHEASTCON '86*, Richmond, Va., March 23–25, 1986, pp. 208–212.

6. *Intel Memory Components Handbook*, Intel Corporation, Santa Clara, Calif., 1985.

7. Johnson, B.W. "Fault-tolerant microprocessor-based systems," *IEEE Micro*, Vol. 4, No. 6, December 1984, pp. 6–21.

8. Johnson, B.W., J.H. Aylor, and H.H. Hana. "Efficient use of time and hardware redundancy for concurrent error detection in a 32-bit VLSI adder," *IEEE Journal of Solid-State Circuits*, Vol. 23, No. 1, February 1988, pp. 208–215.

9. Kohavi, Z. *Switching and Finite Automata Theory*, McGraw-Hill, New York, 1978.

10. Lala, P.K. *Fault Tolerant and Fault Testable Hardware Design*, Prentice-Hall, Englewood Cliffs, N.J., 1985.

11. Lin, S., and D.J. Costello, Jr. *Error Control Coding: Fundamentals and Applications*, Prentice-Hall, Englewood Cliffs, N.J., 1983.

12. Losq, J. "A highly efficient redundancy scheme: Self-purging redundancy," *IEEE Transactions on Computers*, Vol. C-25, No. 6, June 1976, pp. 569–578.

13. Nelson, V.P., and B.D. Carroll. *Tutorial: Fault-Tolerant Computing*, IEEE Computer Society Press, Washington, D.C., 1986.

14. Patel, J.H., and L.Y. Fung. "Concurrent error detection in ALUs by recomputing with shifted operands," *IEEE Transactions on Computers*, Vol. C-31, No. 7, July 1982, pp. 589–595.

15. Reynolds, D.A., and G. Metze. "Fault detection capabilities of alternating logic," *IEEE Transactions on Computers*, Vol. C-27, No. 12, December 1978, pp. 1093–1098.

16. Tang, D.T., and R.T. Chien. "Coding for error control," *IBM Systems Journal*, Vol. 8, No. 1, January 1969, pp. 48–86.

17. Taylor, F.J. "Residue arithmetic: A tutorial with examples," *Computer*, Vol. 17, No. 5, May 1984, pp. 50–62.

18. Toy, W., N., "Self-checking arithmetic unit," United States Patent Number 4, 314, 350, Bell Telephone Laboratories, Murray Hill, N.J., Feb. 2, 1982.

Additional Reading

While this chapter has attempted to cover many of the most important techniques in use in today's fault-tolerant designs, the literature contains other approaches that are beyond the intent of this book. For the reader who is interested in studying further, the following list of articles is provided as suggested reading.

> Abdelaziz, M., and A.D. Friedman. "Efficient design of self-checking checker for any m-out-of-n code," *IEEE Transactions on Computers*, Vol. C-27, No. 6, June 1978.
>
> Armstrong, D.B. "A general method of applying error correction to synchronous digital machines," *Bell System Technical Journal*, Vol. 40, No. 3, March 1961.
>
> Armstrong, J.R. and F.G. Gray, "Fault Diagnosis in a Boolean n Cube Array of Microprocessors", *IEEE Transactions on Computers*, Vol. C-30, No. 8, August 1981.
>
> Clark, E.M., and C.N. Nickolaous. "Distributed reconfiguration strategies for fault tolerant microprocessor systems," *IEEE Transactions on Computers*, Vol. C-31, No. 8, August 1982.
>
> Coyne, R. "Dynamic reconfiguration by a local network's operating system," *Data Communications*, December 1981.
>
> Dishon, Y., and C.J. Georgiou. "A highly available storage system using the checksum method," *Proceedings of the Seventeenth International Symposium on Fault-Tolerant Computing*, July 6–8, 1987, Pittsburgh, Pa.
>
> Gannon, T.G., and S.D. Shapiro. "An optimal approach to fault tolerant software systems design," *IEEE Transactions on Software Engineering*, Vol. SE-4, No. 5, September 1978.
>
> Goodenough, J.B., and C.L. McGowan. "Software quality assurance: Testing and validation," *Proceedings of the IEEE*, Vol. 68, No. 9, September 1980.
>
> Hamill, T.G., and R. Phillips. "A fault tolerant reconfigurable multiprocessor system," *Proceedings of the International Conference on Distributed Computer Control*, Birmingham, England, September 1977.
>
> Jarwala, N., and D.K. Pradhan. "Cost analysis of on chip error control coding for fault tolerant dynamic RAMs," *Proceedings of the Seventeenth International Symposium on Fault-Tolerant Computing*, July 6–8, 1987, Pittsburgh, Pa.
>
> Jensen, P.A. "Quadded NOR logic," *IEEE Transactions on Reliability*, Vol. R-12, No. 3, September 1963.
>
> Kameyama, M., and T. Higuchi. "Design of dependent failure tolerant microcomputer system using triple modular redundancy," *IEEE Transactions on Computers*, Vol. C-29, No. 2, February 1980.
>
> Kinney, L.L., and W.Y. Yueh. "An architecture for a VHSIC computer," *Proceedings of the 8th Annual Symposium on Computer Architecture*, May 1981, Minneapolis, Minn.
>
> Klaschka, T.F. "Reliability improvement by redundancy in electronic systems, II: An efficient new redundancy scheme—Radial logic," *Royal Aircraft Establishment*, Ministry of Technology, 69045, Farnborough, United Kingdom, 1969.

Koren, I. "A reconfigurable and fault tolerant VLSI multiprocessor array," *Proceedings of the 8th Annual Symposium on Computer Architecture*, May 1981, Minneapolis, Minn.

Lin, S. *An Introduction to Error Control Coding*, Prentice-Hall, Englewood Cliffs, N.J., 1970.

Petersen, W.W., and E.J. Weldon, Jr. *Error Correcting Codes*, Second Edition, MIT Press, Cambridge, Mass., 1972.

Pierce, W.H. *Fault Tolerant Design*, Academic Press, Orlando, Fla., 1965.

Rao, T.R.N. *Error Coding for Arithmetic Processors*, Academic Press, Orlando, Fla., 1974.

Stiffler, J.J. "Coding for random access memories," *IEEE Transactions on Computers*, Vol. C-27, No. 6, June 1978.

Taylor, D.J., D.E. Morgan, and J.P. Black. "Redundancy in data structures: Improving software fault tolerance," *IEEE Transactions on Software Engineering*, Vol. SE-6, No. 6, November 1980.

Tryon, J.G. "Quadded logic," in *Redundancy Techniques for Computers*, edited by Wilcox and Mann, Spartan Books, Washington, D.C., 1962.

Wakerly, J.F. *Error Detecting Codes, Self-Checking Circuits, and Applications*, Elsevier North-Holland, New York, 1978.

Wakerly, J.F. "Transient failures in triple modular redundancy," *IEEE Transactions on Computers*, Vol. C-24, No. 5, May 1975.

Weinstock, C.B., and M.W. Green. "Reconfigurable strategies for the SIFT fault tolerant computer," *Proceedings of Compsac 78*, Chicago, November 1978.

Wright, R.S., S.C. Crist, and M. Arozullah. "Fault tolerant techniques for a multiple microprocessor-based space borne packet switch," *Proceedings of the National Telecommunications Conference*, New Orleans, November 1981.

Problems

3.1. Most helicopters use a stabilizer to level the aircraft during flight so that the air drag is significantly reduced. Typically, the stabilizer is moved by opening a hydraulic valve, and movement is stopped when the valve is closed. The valve is controlled by a digital circuit that issues a 1-bit command to a relay that controls the valve. The command originates from triply redundant switches activated by the pilot. Design a TMR circuit to generate the correct command after any single fault. The voter in your circuit will clearly be the weak point. Make sure that your design can at least detect faults that occur within the voter.

3.2. The company that you work for is designing an industrial controller that maintains the temperature of a process during a chemical reaction. The nonredundant controller is fairly simple and consists of an analog-to-digital (A/D) converter, processor, and a digital-to-analog (D/A) converter. You have been asked to develop at least two alternatives for making the controller tolerant of any two component failures. (The term *component* here means an A/D, processor, or D/A.) Show block diagrams of your approaches and compare them. Which approach would you recommend and why?

3.3. A communications link between two processors consists of transmitters, physical media, receivers, buffers, and a memory, as shown in Fig. 3.76. The buffers are first-in, first-out (FIFO) devices and are constructed using chips that are 4-bits wide. The communications link is 40 bits wide, so there are ten such chips in parallel to accommodate the 40 bits. Parity is to be used as a means of error detection with the parity being generated prior to transmission and checked as soon as the data exits the FIFOs. If only 8 of the 40 bits are available for parity bits, devise a parity organization that provides what you feel is the best protection. Justify your solution. One last constraint is that the data is transmitted over the bus at the rate of one word every 10 nanoseconds. Assuming that a two-input logic gate has a propagation delay of 0.5 nanoseconds, make sure that your approach is feasible from a time standpoint.

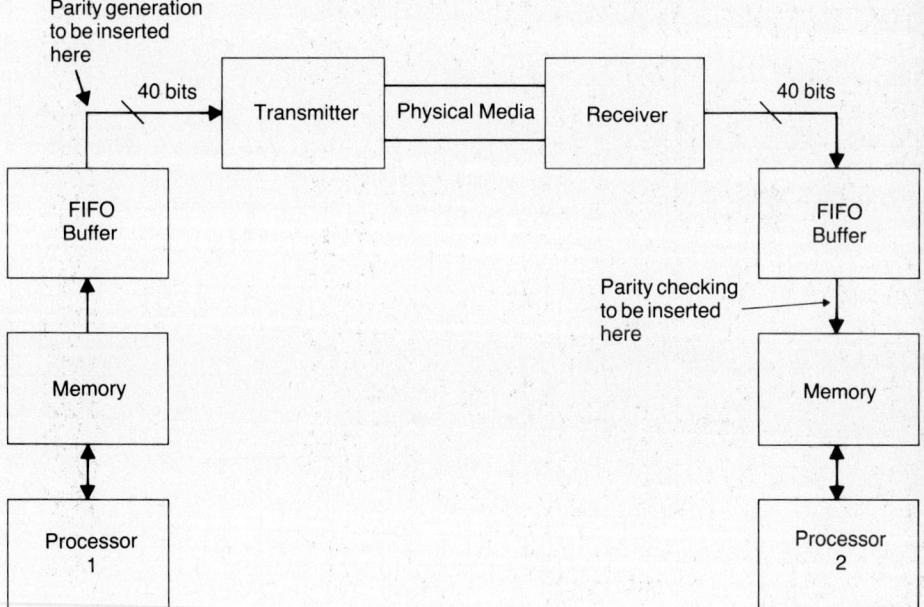

Fig. 3.76 Architecture of system for Problem 3.3.

3.4. As a new employee, you have been assigned to the Fault-Tolerant Systems Branch of your company. Your first assignment has been to design a fault-tolerant network for interconnecting multiple processors. The processors form the core of a transactions processing system for a bank with several branches. Each branch is a node in the network. The basic structure of the nonredundant network is that of a unidirectional ring, as shown in Fig. 3.77. Data is transferred from one node to another around the ring in packets of 20 16-bit words. Develop an architecture for the ring that allows any node failure to be toler-

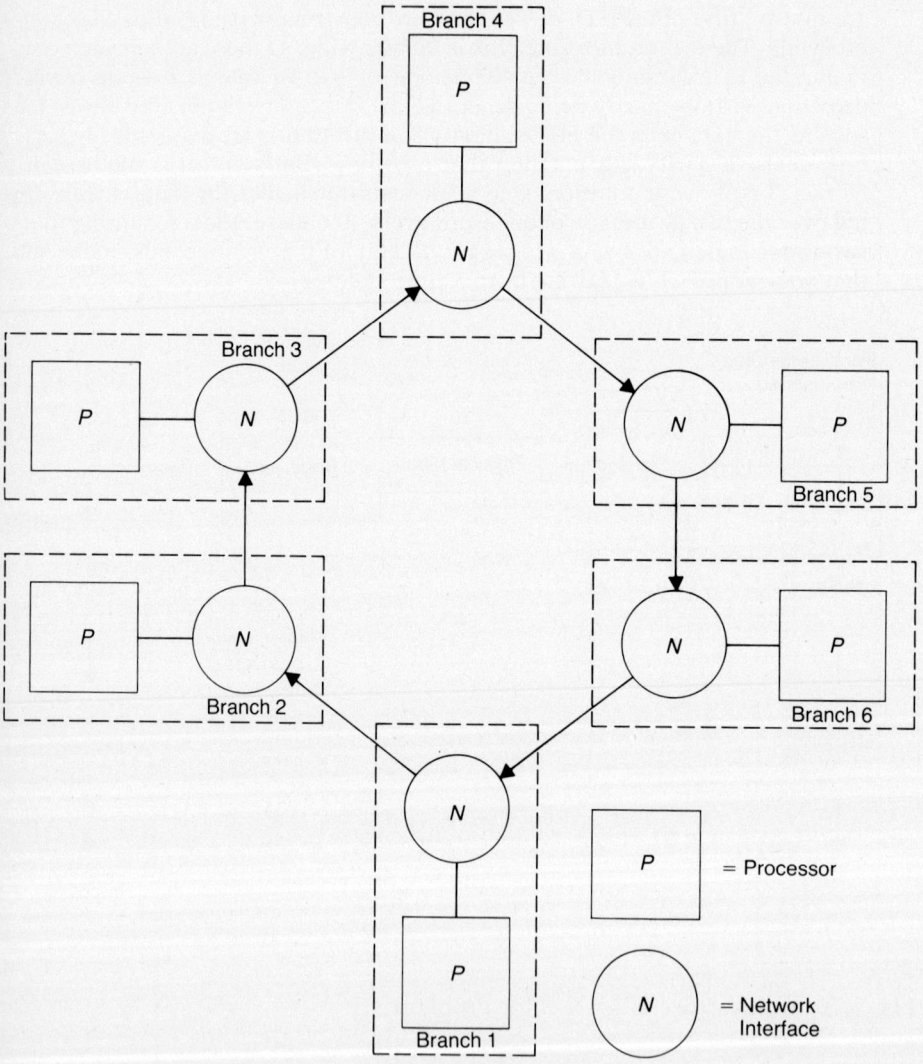

Fig. 3.77 Architecture of the bank transactions processing system for Problem 3.4. The nonredundant network is a unidirectional ring.

ated. Also, describe the techniques that you would use to detect errors in the data as it is transferred around the ring.

3.5. Show the organization of an 8-bit memory with Hamming single-error correction and double-error detection. Be explicit in your descriptions and show the parity groups that result. Associate the syndromes with the particular bit that they identify as erroneous.

3.6. Design a simple, nonredundant circuit to detect erroneous codes in a 4-bit binary coded decimal (BCD) representation. The circuit should produce an output of 1 when the four bits at the inputs do not represent a valid BCD code. Now show the implementation of a self-purging version of that circuit that is capable of tolerating any two faults that can occur.

3.7. Show the sift-out modular redundancy implementation of the combinational circuit of Problem 3.6. Once again, the circuit should tolerate any two faults that occur.

3.8. Show the separable and nonseparable cyclic code words that result for 4-bit information words when the generator polynomial is $G(X) = 1 + X + X^4$. Also, develop a circuit that is capable of encoding the original information and a second circuit that is capable of decoding the code words.

3.9. Show the separable residue and the separable inverse residue codes for the 4-bit, BCD representations of the numbers 0 through 9. Use a modulus of 5.

3.10. Using the Intel 8206 error detection and correction unit, design a 16-bit memory capable of correcting any single error and detecting any double error. Assume that you are using any standard memory chip that is four bits wide. Show the complete block diagram of your memory.

3.11. Investigate the error detection capability of the following time redundancy approach when used on a ripple-carry adder. During the first addition, the operands are encoded using a $3N$ arithmetic code. During the second addition, the operands are encoded using a $5N$ arithmetic code. Will this scheme detect any single error that can occur in the adder? Will the approach detect any double error that can occur in the adder?

3.12. Design an 8-bit adder using the RESWO technique. Make sure that you provide the ability to swap the operands and to compare the results that are computed at the two different points in time.

3.13. Prove or disprove that the RESWO technique is capable of detecting any single error that can occur in a lookahead-carry adder. Assume for simplicity that the adder is eight bits and the lookahead operation is performed over four bits.

3.14. A digital filtering application requires the implementation of the equation, $y((n + 1)T) = Ay(nT) + Bu(nT)$, where A and B are constants, $u(nT)$ is the input at time nT, and $y(nT)$ is the output of the filter at time nT. The input is limited, because of the analog-to-digital converter, to an 8-bit representation. Your design implements the filter in software using a 16-bit processor. Design a simple processing system, including the structure of the software, to provide as much error detection capability as possible without using more than one of any of the hardware components. You may resort to time, information, or software redundancy but not hardware redundancy.

3.15. Figure 3.78 shows the architecture of a simple microprocessor, which has four fundamental components: the arithmetic logic unit (ALU), memory, control logic, and an internal bus. The processor is organized as an 8-bit machine, so each of the ALU, memory, and bus is an 8-bit unit. For the purpose of this design, assume that the control logic is simply a memory capable of storing 16 4-bit words. To simplify matters even further, assume that the ALU can only perform addition. The bus is an 8-bit, parallel bus. Finally, the memory can store up to 32 8-bit words. Develop a design that incorporates redundancy into the various units to achieve fault detection. Your design should be capable of detecting all single faults. You may use any type of redundancy, but you should show an analysis of the extra hardware, time, information, or software that your design requires. Explain, in detail, why you selected your specific approach.

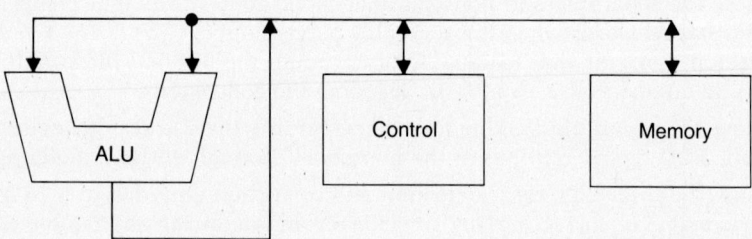

Fig. 3.78 Architecture of a simple microprocessor to be studied in Problem 3.15.

3.16. A common interconnection technique is a crossbar switch, an implementation of which is illustrated in Fig. 3.79. The implementation of Fig 3.79 uses a two-input/two-output switching cell, which is also shown in the figure. The switching cell has two modes of operation: the crossing mode and the bending mode. In the crossing mode, input 1 is connected to output 2 and input 2 is connected to output 1. In the bending mode, input 1 is connected to output 1 and input 2 is connected to output 2. The crossbar switch allows any processor to be directly connected to any memory, if desired. The crossbar switch has been used extensively in interconnection networks for transactions processing systems. The problem is that a single switch failure can make it impossible for certain connections to be realized. Assume that each switching cell can fail in only one of two ways: the cell is stuck in the bending mode or the cell is stuck in the crossing mode. In both cases, the data is not corrupted, but the cell simply loses its switching capability. Develop a design for a crossbar switch that uses the basic switching cell and is capable of tolerating any single switch failure. Assume that your crossbar switch must connect four processors to four memories with each processor being able to connect, one at a time, to any of the memories.

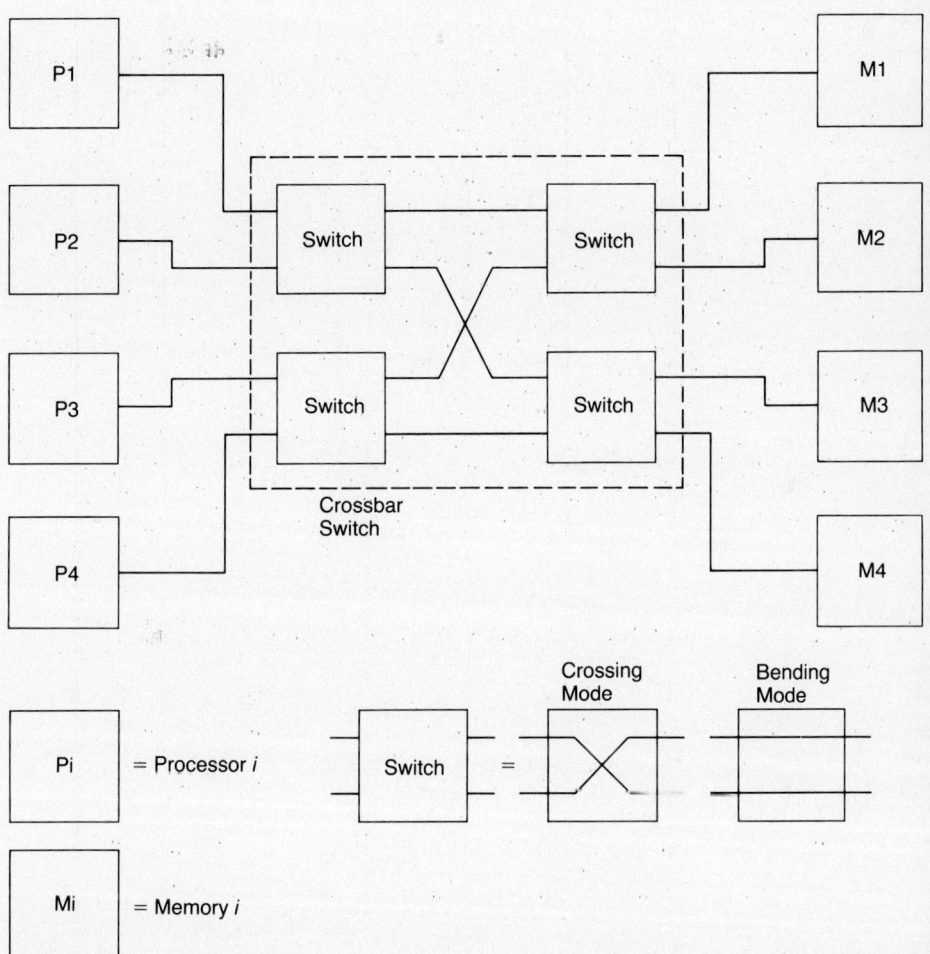

Fig. 3.79 Implementation of a crossbar switch using basic switching elements.

4

Evaluation Techniques

4.1 Introduction
4.2 Quantitative Evaluation Methods
4.3 Reliability Modeling
4.4 Safety Modeling
4.5 System Comparisons
4.6 Availability Models
4.7 Maintainability Models
4.8 Redundancy Ratios
4.9 Qualitative Methods
4.10 Tradeoff Analysis Example
Summary
References
Additional Reading
Problems

4.1 Introduction

The techniques presented in the previous chapter form a collection of approaches that can be used to achieve fault tolerance. The specific techniques used in a given application depend on not only the application, but also on the ideas and philosophies of the designers. One team of designers might develop one fault-tolerant design, whereas another team would choose a completely different approach for the same application. It is important to be able to compare two or more approaches for a particular application. The process of comparison is actually a critical part of the design process because it leads to tradeoffs and modifications of the design. It is through such tradeoffs that the most optimal design is developed.

The methods for evaluating fault-tolerant systems can be divided into two major categories: *quantitative* and *qualitative*. Qualitative measures are typically subjective in nature and describe the benefits of one design over another. Examples include the flexibility of a particular design and the degree to which the fault tolerance techniques are transparent to the user of a system. Quantitative evaluation techniques produce numbers that can be used to compare two or more systems. Examples include specific numbers for the reliability, availability, maintainability, mission life, or fault coverage of a system.

The purpose of this chapter is to examine, in detail, the techniques that are available for evaluating fault-tolerant systems. The final section contains a specific tradeoff example that shows how the evaluation techniques can be used to make design decisions and tradeoffs.

4.2 Quantitative Evaluation Methods

As previously mentioned, the purpose of quantitative evaluation methods is to assign a number to some attribute of a system such that the attribute can be compared among systems. For example, the reliability of one system may be greater than that of another, or the weight and cost of one approach may be less than that of another. Reliability, weight, and cost are three examples of quantitative measures that can be used to evaluate systems. Certainly, the weight and cost of a system are extremely important, and in some cases can be more important than reliability or fault tolerance.

In this chapter, we consider several approaches to quantitative evaluation, including the failure rate, mean time to failure (MTTF), mean time between failure (MTBF), fault coverage, reliability analysis, safety analysis, availability analysis, maintainability analysis, redundancy ratios, and costs. Several techniques for generating the reliability of a system are presented.

4.2.1 Failure Rate and the Reliability Function

Intuitively, the **failure rate** is the expected number of failures of a type of device or system per a given time period [Shooman 1968]. For example, if a computer fails, on the average, once every 2000 hours, the computer has a failure rate of one failure per 2000 hours, or 1/2000 failures/hour. The failure rate is typically denoted as λ. The failure rate is one measure that can be used to compare systems or components. In selecting a computer for a banking application, one would like to select a computer that fails as infrequently as possible. If redundancy has been incorporated as a means of achieving fault tolerance, the failure rate of the redundant system should be lower than the failure rate of a similar, nonredundant system.

To more clearly understand the mathematical basis for the concept of a failure rate, recall the definition of the reliability function. The reliability $R(t)$ of a component, or a system, is the conditional probability that the component operates correctly throughout the interval $[t_0, t]$ given that it was operating correctly at time t_0. Suppose that we test N identical components by placing all N components in operation at time t_0 and recording the number of failed and working components at time t. Let $N_f(t)$ be the number of components that have failed at time t and $N_o(t)$ be the number of components that are operating correctly at time t. It is assumed that once a component fails it remains failed indefinitely. The *reliability* of the components at time t is given by

$$R(t) = \frac{N_o(t)}{N} = \frac{N_o(t)}{N_o(t) + N_f(t)}$$

which is simply the probability that a component has survived the interval $[t_o, t]$. The probability that a component has not survived the time interval is called the *unreliability* and is given by

$$Q(t) = \frac{N_f(t)}{N} = \frac{N_f(t)}{N_o(t) + N_f(t)}$$

Note that at any time t, $R(t) = 1.0 - Q(t)$ because

$$R(t) + Q(t) = \frac{N_o(t) + N_f(t)}{N_o(t) + N_f(t)} = 1.0$$

If we write the reliability function as

$$R(t) = 1.0 - \frac{N_f(t)}{N}$$

and differentiate $R(t)$ with respect to time, we obtain

$$\frac{dR(t)}{dt} = -\frac{1}{N} \frac{dN_f(t)}{dt}$$

which can be written as

$$\frac{dN_f(t)}{dt} = -N \frac{dR(t)}{dt}$$

The derivative of $N_f(t)$, $dN_f(t)/dt$, is simply the instantaneous rate at which components are failing. At time t, there are still $N_o(t)$ components operational. Dividing $dN_f(t)/dt$ by $N_o(t)$ we obtain

$$z(t) = \frac{1}{N_o(t)} \frac{dN_f(t)}{dt}$$

$z(t)$ is called the *hazard function*, *hazard rate*, or *failure rate function*. The units for the failure rate function are failures per unit of time.

The failure rate function can be expressed in different ways. For example, $z(t)$ can be written strictly in terms of the reliability function $R(t)$ as

$$z(t) = \frac{1}{N_o(t)} \frac{dN_f(t)}{dt} = \frac{1}{N_o(t)} \left[-N \frac{dR(t)}{dt} \right] = -\frac{\frac{dR(t)}{dt}}{R(t)}$$

Similarly, $z(t)$ can be written in terms of the unreliability $Q(t)$ as

$$z(t) = -\frac{\frac{dR(t)}{dt}}{R(t)} = \frac{\frac{dQ(t)}{dt}}{1 - Q(t)}$$

The derivative of the unreliability $dQ(t)/dt$ is called the *failure density function*.

The failure rate function clearly depends on time because the value of $N_o(t)$ and the value of $dN_f(t)/dt$ change as functions of time. However, experience has shown that the failure rate function for electronic components does have a period where the value of $z(t)$ is approximately constant. The commonly accepted relationship between the failure rate function and time for electronic components is called the **bathtub curve** and is illustrated in Fig. 4.1. The bathtub curve assumes that during the early life of systems failures occur frequently due to substandard or weak components. The decreasing part of the bathtub curve is called the *early-life* or *infant mortality* region. At the opposite end of the curve is the wear-out region where systems have been functional for a long period of time and are beginning to experience failures due to the physical wearing of electronic or mechanical components. The increasing part of the bathtub curve is called the *wear-out phase*. During the intermediate region, the failure rate function is assumed to be a constant. The constant portion of the bathtub curve is called the *useful life* phase of the system, and the failure rate function is assumed to have a value of λ during that period. (λ is referred to as the failure rate and is normally expressed in units of failures per hour.)

The period of a constant failure rate is typically the most useful portion of a system's life. During the useful-life phase, the system is providing its most predictable service to its users. We usually attempt to get a system beyond the infant mortality stage by using the concept of *burn-in* to remove weak components. Burn-in implies operating a system, often at an acceler-

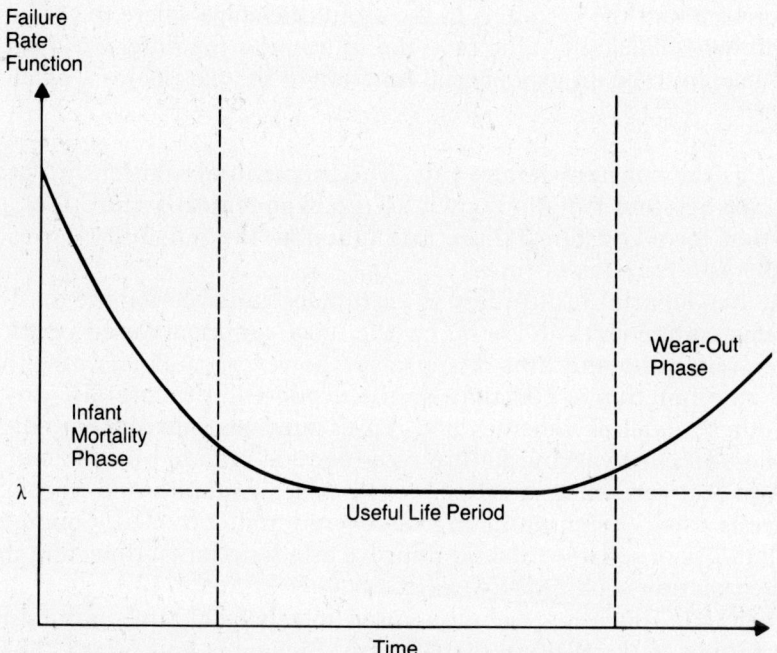

Fig. 4.1 Bathtub curve relationship between the failure rate function and time—the failure rate function is constant during the useful life period.

ated pace, prior to placing the system into service to get the system to the beginning of the useful-life period. In addition, the system is normally replaced before it enters the wear-out phase of its life. Thus, the primary interest is the performance of the system during the useful-life phase.

As noted previously, the failure rate function can be related to the reliability function as

$$z(t) = \frac{1}{N_o(t)} \frac{dN_f(t)}{dt} = -\frac{N}{N_o(t)} \frac{dR(t)}{dt}$$

We know, however, that the quantity $N/N_o(t)$ is the inverse of the reliability function $R(t)$ so we can write

$$z(t) = -\frac{1}{R(t)} \frac{dR(t)}{dt}$$

The result is a differential equation of the form

$$\frac{dR(t)}{dt} = -z(t)R(t)$$

If we assume that the system is in the useful-life stage where the failure rate function has a constant value of λ, the solution to the differential equation is well known to be an exponential function of the parameter λ given by

$$R(t) = e^{-\lambda t}$$

where λ is the constant failure rate. The exponential relationship between the reliability and time is known as the **exponential failure law,** which states that for a constant failure rate function, the reliability varies exponentially as a function of time.

The exponential failure law is extremely valuable for the analysis of electronic components and is by far the most commonly used relationship between reliability and time. Many cases, however, cannot assume that the failure rate function is constant, so the exponential failure law cannot be used; other modeling schemes and representations must be employed. An example of a time-varying failure rate function is found in the analysis of software. Software failures are the result of design faults, and as a software package is used, design faults are discovered and corrected. Consequently, the reliability of software should improve as a function of time, and the failure rate function should decrease.

A common modeling technique used to represent time-varying failure rate functions is the **Weibull distribution** [Siewiorek and Swarz 1982]. The failure rate function associated with the Weibull distribution is given by

$$z(t) = \alpha \lambda (\lambda t)^{\alpha - 1}$$

where α and λ are constants that control the variation of the failure rate function with time. The failure rate function given by the Weibull distribution is intuitively appealing. For example, if the value of α is 1, $z(t)$ is simply the constant λ. If α is greater than 1, $z(t)$ increases as time increases; if α is less than 1, $z(t)$ decreases as time increases. Consequently, we can envision modeling software using the Weibull distribution with the constant α being less than 1.

The reliability function that results from the Weibull distribution is the solution to the differential equation

$$\frac{dR(t)}{dt} = -z(t)R(t) = -\alpha \lambda (\lambda t)^{\alpha - 1} R(t)$$

and is given by

$$R(t) = e^{-(\lambda t)^{\alpha}}$$

The expression for $R(t)$ can be verified by calculating the derivative $dR(t)/dt$. Specifically,

$$\frac{dR(t)}{dt} = -e^{-(\lambda t)^{\alpha}} \alpha (\lambda t)^{\alpha - 1}(\lambda) = -\alpha \lambda (\lambda t)^{\alpha - 1} e^{-(\lambda t)^{\alpha}} = -z(t)R(t)$$

As can be seen, certain values of α result in a reliability function that increases as time increases. For example, if $\alpha = -1$, the reliability is given by

$$R(t) = e^{-1/\lambda t}$$

which approaches 1 as t approaches infinity and is 0 when t is 0. Also note that for $\alpha = 1$, the reliability function is identical to the exponential failure law.

Although time-varying failure rate functions are important in the analysis of software and other systems, by far the most common analysis is performed assuming a constant failure rate function and the exponential failure law. The remainder of this chapter assumes the exponential failure law.

4.2.2 Failure Rate Calculation

An important aspect in the analysis of systems is the estimation of the failure rate of specific components. The most common technique for estimating the failure rate is the United States Department of Defense (USDOD) MIL-HDBK-217 standard ([USDOD 1965], [USDOD 1974], and [USDOD 1979]). Several versions of the standard have been published, including the original standard, MIL-HDBK-217 [USDOD 1965], as well as several revisions, which include, for example, MIL-HDBK-217C [USDOD 1979]. In each version of the standard, the objective has been to develop a model for the failure rate of electronic components using experimental data obtained by analyzing the failures of actual devices. Here, we only summarize the model and the important parameters that are used in calculating the failure rate.

The MIL-HDBK-217B ([Siewiorek and Swarz 1982] and [USDOD 1974]) model predicts the constant failure rate of an integrated circuit (IC) as

$$\lambda = \pi_L \pi_Q (C_1 \pi_T + C_2 \pi_E) \pi_P \quad \text{failures per million hours}$$

where π_L is a *learning* factor, π_Q is a *quality* factor, π_T is a *temperature* factor, π_E is an *environmental* factor, π_P is a *pin* factor, and C_1 and C_2 are *complexity* factors.

The learning factor π_L represents the overall maturity of the fabrication process used to produce the IC. Devices produced using a new, and as yet unproven, manufacturing process are assigned a learning factor of 10, while those produced using a proven process are assigned a learning factor of 1. In other words, the learning factor represents the overall confidence in the ability of the fabrication process to produce devices that will fail infrequently.

The quality factor π_Q represents the amount of device screening that occurs. Device screening is simply the testing that a device goes through prior to being sold by a manufacturer. The lowest level of screening implies that no testing is performed. In other words, the manufacturer simply produces and sells the IC without verifying that it is operational. At higher levels of screening, the manufacturer randomly selects ICs from a manufacturing run and subjects the selected ICs to certain tests. At even higher levels of screening, the manufacturer thoroughly tests each IC produced. In MIL-HDBK-217B, the quality factor varies from 1 to 300, depending on the level of screening.

The four primary screening levels for ICs are Class A, Class B, Class C, and Class D. Classes A and B are the highest screening levels and are used typically in military applications. The quality factor is 1 for Class A and 2 for Class B. Class C components are representative of high-quality commercial components and have a quality factor of 16. Finally, Class D represents a standard, hermetically sealed commercial component and has a quality factor of 150.

The temperature factor π_T is a function of the device technology, operating temperature, device packaging technology, and power dissipation. The specific equations used for the temperature factor are

$$\pi_T = 0.1 e^{(-8121[(1/(T_j+273))-(1/298)])}$$

for linear circuits and

$$\pi_T = 0.1 e^{(-4794[(1/(T_j+273))-(1/298)])}$$

for digital bipolar circuits. T_j is the junction temperature and is expressed in degrees Celsius. The second equation given above for π_T is used for transistor-transistor logic (TTL) circuits. As an example, the calculation for a TTL circuit with a junction temperature of 25 degrees Celsius is

$$\pi_T = 0.1 e^{(-4794[(1/(25+273))-(1/298)])} = 0.1 e^{(0.0)} = 0.1$$

The environmental factor π_E is a function of the harshness of the environment. For example, components operated in an air-conditioned computer room have a much lower environmental factor than those operated on a typical factory floor or in an airborne application. Typical values of the environmental factor vary in the MIL-HDBK-217B standard from 0.2 to 10.0. For example, components located in a computer room have an environmental factor of 0.2; components located in an uninhabited airborne environment have an environmental factor of 6.0; components in a launched missile have an environmental factor of 10.0.

The pin factor π_P is a function of the number of pins on the IC package. In MIL-HDBK-217B, the pin factor ranges from 1.0 to 1.2, for large-scale integration (LSI) technology, as the number of pins increases from 1 to greater than 64. LSI logic is normally defined as an IC having between 100 and 1000 logic gates. For LSI devices, the pin factor is 1.0 if the IC has 25 or fewer pins, 1.1 if the IC has between 26 and 64 pins, and 1.2 for an LSI device having more than 64 pins.

The final factors included in the failure rate model are the complexity factors C_1 and C_2. The complexity factors are a function of the number of gates for logic circuits, the number of transistors for linear circuits, and the number of bits for memories. The complexity factors for ICs having between 100 and 1300 gates are

$$C_1 = (0.0187)e^{(0.00471)(N_g)}$$
$$C_2 = (0.013)e^{(0.00423)(N_g)}$$

where N_g is the number of gates on the IC. The complexity factors for logic having fewer than 100 gates are

$$C_1 = (0.00129)N_g^{(0.677)}$$
$$C_2 = (0.00389)N_g^{(0.359)}$$

where N_g is the number of gates on the IC. The complexity factors for linear circuits are

$$C_1 = (0.00056)N_t^{(0.763)}$$
$$C_2 = (0.0026)N_t^{(0.547)}$$

where N_t is the number of transistors on the IC. The complexity factors for read-only memory (ROM) are

$$C_1 = (0.00114)B^{(0.603)}$$
$$C_2 = (0.00032)B^{(0.646)}$$

where B is the total number of bits in the memory. Finally, the complexity factors for random access memory (RAM) are

$$C_1 = (0.00199)B^{(0.603)}$$
$$C_2 = (0.00056)B^{(0.644)}$$

where B is the total number of bits in the memory.

As an example, consider the calculation of the failure rate for a device having 24 pins and 500 logic gates. We will assume a learning factor of 1.0, a quality factor of 16, a temperature factor of 0.35, and an environmental factor of 0.2. The pin factor is 1.0 for devices with 25 or fewer pins. The complexity factors are calculated as

$$C_1 = 0.0187e^{(0.00471)(500)} = 0.19706$$
$$C_2 = 0.013e^{(0.00423)(500)} = 0.10776$$

and the resulting failure rate is

$$\lambda = \pi_L \pi_Q (C_1 \pi_T + C_2 \pi_E) \pi_P =$$
$$(1.0)(16)[(0.19706)(0.35) + (0.10776)(0.2)](1.0) = 1.448$$

which is in failures per million hours.

Table 4.1 shows some additional typical values computed using the MIL-HDBK-217B standard.

4.2.3 Mean Time to Failure

In addition to the failure rate, the **mean time to failure (MTTF)** is a useful parameter to specify the quality of a system. The MTTF is the expected time that a system will operate before the *first* failure occurs. For example, if we have N identical systems placed into operation at time $t = 0$, and we measure the time that each system operates before failing, the average time is the MTTF. If each system i operates for a time t_i before encountering the first failure, the MTTF is given by

$$\text{MTTF} = \frac{\sum_{i=1}^{N} t_i}{N}$$

The MTTF can be calculated by finding the expected value of the time of failure. From probability theory, we know that the expected value of a random variable X is

$$E[X] = \int_{-\infty}^{\infty} x f(x)\, dx$$

TABLE 4.1 Typical failure rates calculated using MIL-HDBK-217B ($\pi_L = 1$, $\pi_Q = 16$, $\pi_T = 0.35$, $\pi_E = 0.2$, $\pi_p = 1$)

Number of logic gates	Failure rate (Failures per million hours)
(a) Logic circuits	
50	0.1527
100	0.2312
200	0.3655
500	1.4483
1000	14.4880

(b) Memories (RAM) Number of bits	Failure rate (Failures per million hours)
1024 (1K)	0.8837
2048 (2K)	1.3491
8192 (8K)	3.1453
16,384 (16K)	4.8033
32,768 (32K)	7.3362

where $f(x)$ is the *probability density function*. In reliability analysis we are interested in the expected value of the time of failure (MTTF), so

$$\text{MTTF} = \int_0^\infty tf(t)\,dt$$

where $f(t)$ is the *failure density function*, and the integral runs from 0 to ∞ because the failure density function is undefined for times less than 0. We know, however, that the failure density function is

$$f(t) = \frac{dQ(t)}{dt}$$

so, the MTTF can be written as

$$\text{MTTF} = \int_0^\infty t\frac{dQ(t)}{dt}\,dt$$

Using integration by parts and the fact that $dQ(t)/dt = -dR(t)/dt$, we can show that

$$\text{MTTF} = \int_0^\infty t\frac{dQ(t)}{dt}\,dt = -\int_0^\infty t\frac{dR(t)}{dt}\,dt = [-tR(t) + \int R(t)\,dt]_0^\infty = \int_0^\infty R(t)\,dt$$

The term $-tR(t)$ clearly disappears when $t = 0$; but, it also disappears when $t = \infty$ because $R(\infty) = 0$. Consequently, the MTTF is defined in terms of the reliability function as

$$\text{MTTF} = \int_0^\infty R(t)\,dt$$

which is valid for any reliability function that satisfies $R(\infty) = 0$.

If the reliability function obeys the exponential failure law, the result of calculating the MTTF is given by

$$\text{MTTF} = \int_0^\infty e^{-\lambda t}\,dt = \frac{1}{\lambda}$$

In other words, the MTTF of a system that obeys the exponential failure law is the inverse of the failure rate of the system. Note that the reliability at a time equal to the MTTF for the exponential failure law is

$$R(\text{MTTF}) = R\left(\frac{1}{\lambda}\right) = e^{-\lambda(1/\lambda)} = e^{-1} = 0.3678$$

In other words, a system obeying the exponential failure law has a probability of 0.3678 of *not* experiencing a failure before a time equal to the MTTF, given that the system was perfect at the beginning of that time period.

Stated differently, a system obeying the exponential failure law has a probability of 0.6322 of failing during a time period equal to the MTTF, given that the system was perfect at the beginning of that time period.

4.2.4 Mean Time to Repair

The **mean time to repair (MTTR)** is simply the average time required to repair a system. The MTTR is extremely difficult to estimate and is often determined experimentally by injecting a set of faults, one at a time, into a system and measuring the time required to repair the system in each case. The measured repair times are averaged to determine an average time to repair. In other words, if the ith of N faults requires a time t_i to repair, the MTTR is estimated as

$$\text{MTTR} = \frac{\sum_{i=1}^{N} t_i}{N}$$

Often the estimate of the MTTR is improved by averaging over several repair personnel to account for the differences in the abilities of these personnel. For example, if the set of N faults is repaired by M personnel, each of the personnel has an average time to repair, say, MTTR_i, which is the MTTR for the ith person. The estimate of the overall MTTR is the average of the individual MTTRs. In other terms,

$$\text{MTTR} = \frac{\sum_{i=1}^{M} \text{MTTR}_i}{M}$$

The MTTR is normally specified in terms of a **repair rate** μ, which is the average number of repairs that occur per time period. The units of the repair rate are normally number of repairs per hour. The MTTR and the repair rate μ are related by

$$\text{MTTR} = \frac{1}{\mu}$$

4.2.5 Mean Time Between Failure

It is very important to understand the difference between the MTTF and the **mean time between failure (MTBF)**. Unfortunately, these two terms are often used interchangeably. Although the numerical difference is small in many cases, the conceptual difference is very important. The MTTF is the average time until the *first* failure of a system, whereas the MTBF is the average time *between* failures of a system. As noted in the previous section, we can estimate the MTTF for a system by placing each of a population of N

identical systems into operation at time $t = 0$, measuring the time required for each system to encounter its first failure, and averaging these times over the N systems. The MTBF, however, is calculated by averaging the time between failures, including any time required to repair the system and place it back into an operational status. In other words, each of the N systems is operated for some time T and the number of failures encountered by the ith system is recorded as n_i. The average number of failures is computed as

$$n_{avg} = \sum_{i=1}^{N} \frac{n_i}{N}$$

Finally, the MTBF is

$$\text{MTBF} = \frac{T}{n_{avg}}$$

In other words, the MTBF is the total operation time T, divided by the average number of failures experienced during the time T.

If we assume that all repairs to a system make the system perfect once again just as it was when it was new, the relationship between the MTTF and the MTBF is as illustrated in Fig. 4.2. Once successfully placed into operation, a system operates, on the average, a time corresponding to the MTTF before encountering the first failure. The system then requires some time, MTTR, to repair the system and place it back into operation once again. The system then is perfect once again and will operate for a time corresponding to the MTTF before encountering its next failure. The time between the two failures is the sum of the MTTF and the MTTR and is the MTBF. Thus, the difference between the MTTF and the MTBF is the MTTR. Specifically, the MTBF is given by

$$\text{MTBF} = \text{MTTF} + \text{MTTR}$$

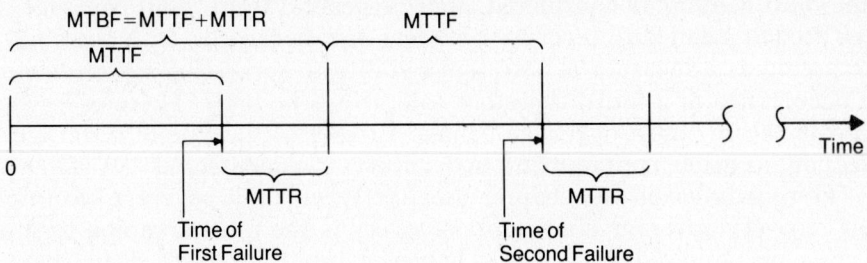

Fig. 4.2 Relationship between the MTBF and the MTTF.

In most practical applications the MTTR is a small fraction of the MTTF, so the *approximation* that the MTBF and MTTF are equal is often quite good. Conceptually, however, it is crucial to understand the difference between the MTBF and the MTTF.

4.2.6. Fault Coverage

An extremely important parameter in the design and analysis of fault-tolerant systems is **fault coverage.** The fault coverage available in a system can have a tremendous impact on the reliability, safety, and other attributes of the system. There are several types of fault coverage, depending on whether the designer is concerned with fault detection, fault location, fault containment, or fault recovery. In addition, there are two primary definitions of fault coverage: one is intuitive, the other is more mathematical.

The intuitive definition is that *coverage* is a measure of a system's ability to perform fault detection, fault location, fault containment, and/or fault recovery. The four primary types of fault coverage are fault detection coverage, fault location coverage, fault containment coverage, and fault recovery coverage. **Fault detection coverage** is a measure of a system's ability to detect faults. For example, a system requirement may be that a certain fraction of all faults be detected; the fault detection coverage is a measure of the system's capability to meet such a requirement. **Fault location coverage** is a measure of a system's ability to locate faults. Once again, it is very common to require a system to locate faults to within easily replaceable modules, and the fault location coverage is a measure of the success with which fault location is performed. **Fault containment coverage** is a measure of a system's ability to contain faults; specifically, the fault containment coverage represents a system's ability to make the *extent* attribute of faults *local* instead of *global*. Finally, **fault recovery coverage** is a measure of a system's ability to recover from faults and maintain an operational status. Clearly, a high fault recovery coverage requires high fault detection, location, and containment coverages.

In the evaluation of fault-tolerant systems, the fault recovery coverage is the most commonly considered, and the general term "fault coverage" is often used to mean fault recovery coverage. In other words, fault coverage is interpreted as a measure of a system's ability to successfully recover after the occurrence of a fault, therefore tolerating the fault. Therefore, when using the term "fault coverage," make sure that the type of coverage—detection, location, containment, or recovery—is understood.

The remainder of this chapter uses the term "fault coverage" to imply fault recovery coverage since fault recovery is the most common form of coverage encountered. In all cases, however, it will be made clear whether detection, location, containment, or recovery coverage is being considered.

Fault coverage is mathematically defined as the conditional probability that, given the existence of a fault, the system recovers [Bouricius, Carter, and Schneider 1969]. In mathematical terms, fault coverage is written as

$$C = P(\text{fault recovery} \mid \text{fault existence})$$

where C is the fault coverage and $P(\text{fault recovery} \mid \text{fault existence})$ is read as the probability of fault recovery *given* the existence of a fault. Recall that fault recovery is the process of maintaining or regaining operational status after a fault occurs.

The fundamental problem with fault coverage is that it is extremely difficult to calculate. Probably the most common approach to estimating fault coverage is to develop a list of all the faults that can occur in a system and to form, from that list, a list of faults that can be detected, a list of faults that can be located, a list of faults that can be contained, and a list of faults from which the system can recover. The fault detection coverage factor, for example, is then computed as simply the fraction of faults that can be detected; that is, the number of faults detected divided by the total number of faults. The remaining fault coverage factors are calculated in a similar manner. As an example, consider the circuit shown in Fig. 4.3 which has fifteen potential sites of stuck-at-1 or stuck-at-0 faults; consequently, there are a total of 30 faults. Table 4.2 shows the input combinations that yield erroneous outputs when certain faults are present, therefore detecting the faults. Note that the circuit performs correctly even if a single stuck-at-0 fault on one of the lines F, G, or M occurs. In other words, a single stuck-at-0 fault on line F, G, or M cannot be detected. As a result, the fault detection coverage for the circuit of Fig. 4.3 is (30−3)/30, or 0.9. In other words, 90% of the stuck-at-1 and stuck-at-0 faults are detected by at least one of the input combinations.

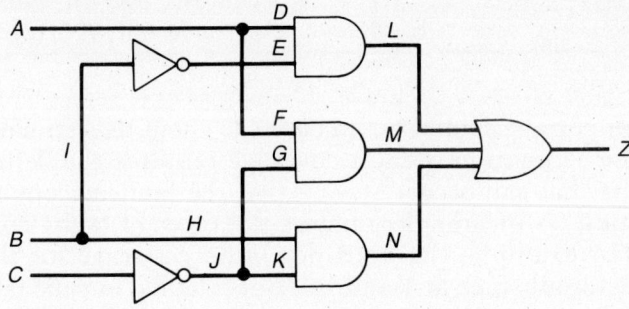

Fig. 4.3 Combinational circuit to illustrate fault detection coverage.

TABLE 4.2 Input patterns capable of detecting faults (test vectors) in the circuit of Fig. 4.3

Fault	Number of test vectors	Test vectors ABC
A_0	2	100, 101
A_1	2	000, 001
B_0	2	010, 111
B_1	2	000, 101
C_0	2	011, 111
C_1	2	010, 110
D_0	1	101
D_1	2	000, 001
E_0	1	101
E_1	1	111
F_0	0	—
F_1	2	000, 101
G_0	0	—
G_1	1	111
H_0	1	010
H_1	1	000
I_0	1	111
I_1	1	101
J_0	2	010, 110
J_1	2	011, 111
K_0	1	010
K_1	2	011, 111
L_0	1	101
L_1	4	000, 001, 011, 111
M_0	0	—
M_1	4	000, 001, 011, 111
N_0	1	010
N_1	4	000, 001, 011, 111
Z_0	4	010, 100, 101, 110
Z_1	4	000, 001, 011, 111

Several important points should be made about the estimation of coverage. First, the estimation of fault coverage requires the definition of the types of faults that can occur. Stating that the fault detection coverage is 0.9, for example, is meaningless unless the types of faults considered are identified. For example, the fault detection coverage for the circuit of Fig. 4.3 is 0.9 for all stuck-at-1 and stuck-at-0 faults, but the fault detection coverage may decrease substantially if stuck-open faults are included.

A second important point about the fault coverage is that it is typically assumed to be a constant. It is easy to envision applications in which the probability of detecting a fault, for example, increases as a function of time, after the occurrence of the fault. However, to simplify the analysis, the various fault coverages are normally assumed to be constants.

4.3 Reliability Modeling

Reliability is perhaps one of the most important attributes of systems. Almost all specifications for systems mandate that certain values for reliability be achieved and in some way proved. We have seen in the previous sections that reliability can be determined experimentally if a set of N systems is operated over a period of time and the number of systems that fail during that time period is recorded. One problem with the experimental approach is the number of systems that would be required to achieve a level of confidence in the experimental results. This is particularly a problem when costs limit the number of systems that can be built. For example, the space shuttle program could not afford to build 1000 of its on-board processing systems such that reliability could be experimentally verified.

A second problem with the experimental approach is the time required to run such experiments. Many systems today are being designed to achieve reliabilities of 0.9_7, or higher, after ten hours of operation. Using the exponential failure law, a reliability of 0.9_7 corresponds to a failure rate of 10^{-8} failures per hour. Therefore, on the average, we would have to wait approximately 100 million hours, or approximately 11,416 years for the first failure to occur. Clearly, we need alternatives to the experimental approach.

The most popular reliability analysis techniques are the analytical approaches. Of the analytical techniques, combinatorial modeling and Markov modeling are the two most commonly used approaches.

4.3.1 Combinatorial Models

Combinatorial models use probabilistic techniques that enumerate the different ways in which a system can remain operational. The probabilities of the events that lead to a system being operational are calculated to form an estimate of the system's reliability.

The reliability of a system is generally derived in terms of the reliabilities of the individual components of the system. The two models of systems that are most common in practice are the series and the parallel. In a **series system,** each element of the system is required to operate correctly for the system to operate correctly. In a **parallel system,** on the other hand, only one of several elements must be operational for the system to perform its functions correctly.

In practice, systems are typically combinations of series and parallel subsystems. Once we have discussed both the series and parallel structures, we will examine techniques for modeling systems that contain *both* series and parallel subsystems.

Series Systems

The series system is best thought of as a system that contains no redundancy; that is, each element of the system is needed to make the system function correctly. For example, a digital filter that contains a microprocessor, an analog-to-digital converter, and a digital-to-analog converter needs each of these elements to perform the digital filtering function; if any one of the three elements fails, the system fails. One way of representing the series system is through the use of **reliability block diagrams.** The reliability block diagram can be thought of as a flow diagram from the input of the system to the output of the system. Each element of the system is a block in the reliability block diagram and, for the series system, the blocks are placed in series to indicate that a path from the input to the output is broken if one of the elements fails.

For example, the generalized reliability block diagram of a series system that contains N elements is shown in Fig. 4.4. Each of the N elements is required for the system to function correctly. The reliability of the series system can be calculated as the probability that none of the elements will fail. Another way to look at this is that the reliability of the series system is the probability that all of the elements are working properly.

Suppose we let $C_{iw}(t)$ represent the event that component C_i is working properly at time t, $R_i(t)$ is the reliability of component C_i at time t, and $R_{\text{series}}(t)$ is the reliability of the series system. Further suppose that the series system contains N series components as shown in Fig. 4.4. The reliability at any time t is the probability that all N components are working properly. In mathematical terms,

$$R_{\text{series}}(t) = P(C_{1w}(t) \cap C_{2w}(t) \cap \cdots \cap C_{Nw}(t))$$

Assuming that the events, $C_{iw}(t)$, are independent, we have

$$R_{\text{series}}(t) = R_1(t)R_2(t)\cdots R_N(t)$$

or

$$R_{\text{series}}(t) = \prod_{i=1}^{N} R_i(t)$$

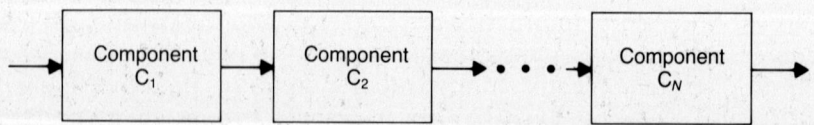

Fig. 4.4 The reliability block diagram of a series system—each element of the system must operate correctly for the system to operate correctly.

An interesting relationship exists in a series system if each individual component satisfies the exponential failure law. Suppose that we have a series system made up of N components, and each component i has a constant failure rate of λ_i. Also assume that each component satisfies the exponential failure law such that the reliability of each component is $R_i(t) = e^{-\lambda_i t}$. The reliability of the series system is given by

$$R_{\text{series}}(t) = e^{-\lambda_1 t} e^{-\lambda_2 t} \cdots e^{-\lambda_N t}$$

or

$$R_{\text{series}}(t) = e^{-\sum_{i=1}^{N} \lambda_i t} = e^{-\lambda_{\text{system}} t}$$

where $\lambda_{\text{system}} = \sum_{i=1}^{N} \lambda_i$ and corresponds to the failure rate of the system. In other words, the failure rate of a series system can be calculated by adding the failure rates of all the components that make up the series system.

As an example of a series system, consider the simple aircraft control system shown in Fig. 4.5. This system contains sensors that are used to measure the roll, pitch, and yaw positions of the aircraft; sensors to measure the crew's desired roll, pitch, and yaw positions; actuators that are used to control the roll, pitch, and yaw; and computers that perform the computations required of the flight control system. The computers receive the sensor values and supply the actuator commands over a serial data bus that connects the sensors, actuators, and the computers. A special high-speed data bus interconnects the computers for the purpose of data transfer among the computers. Each element of the system is required if the system is to perform correctly; there is no redundancy in the system. For example, the failure of any one sensor or computer renders the system unable to operate correctly.

The reliability block diagram of the flight control system is shown in Fig. 4.6. The reliability block diagram illustrates the series nature of the system. For simplicity, assume that all six sensors have the same reliability $R_s(t)$, each of the three actuators has the reliability, $R_{\text{act}}(t)$, and each computer has the reliability $R_c(t)$. Also, let the computer interconnection bus have the reliability $R_{\text{bus1}}(t)$ and the primary control bus have the reliability $R_{\text{bus2}}(t)$. By taking the product of the element reliabilities, we find that the reliability of the system is given by

$$R_{\text{system}}(t) = R_s^6(t) R_{\text{act}}^3(t) R_c^3(t) R_{\text{bus1}}(t) R_{\text{bus2}}(t)$$

Because the failure rates can be added in a series system to obtain the failure rate of the system, we can write

$$\lambda_{\text{system}} = 6\lambda_s + 3\lambda_{\text{act}} + 3\lambda_c + \lambda_{\text{bus1}} + \lambda_{\text{bus2}}$$

where λ_s is the failure rate of one sensor, λ_{act} is the failure rate of one actuator, λ_c is the failure rate of one computer, λ_{bus1} is the failure rate of the com-

188   Evaluation Techniques

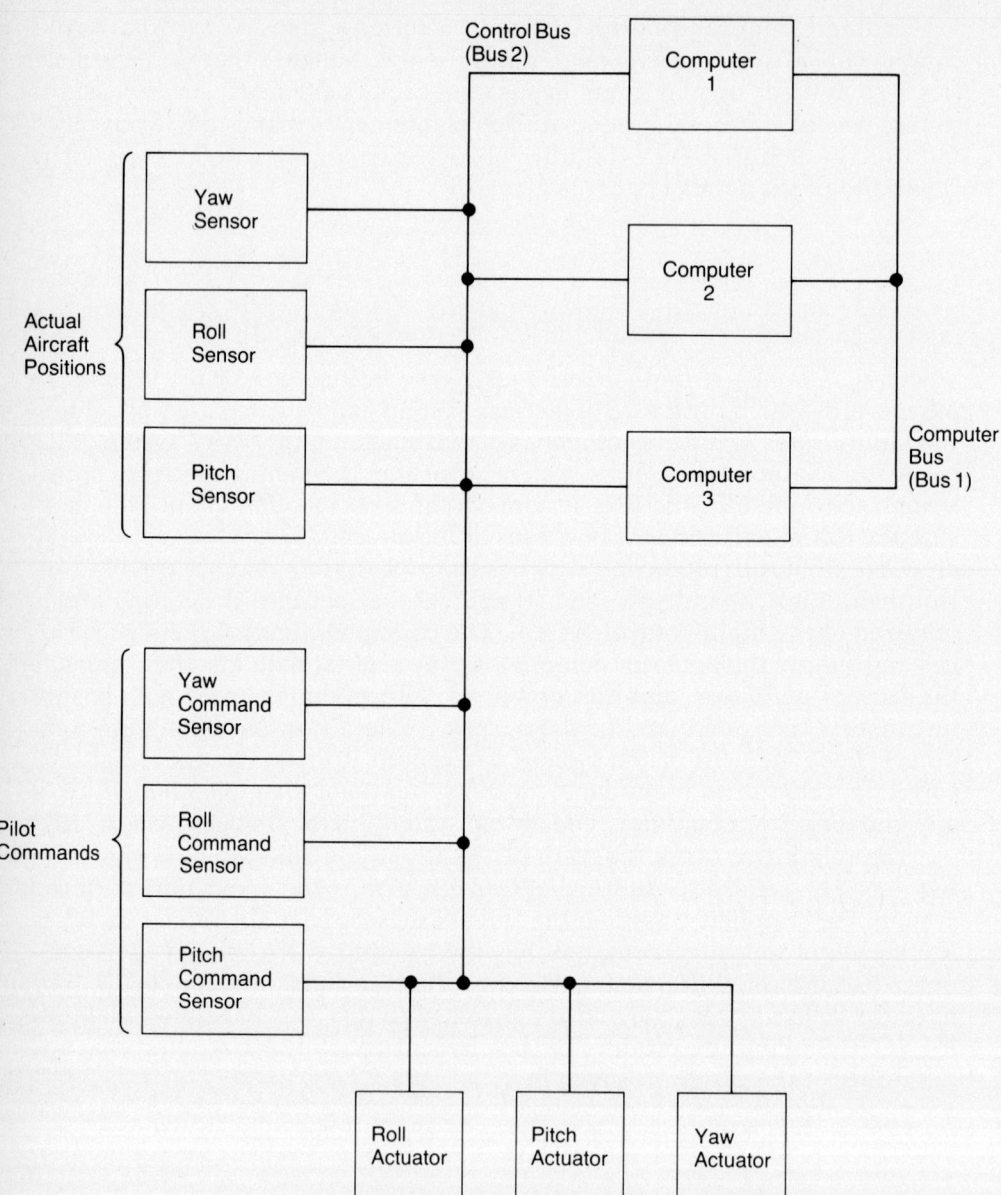

Fig. 4.5 An aircraft control system designed as a series system.

puter interconnection bus, and λ_{bus2} is the failure rate of the primary control bus. λ_{system} is the failure rate of the system. If the failure rates of the system are

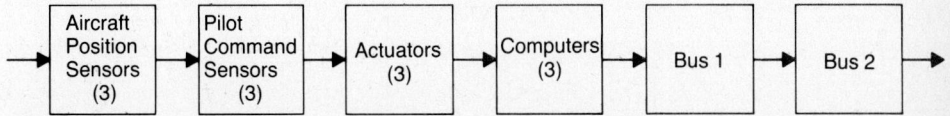

Fig. 4.6 The reliability block diagram of the system in Figure 4.5 illustrates the series nature of the system.

$$\lambda_s = 1 \times 10^{-6} \quad \text{failures per hour}$$
$$\lambda_{act} = 1 \times 10^{-5} \quad \text{failures per hour}$$
$$\lambda_c = 4 \times 10^{-4} \quad \text{failures per hour}$$
$$\lambda_{bus1} = 1 \times 10^{-6} \quad \text{failures per hour}$$
$$\lambda_{bus2} = 2 \times 10^{-6} \quad \text{failures per hour}$$

the system failure rate will be

$$\lambda_{system} = 1.239 \times 10^{-3} \quad \text{failures per hour}$$

The reliability after five hours for this system is approximately 0.995.

Parallel Systems

The distinguishing feature of the basic parallel system is that only one of N identical elements is required for the system to function. For example, many families have two or more cars when one is, in many cases, sufficient to meet the family's needs. The probability of having at least one car working can be determined by modeling the multiple-car family as a parallel system.

The reliability block diagram of the basic parallel system that contains N identical elements is shown in Fig. 4.7. As can be seen, a path exists in the reliability block diagram from input to output as long as one of the N identical elements remains operational. The unreliability of the parallel system can be computed as the probability that all of the N elements fail. Suppose that we let $C_{if}(t)$ represent the event that element i in the parallel system has failed at time t, $Q_{parallel}(t)$ be the unreliability of the parallel system, and $Q_i(t)$ be the unreliability of the ith element. $Q_{parallel}(t)$ can be computed as

$$Q_{parallel}(t) = P(C_{1f}(t) \cap C_{2f}(t) \cap \cdots \cap C_{Nf}(t))$$

or

$$Q_{parallel}(t) = Q_1(t)Q_2(t)\cdots Q_N(t) = \prod_{i=1}^{N} Q_i(t)$$

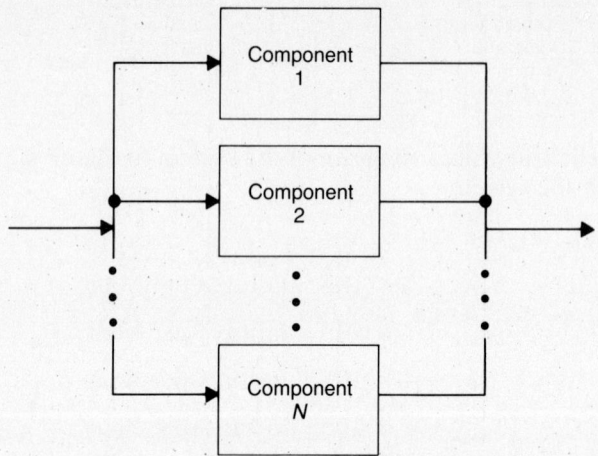

Fig. 4.7 The reliability block diagram of the parallel system—only one of N components must operate correctly for the system to operate correctly.

The reliability of the parallel system can now be computed because we know that the reliability and the unreliability must add to 1.0. Mathematically, we must have $R(t) + Q(t) = 1.0$ for any system. Consequently, we can write

$$R_{\text{parallel}}(t) = 1.0 - Q_{\text{parallel}}(t) = 1.0 - \prod_{i=1}^{N} Q_i(t) = 1.0 - \prod_{i=1}^{N}(1.0 - R_i(t))$$

Note that the equations for the parallel system assume that the failures of the individual elements that make up the parallel system are independent. For random hardware failures, the independence of failures is a good assumption; however, for failures that are the result of items such as external disturbances, the independence assumption is not very good. Therefore, the combinatorial modeling techniques are most often applied to the analysis of random failures in a system's hardware.

To analyze a system that has a parallel structure, consider the system shown in Fig. 4.8. The architecture of the system in Fig. 4.8 is commonly found in aerospace applications. The system consists of two identical computers, two identical interface units, two identical display devices, and two identical communication buses. The system requires that at least one of each unit work properly for the system to perform its functions. Once a particular unit has failed, it is assumed that the second unit of that type automatically assumes the functions of the failed unit.

One important point about the system of Fig. 4.8 is that it has both a series and a parallel structure. It is parallel in the sense that only one of the

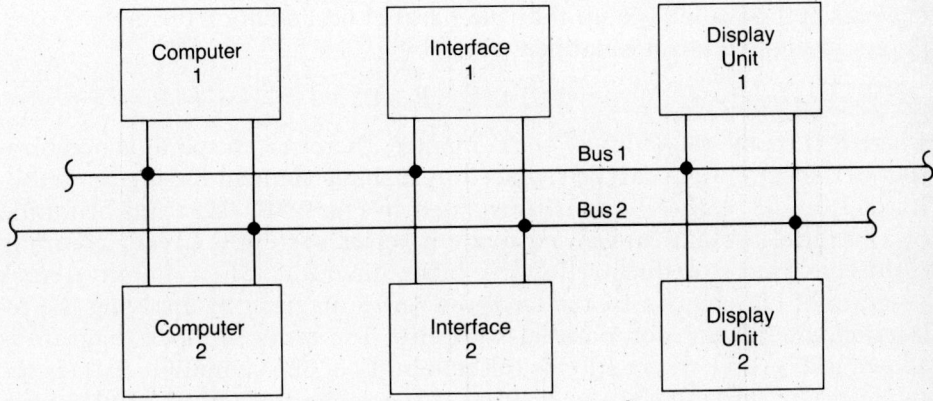

Fig. 4.8 An example computer system having a structure that is a combination of series and parallel.

two computers, for example, must function for the system to function. It is series in the sense that one computer, one interface unit, one display device, *and* one bus must operate for the system to operate. The reliability block diagram of the system of Fig. 4.8 is shown in Fig. 4.9. Note that a path from the input of the diagram to the output exists if and only if enough elements are functioning to allow the system to operate properly.

A reliability block diagram that contains both series and parallel structures can be reduced to a single series diagram by replacing each of the parallel portions of the system with an equivalent, single element that has the same reliability as the parallel structure. For example, we know from the

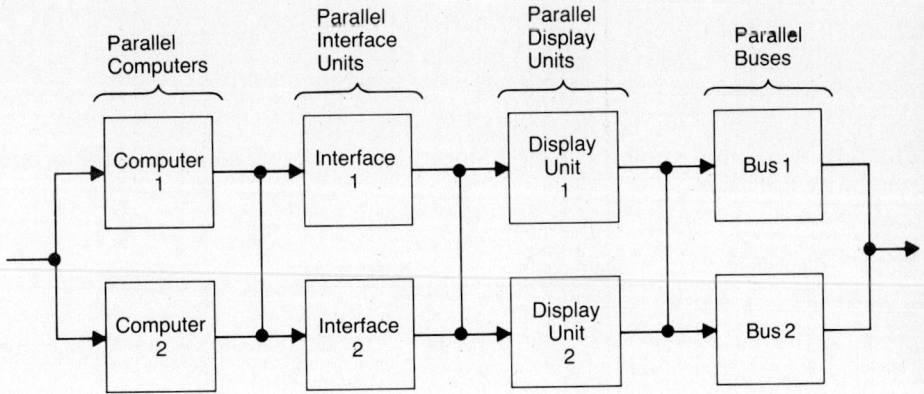

Fig. 4.9 Reliability block diagram of the series/parallel system of Fig. 4.8.

analysis of a parallel system that the parallel organization of the two computers in Fig. 4.9 has a reliability given by

$$1.0 - (1 - R_c(t))^2$$

where $R_c(t)$ is the reliability of one computer. Therefore, the parallel organization of the computers can be replaced by a single element having a reliability of $1.0 - (1 - R_c(t))^2$, as is illustrated in Fig. 4.10. The transformation of a parallel system into an equivalent series system is a very common technique used to reduce reliability block diagrams. The reliability block diagram of Fig. 4.9 can be reduced to a series diagram by applying the reduction concept to each parallel structure. The reduced block diagram is shown in Fig. 4.11 where $R_c(t)$ is the reliability of one computer, $R_{if}(t)$ is the reliability of one interface unit, $R_d(t)$ is the reliability of one display unit, and $R_b(t)$ is the reliability of one bus.

The reliability for the reduced block diagram of Fig. 4.11 can be written as

$$R_{system}(t) = [1 - (1 - R_c(t))^2][1 - (1 - R_{if}(t))^2][1 - (1 - R_d(t))^2] \cdot [1 - (1 - R_b(t))^2]$$

As an example, the reliability of the system after one hour given $R_c(1) = R_{if}(1) = R_d(1) = R_b(1) = .9$ will be

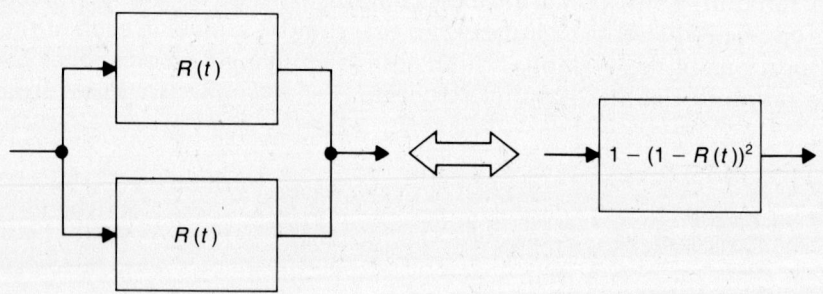

Fig. 4.10 A parallel system can be reduced to a series element with the proper reliability function.

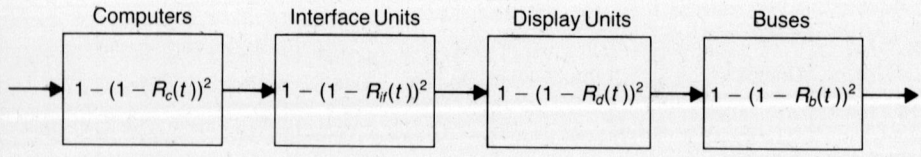

Fig. 4.11 Reduced reliability block diagram for the system of Fig. 4.8.

$R_{system}(1 \text{ hour}) = [1 - (1 - .9)^2][1 - (1 - .9)^2][1 - (1 - .9)^2][1 - (1 - .9)^2] = .96$

Now that we have several tools for investigating the reliability of systems, we can compare the reliability benefits that redundancy can offer. For example, the redundant system just analyzed had a reliability of 0.96 after one hour. The nonredundant system containing one computer, one display device, one bus, and one interface device has a reliability equivalent to the product of the individual element's reliabilities because the nonredundant system is a simple series system. Therefore, the nonredundant system has a reliability of 0.6561 after one hour. As is seen, the incorporation of redundancy has significantly improved the system's reliability.

Note, however, that redundancy does not always improve a system's reliability. Whether or not the reliability is improved depends on the amount of redundancy employed and the reliability of the elements used to construct the system, as well as other factors. For example, if each element of the redundant system in Fig. 4.8 has a reliability of 0.1 at the end of one hour, the redundant system has a reliability of 0.0013 at the end of one hour, and the nonredundant system has a reliability of 0.0001 at the end of one hour. We certainly would hope that elements with a reliability as poor as 0.1 would never be used in a system, but this example does show that the redundancy does not significantly improve the reliability. As we stated in the first chapter, reliability and fault tolerance are not one in the same. As we shall see when we begin to analyze more complex systems, the distinction between fault tolerance and reliability becomes even clearer.

4.3.2 Fault Coverage and Its Impact on Reliability

As defined earlier, fault coverage is a measure of a system's ability to recover from faults. For example, a system with redundant computers that requires reconfiguration before the redundancy can be used depends heavily on good fault coverage. During the analysis of the parallel system, we assumed that the fault coverage was perfect; if we had three computers and needed only one to operate, the reliability was calculated solely as the probability that one of the three computers was operational. Unfortunately, the assumption of perfect fault coverage does not consider that the system may not be able to use the redundancy because it cannot identify that a unit is faulty, remove that faulty unit, and replace it with a fault-free one.

To illustrate the problem, consider a simple parallel system consisting of two identical modules and having the reliability block diagram shown in Fig. 4.12. Assume that module 1 is the primary module and that module 2 is a spare module that is switched on-line in the event of failure of module 1. In other words, the system uses the concept of standby sparing. Under ideal circumstances, the standby sparing system functions correctly as long as one of the two modules functions correctly. In reality, however, the failure

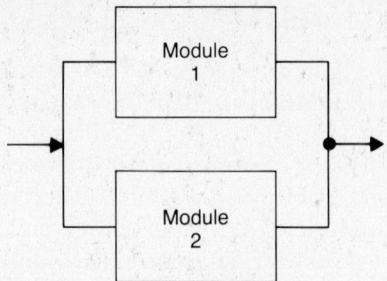

Fig. 4.12 Reliability block diagram of a simple parallel system to illustrate the impact of fault coverage.

of the primary module (module 1, in this case) must be detected and correctly handled before the second module can be used. In other words, the parallel system with two modules functions correctly as long as one of the following two conditions exist:

1. Module 1 is functioning correctly.
2. Module 2 is functioning correctly, module 1 has failed, and the failure was detected and appropriately handled.

The probability that one of these two events will exist can be written in terms of the reliabilities of the modules and the fault coverage as

$$R_{\text{system}}(t) = R_1(t) + (1 - R_1(t))C_1 R_2(t)$$

where C_1 is the fault coverage associated with module 1, $R_1(t)$ is the reliability of module 1, and $R_2(t)$ is the reliability of module 2. The reliability equation enumerates all of the working states of the system. If the reliabilities and the coverage factors of the two modules are identical, the reliability expression reduces to

$$R_{\text{system}}(t) = R(t) + R(t)C(1 - R(t))$$

where $R(t)$ is the reliability of one module and C is the fault coverage. Note that if the fault coverage C, is 1.0, the reliability expression reduces to

$$R_{\text{system}}(t) = 2R(t) - R^2(t) = 1 - (1 - R(t))^2$$

which is the reliability of the perfect parallel system. Also note that if the fault coverage is 0.0, the reliability expression reduces to simply the reliability of one module; therefore, the primary module must function correctly for the system to function correctly.

It is interesting to study the impact of the fault coverage in the parallel system with two modules. Figure 4.13 shows the reliability of the system as a function of fault coverage for a module reliability of R. Note that the

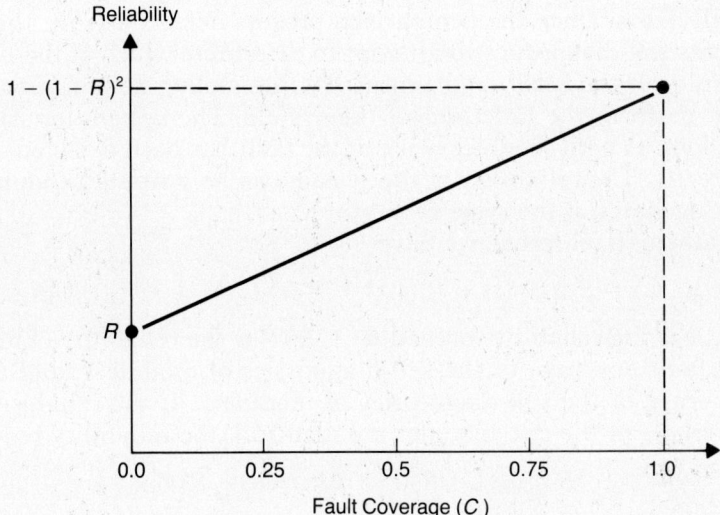

Fig. 4.13 Reliability versus fault coverage for a parallel system with two modules and using standby sparing. Each module has a reliability of R and a fault coverage of C.

reliability of the parallel system with two modules is a linear function of the coverage factor. At a coverage of 1.0, the reliability is $1 - (1 - R)^2$, which is the reliability of the perfect parallel system. At a coverage of 0.0, the reliability is R, which is simply the probability that the primary module will not fail.

One important point about the above analysis is that the failure of the second module is unimportant unless it has replaced the first module. In other words, module 1 is the primary module and as long as it functions correctly, the system functions correctly, even if module 2 fails. In many systems, this may not be true. For example, consider a duplex system that performs comparisons between the two modules as one form of fault detection. Once a fault is detected, the two modules go into more detailed fault analysis routines, often called self-diagnostics, in an attempt to identify which of the two modules is faulty. If the faulty module is identified, the system continues to operate with the one fault-free module, and the comparison mechanism is disabled. If the faulty module cannot be identified, the system discontinues its operation. Therefore, an undetected fault in either of the modules causes the system to fail and must be accounted for in the reliability analysis.

Consider once again the parallel system with two modules as shown in Fig. 4.12, but now perform comparisons between the two modules as one means of fault detection. Assume for now that the comparison is perfect and detects all faults. Once the comparison process detects a fault, the system implements self-diagnostics to attempt to determine which of the two modules is faulty. If the fault can be located successfully, the fault-free module begins to perform the functions of the system. The system functions correctly as long as both modules work or the fault has been detected and handled correctly. The reliability of the system can be written by enumerating the working states of the system.

In mathematical terms, we have

$$R_{system}(t) = R_1(t)R_2(t) + R_1(t)(1 - R_2(t))C_2 + (1 - R_1(t))C_1R_2(t)$$

where $R_1(t)$ is the reliability of module 1, $R_2(t)$ is the reliability of module 2, C_1 is the fault coverage of the self-diagnostics of module 1, and C_2 is the fault coverage of the self-diagnostics of module 2. If the reliabilities and fault coverages of the two modules are identical, the reliability reduces to

$$R_{system}(t) = R^2(t) + 2R(t)C(1 - R(t))$$

For perfect fault coverage, we obtain the same as before; that is, the system has the reliability of the perfect parallel system. If the fault coverage is 0.0, the system has a reliability of $R_{system}(t) = R^2(t)$, which is simply the probability that both modules operate correctly. Figure 4.14 shows the reliability of the system as a function of fault coverage.

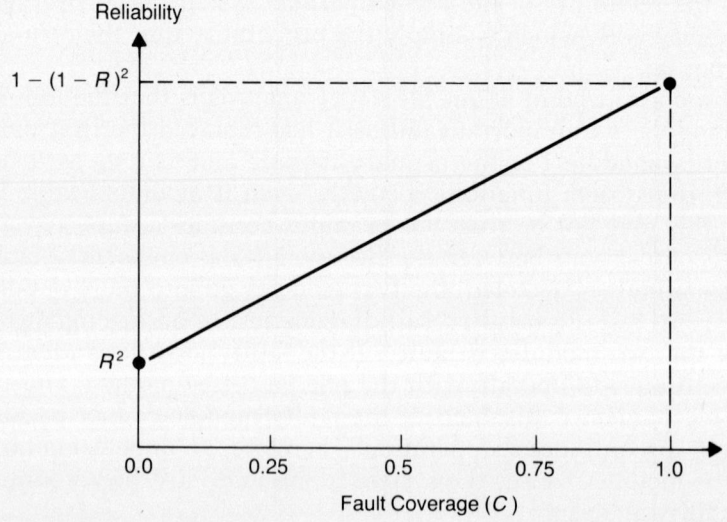

Fig. 4.14 Reliability versus fault coverage for a parallel system with two modules and using comparisons between the two modules. Each module has a reliability of R and a fault coverage of C.

4.3.3 M-of-N Systems

M-of-N systems are a generalization of the ideal parallel system. In the ideal parallel system, only one of N modules is required to work for the system to work. In the M-of-N system, however, M of the total of N identical modules are required to function for the system to function. A good example is the TMR configuration where two of the three modules must work for the majority voting mechanism to function properly. Therefore, the TMR system is a 2-of-3 system.

Consider as an example the TMR system. As seen in the previous sections, we can write the reliability of a system by enumerating all of the possible states in which the system can be functional. Suppose that we have a TMR system with modules 1, 2, and 3 connected in a majority voting arrangement. As long as two of the three modules are functioning correctly, the system will perform correctly. Ignoring the reliability of the voter, the reliability of the TMR system can be written as

$$R_{TMR}(t) = R_1(t)R_2(t)R_3(t) + R_1(t)R_2(t)(1 - R_3(t)) \\ + R_1(t)(1 - R_2(t))R_3(t) + (1 - R_1(t))R_2(t)R_3(t)$$

where $R_i(t)$ is the reliability of the ith module. If $R_1(t) = R_2(t) = R_3(t) = R(t)$, the reliability of the TMR system reduces to

$$R_{TMR}(t) = R^3(t) + 3R^2(t)(1 - R(t)) = 3R^2(t) - 2R^3(t)$$

Now that we have the expression for the reliability of the TMR system, it is interesting to examine the reliability improvements that can be obtained through the use of TMR. Figure 4.15 shows a plot of the reliability of a TMR arrangement as a function of the reliability of the modules that compose the TMR system. In other words, Fig. 4.15 simply shows a plot of the equation $R_{TMR} = 3R^2 - 2R^3$ versus R. As can be seen, there is a point at which the reliability of the TMR system and the reliability of the single module cross. The crossover point is easily found by setting the reliability of the TMR system equal to the reliability of the single module and solving the resulting quadratic equation.

In mathematical terms, we have

$$R_{TMR} = 3R^2 - 2R^3 = R$$

or

$$3R - 2R^2 = 1$$

which implies the quadratic equation

$$R^2 - \frac{3}{2}R + 0.5 = 0$$

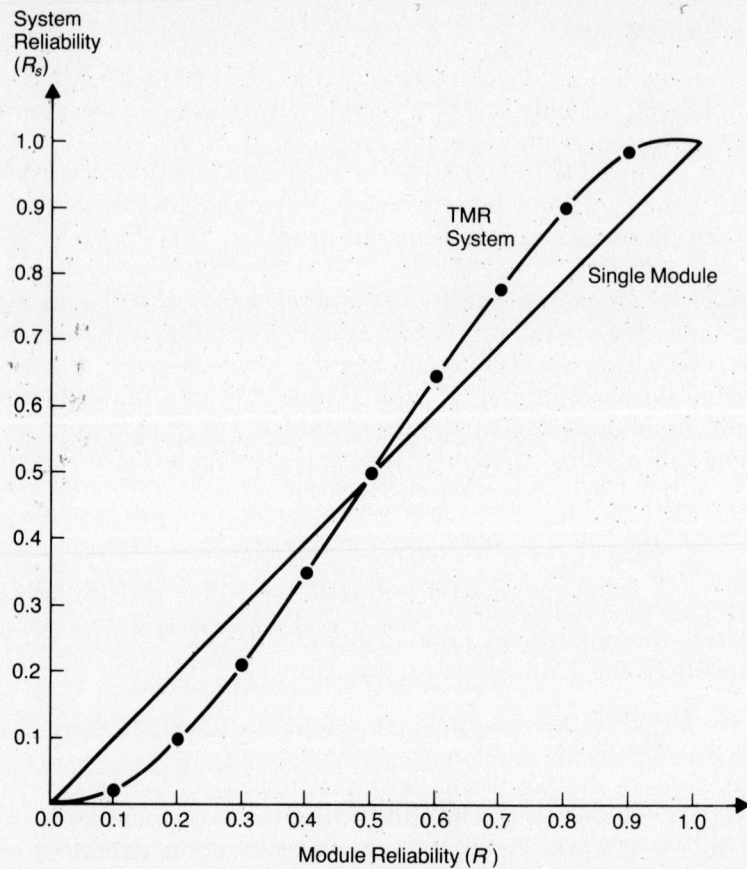

Fig. 4.15 Comparison of the reliability of a TMR system composed of three identical modules with the reliability of a single module.

The two solutions to the quadratic equation are 0.5 and 1.0, which implies that the reliability of the TMR system is equal to that of the corresponding nonredundant system when the reliability of the single module is 0.5 or the module is perfect ($R = 1$).

This further illustrates a point that we made when we defined fault tolerance and reliability. A system can be tolerant of faults and still have a low reliability. For example, a TMR system constructed from modules that have individual reliabilities of 0.5 can tolerate a fault in one of those modules, but the reliability of the TMR system is the same as the reliability of a single module. Conversely, a system can achieve a high reliability without being fault tolerant. Certainly, a system that consists of a perfect module will have the highest possible reliability but will not possess, or need, the attribute of fault tolerance. This is, of course, an unrealistic example, but, in general, as the reliability of the components of a system increases, the reliability of the system also increases. It is possible for the reliability of a nonredundant system to approach that of a redundant system constructed from the same modules. The nonredundant system, however, will not be fault tolerant.

In many cases, we have systems that are of the M-of-N structure but are not TMR; the general NMR system is a good example. In general, if there are N identical modules and M of those are required for the system to function properly, the system can tolerate $N - M$ module failures. The expression for the reliability of an M-of-N system can be written as

$$R_{M\text{-of-}N}(t) = \sum_{i=0}^{N-M} \binom{N}{i} R^{N-i}(t)(1 - R(t))^i$$

where

$$\binom{N}{i} = \frac{N!}{(N-i)!\,i!}$$

For example, the TMR system reliability is given by

$$R_{\text{TMR}}(t) = \sum_{i=0}^{1} \binom{3}{i} R^{3-i}(t)(1 - R(t))^i$$

which reduces to

$$R_{\text{TMR}}(t) = 3R^2(t) - 2R^3(t)$$

which is identical to the expression derived earlier.

4.3.4 Markov Models

The primary difficulty with the combinatorial models is that many complex systems cannot be modeled easily in a combinatorial fashion. The reliability block diagrams can be extremely difficult to construct, and the resulting reliability expressions are often very complex. In addition, the fault coverage that we have seen to be extremely important in the reliability of a system is sometimes difficult to incorporate into the reliability expression in a combinatorial model. Finally, the process of repair that occurs in many systems is very difficult to model in a combinatorial fashion. For these reasons, we often use **Markov models.**

The purpose of the presentation in this text is not to delve into the mathematical details of Markov models but to understand how to use Markov models. For more explicit mathematical details, refer to the references ([Shooman 1968] and [Trivedi 1982]). The discussions here will provide sufficient mathematical background to apply the Markov model but will not pursue various techniques for solving the models.

The two main concepts in the Markov model are the **system state** and the **state transition.** The state of a system represents all that must be known to describe the system at any given instant of time. For reliability models, each state of the Markov model represents a distinct combination of faulty and fault-free modules. For example, suppose we have a TMR system with three identical computers in a majority voting arrangement with a perfect voter. We can define the state of this system as $S = (S_1, S_2, S_3)$ where $S_i = 1$ if module i is fault free and $S_i = 0$ if module i is faulty. The TMR system has eight distinct states in which it can operate: (000), (001), (010), (011), (100), (101), (110), and, (111). Each state represents a unique combination of faulty and fault-free modules within the system. For TMR, we know that at least two of the modules must be fault free for the system to operate correctly. Therefore, the states (000), (001), (010), and (100) represent states in which the system has ceased to function correctly. The remaining states are those in which the system is functioning correctly.

The state transitions govern the changes of state that occur within a system. As time passes and failures and reconfigurations occur, the system goes from one state to another. For example, if the TMR system starts its operation in state (111) and at some time t module 1 fails, the system transitions to state (011). The state transitions are characterized by probabilities such as the probability of failure, fault coverage, and the probability of repair.

As an example of the state transitions that can occur, consider the TMR system. We have already defined the states that can exist in the system; now let us define the transitions that can occur. We construct our transitions using several assumptions. First, we assume that the system does not contain repair. In other words, once a module has failed, it remains failed permanently. Second, we assume that only one failure will occur at a time. In a TMR system, the single failure assumption implies that the system cannot go directly from the state corresponding to all modules operating correctly to a state that corresponds to the system having failed. In other words, no single failure can cause the complete TMR system to fail. Finally, we assume that the system starts in the perfect state (111) where all of the system's modules are operating correctly.

The state diagram that results for the TMR system is shown in Fig. 4.16. As can be seen, the system begins in state (111) and, upon the first module failure, transitions to state (110), (101), or (011), depending on whether module 1, 2, or 3 is the module that fails. Note that the transition exists for the module to remain in a state if a module failure does not occur. The state diagram shown in Fig. 4.16 is analogous to the state diagram of a synchronous digital circuit. When some event, a module failure in the case of the reliability model and the occurrence of a clock signal in the case of a synchronous machine, occurs, the system transitions from one state to another.

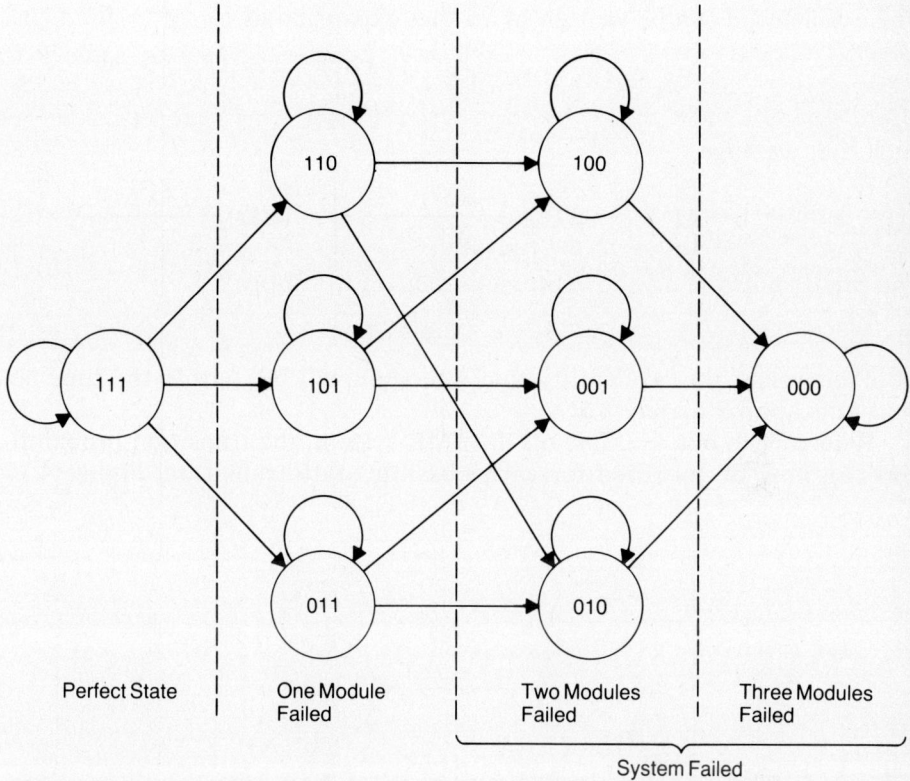

Fig. 4.16 State diagram showing possible states and state transitions for a TMR system.

The states in the diagram shown in Fig. 4.16 can be partitioned into three major categories: the *perfect* state (111) in which all modules function correctly; the *one-failed* states (110), (101), and (011) in which a single module has failed, and the *system-failed* states (100), (001), (010), and (000), in which enough modules have failed to cause the system to fail. State partitioning will be useful later when we attempt to reduce the Markov model.

As stated earlier, each state transition has associated with it a transition probability that describes the probability of that state transition occurring within a specified period of time. In the case of the TMR example that we have been considering, each transition represented in Fig. 4.16 is the result of a single module failure. If we assume that each module in the TMR system obeys the exponential failure law and has a constant failure rate of λ, the probability of a module being failed at some time $t + \Delta t$, given that the module was operational at time t, is given by

$$1 - e^{-\lambda \Delta t}$$

The exponential can be written in a series expansion as

$$e^{-\lambda \Delta t} = 1 + (-\lambda \Delta t) + \frac{(-\lambda \Delta t)^2}{2!} + \cdots$$

such that we have

$$1 - e^{-\lambda \Delta t} = 1 - \left[1 + (-\lambda \Delta t) + \frac{(-\lambda \Delta t)^2}{2!} + \cdots \right] = (\lambda \Delta t) - \frac{(-\lambda \Delta t)^2}{2!} - \cdots$$

For small values of Δt, the expression reduces to simply

$$1 - e^{-\lambda \Delta t} \approx \lambda \Delta t$$

In other words, the probability that a module will fail within the time period Δt is approximately $\lambda \Delta t$.

Referring to our example on the TMR system, the transition probabilities can now be specified for each possible state transition. Figure 4.17

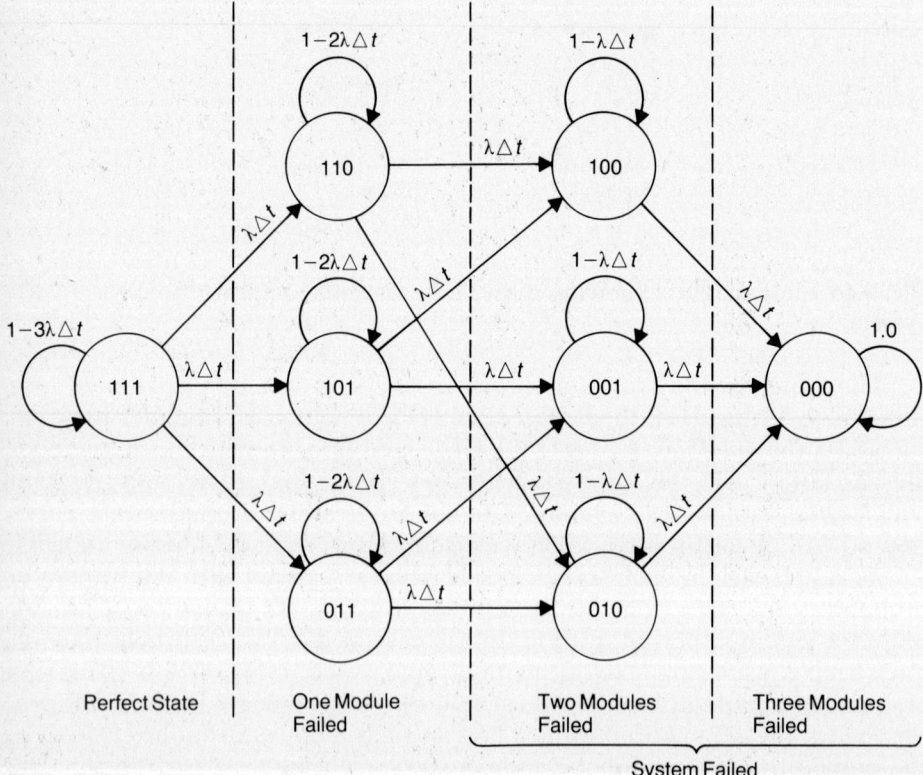

Fig. 4.17 Markov model of the TMR system showing possible states, state transition, and state transition probabilities.

shows the resulting state diagram for the Markov model of the TMR system, including the specification of each state transition probability. It is possible, however, to reduce the Markov model of Fig. 4.17. As mentioned earlier, the states of the TMR model can be partitioned into three major classes: the perfect state, the one-failed state, and the system-failed state. If we appropriately define the state transition probabilities, the several states within the TMR model can be combined.

Suppose that we let state 3 correspond to the state in which all three modules in the TMR system are functioning correctly; state 2 is the state in which two modules are working correctly; state F is the failed state in which two or more modules have failed. The resulting Markov model can be illustrated as shown in Fig. 4.18. The state transition probabilities shown in Fig. 4.18 have been derived to account for one of several failures occurring. For example, the probability of transitioning from state 3 to state 2 depends on the probability of any one of three modules failing. Consequently, the transition probability assigned to the transition from state 3 to state 2 is $3\lambda(\Delta t)$. Likewise the transition probability assigned to the state transition from state 2 to state F is $2\lambda(\Delta t)$.

The equations of the Markov model of the TMR system can be written easily from the state diagram shown in Fig. 4.18. The probability of the system being in any given state S at some time $t + \Delta t$ depends on the probability that the system was in a state from which it could transition to state S *and* the probability of that transition occurring. For example, the probability that the TMR system will be in state 3 at time $t + \Delta t$ depends on the probability that the system was in state 3 at time t (since the system can only transition to state 3 from state 3) and the probability of the system transitioning from state 3 back into state 3.

In mathematical terms, we have

$$p_3(t + \Delta t) = (1 - 3\lambda \Delta t)p_3(t)$$

where $p_3(t)$ is the probability of being in state 3 at time t and $p_3(t + \Delta t)$ is the probability of being in state 3 at time $t + \Delta t$. In a similar fashion, the equations for the remaining two states can be written as

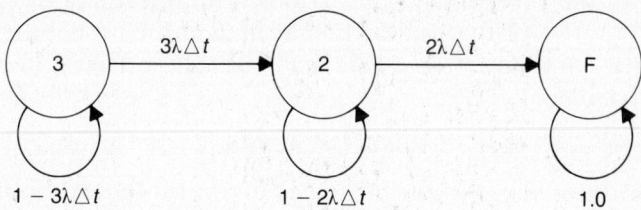

Fig. 4.18 Reduced Markov model of the TMR system with a minimal number of states.

$$p_2(t + \Delta t) = (3\lambda \Delta t)p_3(t) + (1 - 2\lambda \Delta t)p_2(t)$$
$$p_F(t + \Delta t) = (2\lambda \Delta t)p_2(t) + p_F(t)$$

assuming that the system will remain in the failed state if it ever enters the failed state. $p_2(t + \Delta t)$ is the probability of being in state 2 at time $t + \Delta t$; $p_2(t)$ is the probability of being in state 2 at time t; $p_F(t + \Delta t)$ is the probability of being in state F at time $t + \Delta t$; and $p_F(t)$ is the probability of being in state F at time t.

The equations of the Markov model of the TMR system can be written in matrix form as

$$\begin{bmatrix} p_3(t + \Delta t) \\ p_2(t + \Delta t) \\ p_F(t + \Delta t) \end{bmatrix} = \begin{bmatrix} (1 - 3\lambda \Delta t) & 0 & 0 \\ 3\lambda \Delta t & (1 - 2\lambda \Delta t) & 0 \\ 0 & 2\lambda \Delta t & 1 \end{bmatrix} \begin{bmatrix} p_3(t) \\ p_2(t) \\ p_F(t) \end{bmatrix}$$

The resulting matrix equation can be written in a condensed form as

$$\mathbf{P}(t + \Delta t) = \mathbf{A}\mathbf{P}(t)$$

where

$$\mathbf{P}(t + \Delta t) = \begin{bmatrix} p_3(t + \Delta t) \\ p_2(t + \Delta t) \\ p_F(t + \Delta t) \end{bmatrix}$$

$$\mathbf{A} = \begin{bmatrix} (1 - 3\lambda \Delta t) & 0 & 0 \\ 3\lambda \Delta t & (1 - 2\lambda \Delta t) & 0 \\ 0 & 2\lambda \Delta t & 1 \end{bmatrix}$$

$$\mathbf{P}(t) = \begin{bmatrix} p_3(t) \\ p_2(t) \\ p_F(t) \end{bmatrix}$$

$\mathbf{P}(t)$ is the probability state vector at time t, $\mathbf{P}(t + \Delta t)$ is the probability state vector at time $t + \Delta t$, and $\mathbf{A}$ is the transition matrix.

The matrix equations for the Markov model can be viewed as a difference equation for the purpose of obtaining a solution. By assuming some initial value of the probability state vector, $\mathbf{P}(0)$, the value of $\mathbf{P}(\Delta t)$ can be obtained as $\mathbf{P}(\Delta t) = \mathbf{A}\mathbf{P}(0)$. Similarly, the value of the probability state vector at time $2\Delta t$ can be written as $\mathbf{P}(2\Delta t) = \mathbf{A}\mathbf{P}(\Delta t) = \mathbf{A}^2\mathbf{P}(0)$. In general, the solution is given as

$$\mathbf{P}(n\Delta t) = \mathbf{A}^n\mathbf{P}(0)$$

The probability of the system failing is given by the probability of the system being in the failed state. For example, in the TMR illustration, the probability of the system failing is the element of $\mathbf{P}(t)$ given by $p_F(t)$. The reliability of the TMR system can be written as

$$R_{\text{TMR}}(t) = 1 - p_F(t) = p_3(t) + p_2(t)$$

The Markov models considered thus far have been *discrete-time* models in which state transitions occur at fixed intervals Δt. It is possible to model systems using *continuous-time* Markov models in which state transitions can occur at any point in time [Nelson 1986]. The continuous-time equations can be derived from the discrete-time equations by allowing the time interval Δt to approach zero. For example, the equations of the discrete-time Markov model for the TMR system can be written as

$$\frac{p_3(t + \Delta t) - p_3(t)}{\Delta t} = -3\lambda p_3(t)$$

$$\frac{p_2(t + \Delta t) - p_2(t)}{\Delta t} = 3\lambda p_3(t) - 2\lambda p_2(t)$$

$$\frac{p_F(t + \Delta t) - p_F(t)}{\Delta t} = 2\lambda p_2(t)$$

through simple algebraic manipulations. Taking the limit as Δt approaches zero results in a set of differential equations given by

$$\frac{dp_3(t)}{dt} = -3\lambda p_3(t)$$

$$\frac{dp_2(t)}{dt} = 3\lambda p_3(t) - 2\lambda p_2(t)$$

$$\frac{dp_F(t)}{dt} = 2\lambda p_2(t)$$

The simultaneous differential equations can be solved using a number of techniques. For example, if Laplace transforms are used, we have

$$sP_3(s) - p_3(0) = -3\lambda P_3(s)$$
$$sP_2(s) - p_2(0) = 3\lambda P_3(s) - 2\lambda P_2(s)$$
$$sP_F(s) - p_F(0) = 2\lambda P_2(s)$$

where $P_3(s)$ is the Laplace transform of $p_3(t)$, $P_2(s)$ is the Laplace transform of $p_2(t)$, $P_F(s)$ is the Laplace transform of $p_F(t)$, $p_3(0)$ is the initial value of $p_3(t)$ at time $t = 0$, $p_2(0)$ is the initial value of $p_2(t)$ at time $t = 0$, and $p_F(0)$ is the initial value of $p_F(t)$ at time $t = 0$. We assume in the analysis, however, that the system starts in the perfect state at time $t = 0$, so $p_3(0) = 1$, $p_2(0) = 0$, and $p_F(0) = 0$. Consequently, the Laplace transform equations can be written as

$$P_3(s) = \frac{1}{s + 3\lambda}$$

$$P_2(s) = \frac{3\lambda}{(s + 2\lambda)(s + 3\lambda)}$$

$$P_F(s) = \frac{6\lambda^2}{s(s + 2\lambda)(s + 3\lambda)}$$

which can be rewritten as

$$P_3(s) = \frac{1}{s + 3\lambda}$$

$$P_2(s) = \frac{3}{(s + 2\lambda)} + \frac{-3}{(s + 3\lambda)}$$

$$P_F(s) = \frac{1}{s} + \frac{-3}{(s + 2\lambda)} + \frac{2}{(s + 3\lambda)}$$

Taking the inverse Laplace transform results in the solution given by

$$p_3(t) = e^{-3\lambda t}$$

$$p_2(t) = 3e^{-2\lambda t} - 3e^{-3\lambda t}$$

$$p_F(t) = 1 - 3e^{-2\lambda t} + 2e^{-3\lambda t}$$

Recall that the reliability of the TMR system is the probability of being in either state 3 or state 2, so

$$R_{TMR}(t) = p_3(t) + p_2(t) = e^{-3\lambda t} + 3e^{-2\lambda t} - 3e^{-3\lambda t} = 3e^{-2\lambda t} - 2e^{-3\lambda t}$$

which is exactly the same result obtained using the combinatorial techniques. Also note that the sum of $p_3(t)$, $p_2(t)$, and $p_F(t)$ is 1, as expected.

It is interesting to verify that the computer solution of the discrete-time Markov model yields the same results as the equations from the combinatorial model and the continuous-time model. For example, suppose we consider the TMR system. The primary reason for using the TMR system as an example is the relative ease with which both the Markov and the combinatorial models of the TMR system can be constructed. Recall that the combinatorial model of the TMR system that obeys the exponential failure law produces the reliability function

$$R(t) = 3e^{-2\lambda t} - 2e^{-3\lambda t}$$

Table 4.3 shows the values obtained from the combinatorial model of the TMR system at various points in time compared to the values obtained from the computer solution of the discrete-time Markov model of the TMR system. The Markov model was solved by assuming an initial state vector of $P(0) = (100)$ and using a time step, Δt of 0.1 seconds. The failure rate λ has been chosen as 0.1 failures per hour. The differences that exist between the

TABLE 4.3 Comparison of results from computer solution of the discrete-time Markov and the combinatorial model for the TMR system

	Reliability	
Time (t) in minutes	Combinatorial results	Markov results
1	0.99999177	0.99999171
2	0.99996674	0.99996686
3	0.99992549	0.99992561
4	0.99986792	0.99986809
5	0.99979424	0.99979442
6	0.99970472	0.99970472
7	0.99959898	0.99959916
8	0.99947786	0.99947786
9	0.99934101	0.99934095
10	0.99918842	0.99918854

Failure rate λ is 0.1 failures per hour, and time step Δt is 0.1 seconds.

two sets of numbers are within the computational accuracy used to create the numbers.

We have seen how the Markov model can be used to model systems that do not depend on fault coverage or a repair process. Now we want to examine the process of developing a Markov model that depends on the coverage factor. After examining coverage, we will investigate the Markov model of a system with repair.

The system that we wish to model is a triply redundant system that uses fault detection techniques to detect the occurrence of a fault within one of the three independent modules. The modules provide their outputs to a flux-summer such that only one of the three modules must perform correctly for the system to function correctly. Consequently, the system can tolerate as many as two module failures provided that the failures are handled appropriately. The correct way for a module failure to be handled is for the affected unit to be removed from the flux-summing operation by opening a switch. As long as the switch is closed, the associated module provides current to the flux-summer. Once the switch is opened, however, the module is completely disconnected from the flux-summer and no longer affects the operation of the system. The probability that a failure will be correctly handled is the fault coverage and is denoted as C. The basic structure of the system to be analyzed is shown in Fig. 4.19.

The Markov model of this system is similar to that of the basic TMR system with majority voting. In fact, if the coverage factor C becomes zero,

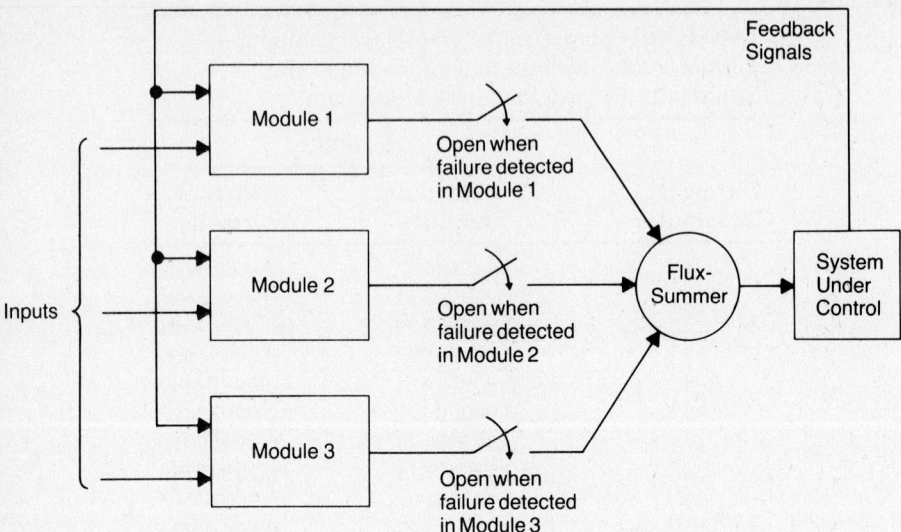

Fig. 4.19 An example hybrid redundancy technique to be used to illustrate the development of a Markov model that includes coverage.

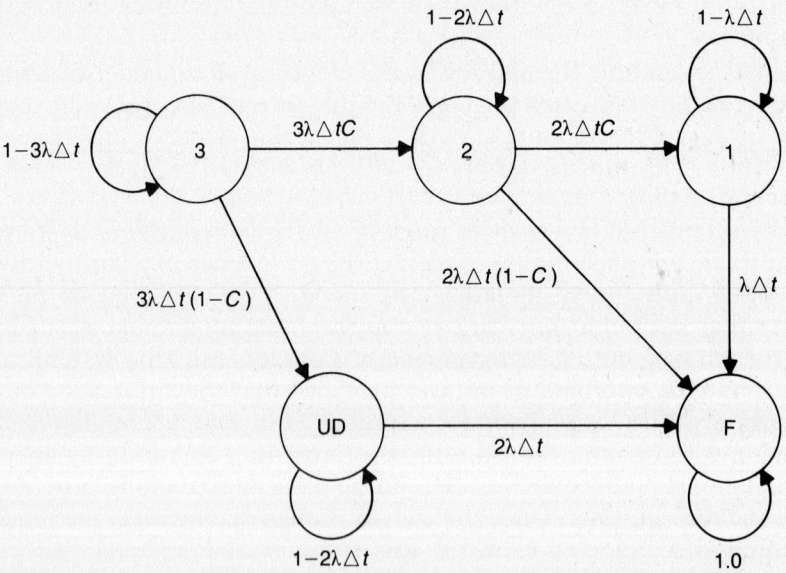

Fig. 4.20 Discrete-time Markov model of the system shown in Fig. 4.19.

the system reduces to the basic TMR system. The Markov model of the system is shown in Fig. 4.20. The system is assumed to begin in the fault-free state, which is labeled as state 3. There are two paths by which the system can exit state 3. The first is shown as a transition to state 2 and corresponds to the failure of one of the three modules and the correct handling of that failure. The second transition that can occur from state 3 is the transition to state UD, which corresponds to one of the three modules failing and the failure going undetected or being handled inappropriately. Once the system enters state UD, it becomes equivalent to the basic TMR system with majority voting; the system cannot tolerate the second failure if the first failure is not handled correctly. The same types of transitions exist from state 2 to state 1 and the failed state. State 2 corresponds to the system having had one module to fail and having handled that failure correctly. While in state 2, the system can tolerate a failure and transition to state 1 provided that the failure is detected and handled correctly. Any undetected failures, however, take the system from state 2 to the failed state. Finally, any failures that occur while the system is in state 1 cause the system to transition immediately to the failed state.

While in any state, there is a nonzero probability that the system will remain in that particular state. For example, if the system is currently in state 3, the system remains in state 3 as long as a failure does not occur. Likewise, the system remains in states 2, 1, and UD if the system is presently in those states and new failures do not occur. The probability of being in the same state at the end of a Δt time period as at the beginning of that time period is calculated as $1 - p_{\text{exit}}(\Delta t)$, where $p_{\text{exit}}(\Delta t)$ is the probability of exiting a state during the Δt time period. For example, the probability of exiting state 3 during a Δt time period is the probability that any one of the three modules will fail. In other terms, the probability of exiting state 3 is $3\lambda \Delta t$. The probability of not exiting state 3 is, therefore, $1 - 3\lambda \Delta t$.

The equations of the Markov model for the system of Fig. 4.19 are developed as they were for the basic TMR system. The probability of being in state i at time $t + \Delta t$ depends on: (1) the probability of being in a state at time t from which the system can transition to state i, and (2) the associated transition probabilities. For example, the system can go to state 2 during a Δt time period if and only if it is in either state 3 or state 2. Therefore, the probability of being in state 2 at time $t + \Delta t$ is

$$p_2(t + \Delta t) = 3\lambda \Delta t C p_3(t) + (1 - 2\lambda \Delta t) p_2(t)$$

The complete set of equations for the Markov model of the system of Fig. 4.19 can be written as

TABLE 4.4 Reliability as a function of fault coverage for the system modeled using the Markov model of Fig. 4.20.

Fault coverage	Reliability (after 1 hour)
0.0	0.97460
0.1	0.97484
0.2	0.97558
0.3	0.97680
0.4	0.97852
0.5	0.98073
0.6	0.98343
0.7	0.98662
0.8	0.99030
0.9	0.99448
1.0	0.99914

Failure rate λ is 0.1 failures per hour, and time step Δt is 0.1 seconds.

$$\begin{bmatrix} p_3(t+\Delta t) \\ p_2(t+\Delta t) \\ p_1(t+\Delta t) \\ p_{UD}(t+\Delta t) \\ p_F(t+\Delta t) \end{bmatrix} = \begin{bmatrix} 1-3\lambda\Delta t & 0 & 0 & 0 & 0 \\ 3\lambda\Delta tC & 1-2\lambda\Delta t & 0 & 0 & 0 \\ 0 & 2\lambda\Delta tC & 1-\lambda\Delta t & 0 & 0 \\ 3\lambda\Delta t(1-C) & 0 & 0 & 1-2\lambda\Delta t & 0 \\ 0 & 2\lambda\Delta t(1-C) & \lambda\Delta t & 2\lambda\Delta t & 1 \end{bmatrix} \begin{bmatrix} p_3(t) \\ p_2(t) \\ p_1(t) \\ p_{UD}(t) \\ p_F(t) \end{bmatrix}$$

The reliability of the system described by the Markov model of Fig. 4.20 is the probability of being in states 3, 2, 1, or UD. In other words, the reliability can be written as

$$R(t) = p_3(t) + p_2(t) + p_1(t) + p_{UD}(t)$$

It is interesting to note the effect that fault coverage has on the reliability of this system. Table 4.4, for example, shows the reliability of the system after one hour as a function of the coverage factor. The time step Δt has been selected as 0.1 seconds and the failure rate λ has been chosen as 0.1 failures per hour. For perfect coverage, the system achieves a reliability of approximately 0.99914, which is the same reliability that can be achieved by the perfect parallel system with three identical modules. When the coverage is zero, the system is identical to the basic TMR system with majority voting. In other words, the system will be a 2-of-3 system where two of the three modules must work for the system to work. The reliability achieved by this system when the coverage is zero is approximately 0.9746. Note in Table 4.4 that the impact of changes in the fault coverage is more significant at the higher values of fault coverage.

We now consider systems that incorporate repair as a form of recovery. For example, many applications require that the repair rate's effect on a system be modeled. A system that possesses a poor repair rate can be required to have fault tolerance to the extent that the system can function while elements are being repaired. The Markov model is an extremely useful tool for analyzing the effect that repair has on a system.

Consider the Markov model of a simple system consisting of one computer and no redundancy. The single computer might be a banking system, for example, and we wish to model the failure and recovery process of this single computer. Further assume that the computer has a constant failure rate λ and a constant repair rate μ. During the time interval Δt, the computer will have a probability of failure given by $\lambda \Delta t$. Since the repair rate is analogous to the failure rate and represents the number of repairs that are expected to occur in a specific time period, the probability of a repair occurring within the time period Δt is $\mu \Delta t$. Using this information allows us to formulate the simple Markov model shown in Fig. 4.21 for the computer. State O represents the condition of the computer being completely operational, and state F represents the failed condition. If the computer system is in state O, the probability of the system transitioning to state F during the time period Δt is $\lambda \Delta t$. Likewise, if the system is in state F, the probability of transitioning to state O is $\mu \Delta t$. As we discovered during previous examples, if the system is in state O and a failure does not occur, the system remains in state O. Similarly, if the system is in state F and a repair does not occur, the system remains in state F.

The equations for the Markov model of Fig. 4.21 can be written as

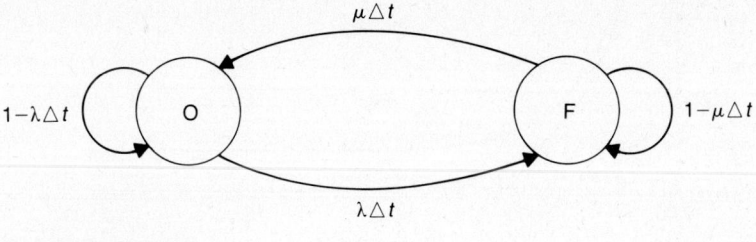

O = Operational State
F = Failed State

Fig. 4.21 Markov model of a simple nonredundant system with repair.

$$\begin{bmatrix} p_O(t + \Delta t) \\ p_F(t + \Delta t) \end{bmatrix} = \begin{bmatrix} 1 - \lambda\Delta t & \mu\Delta t \\ \lambda\Delta t & 1 - \mu\Delta t \end{bmatrix} \begin{bmatrix} p_O(t) \\ p_F(t) \end{bmatrix}$$

It is interesting to solve this Markov model to determine the effect that the repair rate has on the probability of the system being operational. Figure 4.22 shows the plot of the probability of the system being in state O versus the repair rate. The failure rate λ was selected as 0.1 failures per hour, and the time step Δt was chosen as 0.1 seconds for this example. The system was assumed to start in the operational state.

It is also instructive to determine the continuous-time solution for the model shown in Fig. 4.21. Using a procedure identical to that used to determine the continuous-time equations for the TMR system, the equations from the discrete-time Markov model are manipulated algebraically to obtain

$$\frac{p_O(t + \Delta t) - p_O(t)}{\Delta t} = -\lambda p_O(t) + \mu p_F(t)$$

$$\frac{p_F(t + \Delta t) - p_F(t)}{\Delta t} = \lambda p_O(t) - \mu p_F(t)$$

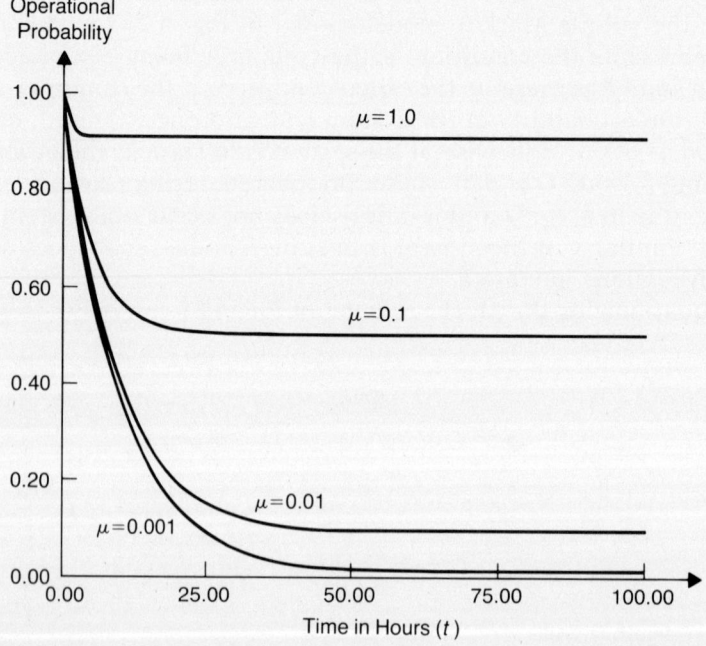

Fig. 4.22 Probability of remaining operational for the system described by the Markov model of Fig. 4.21.

Taking the limit as Δt approaches zero results in the differential equations given by

$$\frac{dp_O(t)}{dt} = -\lambda p_O(t) + \mu p_F(t)$$

$$\frac{dp_F(t)}{dt} = \lambda p_O(t) - \mu p_F(t)$$

Assuming that the initial conditions are $p_O(0) = 1$ and $p_F(0) = 0$ and using Laplace transforms results in the simultaneous equations

$$sP_O(s) = 1 - \lambda P_O(s) + \mu P_F(s)$$

$$sP_F(s) = \lambda P_O(s) - \mu P_F(s)$$

Solving the simultaneous equations for $P_O(s)$ and $P_F(s)$ yields

$$P_O(s) = \frac{1}{s + (\lambda + \mu)} + \frac{\mu}{s(s + (\lambda + \mu))}$$

$$P_F(s) = \frac{\lambda}{s(s + (\lambda + \mu))}$$

which can be rewritten as

$$P_O(s) = \frac{\frac{\mu}{\lambda + \mu}}{s} + \frac{\frac{\lambda}{\lambda + \mu}}{s + (\lambda + \mu)}$$

$$P_F(s) = \frac{\frac{\lambda}{\lambda + \mu}}{s} - \frac{\frac{\lambda}{\lambda + \mu}}{s + (\lambda + \mu)}$$

Taking the inverse Laplace transform results in the time-domain solution given by

$$p_O(t) = \frac{\mu}{\lambda + \mu} + \frac{\lambda}{\lambda + \mu} e^{-(\lambda + \mu)t}$$

$$p_F(t) = \frac{\lambda}{\lambda + \mu} - \frac{\lambda}{\lambda + \mu} e^{-(\lambda + \mu)t}$$

Several interesting features are apparent in the time-domain expressions for $p_O(t)$ and $p_F(t)$. For example, as time approaches infinity, $p_O(t)$ approaches the constant value of

$$p_O(\infty) = \frac{\mu}{\lambda + \mu} = \frac{1}{\frac{\lambda}{\mu} + 1}$$

and $p_F(t)$ approaches

$$p_F(\infty) = \frac{\lambda}{\lambda + \mu} = \frac{1}{\frac{\mu}{\lambda} + 1}$$

As we will discover when discussing availability modeling, the value of $p_O(t)$ as time approaches infinity is the steady-state availability.

4.4 Safety Modeling

The safety of a system, as defined in Chapter 1, is the probability that the system will *either* perform correctly *or* will fail in a safe manner. The concepts of *safe* and *unsafe* are highly dependent upon the application. In many cases, for example, a safe course of action is to simply turn the system off after a failure occurs. In some applications, however, turning the system off can be a disastrous course of action. In any case, however, the fundamental concept of safety analysis is that a system will possess two different ways in which it can fail: one system failure is defined as *safe*, and the other is categorized as *unsafe*. The definition of safe and unsafe failures must be created uniquely for each application.

Safety can be modeled using Markov models by splitting the system failed state into two separate states. One failed state is normally labeled *FS* for *failed safe*, and the other failed state is labeled as FU for *failed unsafe*. A Markov model for a simple system containing one hardware module with a failure rate of λ and self-diagnostics with a fault detection coverage of C is shown in Fig. 4.23. Safe failures are defined in this example as those that are detected by the self-diagnostics. Consequently, unsafe failures are defined as those that are not detected by the self-diagnostics. If a failure occurs, the system transitions to either state *FS* or *FU* depending on whether or not the condition is detected.

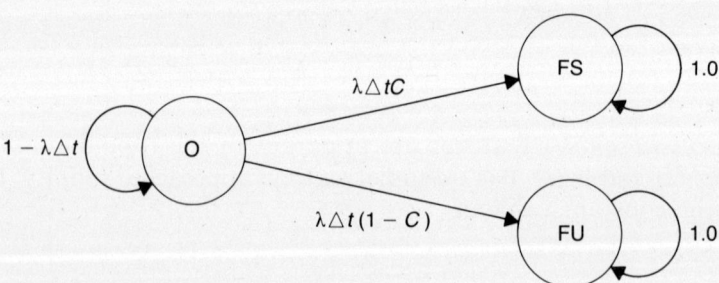

Fig. 4.23 Three-state Markov model allowing safety calculations.

4.4 ■ Safety Modeling

The safety of the system described by the Markov model of Fig. 4.23 can be written as

$$S(t) = p_O(t) + p_{FS}(t)$$

where $S(t)$ is the safety, $p_O(t)$ is the probability of being in the operational state at time t, and $p_{FS}(t)$ is the probability of being in the failed safe state at time t. The complete equations of the discrete-time Markov model can be written as

$$p_O(t + \Delta t) = (1 - \lambda \Delta t)p_O(t)$$
$$p_{FS}(t + \Delta t) = \lambda \Delta t C p_O(t) + p_{FS}(t)$$
$$p_{FU}(t + \Delta t) = \lambda \Delta t (1 - C) p_O(t) + p_{FU}(t)$$

As we have done previously, the differential equations of the corresponding continuous-time Markov model can be written as

$$\frac{dp_O(t)}{dt} = -\lambda p_O(t)$$

$$\frac{dp_{FS}(t)}{dt} = \lambda C p_O(t)$$

$$\frac{dp_{FU}(t)}{dt} = \lambda (1 - C) p_O(t)$$

Taking the Laplace transform results in

$$P_O(s) = \frac{p_O(0)}{s + \lambda}$$

$$P_{FS}(s) = \frac{\lambda C p_O(0)}{s(s + \lambda)} + \frac{p_{FS}(0)}{s}$$

$$P_{FU}(s) = \frac{\lambda (1 - C) p_O(0)}{s(s + \lambda)} + \frac{p_{FU}(0)}{s}$$

where $p_O(0)$, $p_{FS}(0)$, and $p_{FU}(0)$ are the intial values of the respective state probabilities. If we assume that the system begins in state O such that $p_O(0) = 1$, $p_{FS}(0) = 0$, and $p_{FU}(0) = 0$, we obtain

$$P_O(s) = \frac{1}{s + \lambda}$$

$$P_{FS}(s) = \frac{\lambda C}{s(s + \lambda)} = \frac{C}{s} - \frac{C}{s + \lambda}$$

$$P_{FU}(s) = \frac{\lambda (1 - C)}{s(s + \lambda)} = \frac{(1 - C)}{s} - \frac{(1 - C)}{s + \lambda}$$

The time domain solutions can now be written as
$$p_O(t) = e^{-\lambda t}$$
$$p_{FS}(t) = C - Ce^{-\lambda t}$$
$$p_{FU}(t) = (1 - C) - (1 - C)e^{-\lambda t}$$

Intuitively, the equations are as expected. For example, the reliability of the system is
$$R(t) = p_O(t) = e^{-\lambda t}$$
and the probability of being in one of the two failed states is
$$p_{FS}(t) + p_{FU}(t) = 1 - e^{-\lambda t} = 1 - R(t)$$

The safety of the system is written as
$$S(t) = p_O(t) + p_{FS}(t) = C + (1 - C)e^{-\lambda t}$$

At time $t = 0$, the safety of the system is 1, as expected. As time approaches infinity, however, the safety approaches
$$S(\infty) = C$$

In other words, if the fault detection coverage is perfect ($C = 1$), the system has perfect safety. However, if the fault detection coverage is nonexistent ($C = 0$), the system will eventually fail in an unsafe manner. The safety of the system, in this example, is directly dependent upon the fault detection coverage. In subsequent sections, we investigate the safety of more complex systems.

4.5 System Comparisons

Now that we have several tools at our disposal, we can begin to examine the process of comparing two or more systems. When we make comparisons, we must be careful about the parameters that we choose to compare. For example, if we choose to compare the MTTF of two systems or the reliability of the systems, the results can be surprising. Suppose that we wish to compare a simplex system consisting of a single computer to a TMR system with three computers in a majority voting arrangement. Assume for simplicity that the majority voter is perfect. The two systems are shown in Fig. 4.24. The computers in each system are identical and are assumed to obey the exponential failure law.

Recall that the MTTF of a system is defined as
$$\text{MTTF} = \int_0^\infty R(t)\,dt$$

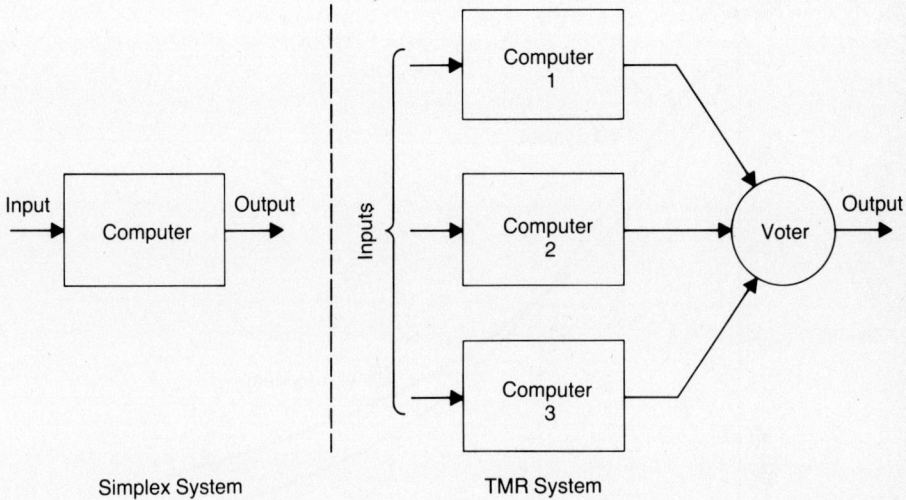

Fig. 4.24 Simplex and TMR systems to be compared to assess the benefits of redundancy.

where $R(t)$ is the reliability function of the system. The reliability function of the simplex system is

$$R_{\text{simplex}}(t) = e^{-\lambda t}$$

whereas that of the TMR system is

$$R_{\text{TMR}}(t) = 3e^{-2\lambda t} - 2e^{-3\lambda t}$$

If we integrate $R_{\text{simplex}}(t)$ and $R_{\text{TMR}}(t)$, we find that the MTTF of each system is given by

$$\text{MTTF}_{\text{simplex}} = \int_0^\infty e^{-\lambda t}\,dt = \frac{1}{\lambda}$$

$$\text{MTTF}_{\text{TMR}} = \int_0^\infty (3e^{-2\lambda t} - 2e^{-3\lambda t})\,dt = \frac{5}{6\lambda}$$

Thus, the MTTF of the TMR system is lower than the MTTF of the simplex system.

Based on these calculations, we might conclude that the TMR system is not as good as the simplex system. This may or may not be a correct conclusion depending on the application and the length of time the system is expected to operate correctly. Figure 4.25 provides a good insight into the reason why the MTTF of the TMR system is less than that of the simplex system. Figure 4.25 shows the reliability of the simplex and the TMR sys-

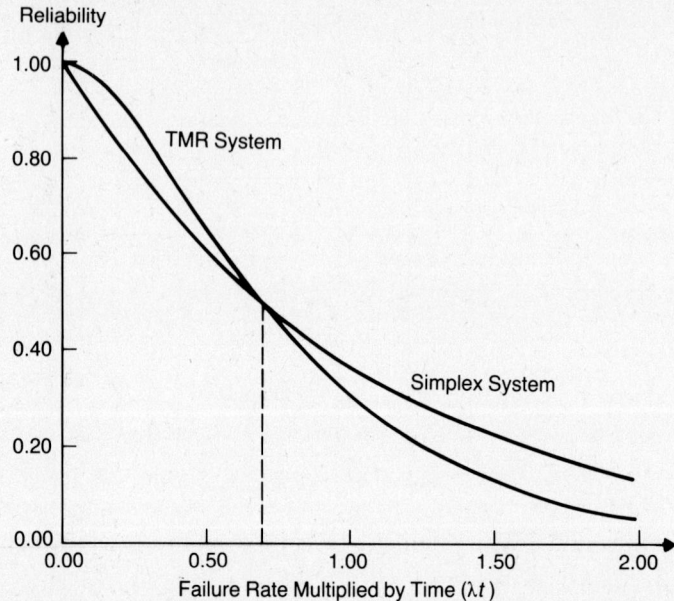

Fig. 4.25 Simplex and TMR reliabilities versus λt—a crossover point occurs where the two reliabilities are equal.

tems as functions of λt. As can be seen, a point in time is reached when the TMR system becomes less reliable than the simplex system. The MTTF is the area under the reliability curve, and that area is larger for the simplex than for the TMR configuration. The point of this discussion is that the MTTF can sometimes misrepresent the quality of a system. For certain values of the product λt, the TMR system's reliability will be superior to that of the simplex system. Regardless of the reliability, if fault tolerance is necessary, the TMR system will be superior to the simplex system.

Sometimes, a single parameter comparison that is better than the MTTF is the **mission time,** denoted as MT[r]. The mission time is the time at which the reliability of a system falls below the level r. For example, a simplex system that obeys the exponential failure law has a reliability of r when

$$r = e^{-\lambda t}$$

The time at which the reliability value of r occurs can be found by taking the natural logarithm of both sides of the preceding equation and solving the resulting equation. The solution yields

$$\text{MT}[r] = \frac{-\ln(r)}{\lambda}$$

A **mission time improvement** can be calculated as the ratio of the mission times of the two systems being compared. For example, suppose that we wish to compare two computer systems: (1) a simplex computer system with a single computer that has a failure rate of 0.01 failures per hour and (2) a TMR system constructed using three of that same computer. The computers are assumed to obey the exponential failure law. We wish to determine the mission time improvement of the TMR system over the simplex system for a reliability of 0.86.

The $MT_{simplex}[0.86]$ is fairly easy to calculate from the exponential reliability function as

$$MT_{simplex}[0.86] = \frac{-\ln(r)}{\lambda} = \frac{-\ln(0.86)}{0.01} = 15.08 \quad \text{hours}$$

The $MT_{TMR}[0.86]$ is found from the solution of the equation

$$R_{TMR}(t) = 3e^{-2 \times 0.01 t} - 2e^{-3 \times 0.01 t} = 0.86$$

for the time t, which results in $MT_{TMR}[0.86] = 27$ hours. The mission time improvement of the TMR system over the simplex system is approximately 1.8. In other words, the TMR system, in this example, can operate 1.8 times as long as the simplex system while still maintaining a reliability of greater than 0.86. The graph shown in Fig. 4.26 illustrates the concepts of the mission time and the mission time improvement for the comparison of the simplex and the TMR systems with the failure rate of 0.01 failures per hour.

4.6 Availability Models

Thus far, we have considered only the modeling of the reliability of a system. However, we have seen in the discussions of Chapters 1 and 2 that parameters such as availability and maintainability are also important in the analysis of fault-tolerant systems. Many computer companies are concerned more with the probability of their systems being available when their customers want to use them (availability) rather than with the length of time the system can operate without failure (related to reliability). As a result, the rate at which a system can be repaired becomes a critical part of the design. The repair rate can dramatically affect the availability of a system.

Recall that the availability $A(t)$ of a system is defined as the probability that a system will be available to perform its tasks at the instant of time t. Intuitively, we can see that the availability can be approximated as the total time that a system has been operational divided by the total time elapsed since the system was initially placed into operation. In other words, the availability is the percentage of time that the system is available to per-

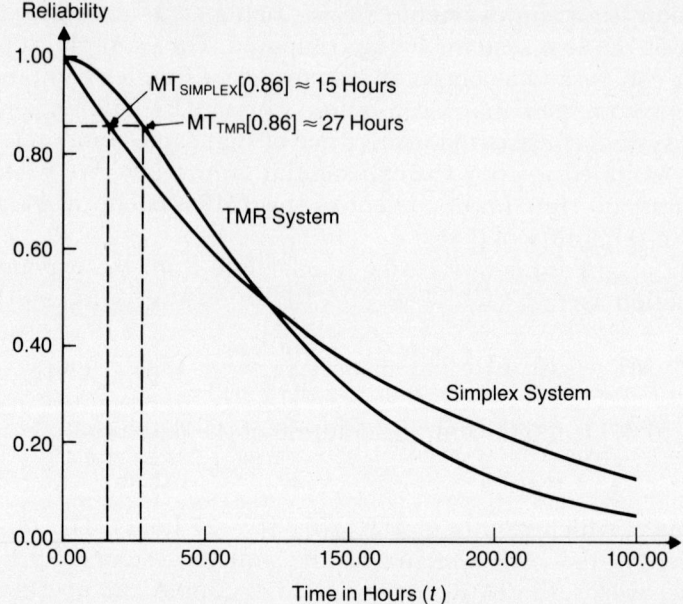

Fig. 4.26 Mission time improvement comparison between simplex and TMR.

form its expected tasks. Suppose that we place a system into operation at time $t = 0$. As time moves along, the system performs its functions, perhaps fails, and is repaired. At some time $t = t_{current}$, suppose that the system has operated correctly for a total of t_{op} hours and has been in the process of repair or waiting for repair to begin for a total of t_{repair} hours. The time $t_{current}$ is then the sum of t_{op} and t_{repair}. The availability can be determined as

$$A(t_{current}) = \frac{t_{op}}{t_{op} + t_{repair}}$$

where $A(t_{current})$ is the availability at time $t_{current}$.

The preceding expression lends itself well to the experimental evaluation of the availability of a system; we can simply place the system into operation and measure the appropriate times required to calculate the availability of the system at a number of points in time. Unfortunately, the experimental evaluation of the availability is often not possible because of the time and expense involved. Also, we would like to have some means of estimating the availability before we actually build the system so that availability considerations can be factored into the design process. We will consider two approaches. The first is based on the single parameter measures such as MTTF and MTTR and yields what is typically called the

steady-state availability A_{ss}. The second approach uses the failure rates and the repair rates in a Markov model to calculate the availability as a function of time.

We have seen that availability is basically the percentage of time that a system is operational. Using knowledge of the statistical interpretation of the MTTF and the MTTR, we expect that, on the average, a system will operate for MTTF hours and then encounter its first failure. Once the failure has occurred, the system will then, again on the average, require MTTR hours to be repaired and placed into operation once again. The system will then operate for another MTTF hours before encountering its second failure. This concept has been illustrated in Fig. 4.2.

If the average system experiences N failures during its lifetime, the total time that the system is operational is $N(\text{MTTF})$ hours. Likewise, the total time that the system is "down" for repairs is $N(\text{MTTR})$ hours. In other words, the operational time t_{op} is $N(\text{MTTF})$ hours and the down-time t_{repair} is $N(\text{MTTR})$ hours. The average, or steady-state, availability is

$$A_{ss} = \frac{N(\text{MTTF})}{N(\text{MTTF}) + N(\text{MTTR})} = \frac{\text{MTTF}}{\text{MTTF} + \text{MTTR}}$$

We know, however, that the MTTF and the MTTR of a simplex system are related to the failure rate and the repair rate, respectively, as

$$\text{MTTF} = \frac{1}{\lambda}$$

$$\text{MTTR} = \frac{1}{\mu}$$

Therefore, the steady-state availability of a simplex system is given by

$$A_{ss} = \frac{\frac{1}{\lambda}}{\frac{1}{\lambda} + \frac{1}{\mu}} = \frac{1}{1 + \frac{\lambda}{\mu}}$$

Recall that the repair rate μ is expressed in repairs per hour, whereas the failure rate λ is in failures per hour. We would expect that if the failure rate goes to 0, implying that the system never fails, or the repair rate goes to infinity, implying that no time is required to repair the system, the availability will go to 1. Looking at the expression for the steady-state availability, we can see that this is true.

As an example calculation, consider a computer system that has a failure rate of one failure every 100 hours and a repair rate of one repair every 10 hours. The failure rate of this system is $\lambda = 0.01$ failures per hour and the repair rate is $\mu = 0.1$ repairs per hour. The steady-state availability is calculated as

$$A_{ss} = \frac{1}{1 + \frac{0.01}{0.1}} = 0.90909$$

This implies that the system is available for use an average of slightly more than 90% of the time.

Now suppose we investigate the use of the Markov model as a means of determining the availability of a system. We already have the necessary tools to accomplish this task. The Markov model of a system with repair is in fact the model required to calculate the availability of a system. Recall the two-state model of a simple system with repair and having a failure rate of λ and a repair rate of μ. The state diagram of this model is repeated in Fig. 4.27 for convenience. State O represents the state in which the system is completely operational, whereas state F is the state in which the system has failed and is in the process of being repaired. The probability of the system failing during the time interval Δt is given by $\lambda \Delta t$, whereas the probability of the system being repaired during the time interval Δt is $\mu \Delta t$.

The equations of the Markov model are given by

$$\begin{bmatrix} p_O(t + \Delta t) \\ p_F(t + \Delta t) \end{bmatrix} = \begin{bmatrix} 1 - \lambda \Delta t & \mu \Delta t \\ \lambda \Delta t & 1 - \mu \Delta t \end{bmatrix} \begin{bmatrix} p_O(t) \\ p_f(t) \end{bmatrix}$$

where $p_O(t)$ is the probability, at time t, that the system is in the operational state and is, therefore, available to perform its tasks. Consequently, $p_O(t)$ is the availability of the system.

As an example, the Markov model shown in Fig. 4.27 has been solved for the failure rate of $\lambda = 0.01$ failures per hour and the repair rate of $\mu = 0.1$ repairs per hour. The plot of the resulting availability is shown in Fig. 4.28. Note that the availability approaches the value of 0.90909 that was previ-

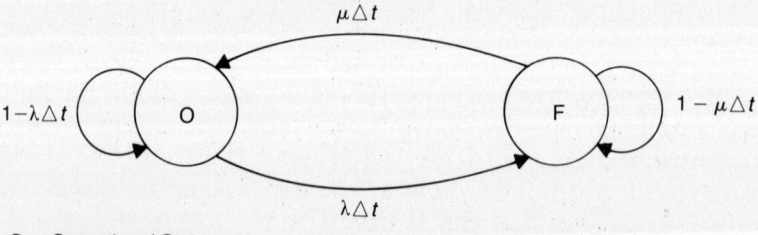

O = Operational State
F = Failed State

Fig. 4.27 Markov model of a simple nonredundant system with repair.

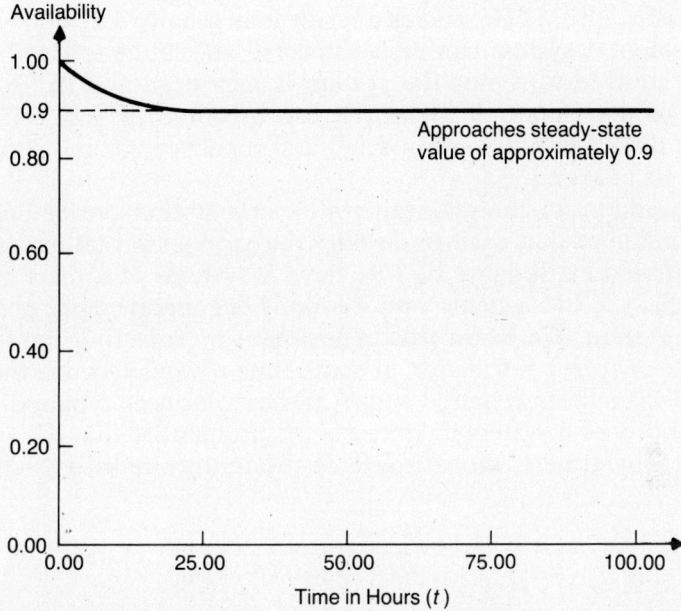

Fig. 4.28 Availability versus time for a simple nonredundant system with $\lambda = 0.01$ failures per hour and $\mu = 0.1$ repairs per hour.

ously determined as the steady-state availability of a system with this particular failure rate and repair rate.

4.7 Maintainability Models

As defined in Chapter 1, the maintainability is the probability that a failed system will be restored to working order within a specified time. We will use the notation that $M(t)$ is the maintainability for time t. In other words, $M(t)$ is the probability that a system will be repaired in a time less than or equal to t.

An important parameter in maintainability modeling is the repair rate μ. The repair rate is the average number of repairs that can be performed per time unit. The inverse of the repair rate is the MTTR, which is the average time required to perform a single repair. Mathematically, the relationship between the repair rate and the MTTR is given by

$$\text{MTTR} = \frac{1}{\mu}$$

In industry, the MTTR, and as a result μ, is usually derived in an experimental fashion. A system can be constructed and faults injected; the average time required to repair the system is measured and recorded as the MTTR. A good estimate of the MTTR can be obtained only if a sufficient number of different faults are injected and repair personnel with a variety of skill levels are used.

An expression for the maintainability of a system can be derived in a manner similar to that used to develop the exponential failure law for the reliability function. Suppose that we have N systems. We inject one unique fault into each of the systems, and we allow one maintenance person to repair each system. We begin this experiment by injecting the faults into the systems at time $t = 0$. Later, at some time t, we determine that $N_r(t)$ of the systems have been repaired and $N_{nr}(t)$ have not been repaired. Since the maintainability of a system at time t is the probability that the system can be repaired by time t, an estimate of the maintainability can be computed as

$$M(t) = \frac{N_r(t)}{N} = \frac{N_r(t)}{N_r(t) + N_{nr}(t)}$$

If we differentiate $M(t)$ with respect to time, we obtain

$$\frac{dM(t)}{dt} = \frac{1}{N} \frac{dN_r(t)}{dt}$$

which can also be written as

$$\frac{dN_r(t)}{dt} = N \frac{dM(t)}{dt}$$

The derivative of $N_r(t)$ is simply the rate at which components are repaired at the instant of time t.

At time t, we have $N_{nr}(t)$ systems that have not been repaired. If we divide $dN_r(t)/dt$ by $N_{nr}(t)$, we obtain

$$\frac{1}{N_{nr}(t)} \frac{dN_r(t)}{dt}$$

which is called the *repair rate function* and is assumed to have a constant value of μ, the repair rate; μ has the units of repairs per unit of time.

Using the expression for the repair rate and the expression for the derivative of $N_r(t)$, we can write

$$\mu = \frac{1}{N_{nr}(t)} \frac{dN_r(t)}{dt} = \frac{N}{N_{nr}(t)} \frac{dM(t)}{dt}$$

which yields a differential equation of the form

$$\frac{dM(t)}{dt} = \mu \frac{N_{nr}(t)}{N}$$

We know, however, that $N_{nr}(t)/N$ is $1 - M(t)$, so we can write

$$\frac{dM(t)}{dt} = \mu(1 - M(t))$$

The solution to the differential equation is well known and is given by

$$M(t) = 1 - e^{-\mu t}$$

The relationship developed for $M(t)$ has the desired characteristics. First, if the repair rate is zero, the maintainability is also zero since the system cannot be repaired in any length of time. Second, if the repair rate is infinite, the maintainability is one since repair can be performed in zero time. A final interesting feature of the maintainability function is its value at a time corresponding to the MTTR. At $t =$ MTTR, the maintainability function will be

$$M(t=\text{MTTR}) = 1 - e^{-\mu 1/\mu} = 1 - e^{-1} = .632$$

which implies that there is a probability of 0.632 that a system will be repaired in a time less than or equal to its MTTR.

As we have seen, the repair rate plays a crucial role in the maintainability of a system. The repair rate can differ depending on the type of repair that must be performed. For example, a banking system that can be repaired on location by a local maintenance person will have a better repair rate than one that must be returned to the factory or some third party for repair. In addition, certain types of faults can be easily repaired on location, whereas others require facilities that are not practical to bring to the location of the electronic system. For example, the replacement of a memory card can be performed easily at the site of the system, but the replacement of the power supply and cooling system can be much more difficult.

Because of the preceding issues, the repair rate is typically specified for several levels of repair. The most common partitioning is to provide three levels of repair. The first is called the *organizational level* and consists of all repairs that can be performed at the site where the system is located. Organizational repairs typically include all faults that can be located to specific circuit cards such that the cards can simply be replaced and the system made operational once again. For example, if an aircraft can be repaired without bringing it off the runway, it is considered an organizational level repair. The key to organizational repairs is the ability to locate the fault. It is seldom feasible to bring sophisticated fault detection and location equipment to the site of the system. Repairs at the organizational level must often depend on the built-in test provided by the system to locate the specific problem.

The second level of repair is called the *intermediate level*. Intermediate level repairs cannot be performed at the organizational level, but they can be performed in the immediate vicinity of the system. For example, a computer firm can have a local repair facility to which the faulty pieces of equipment are taken for repair. Intermediate level repair is not as good as being able to perform the repair on site, but it is better than having to return a piece of equipment to the factory. In the case of an airplane, for example, an intermediate level repair might be made in the hangar as opposed to on the runway.

The final level of repair is called the *depot level* or the *factory level*. In depot level repairs, the equipment must be returned to a major facility for the repair process. For example, if a calculator cannot be repaired at home (organizational level), it is taken to the store from which it was purchased (intermediate level). If the store is unable to perform the repair, they send the calculator to a site designated by the manufacturer as a major repair facility (depot level). The length of time required to perform the repair depends on the level at which it is performed. It may take less than an hour to repair a device at the organizational level, several hours or perhaps days at the intermediate level, and as much as several weeks or months at the depot level.

As an example, assume that the MTTR for a computer system is 2.0 hours at the organizational level, 8.0 hours at the intermediate level, and one week (168 hours) at the depot level. The resulting maintainability functions are plotted versus time in Fig. 4.29. Note the tremendous difference that exists between the maintainability of the system for the different levels of repair.

4.8 Redundancy Ratios

A system that is more reliable or more available than another is better with regards to that one attribute. However, to achieve the improved reliability, availability, or maintainability, the system may contain an excessive amount of redundancy. The cost of the extra redundancy will appear in the weight, size, power consumption, volume, and financial costs of the improved system. In many applications, the improvements in the reliability, for example, may not be worth the extra weight that the system contains.

One good measure of the impact that improvements in reliability, availability, and maintainability have on a system is the **redundancy ratio**. The redundancy ratio is defined simply as the amount of hardware, information, time, or software that the redundant system requires divided by the amount required in a nonredundant system that performs the same function. The redundancy ratio can be specified for each type of hardware component; for example, the processors, memory, buses, interface units, power supplies, and displays. The redundancy ratio gives a measure of the extra resources required for a given application.

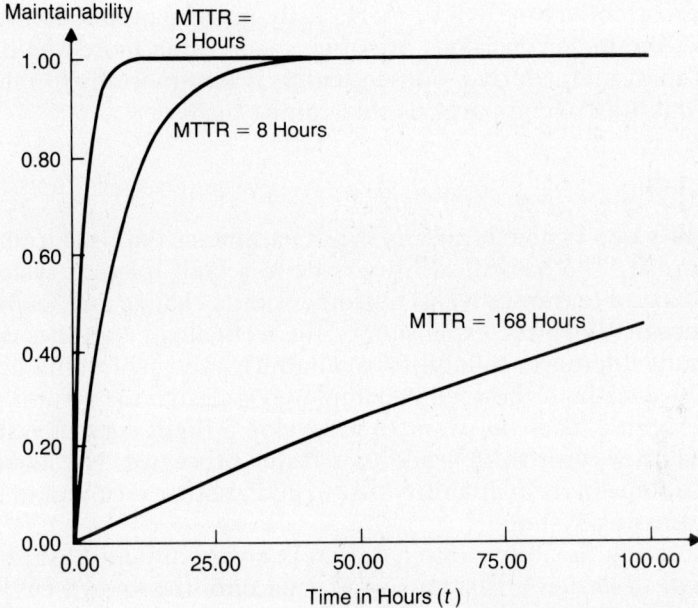

Fig. 4.29 Maintainability $M(t)$ versus time for example values of the MTTR.

A good example of the redundancy ratio can be found in a TMR processor system. If the nonredundant system requires only a single processor, the TMR system will clearly require three and will have a redundancy ratio for the processors of three. Redundancy ratios are equally applicable to software as well as hardware. If a nonredundant system requires 30,000 lines of code and the redundant system requires 40,000 lines, the software redundancy ratio is 1.33. In other words, one-third more software is required to implement the redundant system than is required in the nonredundant system.

4.9 Qualitative Methods

Thus far, we have discussed methods of evaluation that generate specific numbers to use to compare two or more systems. For example, one system may have a higher reliability, another may be less expensive, a third may weigh less, and still another may consume less power. Often, we find that certain attributes of a system that enter into the design process are extremely difficult to quantify. We may anticipate a drop in the prices of some hardware at a future date, or we may feel that the production plan on a given system architecture will fit more easily into the production resources. In practice, the major decisions are often made using more qualitative information than quantitative. Consequently, it is important to understand the many qualitative comparisons that can be made.

4.9.1 Flexibility

The **flexibility** of a system is an important parameter that is extremely difficult to quantify. The basic flexibility issue in a fault-tolerant system is the ability to expand and improve as customer needs change and technological advances occur. When the expansion or the technology upgrades occur, the system's fault tolerance, reliability, availability, and maintainability must be maintained without the need to completely redesign the system. The military, for example, does not want to redevelop a flight control system simply because they wish to upgrade to a faster processor. Nor does a bank wish to lose some of its availability just to add another terminal to its transactions processing system.

Incorporating flexibility into a design is a very difficult task; it is often hard enough to design a system considering only the known environment and technology. To consider possibly unknown circumstances and capabilities to achieve a desired level of flexibility makes the design process even more difficult.

4.9.2 Technology Dependence

The dependence of a system on the technology employed is closely related to the flexibility of the system, but it must often be considered as a separate entity. Many systems are closely tied to the technology. For example, many aircraft flight control systems must use *flight-qualified* technology. In many cases, the technology is five to ten years old before it can meet the requirements necessary to become flight-qualified. Therefore, the designs developed using flight-qualified technology may lack the capability that can be obtained with other technology. **Technology dependence** must be considered when evaluating fault-tolerant systems. A system that is significantly more reliable, lighter, less expensive, and consumes less power may not be selected because it cannot be developed using currently qualified technology. The system's developers may not be willing to take a risk on the required technology becoming qualified before the system must be placed into operation.

4.9.3 Transparency to the User

A factor that is often overlooked in the design of a system is the impact that the system's characteristics have on the user. A system that is difficult to use will be doomed by end users' complaints. The attribute of fault tolerance often affects the end user of a system. For example, if the user's programs on a fault-tolerant system are impacted by the fact that there are redundant memories in the system, the inclusion of fault tolerance has had a tremendous effect on the user. For example, users of the system typically do not want to have to learn much about the system's fault tolerance characteristics in order to effectively use the system. Instead, the user wants to, for example, develop programs, send electronic mail, and perform database searches without any knowledge of the fault detection, voting, or error correction that is occurring within the system.

4.9.4 Testability

Testing is sometimes a painful process because of the lack of preparation for the test procedure. Experience over the past two decades has shown that the test procedure and the design process cannot be independent; instead, the two procedures must be coupled tightly to allow the design to be developed with **testability** in mind.

In a fault-tolerant system, testing is particularly important. Many fault tolerance techniques are designed to hide the occurrence of faults or errors. The test process, however, has completely the opposite goal. Testing attempts to make errors appear at an observable output of a system. In a fault-tolerant design, some means of easily testing the system must be developed. The solutions can be very simple in nature. For example, fault masking makes the TMR approach to achieve fault tolerance difficult to test. Even if one fault occurs, the TMR system still continues to produce the correct output because of the fault masking that is inherent in the majority voting. Attempts to test a TMR system by simply looking at the output as a function of the input can be extremely complicated or impossible. One approach to overcome the testing problem of the TMR system is to simply provide the inputs to the majority voter as primary outputs of the system. As a result, each of the three modules of the TMR system can be tested independently. The inputs to the voter should also be controllable by an external source to allow the voter to be tested independently of the remainder of the system.

4.10 Tradeoff Analysis Example

As with any new material, it is important to see the material applied before a true understanding can be obtained. To accomplish this goal, we examine a practical example that uses some of the evaluation tools that have been presented in this chapter [Johnson and Aylor 1986]. The example involves determining the benefits of including redundancy in the design of an electronic controller. We are interested in obtaining reliability, availability, and safety improvements. We will look at two aspects of the problem. In the first, we use combinatorial modeling techniques to estimate the reliability and availability improvements of including redundancy in the system. In the second, we use Markov models to explore the safety aspects of two architectures for designing the redundant system. We begin our discussions with a brief description of the system.

The electronic controller that is to be designed controls the velocity of a direct current (dc) motor. The structure of the system is similar to that found in electric vehicle control systems and other applications that have a human providing commands to a system and a digital controller modifying those commands to guarantee the stability and acceptable response of the system. Example applications include electric wheelchairs, process control systems, robotic systems, remotely controlled vehicles, and certain aircraft control systems. The digital electronics are critical to the performance of the system, and a failure of the electronics can result in harm to either the system under control or the human.

The control system consists of five essential components: the user command sensors, controller electronics, motor drive electronics, motor feedback sensors, and the electric motors. The command sensors are typically potentiometers contained within joysticks. The velocity feedback sensors can be tachometers, back electromagnetomotive force (EMF) sensors, or other motor speed detection devices. The controller electronics use the driver commands and the feedback signals to develop input signals for the motor drive electronics. The motor drives are typically transistor or relay bridges that control the magnitude and direction of current flow to the direct current (dc) motors. A block diagram of the basic control system is shown in Fig. 4.30.

The core of the control system is the electronic controller. Any signal processing required within the system is normally performed by the controller electronics. For example, the feedback or command signals often require filtering to remove unwanted, and possibly detrimental, noise. Also, any velocity feedback that is implemented will be a part of the controller electronics. All of the control applications for this system use some form of velocity or position feedback to improve their response.

In the design of the electronic controller, safety, reliability, and availability are the three issues of primary importance. High availability is required to ensure that the system is operational for a very high percentage of the time. The user of the control system wants it to be functional and *ready to run* when its services are needed.

The electronic controller must also be safe. A failure of the controller must not result in the user, or the system itself, being hurt in any way. Faults that result in system failures have the potential to compromise the safety of the system, the user, or both. If a system is nonredundant, all failures have the potential to compromise the safety of the system. A system that contains fault detection capability, however, may be able to maintain

4.10 ■ Tradeoff Analysis Example

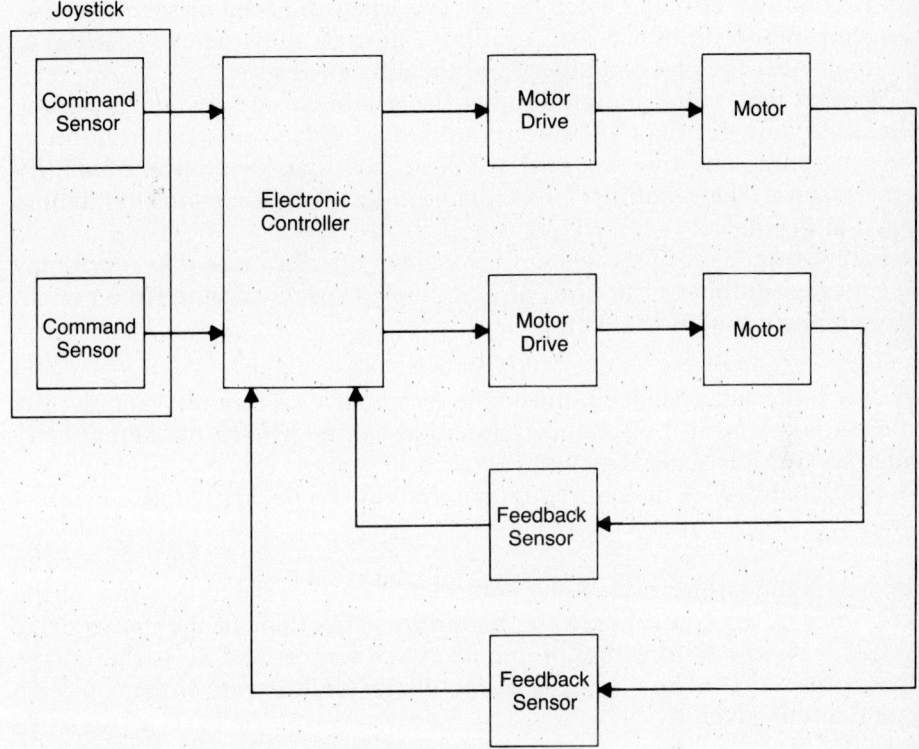

Fig. 4.30 The block diagram of the control system.

the safety of the system even though the fault is not tolerated. In essence, safety is the ability of a system to be fail-safe. Even if a fault results in the controller becoming inoperative, the fault should not cause the user to lose control of the system.

The reliability of the system is important because of the harm that can result if the system performs incorrectly. The manufacturer of the system depends on the analysis for some indication of the probability of system failure.

Specific requirements have not yet been derived for the reliability, availability, and safety of the system. The designer of the system is more interested, at this time, in the difference in these attributes between the redundant and the nonredundant systems.

The issues of reliability and availability are certainly not new ones. As we have seen in this chapter, commercial banking systems, industrial controllers, and military applications have always had stringent availability and reliability requirements. Military aircraft, for example, must have a high probability of being available for use when the need arises. Likewise, the electronics found on board a military aircraft must be extremely reliable to protect the crew members and the aircraft itself.

We first look at the computation of the reliability and availability of the controller that does not include redundancy. We can use our reliability block diagrams and the exponential failure law to calculate the reliability of the system. The reliability block diagram for the basic control system is shown in Fig. 4.31.

Each component of the system is assigned a failure rate that represents the expected number of failures of that element per a specific time period. The reliability $R(t)$ of that component is then

$$R(t) = e^{-\lambda t}$$

where λ is the component's failure rate. As we have seen in this chapter, the exponential form of the reliability function can be proved mathematically under several fairly loose assumptions.

The reliability of the nonredundant system can be written as

$$R_{\text{system}}(t) = e^{-\lambda_i t} e^{-\lambda_i t} e^{-\lambda_e t} e^{-\lambda_d t} e^{-\lambda_d t} e^{-\lambda_m t} e^{-\lambda_m t} e^{-\lambda_f t} e^{-\lambda_f t} = e^{-\lambda_{\text{system}} t}$$

where λ_i is the failure rate of one potentiometer, λ_e is the failure rate of the electronics of the controller, λ_d is the failure rate of one of the motor drive circuits, λ_f is the failure rate of one feedback sensor, and λ_m is the failure rate of one of the motors. λ_{system} is the equivalent failure rate of the complete system and is given by

$$\lambda_{\text{system}} = 2\lambda_i + \lambda_e + 2\lambda_d + 2\lambda_m + 2\lambda_f$$

To allow specific data to be examined, consider a microprocessor-based controller. Assume that the controller consists of one 16-bit processor (such as an Intel 8086, Zilog Z8000, or Motorola 68000), 6K of random access memory (RAM), 64K of erasable, programmable read-only memory (EPROM), a 16-channel multiplexer and analog-to-digital converter, and two channels of digital-to-analog conversion. The motor drives are assumed to be transistor bridge circuits, and the motors are 24-volt, permanent-magnet, direct current motors. The input sensors are analog potentiometers, and the feedback sensors are analog circuits that measure the back EMF.

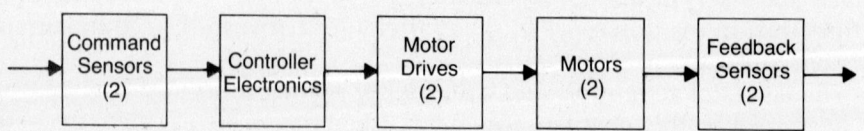

Fig. 4.31 Reliability block diagram for the nonredundant controller of Fig. 4.29.

The failure rates of the individual components are:

$$\lambda_i = 10^{-6} \text{ failures per hour}$$
$$\lambda_e = 3.5 \times 10^{-4} \text{ failures per hour}$$
$$\lambda_d = 5 \times 10^{-5} \text{ failures per hour}$$
$$\lambda_m = 10^{-6} \text{ failures per hour}$$
$$\lambda_f = 10^{-6} \text{ failures per hour}$$

Using the given failure rate data results in a reliability function for the system of

$$R(t)_{system} = e^{-4.56 \times 10^{-4} t}$$

A plot of the reliability function versus time is shown in Fig. 4.32. Figure 4.32 shows the probability that the control system will *not* fail as a function of time. For example, if no maintenance is performed on the system during one month of continuous operation, the probability that a failure will not occur within that month is approximately 0.72. Stated differently, the user has a probability of 0.28 of the system failing within a month of continuous operation.

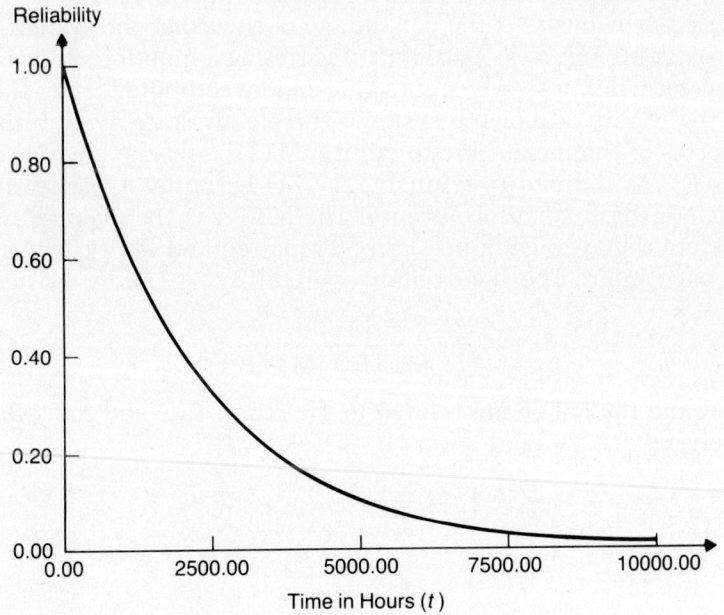

Fig. 4.32 Reliability versus time for the nonredundant system ($R(t) = e^{-4.56 \times 10^{-4} t}$).

The times shown in Fig. 4.32 are actual operation times. It is assumed that the system has a zero probability of failing while it is not operating. If we assume that the system is used approximately four hours per day, the reliability figures indicate that the user will have about 382 days of use before the probability of failure rises above 0.5.

Another way of looking at the reliability of the system is through the mean time to failure (MTTF). Recall that the MTTF is the expected time that elapses before the first failure occurs, and it is defined in terms of the reliability function as

$$\text{MTTF} = \int_0^\infty R(t)\,dt$$

where $R(t)$ is the reliability function. For a simple system with an exponential reliability function, the MTTF can be written in terms of the failure rate as

$$\text{MTTF} = \frac{1}{\lambda}$$

For the system under consideration with the failure rate of $\lambda_{\text{system}} = 4.56 \times 10^{-4}$, the MTTF is approximately 2192 hours. In other words, the user can expect, on the average, to operate the system for only 2192 hours before it fails while in use.

The second crucial issue of the analysis is availability. The availability, as mentioned previously, is the percentage of time that the system is available to provide its services. Availability varies as a function of time, but, as we have seen earlier, a steady-state value can be computed using the failure rate and the repair rate of the system. The steady-state availability is defined in terms of the mean time to failure (MTTF) and the mean time to repair (MTTR). As defined previously, the MTTF is the average time that elapses before the first failure occurs. The MTTR is the average time that elapses before the average failure can be repaired and the system made operational once again. The steady-state availability A_{ss} can be written as

$$A_{ss} = \frac{\text{MTTF}}{\text{MTTF} + \text{MTTR}}$$

The MTTR and the MTTF are related to the repair rate and the failure rate, respectively, as

$$\text{MTTF} = \frac{1}{\lambda}$$

$$\text{MTTR} = \frac{1}{\mu}$$

Therefore, A_{ss} can be written as

$$A_{ss} = \frac{1}{1 + \dfrac{\lambda}{\mu}}$$

The steady-state availability is an estimate of the probability that the system is going to operate correctly when requested to do so. Using the system failure rate of 4.56×10^{-4}, Fig. 4.33 shows the steady-state availability of the system as a function of the MTTR. If one week is required, on the average, for repair then the steady-state availability is slightly higher than 0.92.

Unfortunately, the designers of the system in the present example have little control over the repair rate. The repair rate depends on the capabilities of the user to diagnose the problem, or the proximity of a facility where qualified personnel are employed to aid individuals in repairs. If the user is forced to get the manufacturer to repair the system, the repair period could easily extend beyond a week, particularly if the controller or other component must be mailed back to the factory for diagnosis and repair. To over-

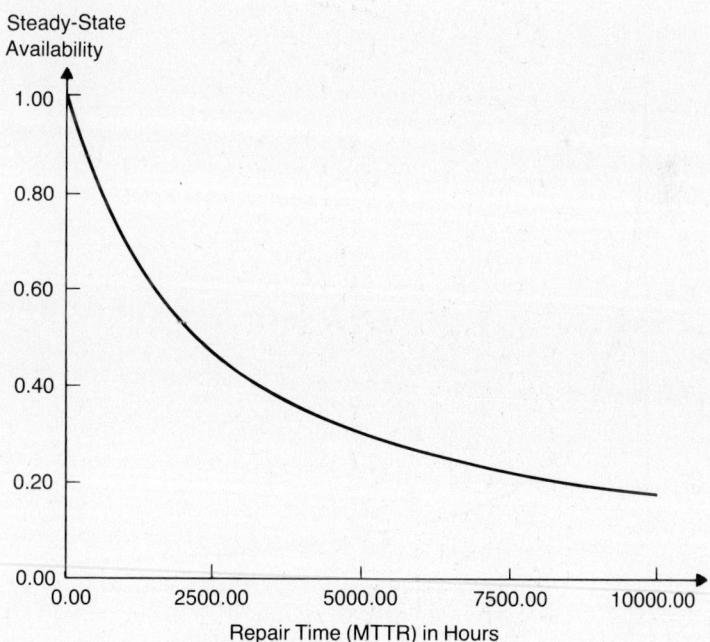

Fig. 4.33 Steady-state availability versus the repair time (MTTR) in hours for the nonredundant system ($\lambda = 4.56 \times 10^{-4}$ failures/hour).

come such problems, many users try to accumulate spare parts of their own such that the system can be operated while the repairs are being made. This is an expensive technique that could possibly be overcome through the development of designs that possess lower system failure rates or ones that use built-in redundancy to accommodate faults until the faulty components can be replaced. Figure 4.34 shows the improvement in availability as the failure rate is improved for a fixed repair rate of one repair per week.

Now we wish to examine the improvements that can be obtained through the use of redundancy to achieve some level of fault detection or fault tolerance. Fault tolerance can improve the reliability by preventing faults from resulting in a failure of the system. For example, if a power supply fluctuation occurs, the controller can be designed to automatically reset, prevent erroneous values from being sent to the motor drive circuits, and recover from the condition. In addition, conditions such as a run-away processor can be detected and a recovery implemented before catastrophic effects result.

Availability can be improved by using redundancy to keep the system functioning while repairs are made. For example, duplicate processors can

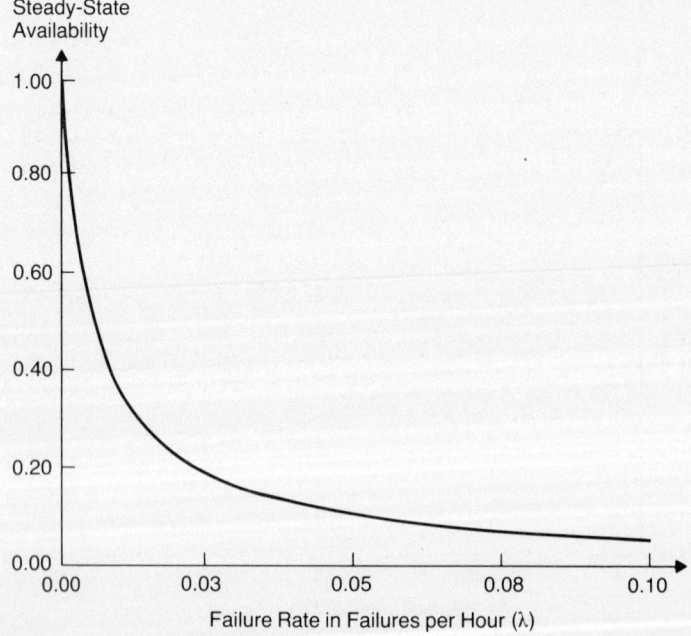

Fig. 4.34 Steady-state availability versus failure rate λ for the nonredundant system (MTTR = 1 week).

4.10 ■ Tradeoff Analysis Example

be incorporated into the controller; when the failure of one processor occurs, the second processor assumes the control functions until the first processor is repaired or replaced. Now, instead of the system being inoperative while being repaired, the redundant unit can perform the necessary computations and keep the system functioning properly. The necessary reconfiguration can be performed automatically to prevent the user from being required to take any action.

Safety can be improved through the incorporation of features that provide for the fail-safe operation of the system. For example, if a fault occurs, a mechanism must be provided to detect the existence of the fault and to prevent the fault from resulting in an undesired response from the control system.

Because of several practical constraints, it is not feasible to introduce redundancy into the system at any point other than the electronic controller. The motors, motor drives, and the sensors are too big and expensive to replicate. The manufacturer of the system would like to determine the advantage of incorporating redundancy into only the electronic controller.

The architecture of the candidate fault-tolerant controller is shown in Fig. 4.35. The system is completely dual redundant and uses an architecture similar to that of the Tandem NonStop computer system [Katzman 1982]. The processors both perform the same computations, and each is capable of functioning as the controller for the system. Various fault detection features are incorporated into the system to provide fault coverage and reconfiguration capability. At this point in the analysis, we will assume that the redundant system implements a reconfigurable duplication approach to fault detection and reconfiguration. In other words, the processors compare their results as well as use internal diagnostics to perform fault detection. Either processor can perform the functions of the system as long as the faulty processor is detected and removed from the system.

The improvements that can be obtained in the reliability and availability of the system can be assessed by analyzing the redundant controller. The reliability block diagram of the redundant system is shown in Fig. 4.36. By comparing Fig. 4.36 with the reliability block diagram of the nonredundant system shown in Fig. 4.31, we can see that the only difference lies in the controller electronics. As discussed earlier, the controller electronics are a weak link in the system, so improvements in the overall reliability and availability should be obtained by improving this weak link.

The reliability of the redundant system can be written by analyzing the combination of the series and parallel systems contained in the reliability block diagram of Fig. 4.36. The reliability of the system of Fig. 4.36 can be written as

$$R_{\text{system}}(t) = R_i(t)R_p(t)R_d(t)R_m(t)R_f(t)$$

238 Evaluation Techniques

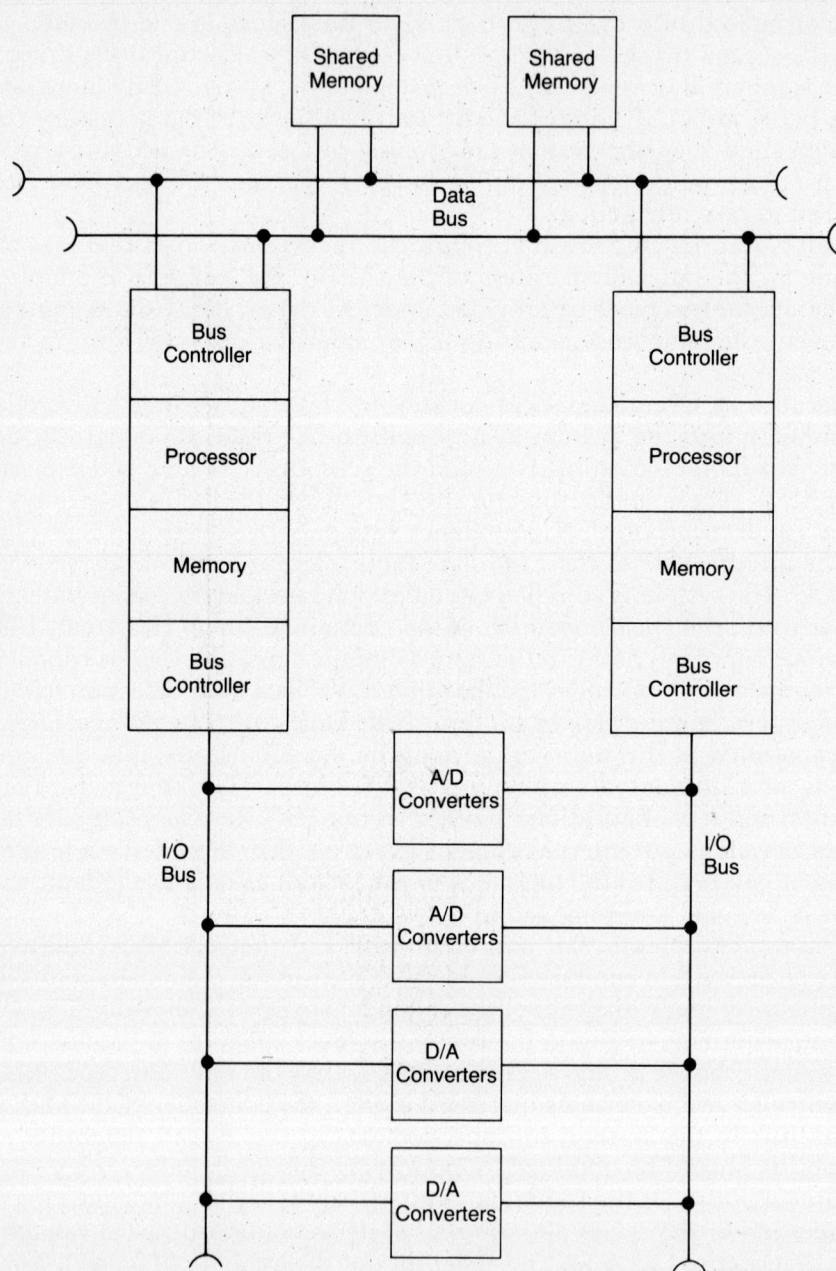

Fig. 4.35 Dual redundant architecture for the electronic controller.

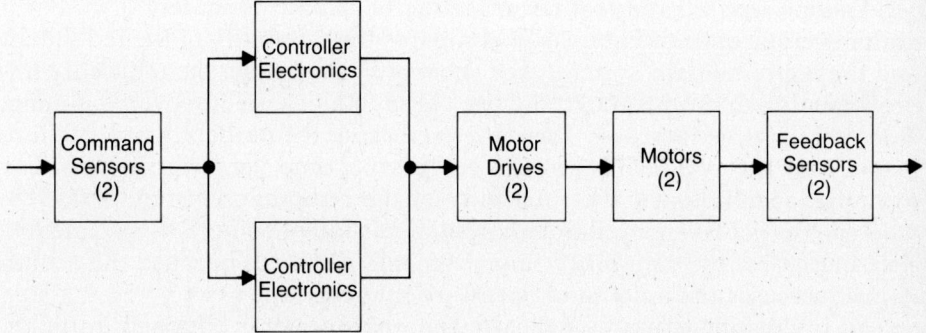

Fig. 4.36 Reliability block diagram of the control system with duplicated controller electronics.

where $R_i(t)$ is the reliability of the series combination of the two input sensors, $R_p(t)$ is the reliability of the parallel combination of the controller electronics, $R_d(t)$ is the reliability of the series combination of the two motor drives, $R_m(t)$ is the reliability of the series combination of the two motors, and $R_f(t)$ is the reliability of the series combination of the two feedback sensors.

The most interesting aspect of the parallel combination of the controller electronics is the fault coverage. We know that the fault coverage plays an important role in the reliability of the parallel system and the overall reliability of the control system. Using the techniques of this chapter, the reliability $R_p(t)$ can be written as

$$R_p(t) = R_e(t)R_e(t) + 2CR_e(t)(1 - R_e(t))$$

where $R_e(t)$ is the reliability of the nonredundant controller electronics.

Since we are assuming that each element of the control system obeys the exponential failure law, the reliability of the complete system can be written as

$$R_{\text{system}}(t) = e^{-\lambda_1 t}[e^{-2\lambda_e t} + 2Ce^{-\lambda_e t}(1 - e^{-\lambda_e t})]$$

where

$$\lambda_1 = 2\lambda_i + 2\lambda_d + 2\lambda_m + 2\lambda_f$$

and λ_e is the failure rate of the nonredundant electronics of the controller. The term C is the fault coverage. If $C = 1$, the system can recover from all faults that can possibly occur. As we have seen earlier in this chapter, the coverage factor can have a substantial impact on the reliability of the system.

240 Evaluation Techniques

Assume that a practical coverage factor is approximately 0.95. For a fault coverage of 0.95, Figure 4.37 compares the reliability of the redundant and the nonredundant systems. For short periods of time, the reliability improvements achieved through the use of redundancy are not overwhelming. However, as time increases, therefore increasing the probability of a failure occurring, the reliability improvements become very significant. For example, at 500 hours, the reliability of the redundant system is 0.9184, whereas that of the nonredundant system has fallen below 0.8. As time further increases, the reliability improvements diminish because the redundancy increases the amount of hardware that can fail.

An interesting way to see clearly the improvements obtained in the redundant system is to look at the time that can elapse before each system's reliability falls below a certain value. In other words, we look at the mission time of the system. For example, the reliability of the nonredundant system falls below 0.9 at approximately 230 hours. The reliability of the redundant system, however, does not fall below 0.9 until about 600 hours. Therefore, the redundancy has increased the time that the reliability remains above 0.9 by a factor of 2.6.

The availability is also improved by the incorporation of redundancy and fault detection techniques into the design of the controller. To deter-

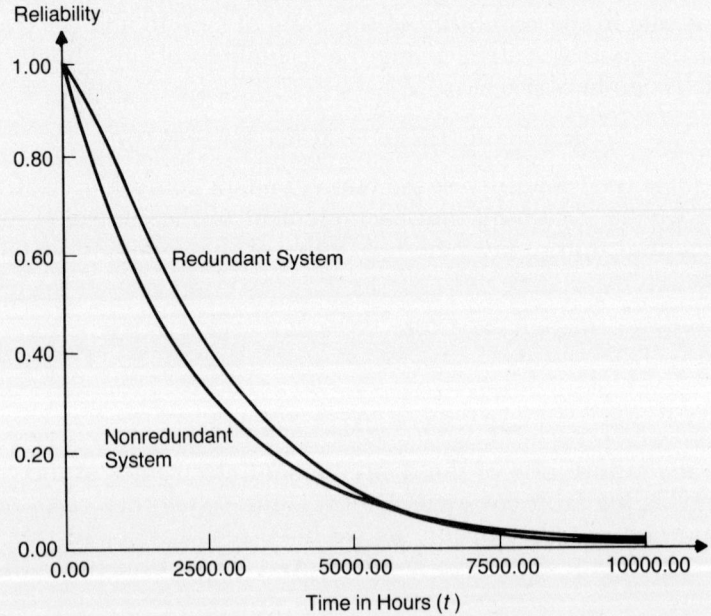

Fig. 4.37 Reliability of the redundant and nonredundant systems versus time.

mine the availability of the redundant system, we first need to compute the MTTF of the redundant system. The MTTF of the redundant system can be calculated by integrating the reliability function of the redundant system. In other words, we have

$$\text{MTTF}_{\text{redundant}} = \int_0^\infty [e^{-\lambda_1 t}[e^{-2\lambda_e t} + 2Ce^{-\lambda_e t}(1 - e^{-\lambda_e t})]]\, dt$$

Evaluating the integral for a coverage factor of 0.95, we find that the MTTF of the redundant system is approximately 3113 hours. Recall that the MTTF of the nonredundant system was approximately 2192 hours, so the redundancy has improved the MTTF by a factor of approximately 1.42. For a coverage factor of 0.95, Fig. 4.38 compares the availability of the redundant and the nonredundant systems.

Now we want to examine some of the specifics of implementing the redundant controller. We have provided some convincing information concerning the benefits of redundancy, but we have not completed the decisions concerning the specifics of the redundant implementation.

Suppose the dual-redundant system can be designed to function in one of two ways. The first is the reconfigurable duplication scheme that we have

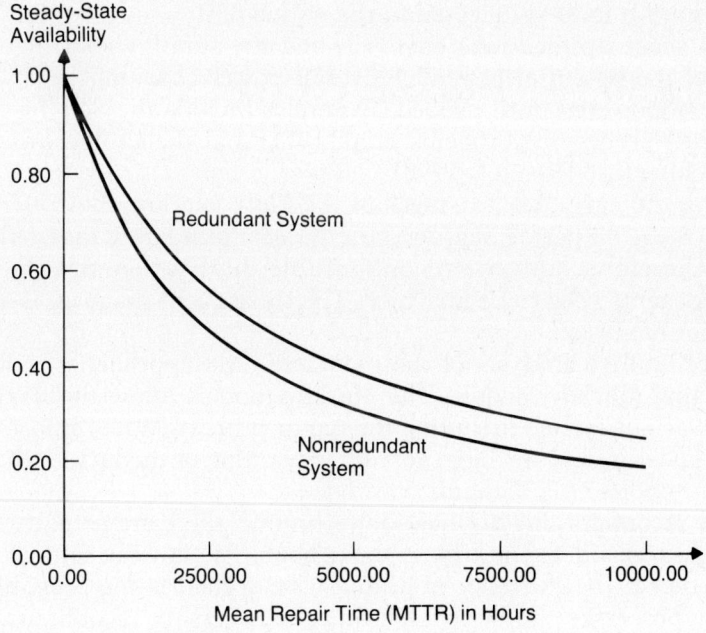

Fig. 4.38 Comparison of steady-state availability as a function of the MTTR.

just considered. In reconfigurable duplication, both processors operate in synchrony while performing the same computations. The results produced by the two processors are compared and any discrepancy is indicative of a fault in some component of the dual-redundant system. Each processor also runs self-diagnostic routines that attempt to locate the source of a fault. If the faulty processor is located, the system is reconfigured to construct a completely functional system using the fault-free processor.

The concept of reconfigurable duplication has been frequently used in industrial applications. Examples include the Bell Electronic Switching [Toy 1978] System (ESS), the COMTRAC railroad traffic control computer [Ihara et al. 1978], the Agusta A129 flight control computer system [Johnson and Julich 1985], and the AXE telephone switching control unit [Ossfeldt and Jonsson 1980]. Reconfigurable duplication is popular because it requires relatively little redundancy and is capable of significantly improving the reliability of a system.

The second possible mode of operation is the use of a standby spare. In standby sparing, one of the two processors is selected as the online unit that performs all of the computations for the system as long as it remains fault free. Concurrent fault detection routines are executed in the online processor to detect faults. If, and only if, a fault is detected in the online unit, the spare processing unit is brought on-line to assume the functions of the failed processor. In many cases, the spare processor might even remain unpowered until a fault is detected in the online unit.

Many space applications have selected the standby sparing approach because of the potential to reduce overall power consumption by keeping the spare unpowered until needed. Examples include the Self-Test And Repair (STAR) processor [Avizienis et al. 1971] and the United Data System (UDS) architecture [Rennels 1978].

During the early design stages of the fault-tolerant controller, certain tradeoffs must be performed on the type of redundancy that will be employed. Therefore, both the reconfigurable duplication and the standby sparing concepts need to be analyzed. The subsequent material is a description of that analysis.

The reliability analysis of the two candidate approaches will be conducted using Markov models. The Markov models allow the flexibility to consider various factors including the repair process. More importantly, the Markov model is easy to construct and solve. One of the factors that is considerably important is the fault coverage.

In the reconfigurable duplication technique, there are actually two fault coverages to consider. The first is the probability that the comparison process will detect the existence of a fault. The second is the probability that the faulty processor can be located and the necessary reconfiguration performed. In the standby sparing approach, only one coverage factor is involved: the probability that the fault will be detected and the spare will successfully replace the online unit.

The Markov model of the reconfigurable duplication system is shown in Fig. 4.39. The model contains four states that represent all configurations of the system. State 2 represents the case where both of the processing modules are operating correctly. Sometimes called the perfect state, this is the state in which the system initially begins operation. State 1 represents the system in which one of the two processing modules has failed and the problem has been successfully detected and appropriately handled. While in state 1, the system has one faulty processor and one fault-free processor, and the fault-free processor is performing the functions of the system. State FS is the fail-safe state. While in the fail-safe state, the system has ceased to

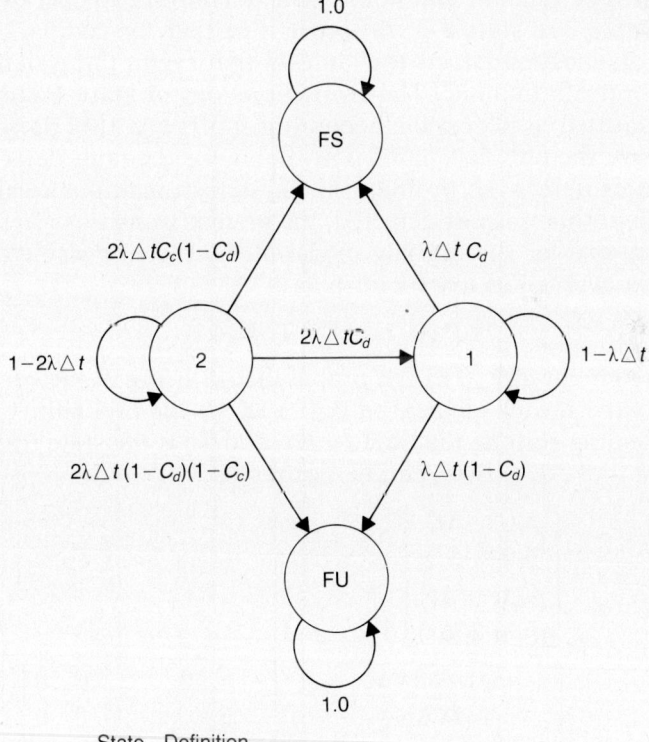

State	Definition
2	Both modules are fault free.
1	One module fault free and one faulty (detected).
FS	System has failed in a safe manner.
FU	System has failed in an unsafe manner.

Fig. 4.39 Four-state Markov model of reconfigurable duplication.

perform its functions, but it has done so in an entirely safe manner. In other words, the system has failed, but the problem has been detected and handled in a safe and effective manner. The final state, state FU, is the failed unsafe state. When in state FU, the system has failed and the problem has not been handled in a manner that guarantees the safe operation of the system. When the system is in state FU, unexpected and perhaps unsafe operations can result.

The transitions illustrated on the diagram of Fig. 4.39 can be explained as follows. The transition from state 2 to state 1 can occur if and only if one of the originally fault-free processors fails and the failure is detected by the self-diagnostics. If the failure is not detected by the self-diagnostics, the system cannot remain operational regardless of whether the comparison process detects the problem. If the comparison process detects the fault and the self-diagnostics do not, the system transitions from state 2 to state FS since the system will be safe but will not be able to continue to operate. The transition from state 2 to state FU will occur if neither the comparison process nor the self-diagnostics detects the fault. In that event, the system will have failed in an unsafe manner. The transitions out of state 1 are the result of the last fault-free processor becoming faulty. At this point, the self-diagnostics are the only mechanism available for the fault detection. If the self-diagnostics detect the problem, the system transitions to state FS. If, however, the problem is not detected, the system transitions to state FU.

The equations for the Markov model of the reconfigurable duplication system can be written in matrix form as

$$\mathbf{P}_{rd}(t + \Delta t) = \mathbf{T}_{rd}\mathbf{P}_{rd}(t)$$

where each element of $\mathbf{P}_{rd}(t)$ is the probability of being in the corresponding state at the time t, each element of $\mathbf{P}_{rd}(t + \Delta t)$ is the probability of being in the corresponding state at the time $t + \Delta t$, and $\mathbf{T}_{rd}$ is the state transition matrix. Each of these quantities can be written as

$$\mathbf{P}_{rd}(t + \Delta t) = \begin{bmatrix} p_2(t + \Delta t) \\ p_1(t + \Delta t) \\ p_{FS}(t + \Delta t) \\ p_{FU}(t + \Delta t) \end{bmatrix}$$

$$\mathbf{T}_{rd} = \begin{bmatrix} 1 - 2\lambda\Delta t & 0 & 0 & 0 \\ 2\lambda\Delta t C_d & 1 - \lambda\Delta t & 0 & 0 \\ 2\lambda\Delta t C_c(1 - C_d) & \lambda\Delta t C_d & 1 & 0 \\ 2\lambda\Delta t(1 - C_d)(1 - C_c) & \lambda\Delta t(1 - C_d) & 0 & 1 \end{bmatrix}$$

$$\mathbf{P}_{rd}(t) = \begin{bmatrix} p_2(t) \\ p_1(t) \\ p_{FS}(t) \\ p_{FU}(t) \end{bmatrix}$$

The solution to the matrix equation of the Markov model can be obtained by selecting values for the initial probability vector **P**(0) and the Δt time increment. The value for **P**(Δt) can be computed by multiplying **P**(0) by the transition matrix. Next, **P**$(2\Delta t)$ can be obtained by multiplying **P**(Δt) by the transition matrix. In general, the solution to the equations of the Markov model is given by

$$\mathbf{P}(n\,\Delta t) = \mathbf{T}^n \mathbf{P}(0)$$

The reliability, at some time t, of the reconfigurable duplication system is the probability that the system operates correctly from the initial time until time t. In reference to the Markov model, the reliability of the reconfigurable duplication system is the probability that the system will be in either state 2 or state 1, which are the only two states in which the system is operating correctly. Therefore, the reliability of the reconfigurable duplication system can be written as

$$R_{rd}(t) = p_2(t) + p_1(t)$$

where $R_{rd}(t)$ is the reliability of the reconfigurable duplication system at the time t.

The Markov model of the standby spare system is constructed in a manner similar to that of the reconfigurable duplication system. The resulting model is shown in Fig. 4.40. The Markov model of the standby spare system

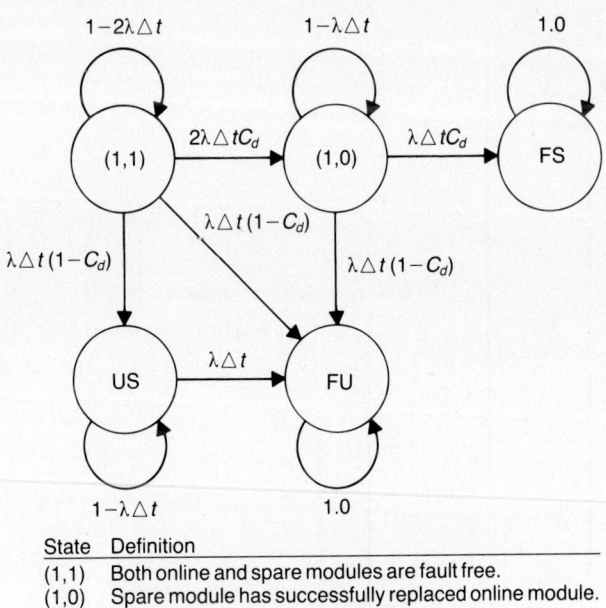

State	Definition
(1,1)	Both online and spare modules are fault free.
(1,0)	Spare module has successfully replaced online module.
US	Undetected failure in spare module.
FS	System has failed in a safe manner.
FU	System has failed in an unsafe manner.

Fig. 4.40 Five-state Markov model of the standby spare system.

contains a total of five states. State $(1,1)$ represents the condition where the system has both the online processor and the spare processor functioning in a fault-free manner. State $(1,0)$ corresponds to the existence of one of two conditions: (1) the online processor has failed and the spare has been successfully brought on-line or (2) the spare has failed, the failure has been detected, and the spare has been taken out of service. State (FS) occurs when both the online processor and the spare processor have failed and both failures have been detected and appropriately handled. While in state (FS), the system cannot function, but the system is safe. State (FU) represents the condition where the system has failed in an unsafe manner. The system can enter state (FU) in two ways: through the undetected failure of the online unit or through the undetected failure of the spare processor and the subsequent use of that spare when the online unit fails. In both cases, the system is operating with processors that possess undetected failures. The final state is state (US). While in state (US), the system continues to perform its functions because the online unit has not failed; however, the spare has failed in a manner that is undetectable. The undetectable failure of the spare results in a faulty spare being substituted for a faulty online unit, in the event that the online unit fails and the failure of the online unit is detected.

The equations for the Markov model of the standby spare system can be written in matrix form as

$$\mathbf{P}_{ss}(t + \Delta t) = \mathbf{T}_{ss} \mathbf{P}_{ss}(t)$$

where

$$\mathbf{P}_{ss}(t) = \begin{bmatrix} p_{(1,1)}(t) \\ p_{(1,0)}(t) \\ p_{(FS)}(t) \\ p_{(FU)}(t) \\ p_{(US)}(t) \end{bmatrix}$$

$$\mathbf{T}_{ss} = \begin{bmatrix} 1 - 2\lambda \Delta t & 0 & 0 & 0 & 0 \\ 2\lambda \Delta t C_d & 1 - \lambda \Delta t & 0 & 0 & 0 \\ 0 & \lambda \Delta t C_d & 1 & 0 & 0 \\ \lambda \Delta t(1 - C_d) & \lambda \Delta t(1 - C_d) & 0 & 1 & \lambda \Delta t \\ \lambda \Delta t(1 - C_d) & 0 & 0 & 0 & 1 - \lambda \Delta t \end{bmatrix}$$

$$\mathbf{P}_{ss}(t + \Delta t) = \begin{bmatrix} p_{(1,1)}(t + \Delta t) \\ p_{(1,0)}(t + \Delta t) \\ p_{(FS)}(t + \Delta t) \\ p_{(FU)}(t + \Delta t) \\ p_{(US)}(t + \Delta t) \end{bmatrix}$$

The standby sparing approach will be completely operational as long as the system is in one of three states: state $(1,1)$, state $(1,0)$, or state (US). Therefore, the reliability of the standby sparing system can be written as

$$R_{ss}(t) = p_{(1,1)}(t) + p_{(1,0)}(t) + p_{(US)}(t)$$

As is evident from the Markov models of the two candidate systems, the reliability depends on several key factors, including the failure rates, coverage factors, and time.

For comparison purposes, assume that the failure rate λ is 3.5×10^{-4}, the coverage provided by the comparison process C_c is perfect, and the self-diagnostics provide a coverage factor C_d of 0.95. Using the Markov models of the two systems allows the reliability of each system to be computed as a function of time. Figure 4.41 compares the reliability of the reconfigurable duplication system and the standby sparing system for the given failure rate and coverage factors. At a reasonably high fault coverage value, the reliability of the two approaches is very similar. The value of the fault coverage factor, however, does have a significant impact on the reliability of the system.

Figure 4.42 shows the reliability of the reconfigurable duplication concept and the standby sparing system as functions of the fault coverage factor C_d. For a given module reliability, the standby sparing approach is capable of achieving a reliability that exceeds that of reconfigurable duplication. The single exception to this fact occurs when the coverage is perfect,

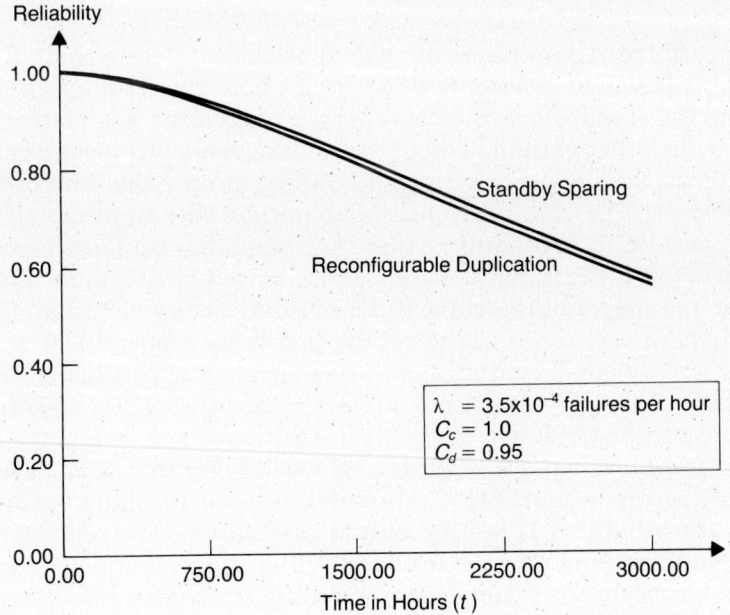

Fig. 4.41 Reliability of standby sparing and reconfigurable duplication versus time.

248 Evaluation Techniques

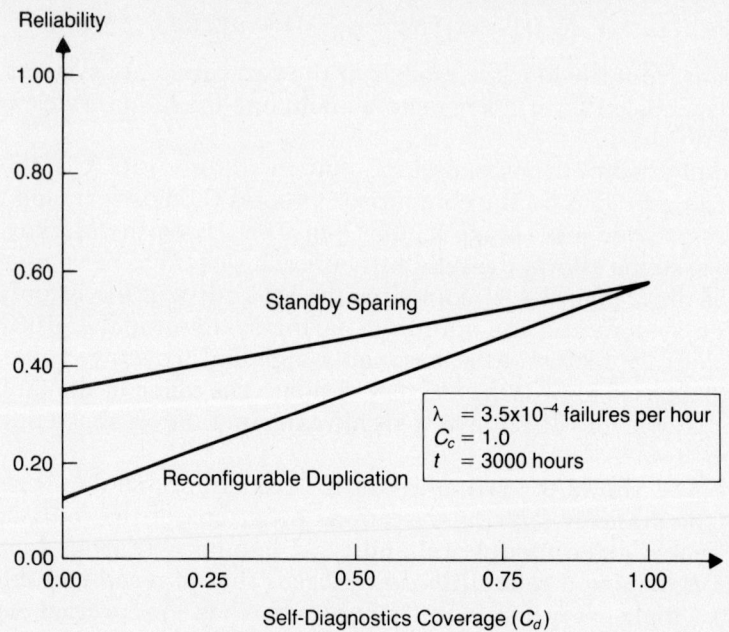

Fig. 4.42 Reliability of standby sparing and reconfigurable duplication versus fault coverage.

in which case the reliability of the two approaches is equivalent. This is as expected because with perfect fault coverage, both the reconfigurable duplication and the standby sparing concepts are equivalent to a perfect parallel system. At the other extreme where the fault coverage factor is zero, the reliability of the standby sparing system will be simply the reliability of the online module. The reliability of the reconfigurable duplication system, however, will be the probability that both modules are fault free because the comparison process can only detect the existence of a fault and cannot determine the source of the fault. If the self-diagnostics have zero fault coverage, any fault will result in the system becoming inoperable.

If one simply conducted a reliability analysis, the conclusion would be that the use of standby sparing is preferred because of the improved reliability that can be obtained. For a given fault coverage factor, the standby sparing approach is capable of achieving a reliability that is at least as high as that of the reconfigurable duplication system. In many applications, however, the reliability is not the only consideration. The reliability of the system simply accounts for the probability that the system will perform its functions correctly. In many cases, the designer is also interested in the probability that if the system does not remain operational it will at least fail in a manner that is safe. For example, the pilot of an airplane can fly the aircraft even if the auto-pilot fails, provided that it fails in such a way that it does not affect the remainder of the system. Likewise, in this control application, the user can survive even if the controller fails, as long as the failure does not result in the system performing some undesired function. This is the concept of *fail-safe operation*. A system fails safely when the failure does not produce an incorrect action from the system; the system may not perform its functions, but it at least does not perform the wrong functions.

Recall that we defined safety as the probability that a system will either perform its functions correctly or will discontinue its functions in a manner that is completely safe. In other words, the system either operates correctly or fails safely.

Referring to the Markov model of the reconfigurable duplication system shown in Fig. 4.39, the system will be safe as long as it is in one of three states: state 2, state 1, or state FS. Therefore, the safety of the system can be written as

$$S_{rd}(t) = p_2(t) + p_1(t) + p_{FS}(t)$$

where $S_{rd}(t)$ is the safety of the reconfigurable duplication system.

The safety of the standby sparing system can be written in a similar manner. Referring to the Markov model of the standby sparing approach, shown in Fig. 4.40, the system will be safe as long as it is in one of four states: state $(1,1)$, state $(1,0)$, state (FS), or state (US). The safety can be written as

$$S_{ss}(t) = p_{(1,1)}(t) + p_{(1,0)}(t) + p_{(FS)}(t) + p_{(US)}(t)$$

It is interesting to compare the reliability and the safety of a particular system. Figures 4.43 and 4.44 show the reliability and safety of both candidate architectures as functions of time. As can be seen, the safety of both systems is considerably higher than the reliability.

To allow the safety feature to be further investigated, the failure rate, time, and the fault coverage of the comparison process were held constant and the fault coverage of the self-diagnostics was allowed to vary. The failure rate is $\lambda = 3.5 \times 10^{-4}$ failures per hour, the comparison fault coverage is $C_c = 1.0$, and the time was selected as 3000 hours. The 3000 hour time period corresponds to approximately eight hours of use per day for one year. For these values, Figs. 4.45 and 4.46 show how the reliability and safety of the two systems vary as the capability of the self-diagnostics varies. As expected, the safety and the reliability of the standby sparing approach are equal when no coverage is provided by the self-diagnostics. The standby sparing system depends on the self-diagnostics to achieve any level of safety.

The safety of the reconfigurable duplication system has several interesting features. As the coverage of the self-diagnostics increases from zero until approximately .5, the safety of the system actually decreases. This can be explained as follows. When the coverage of the self-diagnostics is zero,

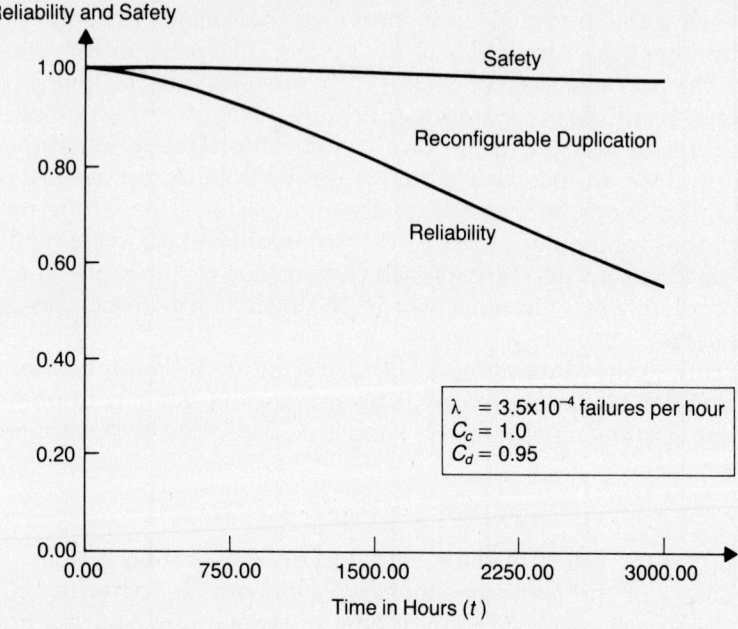

Fig. 4.43 Reliability and safety of reconfigurable duplication versus time.

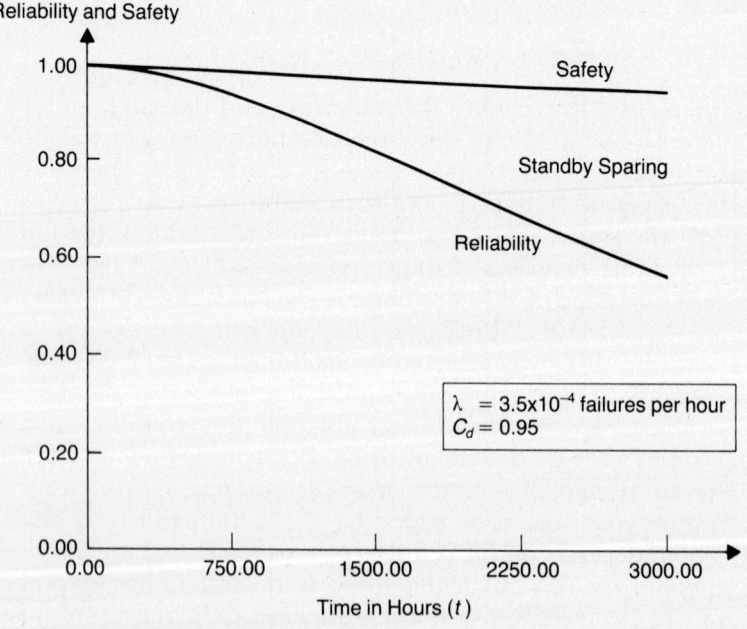

Fig. 4.44 Reliability and safety of standby sparing versus time.

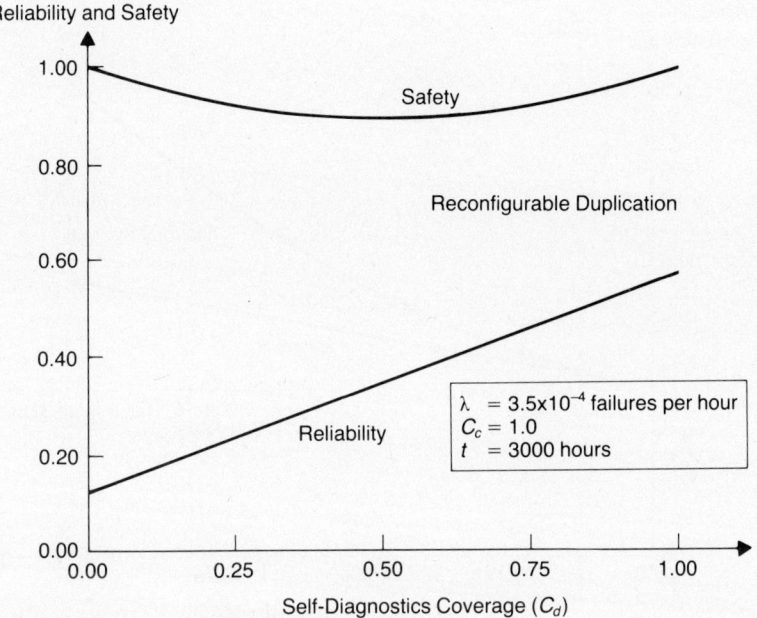

Fig. 4.45 Reliability and safety of reconfigurable duplication versus fault coverage.

the system will never reconfigure because the fault can never be located. The comparison process, in this example, is perfect, therefore, the fault will always be detected and the safety of the system maintained. As the self-diagnostics improve, some of the faults will be located and the system will be reconfigured. Once reconfigured, however, the system must depend on the self-diagnostics to maintain safety. But, when the self-diagnostics are very poor (less than .5 coverage), the safety of the system is compromised.

The coverage of the comparison process has a tremendous impact on the safety of the reconfigurable duplication system. Figure 4.47 shows the variation in the safety of the system as a function of both the coverage of the comparison process and the coverage of the self-diagnostics.

The final comparison is the safety of the reconfigurable duplication system and the standby sparing system. The use of comparison for fault detection cannot be perfect in most applications because of phenomena such as multiple faults and the manner in which the comparison is performed. Seldom is a direct bit-by-bit comparison between two units practical. The units may receive their data through different analog-to-digital converters or from different sensors in a redundant system. Even though the sensors may be functionally equivalent, the outputs may not agree exactly. Small differ-

252 Evaluation Techniques

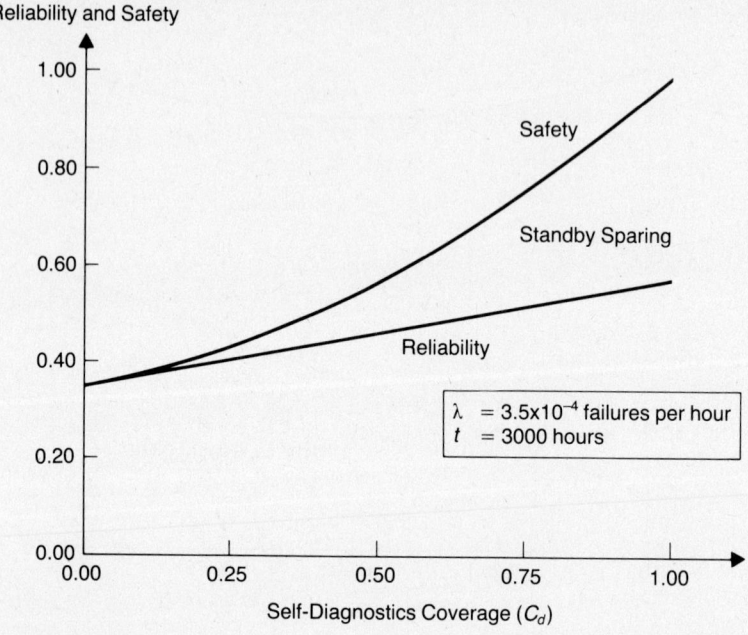

Fig. 4.46 Reliability and safety of standby sparing versus fault coverage.

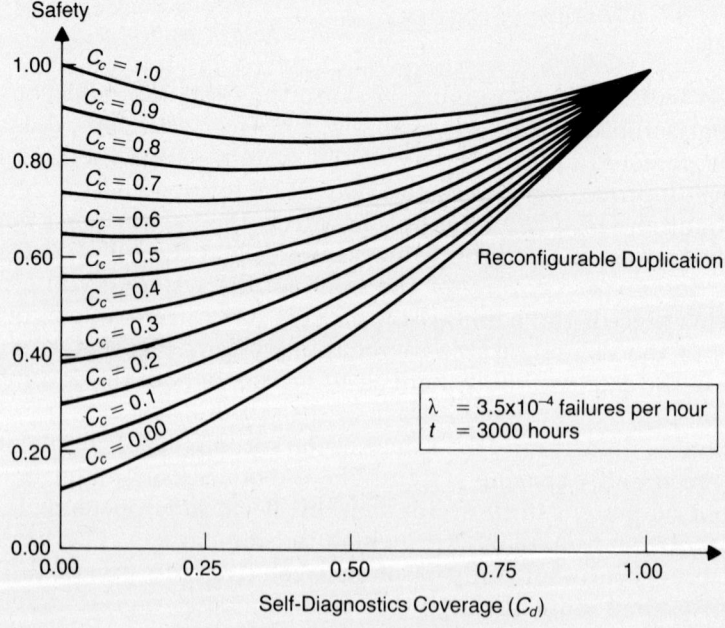

Fig. 4.47 Safety of reconfigurable duplication versus fault coverage for various comparison coverage values.

ences such as these can propagate through calculations to produce larger differences at the outputs of a process. Therefore, the comparison process may be performed by ignoring several of the least significant bits or by computing a difference between the two values and performing a threshold operation on that difference. If the difference exceeds the threshold, a miscompare is indicated. The result is something less than perfect coverage. Practice has shown that the comparison process can achieve a coverage of 0.95 or greater [Johnson and Julich 1985]. Figure 4.48 compares the safety of the reconfigurable duplication system and the standby sparing system for a comparison coverage of 0.95 and for various values of the self-diagnostics coverage factor. As can be seen, the reconfigurable duplication system achieves a much higher safety than the standby sparing approach.

Even though the standby sparing system achieves a higher reliability, for a given self-diagnostics coverage factor, than the reconfigurable duplication concept, the latter is preferable in many applications because of the improved safety. The importance of the system's safety has made the reliability somewhat misleading in that the most reliable system is not necessarily the most desirable one in applications mandating high safety.

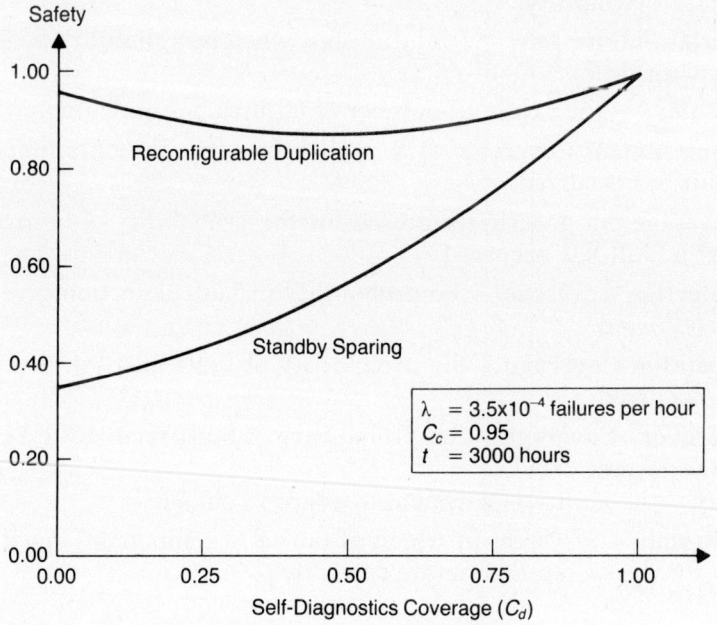

Fig. 4.48 Comparison of standby sparing and reconfigurable duplication safety versus fault coverage.

Summary

This chapter has presented the various analysis techniques that are available to evaluate fault-tolerant systems. Several methods for determining the reliability, availability, and safety of systems have been developed and illustrated, including both combinatorial modeling and Markov modeling approaches. Perhaps one of the most important parts of this chapter has been the example tradeoff analysis. During that analysis, several interesting comparisons were made on the reliability and safety of a practical fault-tolerant computing architecture for a digital control application. The analysis illustrated the importance of considering several factors during the decision making process. With the tools presented in the first four chapters of this book, we are now ready to practice the process of designing fault-tolerant systems. Chapter 5 begins to explore example designs as well as the mechanisms of the design process.

The following list provides a summary of the key concepts presented in this chapter.

Bathtub Curve — the variation of the failure rate with time for electronic components. The failure rate is assumed to be constant during the useful life of the component.

Combinatorial Models — a method of developing an analytical expression for a system's reliability.

Exponential Failure Law — a relationship whereby reliability varies exponentially with time.

Failure Rate, λ — the expected number of failures per unit time.

Fault Containment Coverage — the probability of fault containment given that a fault has occurred.

Fault Coverage — a generic term used for the probability of fault recovery given that a fault has occurred.

Fault Detection Coverage — the probability of fault detection given that a fault has occurred.

Fault Location Coverage — the probability of fault location given that a fault has occurred.

Fault Recovery Coverage — the probability of fault recovery given that a fault has occurred.

Flexibility — the ability of a system to adapt to change.

M-of-N System — a system in which M out of N components must operate correctly for the system to operate correctly.

Markov Model—a method of modeling reliability in terms of system state and state transition.

Mean Time Between Failure (MTBF)—the average time between consecutive failures. The MTBF is the sum of the MTTF and the MTTR.

Mean Time To Failure (MTTF)—the average time that a system will operate before the first failure is encountered.

Mean Time To Repair (MTTR)—the average time required to repair a system.

MIL-HDBK-217—a United States Department of Defense standard that defines a method of calculating failure rates.

Mission Time, MT[r]—the time at which the reliability falls below the value of r.

Mission Time Improvement—the ratio of the mission times of two systems being compared.

Parallel System—a system in which only one of N components must operate correctly for the system to operate correctly.

Redundancy Ratio—the ratio of the amount of a resource (hardware, software, time, or information) used in a redundant system to the amount of the same resource used in a nonredundant system.

Reliability Block Diagram—a graphical method of depicting the elements that must operate correctly for the system to operate correctly.

Repair Rate, μ—the expected number of repairs per unit time.

Series System—a system in which all components must operate correctly for the system to operate correctly.

State Transition—the process of transitioning from one system state to another.

Steady-state Availability—the limiting value of the availability function as time approaches infinity.

System State—a description of the combination of operational and failed modules in a system.

Technology Dependence—the dependence of a system on technological progress.

Testability—the ability to verify that a system is operating correctly.

Transparency to User—the impact of fault tolerance on the user of a system.

Weibull Distribution—a technique used to represent time-varying failure rate functions.

References

1. Avizienis, A., G.C. Gilley, F.P. Mathur, D.A. Rennels, J.A. Rohr, and D.K. Rubin. "The STAR (Self-Testing And Repairing) computer: An investigation of the theory and practice of fault tolerant computer design," *IEEE Transactions on Computers*, Vol. C-20, No. 11, November 1971, pp. 1394–1403.
2. Bouricius, W.G., W.C. Carter, and P.R. Schneider. "Reliability modeling techniques for self-repairing computer systems," *Proceedings of the 24th ACM Annual Conference*, 1969, pp. 295–309.
3. Ihara, H., K. Fukuoka, Y. Kubo, and S. Yokota. "Fault tolerant computer system with three symmetric computers," *Proceedings of the IEEE*, Vol. 66, No. 10, October 1978, pp. 1160–1177.
4. Johnson, B.W., and P.M. Julich. "Fault tolerant computer system for the A129 helicopter," *IEEE Transactions on Aerospace and Electronic Systems*, Vol. AES-21, No. 2, March 1985, pp. 220–229.
5. Johnson, B.W., and J.H. Aylor. "Reliability and safety analysis of a fault-tolerant controller," *IEEE Transactions on Reliability*, Vol. R-35, No. 4, October 1986, pp. 355–362.
6. Katzman, J.A. "A fault-tolerant computing system," in *The Theory and Practice of Reliable System Design* by D.P. Siewiorek and R.S. Swarz, Digital Press, Bedford, Mass. 1982.
7. Nelson, V.P., and B.D. Carroll. *Tutorial: Fault-Tolerant Computing*, IEEE Computer Society Press, Washington, D.C., 1986.
8. Ossfeldt, B.E., and I. Jonsson. "Recovery and diagnostics in the central control of the AXE switching system," *IEEE Transactions on Computers*, Vol. C-29, No. 6, June 1980, pp. 482–491.
9. Rennels, D.A. "Architectures for fault tolerant spacecraft computers," *Proceedings of the IEEE*, Vol. 66, No 10, October 1978, pp. 1255–1268.
10. Shooman, M.L. *Probabilistic Reliability: An Engineering Approach*, McGraw-Hill, New York, 1968.
11. Siewiorek, D.P., and R.S. Swarz, *The Theory and Practice of Reliable System Design*, Digital Press, Bedford, Mass. 1982.
12. Toy, W.N. "Fault tolerant design of local ESS processors," *Proceedings of the IEEE*, Vol. 66, No. 10, October 1978, pp. 1126–1145.
13. Trivedi, K.S. *Probability and Statistics with Reliability, Queuing, and Computer Science Applications*, Prentice-Hall, Englewood Cliffs, N.J., 1982.
14. United States Department of Defense, *Military Standardization Handbook: Reliability Prediction of Electronic Equipment*, MIL-HDBK-217, 1965.
15. United States Department of Defense, *Military Standardization Handbook: Reliability Prediction of Electronic Equipment*, MIL-HDBK-217B, 1974.
16. United States Department of Defense, *Military Standardization Handbook: Reliability Prediction of Electronic Equipment*, MIL-HDBK-217C, 1979.

Additional Reading

For the reader interested in pursuing the topics of this chapter in more detail, the following selection of additional reading is provided. This material covers articles and texts that are specifically devoted to the analysis of digital systems.

Beaudry, M.D. "Performance-related reliability measures for computing systems," *IEEE Transactions on Computers*, Vol. C-27, No. 6, June 1978.

Borgerson, B.R., and R.F. Freitas. "A reliability model for gracefully degrading and standby sparing systems," *IEEE Transactions on Computers*, Vol. C-24, No. 5, May 1975.

Costes, A.C., C. Landrault, and J.C. Laprie. "Reliability and availability models for maintained systems featuring hardware failures and design faults," *IEEE Transactions on Computers*, Vol. C-27, No. 6, June 1978.

Mathur, F.P. "On reliability modeling and analysis of ultra-reliable fault-tolerant digital systems," *IEEE Transactions on Computers*, Vol. C-20, No. 11, November 1971.

Masreliez, C.J., and B.E. Bjurman. "Fault tolerant system reliability modeling/analysis," *Journal of Aircraft*, Vol. 14, No. 8, August 1977.

Meyer, J.F. "On evaluating the performability of degradable computing systems," *IEEE Transactions on Computers*, Vol. C-29, No. 8, August 1980.

Meyer, J.F., D.C. Furchtgott, L.T. Wu. "Performability evaluation of the SIFT computer," *IEEE Transactions on Computers*, Vol. C-29, No. 8, August 1980.

Molloy, M.K. "Performance analysis using stochastic petri nets," *IEEE Transactions on Computers*, Vol. C-31, No. 9, September 1982.

Ng, Y.W., and A.A. Avizienis. "A unified reliability model for fault tolerant computers," *IEEE Transactions on Computers*, Vol. C-29, No. 11, November 1980.

Pedar, A., and V.V.S. Sarma. "Phased-mission analysis for evaluating the effectiveness of aerospace computing systems," *IEEE Transactions on Reliability*, Vol. R-30, No. 5, December 1981.

Siewiorek, D.P., V. Kini, H. Mashburn, S.R. McConnel, and M.M. Tsao. "A case study of C.mmp, Cm*, and C.vmp: Part II- predicting and calibrating reliability of multiprocessor systems," *Proceedings of the IEEE*, Vol. 66, No. 10, October 1978.

Takahashi, K. "Reliability and availability of redundant satellite orbit systems," *IEEE Transactions on Aerospace and Electronic Systems*, Vol. AES-18, No. 3, May 1982.

Varshney, P.K. "On analytical modeling of intermittent faults in digital systems," *IEEE Transactions on Computers*, Vol. C-28, No. 10, October 1979.

Wakerly, J.F. "Microcomputer reliability improvement using triple-modular redundancy," *Proceedings of the IEEE*, Vol. 64, No. 6, June 1976.

Problems

4.1 In most applications, we assume that the reliability of a component obeys the exponential failure law and the failure rate of that component is a constant. The assumption of a constant failure rate is appropriate if we are operating in the flat portion of the bathtub curve. However, if we are in either the infant mortality or wear-out phases, the assumption of constant failure rate is inaccurate. Assume that you are analyzing a component that obeys the exponential failure law, but the failure rate is a time-varying function given by

$$\lambda(t) = e^{kt} \quad \text{failures per million hours}$$

where k is some constant. Positive values of k are used to approximate the wear-out phase of the component's life, and negative values are used to approximate the infant mortality stage. Investigate the reliability of this component for various values of k and as a function of time. Specifically, show plots of the reliability of the component as a function of time and for k equal to 1, -1, 10, -10, 100, and -100.

4.2 Calculate the MTTF of a TMR system that contains three identical modules, each with a failure rate of 0.001 failures per hour. You can assume that the modules obey the exponential failure law. Compare the MTTF of the TMR system with the MTTF of a single module having the specified failure rate.

4.3 The system shown in Fig. 4.49 is a processing system for a helicopter. The system has dual-redundant processors and dual-redundant interface units. Two buses are used in the system and each bus is also dual redundant. The interest-

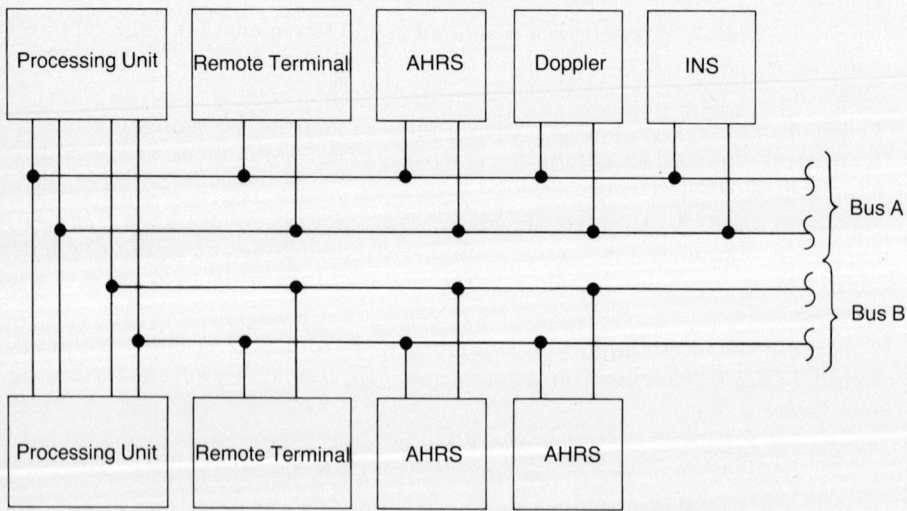

Fig. 4.49 Processing system to be analyzed in Problem 4.3.

ing part of the system is the navigation equipment. The aircraft can be completely navigated using the Inertial Navigation System (INS). If the INS fails, the aircraft can be navigated using the combination of the doppler and the attitude heading and reference system (AHRS). The system contains three AHRS units, of which only one is needed. This is an example of functional redundancy where the data from the AHRS and the doppler can be used to replace the INS, if the INS fails. Because of other sensors and instrumentation, both buses are required for the system to function properly regardless of which navigation mode is being employed. Draw the reliability block diagram of the system. Calculate the reliability of the system for a one-hour flight using the MTTF figures given in Table 4.5. Assume that the exponential failure law applies and that the fault coverage is perfect.

4.4 Repeat Problem 4.3, but this time incorporate a coverage factor for the fault detection and reconfiguration of the processing units. Using the same failure rate data, determine the approximate fault coverage value that is required to obtain a reliability (at the end of one hour) of 0.99999.

4.5 Once again using the system and the data of Problem 4.3, determine the availability of the system when the fault coverage is perfect. Assume that the MTTR varies from one hour to two weeks and plot the availability as a function of the MTTR. Compare this availability to that which could be achieved if the INS was not included in the system and navigation depended completely on the AHRS and the doppler.

4.6 Construct the Markov model of a TMR system with a single spare. Make sure that you incorporate a coverage factor associated with the process of identifying the failed module and switching in the spare. You can assume that the spare is always powered and is just as likely to fail as the primary modules. If the failure rate of each module is 0.001 failures per hour, what is the reliability of the TMR system with the single spare at the end of a one-hour period, as a function of the fault coverage factor? If the coverage is perfect, how does the reliability of the TMR system with the single spare compare with the reliability of a TMR system without a spare (again at the one hour time period)?

4.7 Figure 4.50 shows the architecture of a network of computers in a banking system. The architecture is called a skip-ring network and is designed to allow

TABLE 4.5 MTTF data for use in Problem 4-3.

Equipment	MTTF
Processing Unit	2000 hours
Remote Terminal	3500 hours
AHRS	1000 hours
INS	1000 hours
Doppler	700 hours
Bus	5000 hours

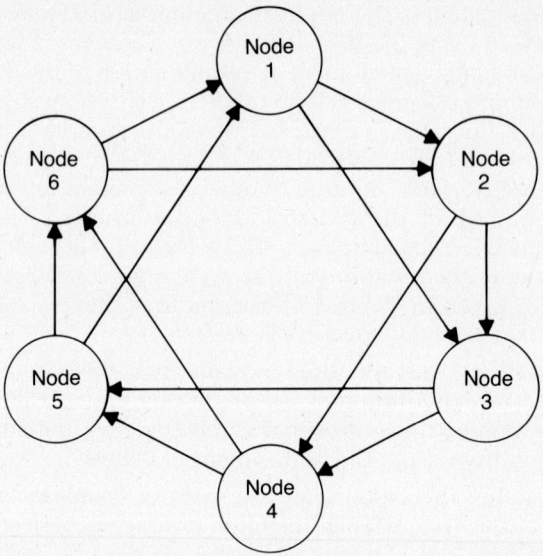

Fig. 4.50 Computer network architecture for a banking system.

processors to communicate even after node failures have occurred. For example, if node 1 fails, node 6 can bypass the failed node by routing data over the alternate link connecting nodes 6 and 2. Assuming that the links are perfect and the nodes each have a reliability of R_m, derive an expression for the reliability of this network. Include in that expression a fault coverage factor that reflects the probability of a node recognizing that a problem exists and appropriately rerouting the data. If R_m obeys the exponential failure law and the failure rate of each node is 0.002 failures per hour, determine the reliability of the system at the end of a ten-hour period for coverage values of 0, 0.5, 0.9, 0.95, and 1.0.

4.8 Figure 4.51 shows the architecture of a process control system. The basic system consists of two unique components; the processors and the interface units. The processors use self-diagnostics to detect and handle failures in the interface units and the processors themselves. A company is trying to determine the maintenance interval to recommend to its customers. The customers want to perform routine maintenance frequently enough to prevent the reliability of the system from falling below 0.995, assuming that once maintenance is performed, the system becomes perfect again. Develop a Markov model of the system and use the model to determine how often the system should be maintained. Assume that the failure rate of each processor is 0.002 failures per hour and the failure rate of each interface unit is 0.0008 failures per hour. Show how the required maintenance interval varies as a function of the fault coverage capability of the system.

4.9 The architecture of a flight control system that you have designed is simply a TMR system that uses flux-summing as the voting mechanism. Each of the

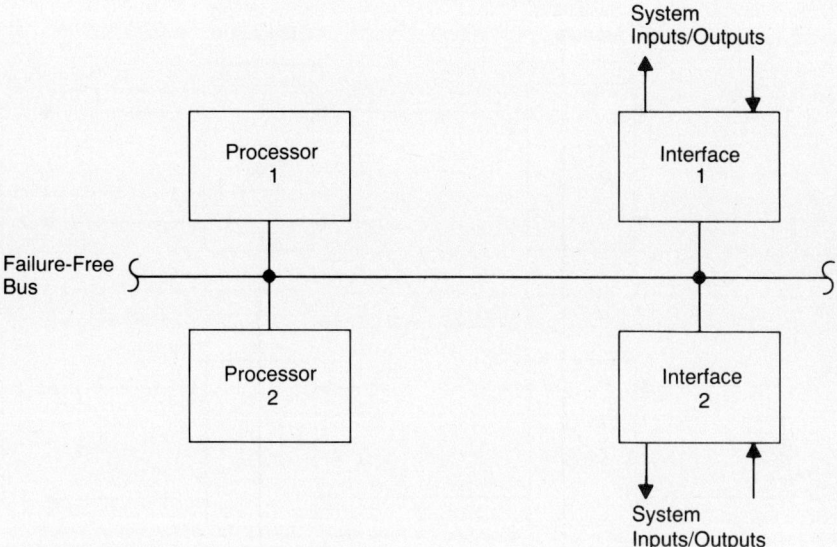

Fig. 4.51 Process control system architecture.

three processors in the system performs self-diagnostics to allow faults to be detected. If a fault is detected, the affected processor removes itself from the flux-summing arrangement. Because of customer concerns about the safety of the system, a safety analysis of the proposed architecture is to be performed. Construct a Markov model of the system and develop a safety analysis. Compare the reliability and safety of the system as functions of the fault coverage factor associated with the self-diagnostics.

4.10 A crossbar switch is a device that is capable of connecting each of n input lines to any one of n output lines. By connecting each of n processors to one input line and one output line, the crossbar switch allows each processor to connect itself to any other processor for the purposes of transferring data. A crossbar switch with four input lines and four output lines is shown in Fig. 4.52. The complete switch is constructed from smaller switching modules that have two inputs and two outputs. Develop an expression for the reliability of the complete switch in terms of the reliability R_s of the smaller switching components.

4.11 Suppose that a simplex (no redundancy) computer system has a failure rate of λ and a fault detection coverage factor of C. The fault detection capability is the result of self-diagnostics that are run continuously. If the self-diagnostics detect a fault, the time required to repair the system is 24 hours because the faulty board is identified, obtained overnight, and easily replaced. If, however, the self-diagnostics do not detect the fault, the time required to repair the system is 72 hours because a repair person must visit the site, determine the problem, and perform the repair. The disadvantage, however, is that the inclusion of self-diagnostics results in the failure rate becoming $\alpha\lambda$. In other words,

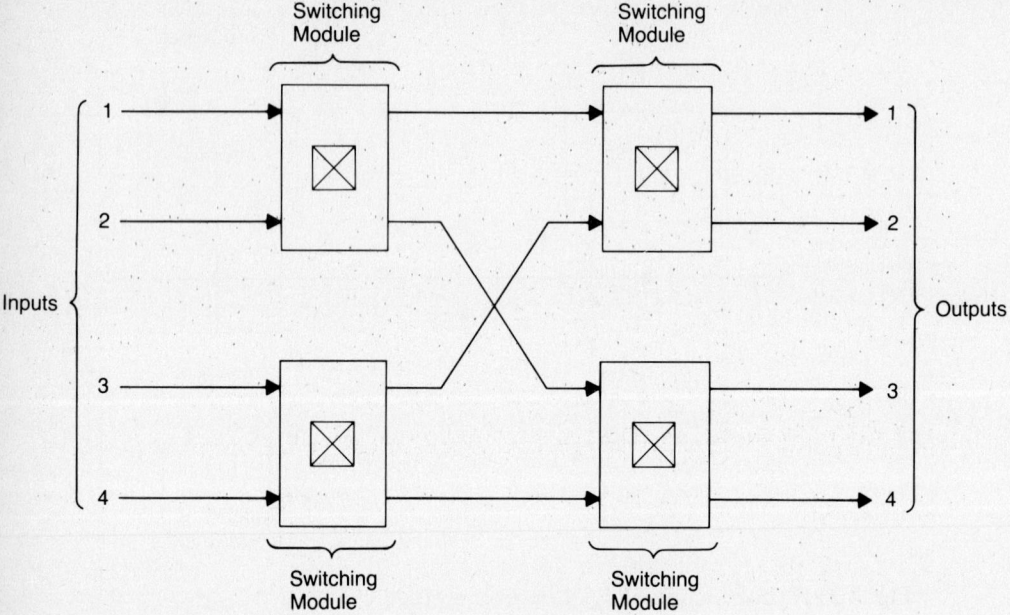

Fig. 4.52 Crossbar switch to be analyzed in Problem 4.10.

the failure rate is increased by a factor of α because of the self-diagnostics. Determine the value of α, for a coverage factor of 0.95, at which including the self-diagnostics begins to degrade the availability of the system.

4.12 Show the three-state, discrete-time Markov model of a nonredundant computer module with failure rate λ, repair rate μ, and a fault coverage factor C. The fault coverage is achieved using self-diagnostics. The three states should include the operational state, failed-safe state, and the failed-unsafe state (where a safe failure is defined as one that is covered by the self diagnostics). The model should account for the repair process. Starting with the difference equations that describe the model, develop the continuous-time expressions for the reliability, availability, and safety of the system.

5

The Design of Practical Fault-Tolerant Systems

5.1 Introduction
5.2 The Design Process
5.3 The Use of Fault Avoidance in the Design Process
5.4 A Sample Design
5.5 Sample Fault-Tolerant Systems
    Summary
    References
    Additional Reading
    Projects

5.1 Introduction

In the first four chapters of this text, we examined numerous *techniques* that can be used in designing and analyzing fault-tolerant digital systems. We have not, however, studied the *complete process* of actually designing practical, fault-tolerant systems. Our knowledge at this point can be considered somewhat as a "bag of tricks" (the techniques of Chapter 3 and the analysis tools of Chapter 4). We are now ready to learn how to apply these "tricks" to the design of fault-tolerant digital systems.
This chapter examines two primary topics:

1. The design process used to create a fault-tolerant system using the techniques presented in the first four chapters.
2. Examples of fault-tolerant systems that have been previously designed and that illustrate many of the techniques we have considered.

This chapter concludes with a selection of suggested projects that can be used to solidify the knowledge of fault-tolerant system design.

The design of a fault-tolerant system is often considered somewhat of an art. There is seldom *one* correct way to solve a problem. A particular application can have numerous approaches that completely satisfy all the requirements. Each approach, however, can have subtle differences that make one approach more attractive than another. For example, one candidate design may be easier to manufacture because it uses a processor that currently exists in a company's manufacturing capability. Another design, however, may be easier to expand because it uses the features of a more advanced processor that requires starting a new manufacturing process. The **tradeoff** in this simple example is the advantage of the increased expansion capability versus the disadvantage of the cost of starting a new manufacturing procedure.

Tradeoffs are extremely important in the design of fault-tolerant systems. Seldom can we design exactly as we wish; we must reach compromise solutions where items such as reliability are *traded* for cost and weight. The process of *compromising* is inherent in the meaning of the term tradeoff, and tradeoffs are an integral part of the design process.

The primary purposes of the design process are to:

1. Identify the candidate approaches.
2. Ascertain the advantages and disadvantages of each approach over the others.
3. Converge to a most nearly optimal solution through a series of design tradeoffs, concessions, and decisions.

As we saw while discussing evaluation methods in Chapter 4, many of the attributes of a system are quantitative, whereas many others are qualitative. In practical designs, qualitative issues such as flexibility and compatibility with previous designs can be far more important than quantitative issues such as reliability and availability. A good engineering design must consider all the attributes of a system.

The design of a fault-tolerant system requires the use of fault-tolerance techniques, fault-avoidance methods, and system evaluation procedures. These three items must be used effectively to meet the overall design objectives. We will first discuss the design process and see how fault-avoidance techniques can be used. Next, we will illustrate with an example the major steps of the design process. Finally, we will examine interesting features that have been used in some previously designed, practical systems. In the discussions of practical systems, we will examine systems from several of the major application areas.

5.2 The Design Process

The design process consists of many phases, each of which may be performed numerous times before an acceptable solution is obtained [Johnson 1987]. Figure 5.1 illustrates the various phases of the design process as well as the iterative nature of that process. The fact that the design process is iterative is extremely important! The process of performing tradeoffs naturally results in modifications during the design process. For example, referring to Fig. 5.1, we see that the derivation of the requirements can cause us to revisit the definition of the problem and perhaps eliminate certain aspects of the design to make the requirements more reasonable.

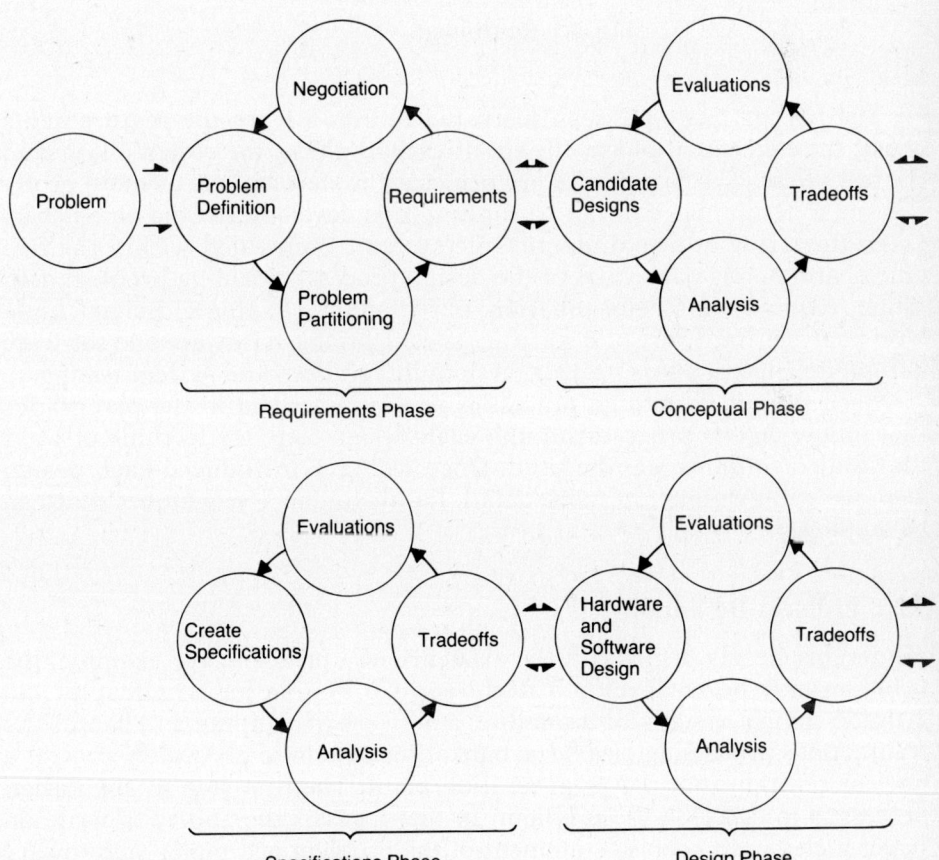

Fig. 5.1 The five primary phases of the design process include: Requirements Phase, Conceptual Phase, Specifications Phase, Design Phase, and Test Phase.

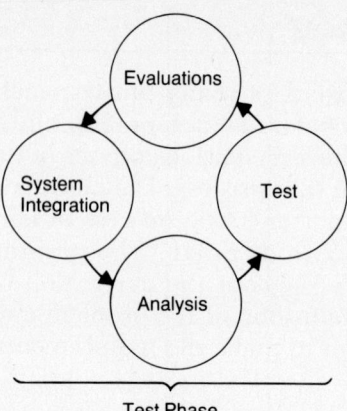

Fig. 5.1 Continued

The five primary phases illustrated in Fig. 5.1 are the **requirements phase,** the **conceptual phase,** the **specifications phase,** the **design phase,** and the **test phase.** The five phases are necessary to develop one or more prototypes of a design; we will not consider in this development the phases necessary to arrive at a manufacturable system. Embedded within the five phases are the primary steps of the design process, including problem definition, requirements determination, partitioning, candidate designs, high-level analysis, hardware and software specifications, hardware and software design, detailed analysis, testing, system integration, and system testing.

Each step of the design process is briefly described in the paragraphs that follow. As we proceed through each design step, try to think of ways that fault avoidance can be used. Once we have introduced each design step, we will examine, in more detail, fault-avoidance techniques that can be employed.

5.2.1 Problem Definition

All designs clearly begin with the existence of a problem. For example, the problem may be to develop a flight-control system for a helicopter, an attitude-control system for a satellite, a network of computers to handle the transactions processing of a large bank, or a system to accurately control a nuclear reaction in a power generation plant. The first step in any design process is to describe the problem. In many cases, the ability to write on paper a clear and concise statement of the problem is a major step toward solving the problem.

The **problem definition** may seem like a trivial phase of the design process, but it is not! Designs have frequently been delayed or performed erroneously because the designers did not clearly understand the problems to be solved. Imagine the embarrassment of preparing a detailed presentation on an advanced auto-pilot only to learn that the aircraft is not going to have an auto-pilot. This is an extreme example, but stranger things have happened.

The major purpose of the problem definition phase of the design process is to make sure that you completely understand the problem, or problems, to be solved. Without this understanding, the design process cannot continue.

5.2.2 System Requirements

The second step in the design process is to extract, or create, a set of **requirements** from the problem description. The system requirements typically consist of requirements on reliability, cost, weight, power consumption, physical size, performance (for example, speed), maintainability, and other system attributes. Many of the requirements (for example, reliability) are quantitative, and others (for example, compatibility with existing designs) are more qualitative. The derivation of the requirements may cause us to revisit the problem description as we learn more about the problem and its specific attributes. Also, the definition of the problem can make certain requirements impossible to attain.

During the determination of requirements, it is extremely important to attempt to be reasonable and to set requirements that can be verified either experimentally or analytically. For example, it may be unreasonable to require that a computer system have a reliability of 0.9999 when the reliability of the motor that it is controlling is only 0.9. Not only might this be unreasonable, it may be extremely expensive as well. Another good example of a requirement is fault coverage. A fault coverage requirement without a statement of the fault model to be used is misleading. If we require a system to have a 95% fault detection capability, we must state whether that is for stuck-type faults only or for any possible fault. If it is the latter, we may never be able to verify that the requirement has been met simply because we may never be able to identify all the faults that can occur. A requirement that we cannot verify that we have successfully met is not a useful requirement!

5.2.3 System Partitioning

Once the system requirements are well defined and understood, the design process requires **partitioning** the problem into manageable subproblems, called subsystems, that can each be handled easily by individual teams of design engineers. For small projects, each team may be a single engineer, or one designer may handle several portions of the total design. In more complicated systems, each design team may consist of tens or hundreds of engineers. The objective of the partitioning process is to divide the complete problem into manageable pieces. The partitioning can be in terms of hardware and software or perhaps a specific division of hardware subsystems. One form of partitioning can be performed based on the system requirements.

A portion of the partitioning process involves categorizing various parts of the system based on the reliability (or availability, maintainability, or some other attribute) requirements. For illustrative purposes, we will consider the reliability. The higher the reliability required of a subsystem, the more redundancy that will be necessary to attain that reliability. For example, a complex system can have one subsystem that is life-critical, such as the flight-control system of an aircraft, whereas another is only a convenience function such as the control of the "No Smoking" light on a commercial airliner. Clearly, it does not seem necessary, nor cost effective, to provide the same level of reliability for the "No Smoking" light that you provide for the flight-control system.

One way to partition a system based on reliability requirements is into categories of varying degrees of criticality [Johnson and Julich 1985]. For example, the aerospace industry classifies functions as either *flight-critical*, *mission-critical*, or *convenience* functions. Flight-critical functions are those functions that, if discontinued or performed incorrectly, could result in the loss of the aircraft or the crew. In simple terms, a flight-critical function is one that is required to keep the aircraft flying. For example, the flight-control system in a fly-by-wire airplane is a flight-critical function. Flight-critical functions are the most important functions and, as a result, usually require the highest reliability.

Mission-critical functions are those functions that are required for the aircraft and its crew to complete its intended mission. For example, an airplane could certainly fly if the radio were to fail, but it is highly unlikely that the crew could complete its intended job. The mission-critical functions usually follow immediately the flight-critical functions in the level of reliability required.

Convenience functions are those functions that are nice, and desirable to have, but that have little impact if discontinued. The "No Smoking" light in the commercial aircraft is one example. Compared to the flight-critical and mission-critical functions, the convenience functions, if discontinued, will produce relatively minor impacts. The electronic maintenance log in a military aircraft is another example of a convenience function. Although the maintenance log is convenient and significantly improves the maintenance process, neither the crew, the mission, nor the aircraft will be endangered during flight if the maintenance log becomes inoperative.

The three levels of functions—flight-critical, mission-critical, and convenience—definitely have differing requirements. As an example, the fault-tolerant computer system described by Johnson and Julich [Johnson and Julich 1985] and applied to the avionics of a combat helicopter has the three distinct sets of requirements—flight-critical, mission-critical, and convenience. The flight-critical functions consist of the fly-by-wire control system for the tail rotor. The mission-critical functions include navigation, communications, weapons delivery, stability augmentation, numerous automatic flight modes such as attitude and altitude hold capabilities, cockpit interfaces, and other similar functions. Finally, the convenience functions include items such as the maintenance log that provides engine and transmission statistics. The reliability required of the flight-critical functions was 0.9_7, whereas the reliability required of the mission-critical functions was 0.995. The convenience functions, however, required a reliability of only 0.95.

The concept of having levels of reliability or availability can be applied to systems other than aerospace ones. For example, a banking system certainly has both mission-critical and convenience functions. Mission-critical functions might include those necessary to manage the financial databases and keep the bank operational and performing its required transactions. A 24-hour teller machine, on the other hand, might be considered a convenience function to some extent. If the teller machine becomes inoperative, numerous customers may be inconvenienced, but the financial operations of the bank can still be performed.

5.2.4 Candidate Designs

One of the most important steps in the design of a fault-tolerant system is the creation of several candidate designs. Certainly, the initial candidates will be significantly deficient in the detail required of a complete design, but they will illustrate a basic approach that can be taken. For example, you may consider triple modular redundancy (TMR) as a candidate approach. You may not have yet defined the mechanism used to perform the voting, the manner in which synchronization is achieved, or even the degree of synchronization required, but the basic features of TMR can be analyzed, in depth, to determine if they are appropriate for the particular application. TMR can be contrasted with standby sparing or the triple-duplex approach to determine, at least initially, which approach, if any, is best for the application at hand.

One key reason for developing several candidate designs is that the process of determining the advantages of one approach will very likely uncover the disadvantages of another approach, and vice versa. The evaluation of the candidate designs is a key step in the overall design process.

5.2.5 High-Level Analysis

Once candidate systems are defined, the next step in the design process is to perform a preliminary analysis of each candidate architecture. A preliminary analysis can consist of a reliability estimate, cost estimates, weight estimates, and so on. At this stage of the design, many candidates that are obviously not suitable for the particular application can be eliminated quickly from consideration and further detailed development. A good, high-level analysis can easily save the designers much expense by significantly decreasing the number of candidates. For example, you might initially consider TMR to be a viable alternative for an industrial control system, but you may eliminate TMR from consideration when a high-level analysis indicates that TMR significantly exceeds the reliability requirements and the weight limitations.

The importance of some type of high-level analysis cannot be overemphasized. Analysis, in general, must be an integral part of the design process if the design is to be successful. The earlier that deficiencies in a candidate design are identified, the more cost effective the design process will be. You do not want a candidate to remain in consideration if significant problems exist in the approach. You also do not want to spend large amounts of time further developing a candidate that clearly cannot meet the requirements.

The difficulty with performing a high-level analysis is that many of the analysis techniques require a substantial amount of information on the specific design before an analysis can be performed. For example, we need some idea of the failure rate of the system's modules before we can determine the system's reliability [Johnson and Julich 1984]. Likewise, if we want to simulate the system as a means of functional evaluation, many of today's simulation tools require that significant detail be available before the simulation can be constructed [Breuer and Friedman 1976]. In many instances, we overcome these problems by analyzing the system for a range of parameters. For example, we can calculate reliability as a function of the failure rate and determine the reliability for a range of failure rates.

The outcome of the high-level analysis will be one or more candidates that will each be designed in more detail. If a single candidate must be selected, the analysis must often be significantly more detailed to allow the selection to be made intelligently. Ideally the selection is not irrevocable. If, in later design phases, it is determined that an inappropriate selection was made, the design can return to the analysis phase, or an even earlier phase, and start over. Practically speaking, however, cost considerations may dictate that the design is committed to a specific candidate once that candidate is selected and further developed.

5.2.6 Hardware and Software Specifications

The term "specification" is often used interchangeably with the term "requirement." In this text, however, a requirement is considered to be some attribute, or quality, that is demanded of a system. A **specification**, on the other hand, is a detailed plan for a design that is capable of meeting certain requirements. For example, the requirement might be to achieve a reliability of 0.995 for a simple digital filter. The specification would be the outline of the design that could meet the requirement.

Once a high-level solution to the problem has been developed, analyzed, and refined, the specifications for the hardware and software to implement the design must be developed. At this point in the design process, it is crucial that much interaction occur between the systems engineers who created the high-level design and the hardware and software design engineers who will actually create the designs. The specifications must achieve a delicate balance necessary to meet the system requirements and, at the same time, produce a practical, implementable, and manageable design.

The importance of the specifications cannot be overemphasized. A system that is specified incorrectly will never perform as desired. Relative to the causes of faults that we have discussed, specification mistakes are very much prevalent in the complex designs encountered in today's technology [Toy 1978].

5.2.7 Hardware and Software Design and Analysis

The result of the high-level analysis should be the selection of a final set of candidate solutions to the original problem. The set of possible solutions can contain only one candidate if all others have been eliminated for one reason or another. On the other hand, the set can contain two or more solutions that will each be carried through the complete design process to allow more detailed comparisons to be conducted. For example, NASA funded the complete development of the Fault-Tolerant Multi-Processor (FTMP) [Hopkins, Smith and Lala 1978] and the Software-Implemented Fault Tolerance (SIFT) concept [Wensley et al., 1978], both of which are solutions to the reliable, commercial aircraft control problem, because the designs used completely opposing approaches. The results of evaluating prototypes of both FTMP and SIFT will prove extremely valuable.

Each candidate that remains after the high-level analysis will be carried through a complete hardware and software design and construction process resulting in both hardware and software prototypes. In most instances, the hardware and software design can be performed in parallel with close interaction between the two groups to ensure that hardware decisions do not negatively impact the software, and, likewise, that software decisions do not negatively impact the hardware design.

It is very important to continue the analysis of the designs in parallel with the actual design and construction to ensure that the system requirements (reliability, availability, maintainability, cost, and so on) are not compromised by design decisions that are made. Many times the analysis portion of the design process is overlooked until the design is complete. At that point, it is often too late to make substantial changes to remedy any design problems. As the hardware is developed, more specific data can be provided to the analysis; for example, the failure rate information and the fault coverage data. Therefore, the analysis can be refined as the design is refined. Once the design is complete, the analysis should also be complete.

5.2.8 Testing

An extremely important part of the design process is the development of a plan for testing the resulting designs and the actual testing itself. A crucial part of the design process is developing an easily testable design. Testing involves searching for faults of all types, including faults resulting from design mistakes, implementation mistakes, and component defects. The overall purpose of the test phase of the design process is to ensure the correct operation of the system. As in all the steps of the design process, the outcome of testing can result in a redesign of the hardware, software, or both. Also, the test process can uncover architectural or algorithmic problems that must be corrected.

At this stage of testing, the hardware and software are usually tested separately. Several small programs can be written to allow the hardware to be tested adequately, but you want to maintain as much independence between the hardware and software as possible to allow problems to be identified easily as either hardware or software problems.

5.2.9 System Integration and Test

Once the hardware and software prototypes have been tested adequately, they must be combined to form a complete, operational system. The process of combining the hardware and the software is usually called **system integration**. The fundamental purpose of system integration is to get all of the subsystems working together to perform the desired functions of the system. Each of the subsystems can work perfectly when tested independent of the remainder of the system; however, when all the subsystems are required to coordinate, interfacing problems can arise.

System integration can be performed first in the laboratory. For example, the nodes of a banking computer network can be interconnected and tested in the laboratory before the equipment is placed in the field.

Once the system integration is complete, the system must be tested. Software faults, for example, that never create errors in the emulators used to develop the software can suddenly produce erroneous results because of hardware idiosyncrasies. Likewise, hardware faults can emerge because the software is exercising the hardware in a manner slightly different than expected. The ultimate outcome of the system test phase will be an operational system. Like system integration, system test can be performed in the laboratory before being performed in the field.

5.3 The Use of Fault Avoidance in the Design Process

The basic goal of fault avoidance is to prevent faults from occurring. For example, fault avoidance can be used to prevent design mistakes or implementation mistakes. The underlying theme of fault avoidance is that you achieve neither fault tolerance, high reliability, high availability, nor any other key system attribute unless you do something during the design process to ensure that the desired attributes are being *designed into* the system. In other words, simply testing for quality at the end of the design process is insufficient! We must incorporate procedures at all phases of the design to guarantee the achievement of a system's requirements.

Several fault avoidance techniques can be applied at different points during the design process. Examples include various types of design reviews, adherence to design rules, shielding against external disturbances, and quality control checks. Figure 5.2 shows the design process that was il-

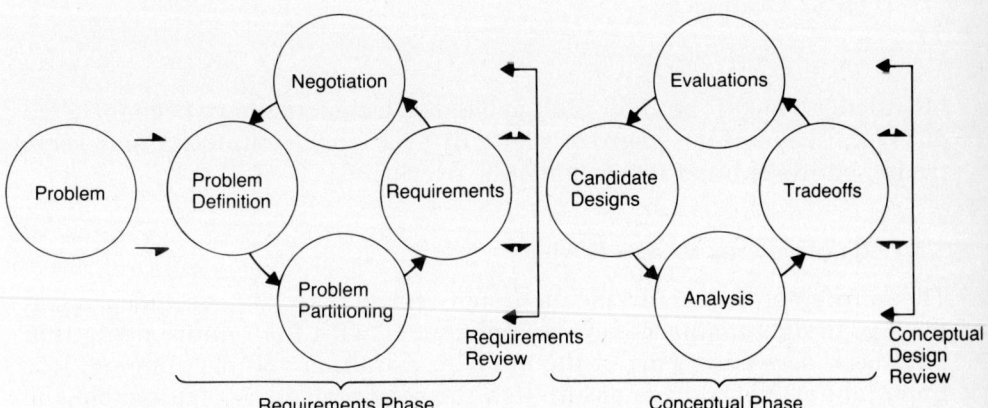

Fig. 5.2 Fault avoidance techniques such as design reviews can be included between the primary design phases.

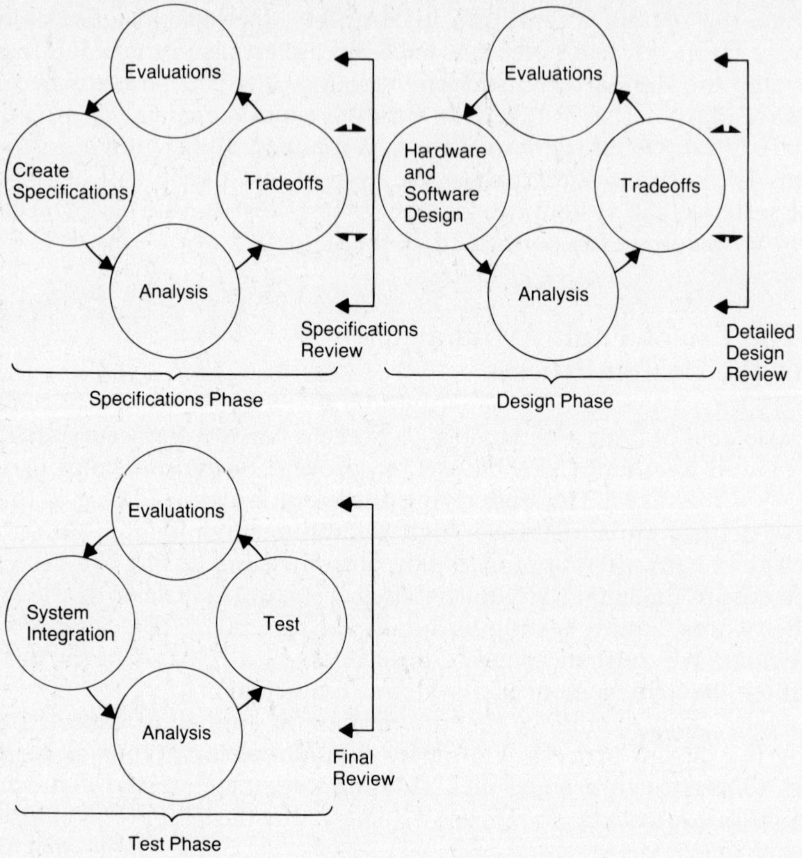

Fig. 5.2 Continued

lustrated in Fig. 5.1, but now fault avoidance techniques have been included at several key points. Many of the fault avoidance techniques may seem trivial, but their importance cannot be overlooked.

5.3.1 Requirements Design Review

The purpose of the **requirements design review** is to have an independent team of design engineers review and concur with the requirements that have been derived as part of the problem definition and partitioning. It is important at this point to ensure that the requirements are reasonable and will be verifiable during the analysis and testing process. It is also important to ensure that all necessary requirements have been identified. For example, if a weight limitation exists but is simply forgotten, the design may have to be modified later on (at perhaps considerably more expense) to reduce weight and meet the limitations.

The requirements design review helps eliminate specification mistakes as a cause of faults. If an erroneous weight requirement is derived, the designers may later specify a hardware design that is too heavy, too slow, or consumes too much power. The designers may actually believe they are fulfilling the requirements, but, in fact, the requirements are wrong and a costly design modification can result.

5.3.2 Conceptual Design Review

The **conceptual design review** occurs immediately after a high-level, candidate design has been developed. The purpose of the conceptual design review is to ensure that the basic concept of the candidate design is correct and meets the requirements that have been developed. During the conceptual design review, the design engineers normally present the results of their initial analysis to justify their belief that the candidate design can fulfill all the system requirements. In addition, the conceptual design review examines candidates that were eliminated from consideration to verify that the reason for elimination is sufficient.

The conceptual design review attempts to eliminate specification mistakes as a cause of faults by ensuring that the concept that is ultimately specified is correct. The review should be performed by a team of independent designers to guarantee that the reviewers are not biased prematurely, in any way, toward or against the proposed candidate solution.

5.3.3 Specifications Design Review

The result of the specifications phase of the design process is a detailed hardware and software specification and a plan for testing the resulting hardware and software. The **specifications design review** attempts to verify the specifications to ensure their validity. Before the hardware and software designs are begun, it is important to make sure that the specifications are both correct and understood completely. Otherwise, faults due to specification mistakes will inevitably occur in the design. Also, it is extremely important to guarantee that the designs will be testable.

As in the other design reviews, the specifications review should be conducted by an independent team of designers capable of completely understanding and critiquing the material. After the completion of the review and any resulting modifications, the specifications for the hardware and software should be ready to pass along to the designers, and the detailed design process can begin.

5.3.4 Detailed Design Review

Detailed design reviews must be performed on both the hardware and software designs before beginning the actual implementations. The purpose of the review is to ensure that the specific designs meet the specifications and can fulfill the system requirements. A detailed design review is one of the most important and difficult reviews that must be performed because the specific details of the design must be reviewed and verified. The detailed design review must encompass both the detailed design and the detailed analysis. For example, once the detailed design is performed, specific failure rates should be known such that the accuracy of the reliability analysis can be refined.

When the detailed design review is completed and any revisions have been made, the design should be ready to implement. The review process is a key method of preventing design mistakes from appearing within the system.

5.3.5 Final Review

The **final review** is intended to be a last checkpoint in the design process. At the time of the final review, a working prototype is often available. The basic purpose of the final review is to examine the performance of the prototype and the results of the final analysis to ensure that specifications have been adhered to and that the system requirements have been met. If the design has been performed correctly and the design reviews have been successful throughout the design process, the final review should not uncover any major problems. The final review should simply be an affirmation of a job well done.

5.3.6 Parts Selection

The selection of parts for a system can be critical to the achievement of reliability, availability, or some other system attribute. There are many levels of quality in the components that are available for use in designs; in many cases, the failure rate of a high-quality part can differ from that of the same part at a lower quality by a factor of 300. As might be expected, however, there is also a tremendous difference in the costs of parts of different qualities. Therefore, it is extremely important to guarantee that the components are selected appropriately. The primary tradeoffs in the parts selection are:

1. The cost of the part versus the failure rate of the part
2. The availability of the part versus the failure rate
3. The cost (both financial and otherwise) of a failure of the part

5.3.7 Design Rules

Design rules often play an extremely important role in the design of a system; the strict adherence to design rules can improve the system substantially. Design rules can address, as examples, packaging, testing, shielding, or circuit layout. For example, a design rule can require that the system be partitioned into subsystems that are no larger than some predefined value. The size of the subsystem can be controlled in terms of the number of logic gates, the physical size of the subsystem, the total number of cards required, or some other metric. By partitioning based on size, each subsystem becomes more manageable, and the probability of design mistakes is significantly decreased.

5.3.8 Documentation

Documentation is another example of an extremely important aspect of fault avoidance. Design projects often require that different phases of the design be handled by different teams of designers. For example, a team of *systems engineers* develops a system's architecture, and then a team of *design engineers* takes the systems engineers' specifications and designs the hardware and software. If the systems engineers incorrectly document the top-level architecture, the final hardware and software designs can be incorrect. The design reviews mentioned previously are mechanisms that can be used to detect documentation errors. It is critical to the success of a project that each stage of the design be described correctly and clearly in the documentation.

5.4 A Sample Design

The following sample design illustrates many of the features of the design process described in the previous sections. It is impossible to provide complete details on a design of any complexity in just a few pages; however, many of the important concepts can be illustrated. We limit ourselves to discussions of: problem definition, system requirements, system partitioning, candidate designs, and high-level analysis. Our design is significantly simplified here so that many of the crucial aspects of the design process can be presented in a concise manner. The projects at the end of this chapter allow investigation, in more detail, of the specifics of each step of the design process.

Our sample design is that of the electronics for an aircraft that uses fly-by-wire control technology. The basic structure of the aircraft is illustrated in Fig. 5.3. Movements of the pilot's yoke, rudder pedals, or the throttle are

278   The Design of Practical Fault-Tolerant Systems

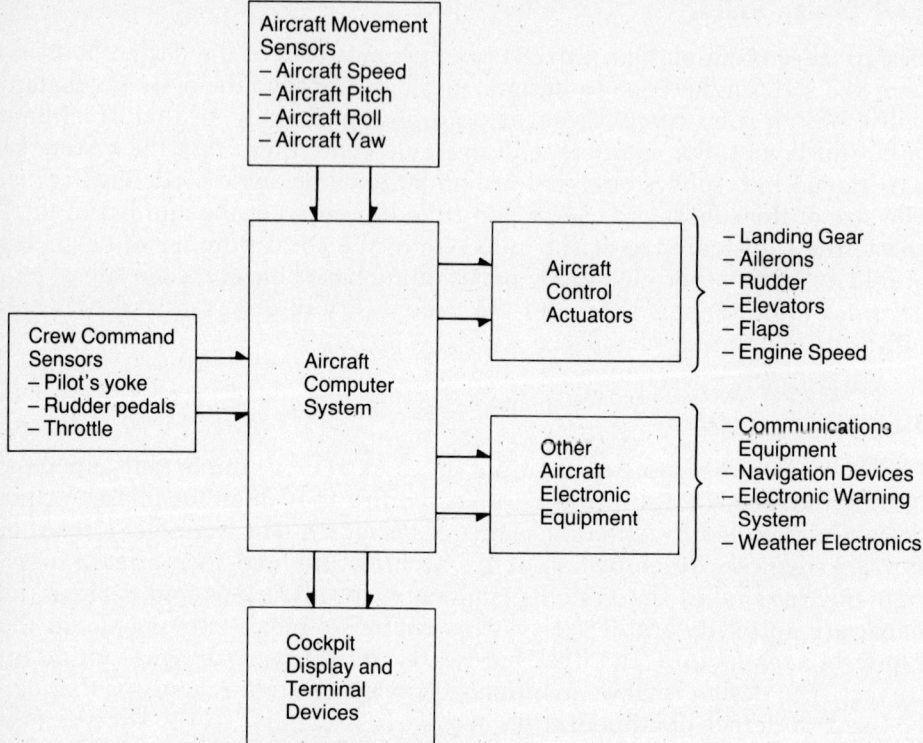

Fig. 5.3 The aircraft computer system must accept crew commands on aircraft flight as well as communications and navigation commands. Based on available data, the computer system controls actuators and all electronic devices.

measured by sensors, which in turn are to be sampled by the flight-control computer system. Based on the crew commands, the control surfaces (rudder, elevators, and ailerons) are to be positioned to appropriately control the aircraft. The throttle specifies jet-engine speed to allow the crew to control the air- and ground-speed of the airplane.

In addition to these basic flight-control functions, the electronics are also responsible for controlling the communications, navigation, cockpit display, and the electronic warning system. The communications equipment on board the aircraft consists of several different types of radios, including a high-frequency one for air-to-ground communications and another for communications among the crew. The electronics must be capable of allowing the crew to specify electronically which radio they wish to use as well as the radio frequency.

The navigation system is responsible for all of the on-board navigation equipment. Examples include the radar altimeter for accurate altitude readings, a weather radar for weather information, distance measuring equipment, an inertial reference system, and an instrument landing system.

Cockpit display devices allow the crew to examine information on a cathode ray tube (CRT) display during a flight. Displayed information might include, for example, engine performance, flight-plan checkpoints, heading and altitude information, landing information, and the locations of other aircraft traffic.

Finally, the electronic warning system is responsible for monitoring the complete system and issuing warnings to the crew in the event that anomalies occur. The warnings may include overheating components, unlocked landing gear, unusually low altitude, and equipment failure.

5.4.1 Problem Definition and Initial Partitioning

The problem, at this point, can be divided into two subproblems:

1. The development of an architecture that can sample the pilot's commands, convert those commands into signals that specify the position of the control surfaces and the amount of thrust, and apply the resulting signals to the actuators that physically perform the control action.
2. The development of an architecture that can support the use of the remaining aircraft equipment, for example, the radios.

The architecture that solves the first problem is the *flight-control system*, whereas the solution to the second problem is the *flight-management system*. Note that this is a natural partitioning of the problem that has arisen. We will focus primarily on the flight-control architecture during the development and analysis of candidate architectures.

We need to investigate in more detail the functions that each of the subsystems must perform before we can proceed past the problem definition stage of the design process. We will first consider the flight-control system. The flight-control system has five major functions:

1. Controlling aircraft's roll motions by adjusting the ailerons on the wings
2. Controlling aircraft's yaw motions by adjusting the rudder
3. Controlling aircraft's angle-of-attack (pitch) via the elevators on the tail
4. Controlling aircraft speed and other related factors by varying engine thrust
5. Controlling lift and speed via the flaps located on both wings

Note, if it is not apparent by now, that if the flight-control system fails to perform its tasks correctly, the crew will not be able to fly the airplane.

The flight-management system has four major functions:

1. Accepting information from the crew to allow, for example, radio selection, equipment initialization, and specification of the flight plan
2. Providing the control stimulus to the appropriate equipment to implement the functions listed in (1)
3. Displaying the results of crew requests and navigation information
4. Monitoring all equipment and immediately reporting all failures to the crew

The item most noted about the flight-management functions is that the airplane is still flyable even if the flight-management functions cease to be performed correctly.

At this point, the problem description is still not very detailed but can be stated succinctly as one of designing both the flight-management system and the flight-control system architectures. Before we further refine the problem statement, we need to investigate the requirements that will be placed upon the system. From this point on, we will consider only the development of the flight-control architecture.

5.4.2 Requirements Definition

In most problems, the derivation of the requirements is one of the most difficult jobs to perform. It is easy to say that we want something to be *highly* reliable, *extremely* light, and consume *little* power, but the words "highly," "extremely," and "little" are not descriptive enough to build a design upon. A highly reliable system to one designer might be marginally reliable to another. Likewise, a system that is considered light, in terms of weight, to the commercial aircraft designer can be considered prohibitively heavy to the designer of a satellite system.

With many systems like the one described in this example, the reliability of the electronics is often established based on the reliability of the mechanical parts such as the aircraft structure and the engines. For example, the probability that the airplane will be capable of flying throughout a given time interval depends on the probability that *both* the electronics and the mechanical elements are operating correctly throughout that same time interval. Assume that the time interval is one hour. If the reliability of the mechanical elements is 0.99, it hardly seems worth it to design the electronics to achieve a reliability of 0.9_7. However, if the reliability of the mechanical elements is 0.9_7 or greater, the electronic system should normally be designed so as to not significantly degrade the overall system reliability.

One of the first items that must be established is the time interval over which correct operation must be guaranteed. As we have previously seen, a satellite may have to operate correctly for five years, whereas an aircraft control system may need to operate correctly for only a few hours. Once again, the physical elements of the system can provide the answer. We probably would not design a flight-control system to have an extremely high reliability over an interval of time significantly longer than the length of the longest possible flight. In this example, we assume that the longest possible flight will be three hours, and the reliability requirements will be taken over a three-hour time interval.

To better understand the requirements that must be placed upon the flight-control system electronics, we will perform a preliminary reliability analysis of the complete system. A simplified reliability block diagram of the aircraft system is shown in Fig. 5.4. The primary use of the reliability block diagram of Fig. 5.4 is to establish, in a very general manner at this point, how the flight-control system electronics affect the overall system reliability. A primary objective in the design of the flight-control system electronics is to prevent the electronics from being the *weak link* in the system. In other words, the reliability of the flight-control system electronics should be at least as high as that of the series combination of the remaining elements, both mechanical and electrical. We assume for simplicity that the aircraft actuators are the limiting factor with a reliability of 0.9_7. Consequently, an initial design goal is to ensure that the flight-control electronics have a reliability of at least 0.9_7 at the end of a three-hour flight.

In addition to being reliable, the flight-control architecture must be fault tolerant. In other words, the flight-control system must continue to operate correctly after the occurrence of at least one fault since a single fault should not result in the loss of the aircraft. The primary question concerns the number of faults that should be tolerated by the system. Is it sufficient to provide for the tolerance of only one fault? Should the flight-control architecture tolerate two, three, or perhaps even four faults? What should be the safety requirements? Is fail-safe operation required after the second fault? Should we allow the system to tolerate the first fault and fail safely after the second fault? These are extremely difficult questions to answer,

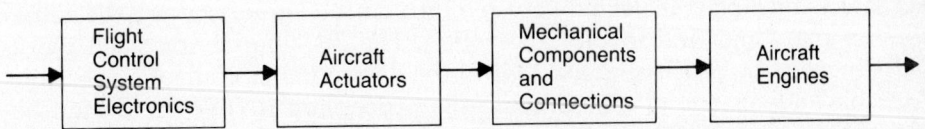

Fig. 5.4 The four primary elements of the flight-control system are: electronics, actuators, mechanical devices and connections, and engines. One objective in the design of the electronics is to prevent the electronics from being the "weak link" in the system.

but the answers must be provided before the design of the system can be continued. We must develop the goals of the design!

In many cases, the design goals must be developed after performing some preliminary analysis of the system. We assume that such an analysis has been performed and that the result has been the requirement that the flight-control architecture must be capable of tolerating a first fault 100% of the time. The second fault must be tolerated a high percentage of the time, but, at this point in the design, it is not known what percentage of second faults must be tolerated. It is considered a part of the design process to establish this requirement.

5.4.3 System Partitioning

The flight-control architecture can be partitioned physically by considering the basic structure of the control system. The flight-control system must accept commands from the crew via the appropriate interfaces, process those commands to determine the desired positions of the control surfaces and the engine thrust, develop commands for the individual actuators, and supply those commands to the actuators.

The control system is a complex feedback system, as illustrated in Fig. 5.5. The system contains both a major feedback loop and several minor feedback loops. The major loop uses the crew's desired speed, roll, pitch, and yaw, and the actual speed, roll, pitch, and yaw to determine the desired positions of the various actuators. The minor loops use the desired and actual positions of the actuators to control the movement of the actuators.

The control system can be implemented in two basic ways. First, all processing, including the major and minor loops, can be performed by a centralized computer that sends drive signals to the actuators on the wings and in the tail section. The second option is to distribute the processing and to locate computers in the wings and the tail sections. The wing and tail computers could perform the processing required of the minor feedback loops. One advantage of distributing the processing in this manner is that the transfer of data over some medium is minimized. In a centralized approach, the drive signals to the actuators and the feedback signals from the actuators must be transferred between the central computers and the actuators on the wings and in the tail section. The distributed approach physically locates the processing as close to the device under control as possible. A second advantage of distributing the processing is that the design becomes partitioned into smaller, and perhaps easier to handle, pieces.

Because of the above-mentioned reasons, the distributed approach will be employed. The basic architecture of the distributed system is shown in Fig. 5.6. A centralized processing unit samples the crew commands and the state of the aircraft and calculates the required actuator positions to move

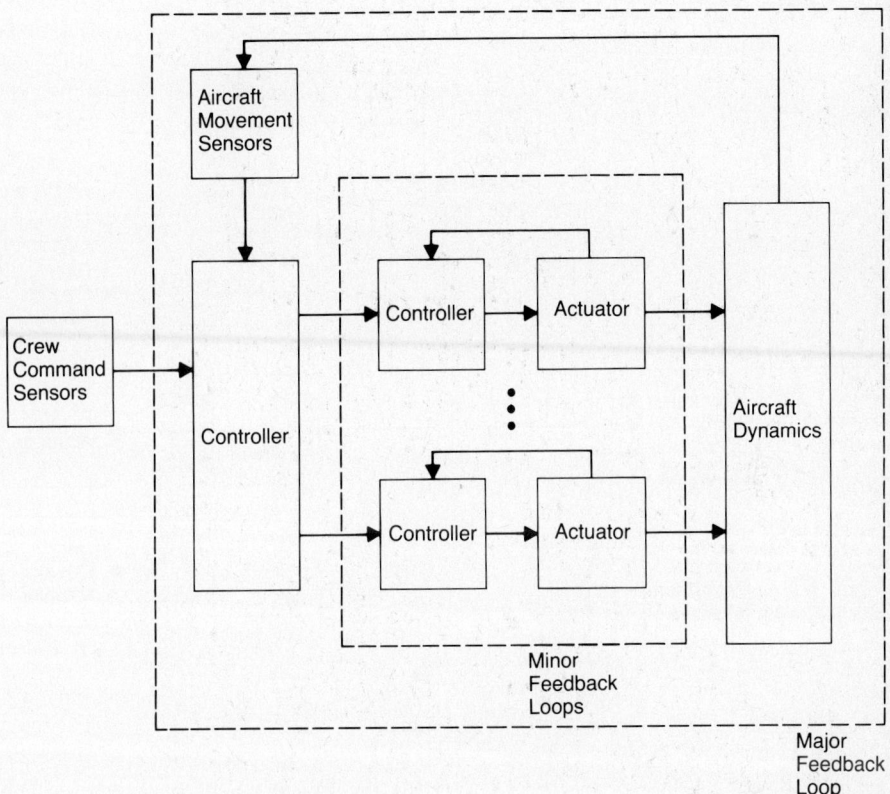

Fig. 5.5 The aircraft control system contains both major and minor feedback loops. The major loop uses actual and desired aircraft movement to generate desired actuator positions. The minor loops control each actuator based on desired and actual actuator positions.

the aircraft state to the desired value. The required actuator positions are transferred from the central computer to a *local* computer at the specific actuator. The local computers use feedback on the actual actuator positions to appropriately control those actuator positions. In other words, the central computer implements the major control loop, and the local computers implement the minor control loops.

Figure 5.6 shows only the functional configuration of the system; there are a number of techniques for interconnecting the central and local computers. Several interconnection options will be considered during the discussion of the candidate designs. It is also important to point out that Fig. 5.6 considers only the control of the ailerons, flaps, rudder, and eleva-

284 The Design of Practical Fault-Tolerant Systems

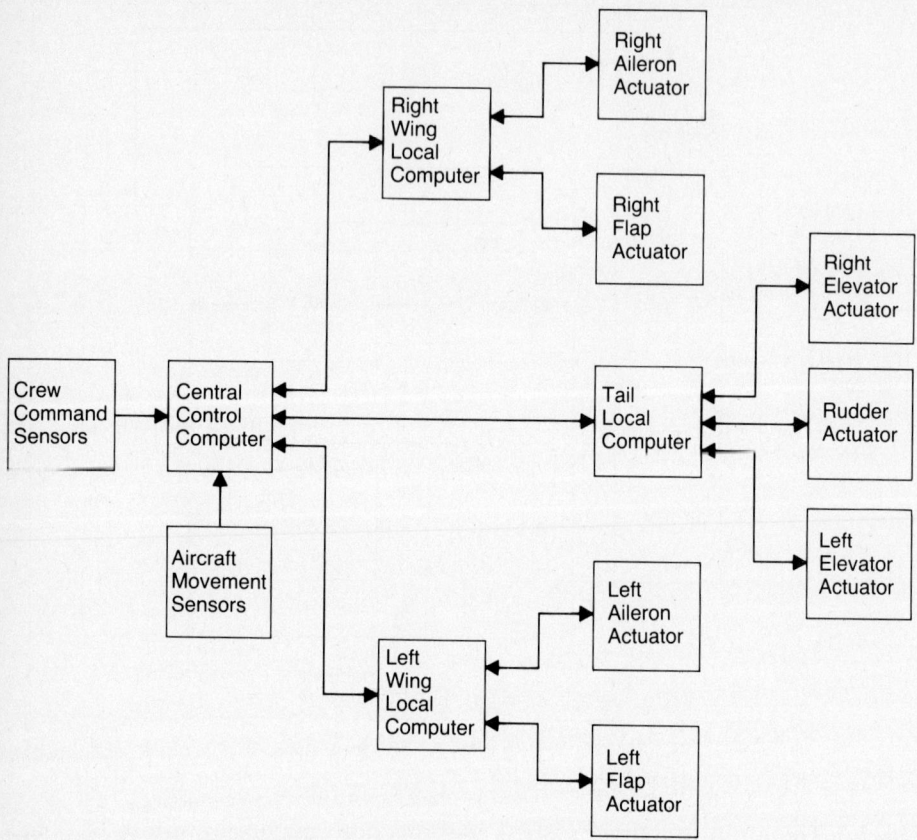

Fig. 5.6 The processing capabilities are distributed throughout the aircraft. A central computer performs the major-loop processing while local computers perform the minor-loop processing.

tors. For simplicity, the control of the landing gear, engine thrust, and other entities has been ignored; however, the distributed structure is also assumed to be applied to their control.

5.4.4 Candidate Designs

The purpose of this section is to develop several candidate approaches for the flight-control architecture. Note that we will not be able to consider all possible approaches. If we were actually given the problem of designing the flight-control system, we would consider a wide range of alternatives before selecting one or two to further develop and analyze. Unfortunately, we must limit the considerations in this text because of time and space restrictions.

5.4 ■ A Sample Design

Before we consider the specific redundancy technique that will be used or the amount of redundancy that is needed, we will examine the interconnection techniques used to allow the central and local computers to communicate. Two approaches are considered. The first is a busing structure that is shown in Fig. 5.7. In the approach of Fig. 5.7, the central and local computers communicate via redundant buses. The central computer can communicate, via any of the redundant buses, with any of the local computers. A second possible approach is to use point-to-point interconnections, as shown in Fig. 5.8. In the structure of Fig. 5.8, the central computer has a dedicated bus for each of the local computers and the sensors. Several

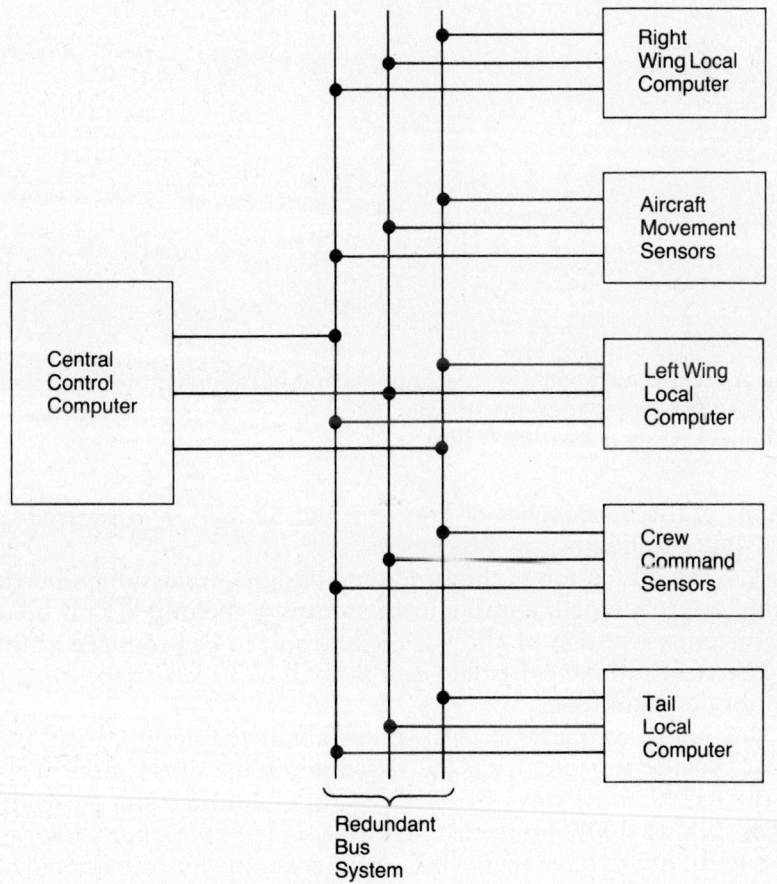

Fig. 5.7 One approach for communicating between central and local computers is to use a redundant bus structure. The failure of any one bus will not prevent the necessary communications from occurring.

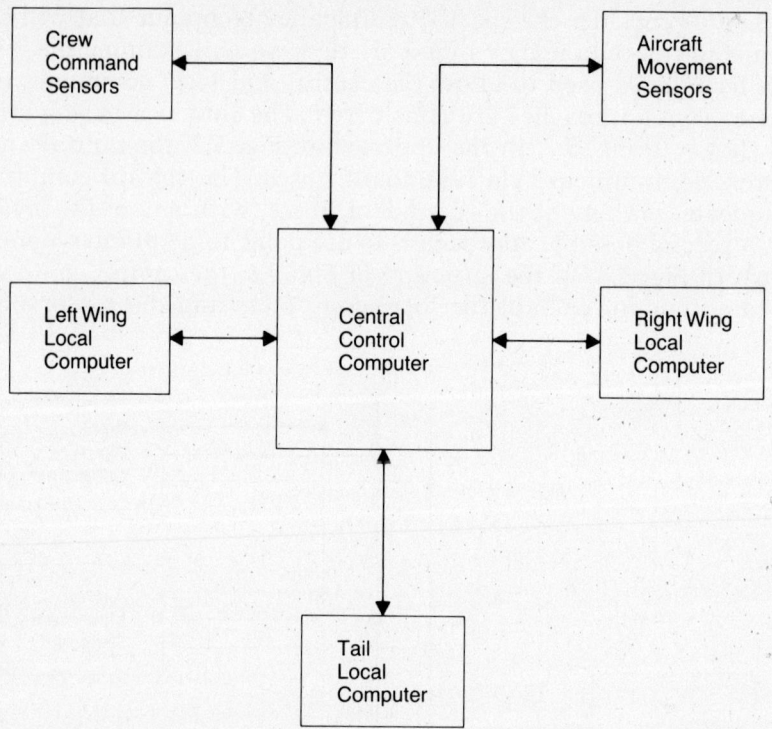

Fig. 5.8 A second approach for communicating between central and local computers is to use point-to-point interconnections. Each interconnection may be redundant to allow tolerating failures.

variations of the approaches of Figs. 5.7 and 5.8 will be discussed in more detail as the candidates are developed.

During the discussion of the candidate designs, we assume that the flux-summing approach or a similar force-summing technique will be used in the flight-control system to allow a single result to be produced at the actuators. Therefore, all the candidate designs will be based on the fundamental concept of flux-summing.

At this point, we know several things about the design. First, we know we must provide for tolerating the first fault 100% of the time. This mandates the use of some form of passive fault tolerance since dynamic approaches require 100% fault coverage to provide 100% fault tolerance for the first fault. As we have seen, 100% fault coverage for all possible faults is impossible to verify. Second, we know that the system must achieve a reliability of 0.9_7 or greater throughout a three-hour time interval. Finally, we know that the processing will be distributed physically throughout the aircraft with the major-loop control performed in a centralized location, whereas the minor-loop control of the individual actuators is performed as close as possible to the actuator.

A natural candidate that arises is triple modular redundancy (TMR). TMR provides for the masking of the first fault, and with the addition of fault detection and reconfiguration capability, TMR can provide some tolerance of the second fault. An advantage of TMR is its relative simplicity when compared to other approaches.

The architecture of a candidate TMR system using triplicated TMR (or TTMR) and a bus structure is shown in Fig. 5.9. There are three central computers that each have a dedicated set of sensors for determining the crew commands and the state of the aircraft. Each central computer uses the crew commands and the aircraft state to determine the desired position of each of the control-surface actuators. The desired positions produced by each central computer are voted upon (using a majority vote) before being transferred to the appropriate local computer via the redundant buses. The values received on the buses at the local computers are voted upon (again, a simple majority vote) before being processed by the local computers. The drive currents produced by the local computers are flux-summed to control the actuators. The system in Fig. 5.9 can tolerate any single faults that occur. The failure of two local computers can be tolerated as long as the affected local computer is identified and removed from the flux-summing process. The failure of two buses or of two central computers, however, cannot be tolerated in this TMR system.

An alternative implementation of the TMR system is shown in Fig. 5.10. The architecture of Fig. 5.10 is referred to simply as TMR. In Fig. 5.10, each of the three redundant central computers communicates with one, and only one, of the local computers at each actuator. In other words, the system consists of three independent channels with each channel containing a central computer, a sensor set, and a local computer for each of the actuators. The major feature of this approach is that the flux-summing is now the only fault masking that occurs within the system. The failure of one of the central computers propagates through each of the local computers to which it is connected. A disadvantage of this approach is that any fault in a channel results in the entire channel becoming inoperative. A strong advantage of the approach is that the system can tolerate faults in more than one channel as long as the faults are detected and the channel is removed from the flux-summing process. Although several other variations of the TMR system could be considered, we will limit our discussions to those presented in Figs. 5.9 and 5.10.

A third candidate is the triple-duplex architecture presented in Chapter 3 and referred to here as TDTMR (triple-duplex TMR). TDTMR is identical to the TMR approach of Fig. 5.10 except that each channel of the system uses

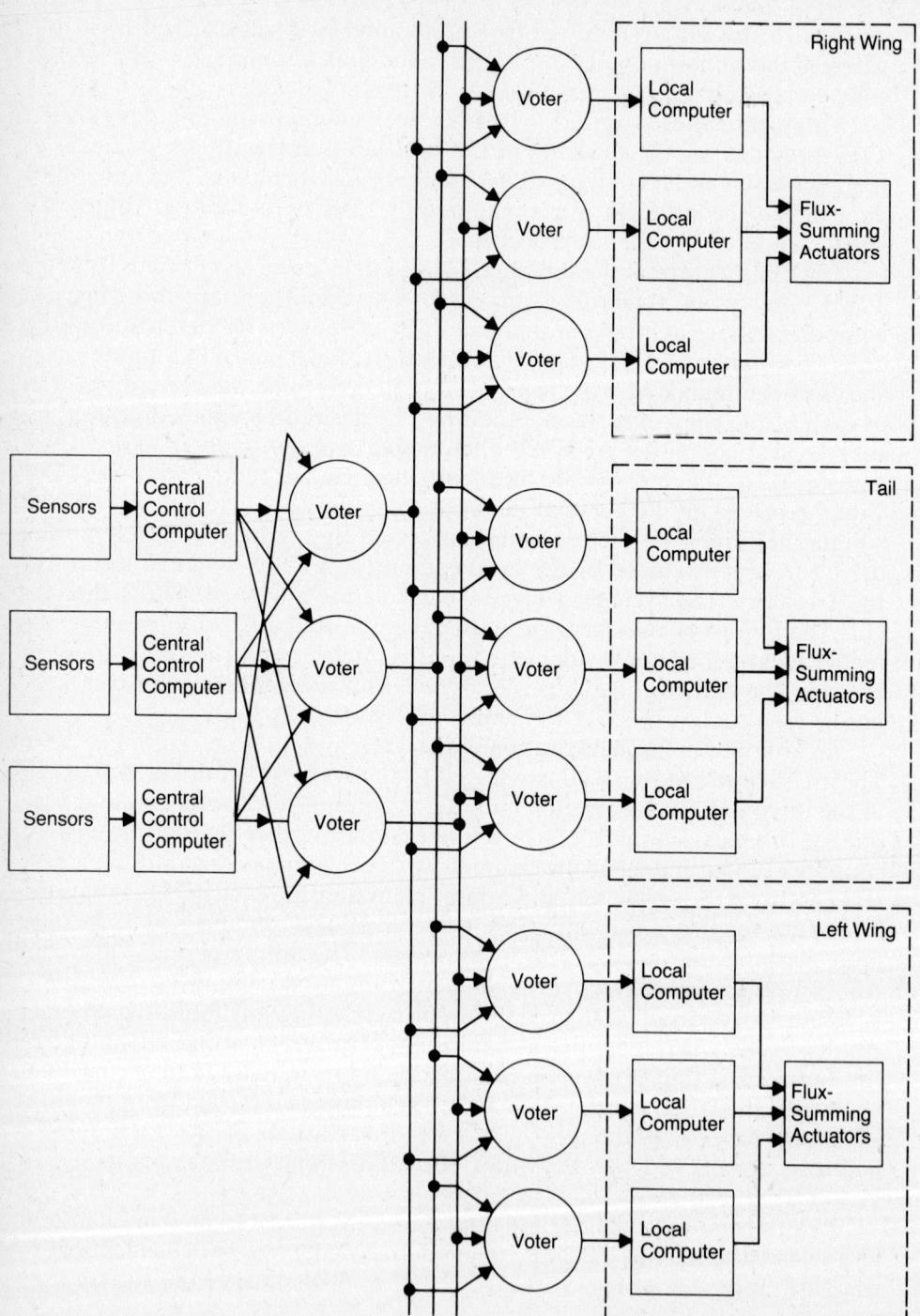

Fig. 5.9 Triplicated TMR (TTMR) uses triple redundancy of all units, including central computers, local computers, voters, and buses.

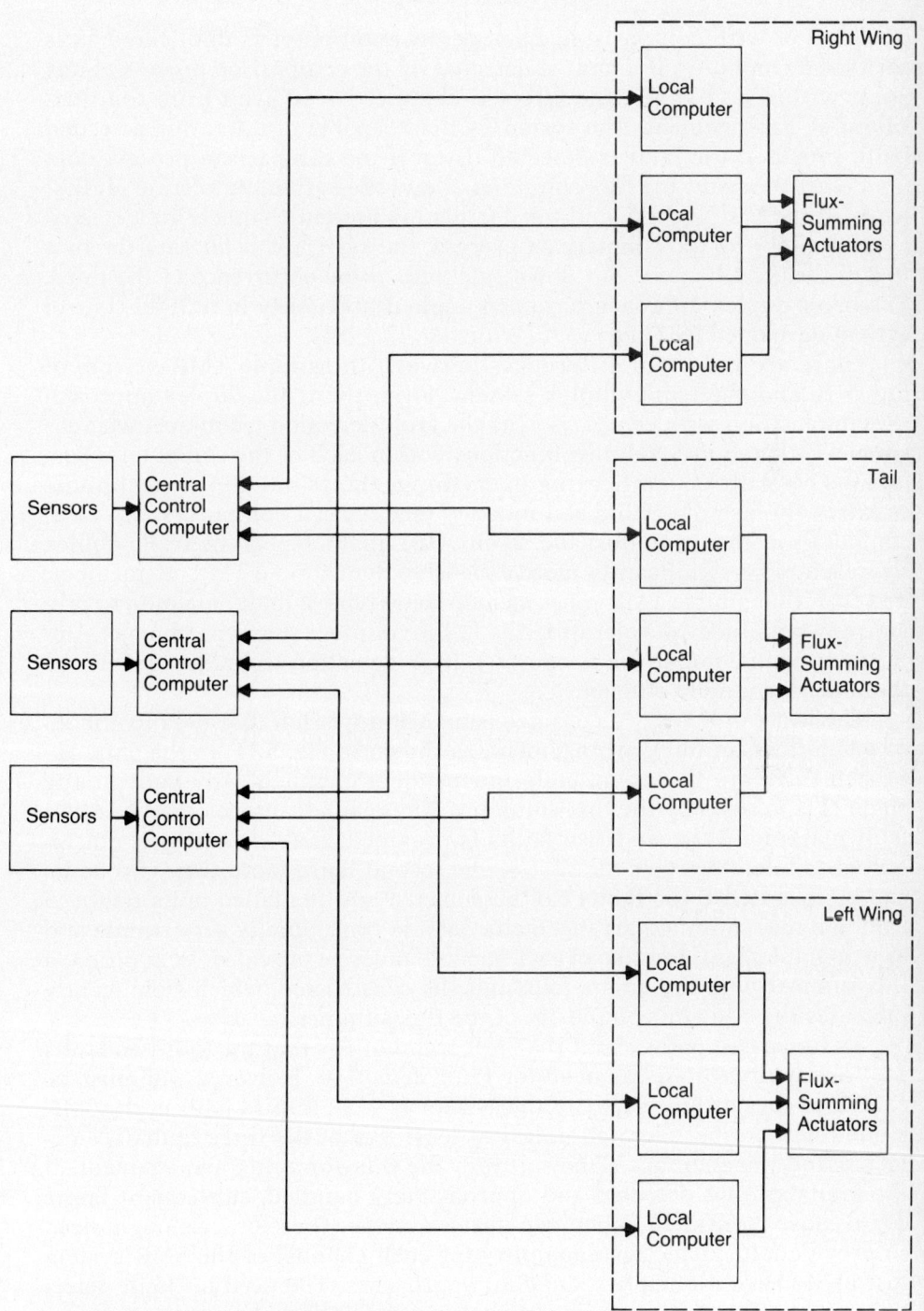

Fig. 5.10 The TMR architecture contains three independent channels. Each channel consists of a sensor set, central control computer, right wing local computer, tail local computer, left wing local computer, and appropriate interconnections.

duplication with comparison. Each central computer is duplicated as is each local computer. If a fault is detected by the comparison process at any point within a channel, the affected channel is removed from the flux-summing arrangement. The system is then capable of tolerating a second fault, provided the fault is detected. Even if the comparison process does not detect the fault, the flux-summing allows the system to tolerate all first faults. If the first, second, and third faults in different channels are detected appropriately by the comparison process, the system can tolerate the first two of those faults and shut down safely upon the occurrence of the third. The triple-duplex approach has been applied previously in a flight-control system developed by Honeywell [White et al., 1981].

There are only two differences between the simple TMR system of Fig. 5.10 and the triple-duplex system. First, the triple-duplex approach uses duplication with comparison as the fault detection technique, whereas simple TMR requires self-test functions within each of the computers. Second, as a result of duplicating everything, the triple-duplex technique requires the use of sixfold redundancy (six central computers, six local computers at each actuator, and so on). One might expect the triple-duplex approach to be significantly more expensive than simple TMR. Remember, however, that simple TMR must include some type of fault detection to provide any possibility of tolerating the failure of more than one channel. Duplication with comparison is one such fault detection technique. It may be that others are more efficient.

The last candidate we consider is a 5MR approach that uses five modules in a flux-summing arrangement, as shown in Fig. 5.11. In the 5MR approach there are five completely independent channels. Any fault in any channel is masked by the flux-summing. The real advantage of 5MR is that fault masking of the first two faults is provided; fault detection is not required to tolerate either the first or the second fault. More than two faults can be tolerated if the faults can be detected and the failed units removed from the flux-summing arrangement. 5MR is conceptually very simple and easy to implement, but five of each module must be provided. In addition, a flux-summer with five input coils must be constructed, which significantly increases the size and complexity of the flux-summer.

An important point about the 5MR technique is that the first two faults are tolerated without the need for fault detection. However, we must be careful to consider the safety of the system as well. A third fault in the 5MR system will not be tolerated unless at least two of the three faults are detected and the channels removed from the flux-summing arrangement. If the faults are not detected and appropriately handled, subsequent faults could cause the system to perform unsafe actions. Therefore, we might need to provide fault detection capability for each channel of the 5MR system just as we have the triply redundant approaches. The need for fault detection might affect the apparent simplicity of the 5MR technique.

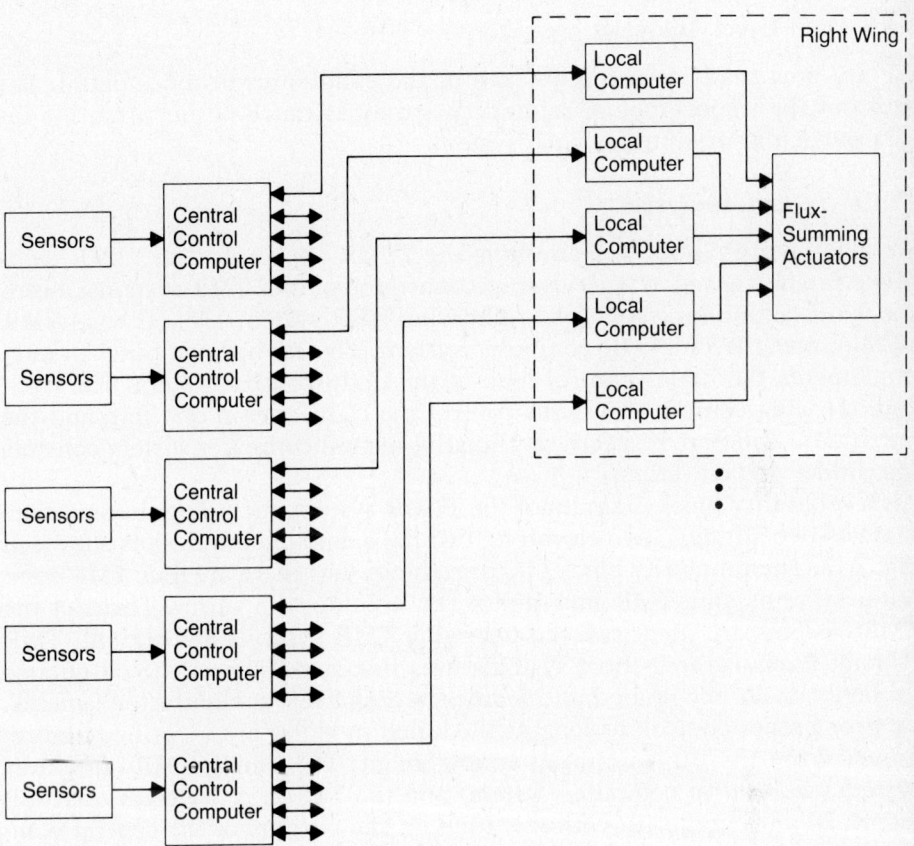

Fig. 5.11 The 5MR approach uses five of each device to provide masking of up to two failures. A local computer structure exists for the right wing (as shown), left wing, and the tail.

At this point we have developed four primary candidates for the flight-control system. Once again we must stress that we are significantly simplifying the problem for illustrative purposes. An actual design has numerous other considerations. For example, the sensors can be arranged such that they are shared with sensor data being voted upon by the computers; we have assumed that each computer has its own sensor set. Likewise, several additional variations of the TMR and triple-duplex approaches could be considered. Finally, architectures like the quad approach employed in the space shuttle [Sklaroff 1976] could also be considered as viable candidates for this system.

292 The Design of Practical Fault-Tolerant Systems

5.4.5 High-Level Analysis

We are now ready to examine each of the candidates in more detail. Because of the importance of reliability, we focus much of our attention on that particular attribute of each system.

TTMR System Analysis

We begin our analysis by examining the TTMR approach. The TTMR technique can be viewed as a series interconnection of four TMR systems: a central TMR computer system, the right wing TMR system, the left wing TMR system, and the tail TMR computer system. The central computer system implements the major control loop of the flight-control system. The right and left wing computer systems control the right aileron and flap and the left aileron and flap, respectively. Finally, the tail computer system controls the rudder and elevators.

A reliability block diagram of the TTMR system, including the seven actuators (two ailerons, two elevators, two flaps, and one rudder), is shown in Fig. 5.12. The reliability block diagram shows that there are four TMR computer systems, one TMR bus, and seven actuators in series. Three of the TMR systems are identical: the two wing TMR systems and the tail TMR system. Each of these three systems uses flux-summing on the outputs so that only one of the redundant modules in each TMR configuration is necessary for proper control, as long as the failed modules are identified and removed from the flux-summing arrangement. The fourth TMR computer system (the central computer system) and the TMR bus are somewhat different. Because majority voting is used on the outputs of the central computers and on the buses, at least two of the redundant modules in each TMR configuration must work properly for the TMR unit to perform properly.

The Markov model of a flux-summing TMR unit is shown in Fig. 5.13. The model has the following five states:

State 3 All three modules in the TMR configuration are operating properly. This is sometimes called the perfect state.

State 2D One module has failed, but the condition has been detected, and the module has been appropriately removed from the

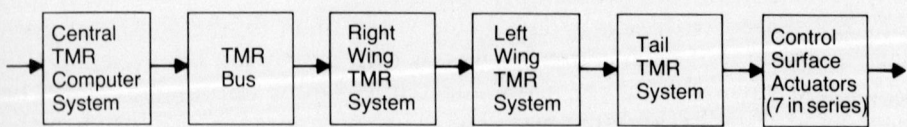

Fig. 5.12 Reliability block diagram of the TTMR system.

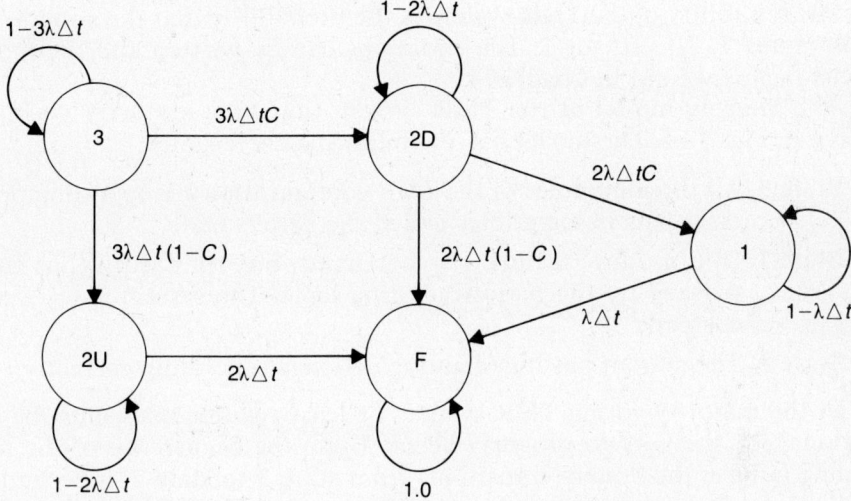

Fig. 5.13 Five-state Markov model of a flux-summing TMR system. The system can operate with two failed modules provided both failures are detected.

flux-summing arrangement. The system is operating correctly.

State 2U One module has failed, but the condition has not been detected. The flux-summing has allowed the failure to be tolerated, but the lack of detection prevents subsequent module failures from being tolerated. The system continues to operate correctly while in state $2U$.

State 1 Two modules have failed, and both failures have been detected and appropriately handled. The system is operating correctly.

State F The system has failed and is no longer operating correctly.

The system will transition from state 3 to state $2D$ upon the occurrence of the first module failure, provided that the failure is detected and appropriately handled. If the first failure is not detected or not appropriately handled, the system continues to operate but will be in state $2U$. Because state $2U$ represents a condition where a module failure has not been handled, the system cannot tolerate subsequent module failures when it is operating in state $2U$. State 1 can be reached only if two modules have failed and both have been detected and handled. Finally, the system transitions to state F if three module failures have occurred, or if two failures, one of which is not handled, have occurred.

294   The Design of Practical Fault-Tolerant Systems

The reliability of the TMR system is the probability that the system will be in states 3, 2D, 2U, or 1. The equations can be written and the model solved just as we did in Chapter 4.

The Markov model of the TMR system that uses majority voting is shown in Fig. 5.14. The model has the following three states:

State 3 All three modules in the TMR configuration are operating properly. This is sometimes called the perfect state.

State 2 One module failure has occurred, but the failure has been masked by the majority-voting logic. The system is operating correctly.

State F The system has failed and is no longer operating correctly.

In the majority-voting TMR system, at least two modules must be operational for the vote to properly occur. Upon the occurrence of the first module failure, the system transitions from state 3 to state 2, independent of any detection process that can be performed. Once the second module failure occurs, the whole system will have failed because at least two good modules are required to implement the majority voting process. The reliability of the system is the probability that it will be in state 3 or 2.

The reliability of the complete TTMR system can be calculated by taking the product of the individual TMR system reliabilities and the actuator reliabilities. The key factors in determining the reliability are the coverage factor and the failure rate of each module. We assume for simplicity that the coverage factor in each flux-summing TMR system is the same and that the failure rate of each module in each TMR system (both flux-summing and majority voting) is the same.

At this point in the design we have not selected a computer, consequently we do not have specific failure rate information. In addition, we have not defined how the fault detection will be performed, so we do not know what fault coverage can be achieved. However, we can still perform important comparisons by examining the systems for several failure rates and coverage factors. Table 5.1, for example, shows the reliability of the sys-

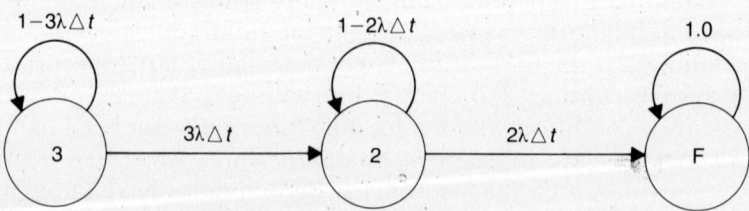

Fig. 5.14 Three-state Markov model of a majority-voting TMR system. Two of the three modules must work for the system to work.

TABLE 5.1 Reliability of the TTMR candidate system as a function of failure rate and fault coverage

		Failure rate (Failures per 10^6 hours)					
		0.1	1.0	10.0	100.0	1000.0	10000.0
	0.0	$0.9_{12}20$	$0.9_{10}19$	0.9_805	0.9_618	0.9_419	0.9_223
	0.1	$0.9_{12}20$	$0.9_{10}20$	0.9_806	0.9_618	0.9_419	0.9_223
	0.2	$0.9_{12}22$	$0.9_{10}21$	0.9_808	0.9_62	0.9_421	0.9_225
	0.3	$0.9_{12}24$	$0.9_{10}24$	0.9_811	0.9_623	0.9_424	0.9_227
	0.4	$0.9_{12}28$	$0.9_{10}28$	0.9_815	0.9_627	0.9_427	0.9_231
Fault coverage	0.5	$0.9_{12}33$	$0.9_{10}32$	0.9_821	0.9_632	0.9_432	0.9_235
	0.6	$0.9_{12}39$	$0.9_{10}38$	0.9_828	0.9_638	0.9_438	0.9_241
	0.7	$0.9_{12}46$	$0.9_{10}45$	0.9_836	0.9_645	0.9_445	0.9_247
	0.8	$0.9_{12}54$	$0.9_{10}53$	0.9_845	0.9_653	0.9_453	0.9_255
	0.9	$0.9_{12}63$	$0.9_{10}62$	0.9_856	0.9_662	0.9_462	0.9_264
	1.0	$0.9_{12}73$	$0.9_{10}73$	0.9_868	0.9_673	0.9_473	0.9_273

tem as a function of the fault coverage factor and for several values of the failure rate λ of one computer module. The reliability shown in Table 5.1 is the reliability of the system after three hours.

TMR System Analysis

The analysis of the TMR system is relatively straightforward and can be considered as simply the analysis of a flux-summing TMR system. The only difference is that each redundant channel of the TMR flight-control system actually contains four computer modules in series. The failure of any one of the four series computer modules results in the loss of that channel of the TMR system. If we select λ as the failure rate of one computer module, as we did in our analysis of the TTMR system, the Markov model of Fig. 5.15 can be constructed. The model of Fig. 5.15 is identical to that of Fig. 5.13 except that we now have three channels, each of which has a failure rate of 4λ. As we found before, the reliability of the system is the probability that the system is operating in states 3, $2D$, $2U$, or 1. Table 5.2 shows the reliability of the system as a function of fault coverage and for several values of the failure rate λ. Once again, the reliability is after three hours.

296   The Design of Practical Fault-Tolerant Systems

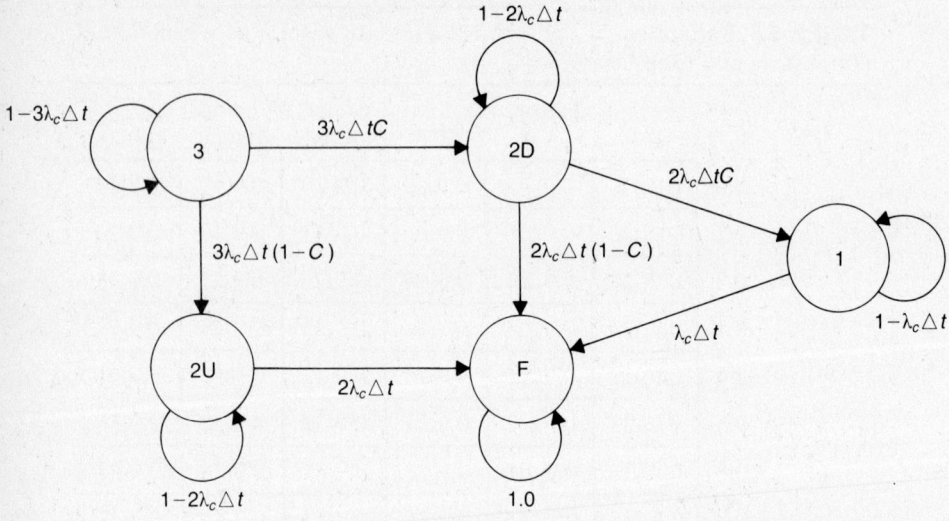

λ_c = failure rate of one channel
$\lambda_c = 4\lambda$

Fig. 5.15 Five-state model of the TMR candidate system. Each of the three channels has a failure rate of $\lambda_c = 4\lambda$. λ is the failure rate of one computer module.

TABLE 5.2 Reliability of the TMR candidate system as a function of failure rate and fault coverage

		Failure rate (Failures per 10^6 hours)					
		0.1	1.0	10.0	100.0	1000.0	10000.0
	0.0	$0.9_{11}56$	0.9_965	0.9_757	0.9_557	0.9_358	0.965
	0.1	$0.9_{11}57$	0.9_965	0.9_758	0.9_558	0.9_359	0.966
	0.2	$0.9_{11}58$	0.9_966	0.9_759	0.9_559	0.9_360	0.967
	0.3	$0.9_{11}60$	0.9_968	0.9_761	0.9_561	0.9_362	0.968
	0.4	$0.9_{11}63$	0.9_970	0.9_764	0.9_564	0.9_365	0.970
Fault coverage	0.5	$0.9_{11}67$	0.9_973	0.9_768	0.9_568	0.9_368	0.973
	0.6	$0.9_{11}72$	0.9_977	0.9_772	0.9_572	0.9_373	0.977
	0.7	$0.9_{11}78$	0.9_982	0.9_778	0.9_578	0.9_378	0.981
	0.8	$0.9_{11}84$	0.9_987	0.9_784	0.9_584	0.9_384	0.986
	0.9	$0.9_{12}18$	$0.9_{10}33$	0.9_819	0.9_619	0.9_419	0.992
	1.0	$0.9_{15}0$	$0.9_{15}0$	$0.9_{11}83$	0.9_883	0.9_583	0.998

TDTMR System Analysis

The TDTMR system can be analyzed in an identical manner to that used in the analysis of the TMR system. The only difference is that each computer module now consists of two computer modules and some comparison circuitry. If we assume that the failure rate of the duplexed computer configuration is approximately twice that of the simplex computer, the Markov model of Fig. 5.16 can be used. Note that the only difference between the models in Figs. 5.16 and 5.15 is the transition probabilities. Since the TDTMR system has twice as much hardware to fail as the TMR system, we expect the reliability to be lowered. However, the duplication with comparison allows a higher fault coverage to be achieved, so that it might be possible to obtain a higher reliability simply by significantly improving the coverage. Table 5.3 shows the reliability, after three hours, of the TDTMR system as a function of λ and the fault coverage factor.

5MR System Analysis

The 5MR system is identical to the TMR system except that five channels are used rather than three. The 5MR approach allows the first two module failures to be masked without the use of fault detection and reconfiguration. However, if fault detection and reconfiguration are provided, the third and

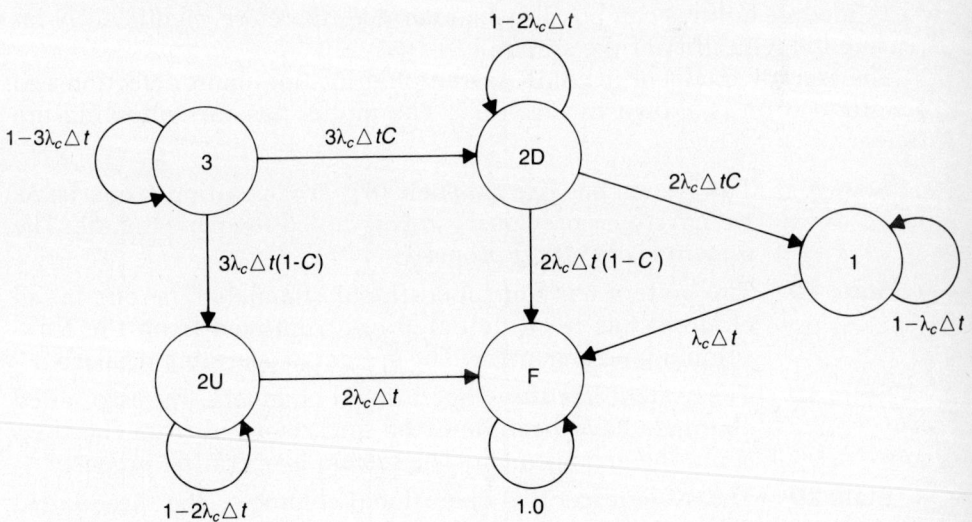

λ_c = failure rate of one channel
$\lambda_c = 8\lambda$

Fig. 5.16 Five-state model of the TDTMR system. Each of the three channels has a failure rate of $\lambda_c = 8\lambda$. λ is the failure rate of one computer module.

TABLE 5.3 Reliability of the TDTMR candidate system as a function of failure rate and fault coverage

		Failure rate (Failure per 10^6 hours)					
		0.1	1.0	10.0	100.0	1000.0	10000.0
	0.0	$0.9_{10}82$	$0.9_8 83$	$0.9_6 83$	$0.9_4 83$	$0.9_2 83$	0.884
	0.1	$0.9_{10}83$	$0.9_8 83$	$0.9_6 83$	$0.9_4 83$	$0.9_2 83$	0.885
	0.2	$0.9_{10}83$	$0.9_8 83$	$0.9_6 83$	$0.9_4 83$	$0.9_2 84$	0.889
	0.3	$0.9_{10}84$	$0.9_8 84$	$0.9_6 84$	$0.9_4 84$	$0.9_2 85$	0.894
	0.4	$0.9_{10}85$	$0.9_8 85$	$0.9_6 85$	$0.9_4 85$	$0.9_2 86$	0.90
Fault coverage	0.5	$0.9_{10}87$	$0.9_8 87$	$0.9_6 87$	$0.9_4 87$	$0.9_2 87$	0.911
	0.6	$0.9_{10}88$	$0.9_8 89$	$0.9_6 89$	$0.9_4 89$	$0.9_2 89$	0.922
	0.7	$0.9_{11}12$	$0.9_9 13$	$0.9_7 13$	$0.9_5 13$	$0.9_3 16$	0.936
	0.8	$0.9_{11}38$	$0.9_9 39$	$0.9_7 39$	$0.9_5 39$	$0.9_3 40$	0.952
	0.9	$0.9_{11}67$	$0.9_9 67$	$0.9_7 67$	$0.9_5 67$	$0.9_3 68$	0.97
	1.0	$0.9_{15}0$	$0.9_{15}0$	$0.9_{10}86$	$0.9_7 86$	$0.9_4 87$	0.99

fourth module failures can possibly be tolerated, therefore, significantly increasing the reliability of the system.

The Markov model of the 5MR system that includes fault detection and reconfiguration is shown in Fig. 5.17. The model has the following ten states:

State 5 The system has five channels that are operating properly. As we have seen previously, this is called the perfect state. The system is operating properly.

State 4D The system has four operational channels. The one failed channel has been detected and removed from the flux-summing arrangement. The system is operating properly.

State 3D The system has three operational channels. The two failed channels have been detected and removed from the flux-summing arrangement. The system is operating properly.

State 2D The system has two operational channels. The three failed channels have been detected and removed from the flux-summing arrangement. The system is operating properly.

State 1 The system has one operational channel. The four failed channels have been detected and removed from the flux-summing. The system continues to operate correctly.

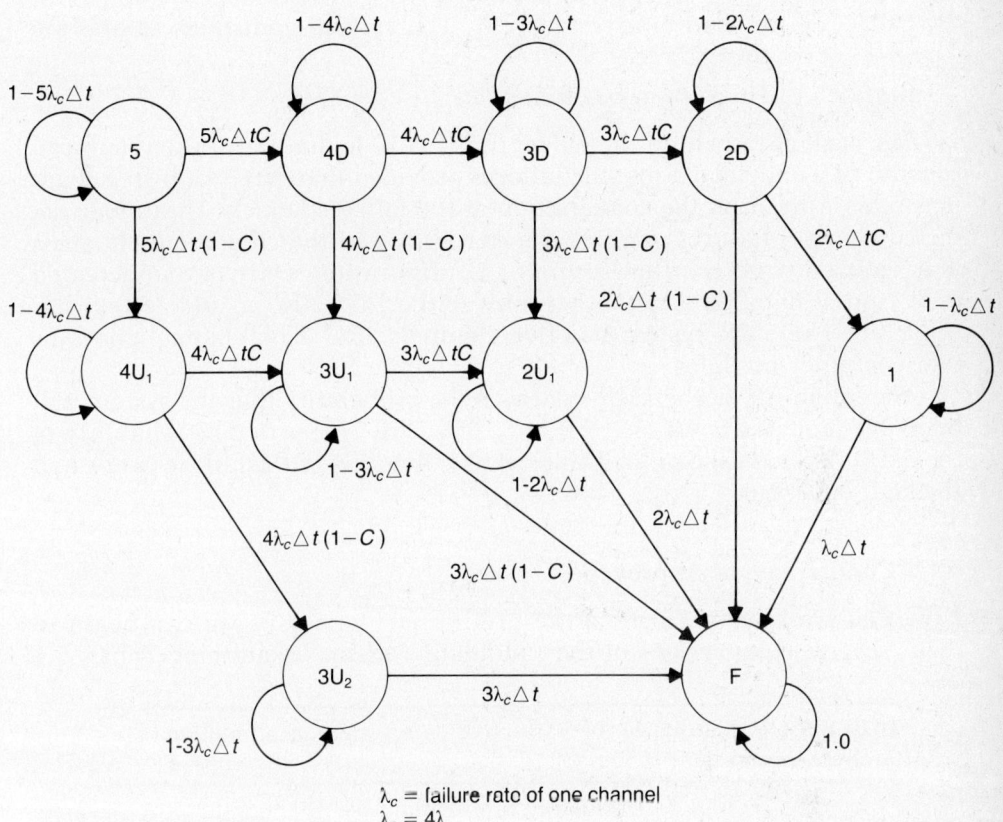

Fig. 5.17 Ten-state Markov model of a flux-summing 5-modular redundancy (5MR) system. Each channel has a failure rate of $\lambda_c = 4\lambda$, where λ is the failure rate of one computer module.

State $4U_1$ The system has four operational channels. The one failed channel has not been detected and removed from the flux-summing arrangement. The system is operating correctly.

State $3U_1$ The system has three operational channels. One of the two failed channels has been removed from the flux-summing, but the other failed channel has not. The system continues to operate correctly.

State $2U_1$ The system has two operational channels. Two of the three failed channels have been removed from the flux-summing, but the other failed channel has not. The system continues to operate correctly.

State $3U_2$ The system has three operational channels. Neither of the two failed channels has been removed from the flux-summing arrangement. The system continues to operate correctly.

State F The system has failed.

The basic idea behind the model in Fig. 5.17 is that the 5MR system can tolerate two undetected module failures provided that three fault-free channels remain to mask the contribution of the failed channels. Up to four detected channel failures can be tolerated provided that the failed channels are removed from the flux-summing configuration. Only one undetected fault can be handled if any other fault (either detected or undetected) has occurred. The 5MR system has five channels, and each channel contains four computer modules.

The reliability of the 5MR system is the probability that the system will be in one of states 5, $4D$, $3D$, $2D$, $4U_1$, $3U_1$, $2U_1$, $3U_2$, or 1. The reliability of the 5MR system is shown in Table 5.4 as a function of the failure rate λ and the fault coverage.

5.4.6 Comparison of Approaches

Now that we have performed the preliminary analysis, we can begin to make several comparisons of the candidate systems. Examining Tables 5.1

TABLE 5.4 Reliability of the 5MR candidate system as a function of failure rate and fault coverage

		Failure rate (Failure per 10^6 hours)					
		0.1	1.0	10.0	100.0	1000.0	10000.0
	0.0	$0.9_{15}0$	$0.9_{13}83$	$0.9_{10}83$	0.9_783	0.9_483	0.998
	0.1	$0.9_{15}0$	$0.9_{13}84$	$0.9_{10}83$	0.9_783	0.9_484	0.988
	0.2	$0.9_{15}0$	$0.9_{13}85$	$0.9_{10}85$	0.9_785	0.9_485	0.989
	0.3	$0.9_{15}0$	$0.9_{13}87$	$0.9_{10}86$	0.9_786	0.9_487	0.990
	0.4	$0.9_{15}0$	$0.9_{13}89$	$0.9_{10}89$	0.9_789	0.9_489	0.992
Fault coverage	0.5	$0.9_{15}0$	$0.9_{14}2$	$0.9_{11}16$	0.9_816	0.9_518	0.993
	0.6	$0.9_{15}0$	$0.9_{14}4$	$0.9_{11}41$	0.9_841	0.9_542	0.995
	0.7	$0.9_{15}0$	$0.9_{14}6$	$0.9_{11}63$	0.9_863	0.9_564	0.997
	0.8	$0.9_{15}0$	$0.9_{14}8$	$0.9_{11}82$	0.9_882	0.9_582	0.998
	0.9	$0.9_{15}0$	$0.9_{15}0$	$0.9_{12}53$	0.9_953	0.9_651	0.9_343
	1.0	$0.9_{15}0$	$0.9_{15}0$	$0.9_{15}0$	0.9_148	0.9_971	0.9_482

through 5.4 we see that each candidate system is capable of achieving the desired reliability of 0.9_7 for certain ranges of the failure rate and the fault coverage. For example, the TTMR approach meets the reliability requirements for all fault coverage values if λ is less than or equal to 10.0 failures per million hours. The TDTMR approach, however, requires a fault coverage of greater than 0.7 to meet the reliability requirements for values of λ less than or equal to 10.0 failures per million hour.

To allow the systems to be compared in more detail, we will make some assumptions about the hardware used to implement the machines. First, we assume that the computer modules are the dominant entities in the system in terms of power consumption, failure rate, weight, size, and so on. We want to compare the candidate systems for reliability, power consumption, and weight to determine which approach yields the most reliable system for the least cost (cost is assumed to be directly related to power consumption and weight). As is evident, this simple analysis ignores development costs and other important factors, but remember that we are performing a high-level analysis to gain initial insight into the relative effectiveness and efficiency of candidate approaches. Later, more detailed analysis would have to be used to make final judgments on the candidate systems.

Each computer module is assumed to consist of one processor card and one memory card. The exception to this assumption is the TDTMR approach where computer modules use duplication. In addition, there is one interface card for each connection required to another computer module. An analog-to-digital converter card and a digital-to-analog converter card are also available to be installed in computer modules as required. For example, both the local and the central computer modules require analog-to-digital conversion; however, only the local computer modules require digital-to-analog conversion. A comparator card is available for use in the TDTMR system. Computer modules are assembled by placing the appropriate cards in a card housing. Three card housings are available: one that holds a total of eight cards, a second that holds twelve cards, and a third that holds fourteen cards. A power-supply card that occupies two card positions is also available. Table 5.5 summarizes the available hardware and gives weights, power consumption, and, for the electronics, failure rates.

Tables 5.6 through 5.9 show the hardware used to construct each candidate system's central and local computers and the associated weight, power, and failure rate figures. For example, the TDTMR system requires two processor cards, two memory cards, and so on, in each local and each central computer. The primary impact of the duplication with comparison is the requirement to use the larger and heavier card housing and the increase in each computer's failure rate.

Using the information from Tables 5.6 through 5.9, one can construct total power and weight estimates, as shown in Table 5.10. Note that the most efficient approach (in terms of power and weight) is also the simplest

TABLE 5.5 Available hardware used to construct each candidate system

Item	Weight (pounds)	Power consumption (watts)	Failure rate (per 10^6 hours)
Majority voter	0.425	18	40
Processor	0.875	21	85
Memory	2.500	35	42
Interface	0.425	11	75
Comparator	0.425	11	60
Analog-to-digital	0.750	15	35
Digital-to-analog	0.750	15	35
Power supply	11.400	110	20
8-card chassis	7.500	—	—
12-card chassis	11.900	—	—
14-card chassis	15.800	—	—

TABLE 5.6 Hardware requirements for each central and local computer of the TTMR candidate system

Item	Weight (pounds)	Power consumption (watts)	Failure rate (per 10^6 hours)
Central control computer			
Processor card (1)	0.875	21	85
Memory card (1)	2.500	35	42
Interface cards (3)	1.275	33	225
A/D card (1)	0.750	15	35
Power supply (1)	11.400	110	20
Voter card (1)	0.425	18	40
12-card chassis (1)	11.900	—	—
Totals	29.125	232	447
Local computer			
Processor card (1)	0.875	21	85
Memory card (1)	2.500	35	42
Interface card (1)	0.425	11	75
Voter card (1)	0.425	18	40
A/D card (1)	0.750	15	35
D/A card (1)	0.750	15	35
Power supply (1)	11.400	110	20
8-card chassis (1)	7.500	—	—
Totals	24.625	225	332

TABLE 5.7 Hardware requirements for each central and local computer of the TMR candidate system

Item	Weight (pounds)	Power consumption (watts)	Failure rate (per 10^6 hours)
Central control computer			
Processor card (1)	0.875	21	85
Memory card (1)	2.500	35	42
Interface cards (3)	1.275	33	225
A/D card (1)	0.750	15	35
Power supply (1)	11.400	110	20
8-card chassis (1)	7.500	—	—
Totals	24.300	214	407
Local computer			
Processor card (1)	0.875	21	85
Memory card (1)	2.500	35	42
Interface card (1)	0.425	11	75
A/D card (1)	0.750	15	35
D/A card (1)	0.750	15	35
Power supply (1)	11.400	110	20
8-card chassis (1)	7.500	—	—
Totals	24.200	207	292

approach (TMR). The most costly candidate is the 5MR approach, however, one might expect the 5MR system to also be the most reliable.

To examine the reliability of each candidate more closely, the failure rate estimates of Tables 5.6 through 5.9 were used in the reliability models. Table 5.11 shows the resulting reliability for each candidate approach. As can be seen, the TMR and 5MR approaches are the only two capable of meeting the reliability requirement. However, the TMR candidate requires nearly perfect coverage, whereas the 5MR system meets the requirement for almost any reasonable value of coverage. On the other hand, 5MR consumes almost 71% more power and weighs nearly 76% more than TMR.

At this point in the development, it appears that TMR and 5MR are the best potential candidates. Considering that all of our failure rates are rough estimates at this point, it is reasonable to keep TMR as a candidate, particularly because of the savings in power and weight. Remember, however, that we want to achieve fault tolerance in addition to high reliability, so we must continue our analysis and design. The first project at the end of this chapter allows flight-control system architectures to be investigated in much more detail and the analysis performed here to be extended to include safety.

TABLE 5.8 Hardware requirements for each central and local computer of the TDTMR candidate system

Item	Weight (pounds)	Power consumption (watts)	Failure rate (per 10^6 hours)
Central control computer			
Processor cards (2)	1.75	42	170
Memory cards (2)	5.00	70	84
Interface cards (6)	2.55	66	450
A/D cards (2)	1.50	30	70
Power supply (1)	11.40	110	20
14-card chassis (1)	15.80	—	—
Totals	38.00	318	794
Local computer			
Processor cards (2)	1.75	42	170
Memory cards (2)	5.00	70	84
Interface cards (2)	0.85	22	150
A/D cards (2)	1.50	30	70
D/A cards (2)	1.50	30	70
Power supply (1)	11.40	110	20
12-card chassis (1)	11.90	—	—
Totals	33.90	304	564

5.5 Sample Fault-Tolerant Systems

We are now in a position to consider practical, fault-tolerant designs that have been created in the past and applied to specific applications. It is very important for us to understand the approaches taken by others in previous designs so that we can more easily generate new ideas for new designs.

The material in this section is partitioned according to the application area. The discussions describe the design requirements, any unique problems associated with the application, and the specific techniques used to meet the design requirements. Because of the complexity of most of the designs, it will be impossible to show all the details of each system. Therefore, the discussions include an overview of the system and then more detail on the aspects of the system that are particularly unique.

5.5.1 Long-Life Applications

We will consider three systems designed for long-life applications: the Self-Testing And Repairing (STAR) computer [Avizienis et al., 1971], the Fault-

TABLE 5.9 Hardware requirements for each central and local computer of the 5MR candidate system

Item	Weight (pounds)	Power consumption (watts)	Failure rate (per 10^6 hours)
Central control computer			
Processor card (1)	0.875	21	85
Memory card (1)	2.500	35	42
Interface cards (5)	2.125	55	375
A/D card (1)	0.750	15	35
Power supply (1)	11.400	110	20
12-card chassis (1)	11.900	—	—
Totals	29.550	236	557
Local computer			
Processor card (1)	0.875	21	85
Memory card (1)	2.500	35	42
Interface card (1)	0.425	11	75
A/D card (1)	0.750	15	35
D/A card (1)	0.750	15	35
Power supply (1)	11.400	110	20
8-card chassis (1)	7.500	—	—
Totals	24.200	207	292

Tolerant Spaceborne Computer (FTSC) [Burchby, Kern and Sturm 1976], and the Fault-Tolerant Building Block Computer (FTBBC) [Rennels, Avizienis, and Ercegovac 1978]. Each system was designed for space applications such as satellites or space exploration vehicles. Recall that the unique requirement for this application area is to achieve correct operation over long periods of time, for example, as many as five or ten years.

Self-Testing And Repairing Computer (STAR)

The Self-Testing And Repairing (STAR) computer was one of the first major efforts in the design of a fault-tolerant system for space applications. Although the design of the STAR computer is almost twenty years old, many of the techniques used in the system are applicable today and are certainly worthy of attention and understanding.

    The STAR computer was developed to satisfy all the expected requirements of a guidance, control, and data acquisition computer for use in unmanned space flights of ten years or longer. It was projected that the space flights would include the exploration of the outer planets of the solar sys-

TABLE 5.10 Total weight and power requirements for each candidate architecture

Candidate system	Total weight (pounds)	Total power consumption (watts)
TTMR	309.00	2721
TMR	290.70	2505
TDTMR	419.10	3690
5MR	510.75	4285

TABLE 5.11 Reliability of each candidate system as a function of fault coverage

Candidate system	Fault coverage					
	0.95	0.96	0.97	0.98	0.99	1.0
TTMR	0.9_535	0.9_537	0.9_539	0.9_542	0.9_544	0.9_546
TMR	0.9_579	0.9_583	0.9_587	0.9_616	0.9_657	0.9_782
TDTMR	0.9_483	0.9_486	0.9_489	0.9_530	0.9_563	0.9_659
5MR	0.9_783	0.9_788	0.9_835	0.9_869	0.9_909	$0.9_{11}21$

tem. It was desired to achieve a probability of 0.95 that the STAR computer would be completely operational at the end of a ten-year period.

The designers of the STAR computer selected the active, or dynamic, redundancy method of achieving fault tolerance. Recall from Chapter 3 that the basic characteristic of active redundancy is that fault tolerance is achieved by first detecting the existence of a fault, then locating the source of the fault, and finally performing some fault recovery. The success of active redundancy techniques depends directly on the system's ability to perform correctly fault detection, fault location, and fault recovery. In other words, fault coverage, as we have already seen, is extremely important in systems that use active redundancy approaches.

The STAR computer uses the concept of standby sparing to achieve fault tolerance. The computer consists of eight basic subsystems including the control processor (COP), the logic processor (LOP), the main arithmetic processor (MAP), read-only memory (ROM), read-write memory (RWM), an input/output processor (IOP), an interrupt processor (IRP), and the test and repair processor (TARP). Each subsystem is supplemented with one or more spares that are used to replace failed subsystems. The spares remain unpowered until a module failure occurs so that total power consumption is minimized. The basic architecture of the STAR computer is shown in Fig. 5.18. Information is exchanged among the different units via two buses; the memory-out bus provides for unidirectional transfers from the memories to the processors, whereas the memory-in bus allows information to be transferred from the processors to the memories. Each bus is four bits wide.

The control processor performs basic control operations such as address translation. The logic processor performs the basic logic operations, and the main arithmetic processor provides all arithmetic operations on the data words. Input/output interfaces are provided by the I/O processor, which contains registers to buffer the flow of data from the STAR computer to peripheral devices, and vice versa. The interrupt processor handles all interrupt requests, and the individual memory units provide for the storage of both data and instructions.

As far as fault tolerance is concerned, the test and repair processor (TARP) is the heart of the STAR computer. The TARP provides the error checking on information transmitted over the buses and reconfigures the system in the event that a unit fails. In general, the TARP is responsible for the error-free performance of the computer.

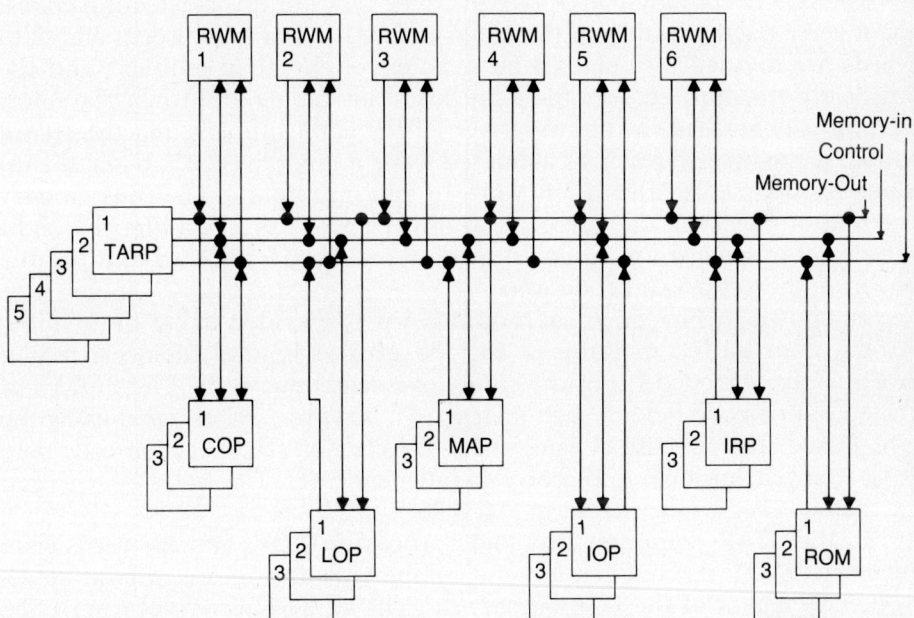

Fig. 5.18 The STAR system uses the concept of standby sparing. Each unit (COP, LOP, MAP, IOP, IRP, and ROM) has spares that are brought into service in the event of a failure. The TARP uses triplication with voting and two spares to achieve fault masking and fault recovery. (From [Avizienis et al., 1971] © 1971 IEEE)

The primary method of providing fault tolerance in the STAR computer is the use of reconfiguration and standby spares. Before the spares can be used, however, faults must be detected and located. To facilitate detection and location, each subsystem is designed with maximum autonomy. In other words, each functional unit of the computer can perform its own fault detection. Each unit sends status information on the results of fault detection tests to the TARP, which then performs a reconfiguration, if required. The TARP reconfigures the system by removing power from a unit that reports itself as failed and applying power to one of the spare units.

The functional units of the STAR computer use such fault detection techniques as error-detecting codes, watchdog timers, comparison, and triplication with voting. Error-detecting codes are used on all data and instruction words and provide for the detection of faults that occur during the storage, transmission, or processing of data or instructions. A watchdog timer requires that each unit of the computer transmit a status message to the TARP at a predefined frequency. Certain units of the STAR computer are duplicated, and a comparison between the duplicated units provides fault detection. Because of the importance of the functions performed by the TARP, triplication and voting is used in its design.

The STAR computer uses two primary types of error-detecting codes: the inverse-residue code and the 2-of-4 code. All instruction words and data words are divided into eight 4-bit entities (often called nibbles), and the words are transmitted over the 4-bit buses one nibble at a time. The information is contained in the first seven of the eight nibbles, and the eighth nibble is the inverse residue calculated using a modulus of 15. Using the inverse residue causes the 32-bit word to be a multiple of 15. Thus an easy method of checking the correctness of each word is available: the eight nibbles of one word are summed using a modulo-15 adder, and a residue of zero indicates a valid code word.

As shown in Fig. 5.19, each data word is encoded using an inverse-residue code with a modulus of 15. The address field of each instruction word is also encoded using the same inverse-residue code. The nibbles of the operation-code field of each instruction, however, are encoded using the 2-of-4 code. The purpose of using the 2-of-4 code on the operation-code field is to allow each nibble to be checked independently. The codes are checked by the bus checker, as shown in Fig. 5.20.

In the STAR computer, the logic processor (LOP) and the read-write memory (RWM) use duplication and comparison as a means of fault detection. Two copies of the LOP operate in tight synchronism, performing the same functions on the same data. The results produced by the two LOPs are compared bit by bit, and any miscompare is reported to the TARP so that reconfiguration can occur. An identical operation occurs for the RWM. When information is read from memory, one copy is obtained from each of two separate memories, and the copies are compared bit by bit. If the two

5.5 ▪ Sample Fault-Tolerant Systems

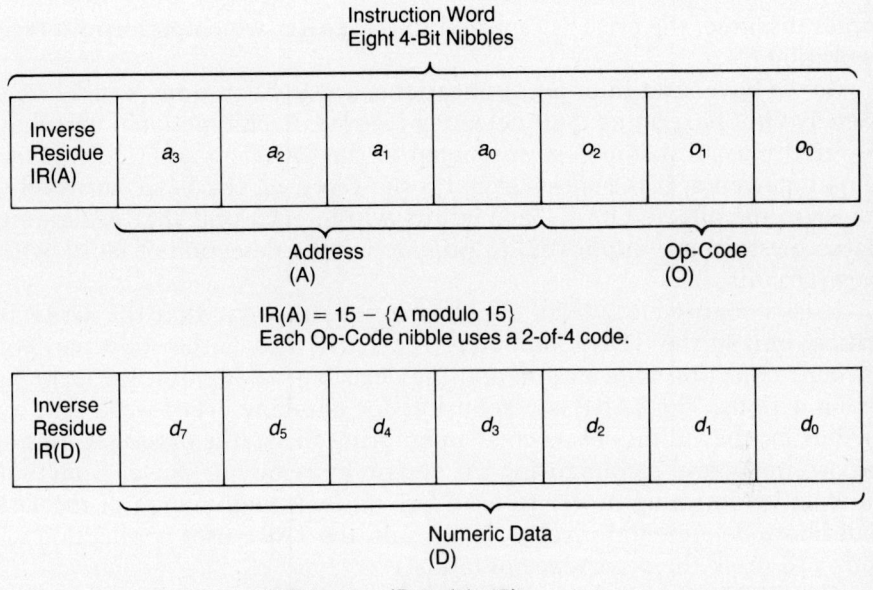

Fig. 5.19 Each data word is encoded in the STAR system. Both inverse residue and 2-of-4 codes are used. (From [Avizienis et al., 1971] © 1971 IEEE)

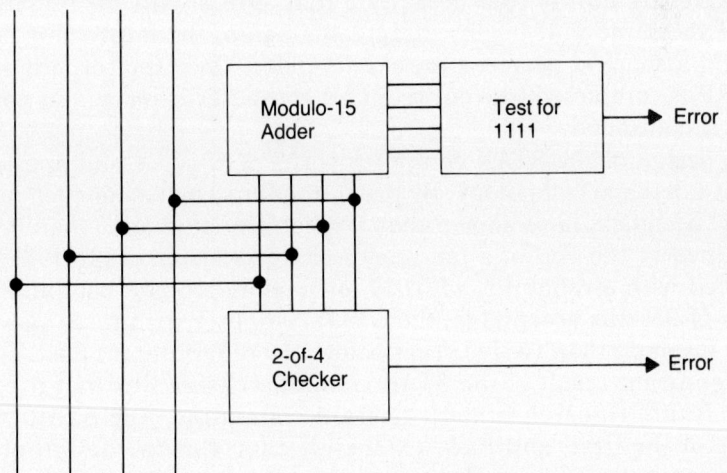

Fig. 5.20 The bus checker consists of a 2-of-4 checker, a modulo-15 adder, and a circuit to test for a 1111 sum from the modulo-15 adder. Op-codes are checked using the 2-of-4 code; all other information is protected with the inverse-residue code. (From [Avizienis et al., 1971] © 1971 IEEE)

copies disagree, the error is reported to the TARP, which performs a reconfiguration.

As we have seen in other applications, a watchdog timer can be an extremely vital part of the fault detection scheme. Each functional unit that is powered must transmit a status signal to the TARP on a periodic basis to remain powered. If the status signal is not received, the TARP turns off the power to the affected unit. The circuits within each unit that generate the status message are duplicated to provide for the detection of faults within those circuits.

It is probably clear from our descriptions thus far that the TARP is a critical unit in the STAR computer. The TARP acts as the collection point for fault detection information and the decision-making unit for reconfiguration actions. The TARP is responsible for checking every word sent over the bus for the validity of its code, monitoring the status messages from the various units, and reconfiguring the system by removing power from faulty units and connecting power to fault-free units. A malfunction of the TARP could have detrimental results. As a result, the TARP uses triplication with voting to mask the occurrence of faults.

The TARP design uses triple modular redundancy with spares. Three identical copies of the TARP are powered at all times, and two unpowered spares are available to replace any one of the three powered units. A 2-of-3 majority vote is used to determine the output of the TARP. The signals produced by the three powered TARPs are also compared so that differences can be detected. In the event that one TARP disagrees with the other two, the disagreeing unit is reset to a state that agrees with the remaining two units. If the same unit begins disagreeing again, the faulty unit is unpowered, and a spare is powered to take its place. Once the reconfiguration of the TARP is completed, the entire STAR computer is reset, and normal operation is continued.

The design of the STAR computer was a very successful application of the fault tolerance technology. By providing three spares for each functional unit, the designers have shown that a reliability of greater than 0.95 could be obtained at the end of a ten-year period [Avizienis et al., 1971]. This is contrasted with a reliability of 0.019 for a nonredundant computer. If a reliability of 0.9 was acceptable, the STAR computer with three spares could operate for more than twelve years before the reliability dropped below 0.9.

A significant result of the STAR computer design was that it paved the way for future research in fault-tolerant computing. The STAR computer was one of the first, and most extensive, efforts in the design of a fault-tolerant computer system. Many of the approaches tried and proven in the STAR computer design remain valid approaches today.

Fault-Tolerant Spaceborne Computer (FTSC)

The development of the Fault-Tolerant Spaceborne Computer (FTSC) was sponsored by the United States Air Force and was intended to produce a design suitable for applications requiring long, unattended service in the space environment [Burchby, Kern, and Sturm 1976]. Two competing architectures were developed under general guidelines and specifications defined by the Air Force. We will discuss the requirements of the system and the basic architecture that evolved.

The FTSC design was intended to support satellite applications requiring continuous, correct operation at extremely high computational loads. The computer system had to support extremely high data rates and function reliably in the harsh space environment. In addition, modularity and flexibility were important attributes. It was desired that the FTSC support a variety of space applications using essentially the same hardware in each. The FTSC concept used a standard central computer designed in a modular fashion to allow flexibility in targeting the FTSC to a particular application or mission.

The general goal specified for the FTSC was to achieve a probability of 0.95 of maintaining complete functional capability over five years in an unattended space environment. This reliability goal was coupled with goals of no more than 30 watts of power consumption, 25 pounds in weight, and the ability to perform 200,000 operations per second.

Two candidate architectures were developed for the FTSC design: one architecture was proposed by Raytheon and the other by Ultrasystems. The final architecture selected by the Air Force combined the advantageous features of both approaches. The architecture proposed by Raytheon was a general-purpose, microprogrammed digital computer that used *subelement* redundancy to improve reliability. Subelement redundancy simply implies that modules within the computer were replicated to provide fault detection and to improve computer reliability. Raytheon's approach relied heavily on the use of duplication with comparison as the primary form of fault detection.

The architecture proposed by Ultrasystems used arithmetic codes to provide fault detection. Recovery was implemented by replacing the faulty module with a spare module. The replacement process was under the control of a configuration control processor (CCP). The Ultrasystems' architecture was similar, in many respects, to the STAR architecture that we have already discussed.

After a comparative study of the competing architecture, the Air Force selected an approach that combined features from both candidates. Duplication, at the module level, was selected for the FTSC because of the simplicity of the concept and the high fault coverage that can be obtained. Arithmetic codes were not selected because of their inability to detect faults that arise in logical operations. Also, the coverage provided by the arithmetic codes was judged to not be as good as that provided by the duplication and comparison concept.

Many of the key features associated with fault detection and fault tolerance in the FTSC were developed in the Sub-Element Redundant Fault-tolerant (SERF) computer [Stiffler 1973]. The basic architecture of the SERF computer is shown in Fig. 5.21. The 32-bit computer consists of an arithmetic processing unit (APU), a control unit, memory, a configuration control unit, internal buses, and external bus interfaces. Each of the basic components relies on some form of hardware redundancy to achieve fault detection and fault tolerance.

The APU uses the *pair and a spare* technique to provide concurrent fault detection and recovery. The 32-bit APU is constructed from four 8-bit APUs, each of which uses the duplication and comparison method. In other words, each 8-bit APU actually contains two 8-bit units that are compared. One duplex, 8-bit APU is provided as a spare to replace a failed unit. The basic structure of the APU is illustrated in Fig. 5.22. Each half of the 8-bit APUs receives identical inputs, and the outputs of the halves are compared bit by bit to determine if faults have occurred. Upon the occurrence of a miscompare, the faulty module is replaced with one of the spares.

The control unit of the SERF computer was a relatively simple unit that could be replicated several times without a significant hardware penalty. Therefore, the control unit was designed using triple modular redundancy with majority voting to provide for tolerating one control unit fault.

One of the most interesting features of the SERF computer is the reconfiguration control logic. Reconfiguration occurs when a fault is identified. The reconfiguration is implemented by the configuration control unit and unique distributed switching networks that the designers called *ripplers*. In one variation of the rippling approach, each module has its own rippler,

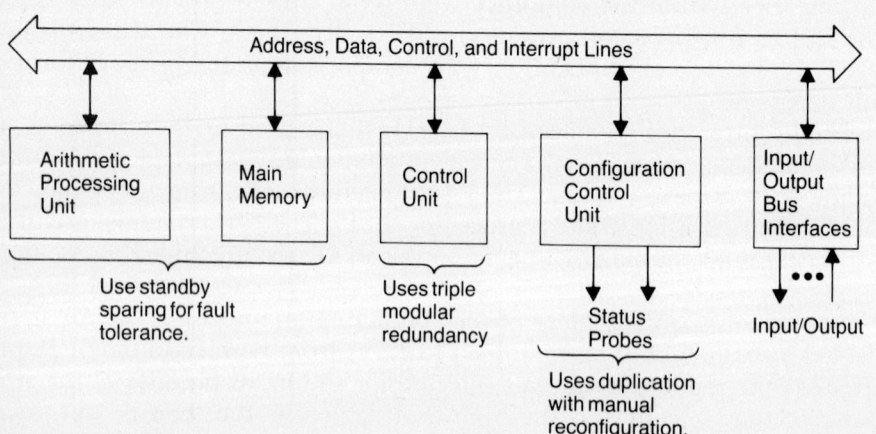

Fig. 5.21 The SERF computer has five primary components: the arithmetic processing unit, main memory, control unit, configuration control unit, and input/output bus interfaces. A variety of fault detection and fault tolerance techniques are used.

5.5 ■ Sample Fault-Tolerant Systems

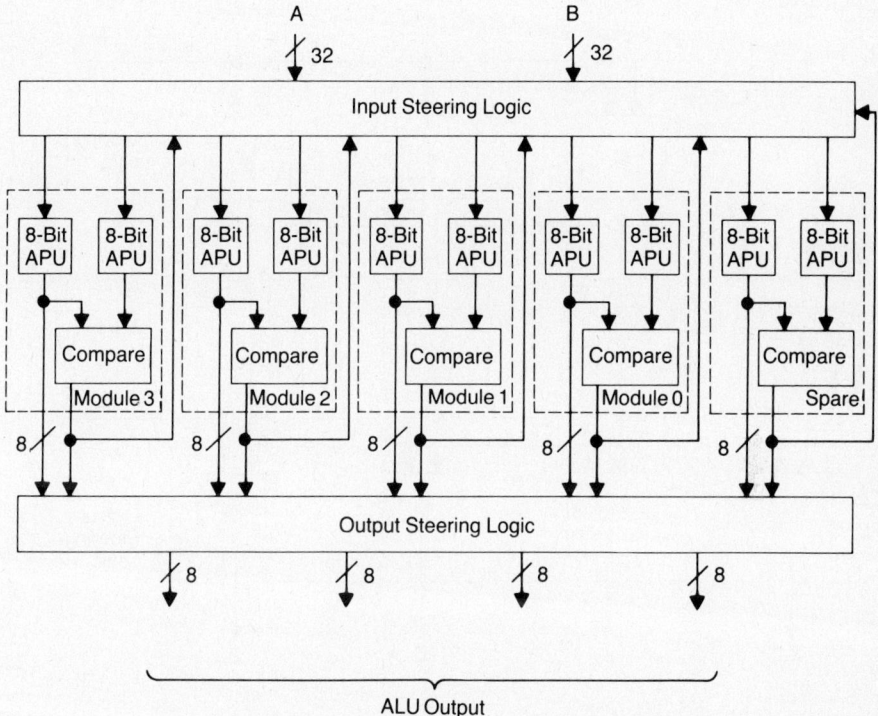

Fig. 5.22 The APU contains five modules, one of which is a spare. Each module has duplicated 8-bit APUs which are compared to detect faults. A failed module causes the spare to be switched in by the steering logic.

and the function of the rippler is to allow any module to be replaced with a neighboring module. The reconfiguration is accomplished with both input and output steering logic. The input steering logic can select either the normal input for its associated module or an input from a previous module. Likewise, the output of a module can be directed to the normal place or an alternate destination.

As an example of the rippling process, consider the simple 4-bit complementation circuit shown in Fig. 5.23. The 4-bit complementer is composed of four inverter cells, and a single spare cell is provided. The input steering logic and output steering logic are simply multiplexers that are set to appropriately control the flow of information around failed cells. During normal operation, the multiplexers are set to route the inputs through cells 0 through 3 and to select the outputs from those same cells. The spare cell is not used during normal operation. If, for example, Inverter 2 fails, the multiplexers are set to allow Inverter 1 to perform the function of Inverter 2.

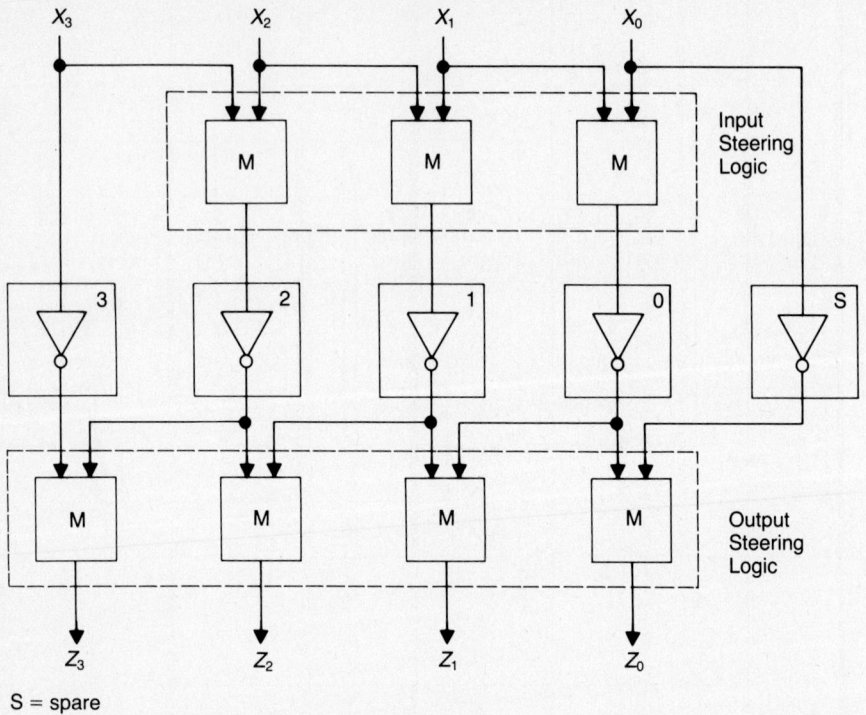

Fig. 5.23 The rippling concept allows neighboring inverter cells to assume the functions of failed cells. Input and output steering logic accomplishes the necessary multiplexing. A single spare is provided in this example.

The function of Inverter 1 is then *rippled* to Inverter 0, and the function of Inverter 0 is rippled to the spare inverter.

In the SERF computer, each module used duplication with comparison to provide for fault detection, and the ripplers and the configuration control unit provided for the insertion of spare modules to replace the failed ones. The advantages of using the ripplers are simplicity, regularity of structure, and the convenience of physically locating the spares. Using the ripplers, spares do not have to be physically located adjacent to the module that they will replace.

The SERF computer was one of the first proposed uses of subelement redundancy to improve the reliability of a computer system. Reliability estimates show that the SERF concept is capable of surpassing the goal of 0.95 reliability at the end of five years based on a medium-scale integration (MSI), metal-oxide semiconductor (MOS) implementation. The designers of SERF reported that a simplex computer with the same functional capability would have a reliability of less than 0.2 at the end of five years.

Fault-Tolerant Building Block Computer (FTBBC)

Like the research on the STAR computer, the work on the Fault-Tolerant Building Block Computer (FTBBC) was performed at the Jet Propulsion Laboratory (JPL) [Rennels, Avizienis, and Ercegovac 1978]. The FTBBC project was begun at JPL as an effort to promote a wider use of fault-tolerant computing techniques. The basic concept was to define a small number of building blocks that could be implemented in VLSI and used to construct various types of fault-tolerant computing systems. The building blocks included a self-checking computer module (SCCM), a memory interface, a bus interface, and a core building block. Each building block used proven techniques such as duplexing and error-detecting and error-correcting codes to provide the required fault tolerance. Many of the techniques used in the building blocks were proven during the development of the STAR computer.

Numerous goals were established for the design of the basic building blocks. In general, the goal was to provide a fundamental set of circuits from which larger, fault-tolerant systems could be constructed. At the same time, however, the designers recognized that complex systems often required various degrees of fault tolerance. For example, as we have seen previously, many systems have a combination of functions ranging from highly critical operations to convenience operations. The building blocks had to allow the designer the flexibility to include the desired amount of redundancy to achieve the design goals in each part of the system.

Seven primary requirements were derived for the building blocks [Rennels, Avizienis, and Ercegovac 1978]. First, the number of distinct building blocks must be kept as small as possible. This requirement is extremely difficult to satisfy because you want enough building blocks to support the design, however, the number must not be so large that the design process is complicated.

Second, the building blocks were developed primarily for use in distributed, multiple-processor systems, so it was desired that the building blocks be usable with off-the-shelf microprocessor and memory circuits. Also, the building blocks were developed to support and encourage widespread use of fault tolerance; therefore, the building blocks needed to be as versatile as possible.

Third, existing standards for busing and input/output operations were to be followed as closely as possible. Once again, the objective of this requirement was to promote compatibility and ease of use to encourage designers to employ the building blocks in their designs.

Fourth, extremely high fault detection coverage must be a characteristic of the building blocks. Fault detection is the cornerstone of fault tolerance, so each building block must be capable of detecting nearly 100% of its own faults. Likewise, the building blocks must allow the designer to achieve nearly perfect recovery and reconfiguration capabilities.

Fifth, the building blocks were developed primarily for applications where periodic human maintenance is possible. During the maintenance process, failed modules will be replaced by spares while the normal operations of the system continue uninterrupted. This requirement particularly affects the interface building blocks; each must be designed to support such repairs by isolating modules from the remainder of the system.

Sixth, the architectures developed using the building blocks must be flexible enough to satisfy the requirements of a number of applications. Remember that an overall objective of the FTBBC project was to promote the use of fault tolerance; therefore, flexibility was mandatory. A part of the flexibility also includes the ability to support various amounts of redundancy. For example, one application may require three spares of each module, whereas another application requires only one. The building blocks should be flexible enough to support either application.

Finally, the building blocks must be conceptually simple and easy to use. Certainly, if the building blocks are complex or difficult to use, the designers will not use them, and the ability to promote widespread usage will be lost.

A key module developed in the FTBBC research was the self-checking computer module (SCCM). The SCCM is a small computer complete with a central processing unit, memory, and input/output interfaces. The key attribute of the SCCM is the ability to detect its own faults during normal operation. The fault detection information can be used to initiate reconfiguration operations.

The basic architecture of the SCCM is shown in Fig. 5.24. The SCCM is constructed using other building blocks, including the memory interface building block, the bus interface building block, the input/output building block, and the core building block. Each of the individual building blocks is responsible for detecting faults within its own circuitry and notifying the core building block of the existence of the fault. The core building block uses the fault detection information provided by the remaining building blocks to decide on the overall condition of the SCCM. If faults are detected, the core building block disables the bus control and input/output functions of the SCCM, thereby isolating the SCCM from the remainder of the system. The core building block can either completely halt the processing of the SCCM or initiate a restart of the processor and attempt a recovery operation.

A primary fault detection mechanism in the SCCM is the use of duplication and comparison on the central processing units (CPUs). Each SCCM contains two CPUs whose results are continuously compared bit by bit. Any differences are reported to the core building block as processor miscompares.

Another primary feature of the SCCM is the error-correcting memory. The memory uses both Hamming error-correction coding and standby

5.5 ■ Sample Fault-Tolerant Systems

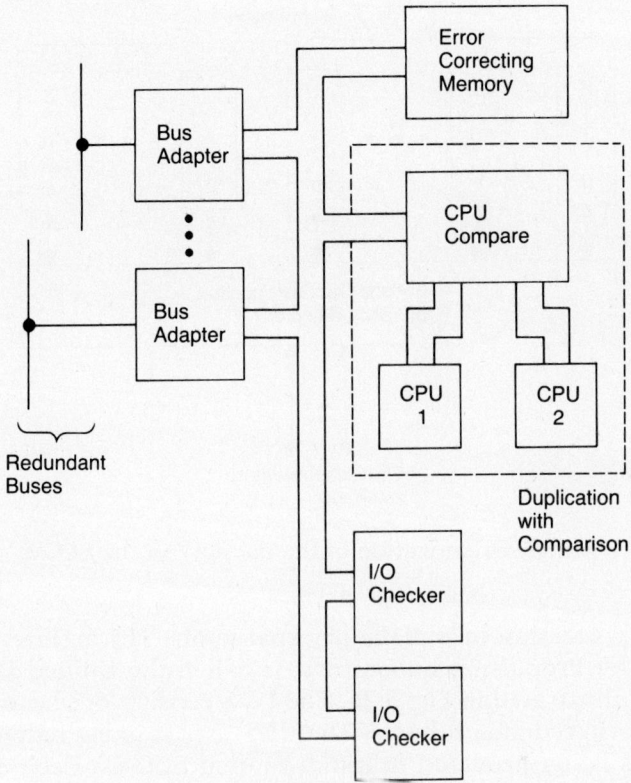

Fig. 5.24 Architecture of the SCCM.

sparing to provide for fault detection, fault masking, and fault recovery. Memory words are 16 bits, and 7 additional bits are provided for the Hamming code (6 bits) and the sparing (1 bit). The organization of the memory is shown in Fig. 5.25.

The 16-bit data words are transferred to and from memory via the internal bus, which is protected by two parity bits—one parity bit for each 8-bit byte. Parity is checked on all words prior to writing the words to memory, and then the appropriate Hamming check bits are generated. The modified Hamming code is used to allow detection of double errors, thus preventing erroneous correction. Upon reading a word from memory, Hamming correction occurs by complementing the erroneous bit, and the parity bits for each byte are generated to provide detection capability for errors that occur during transmission over the internal bus.

The SCCM is used as a computer module building block to develop fault-tolerant architectures. The developers of the SCCM have designed and constructed a sample architecture for the distributed processing application

318   The Design of Practical Fault-Tolerant Systems

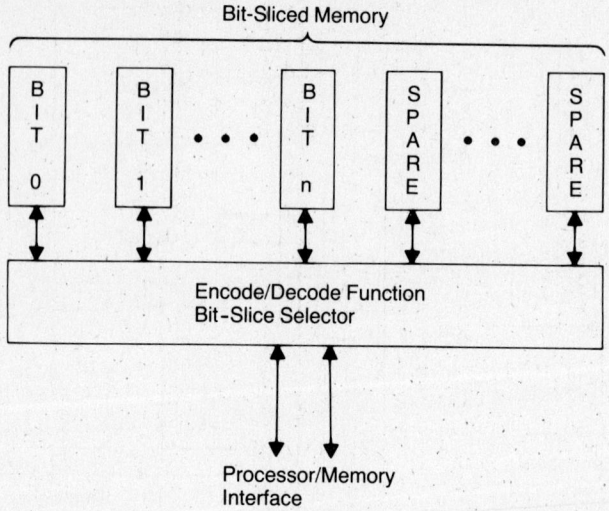

Fig. 5.25 Organization of the memory in the SCCM.

of on-board processing in a planetary spacecraft. The architecture, developed at the Jet Propulsion Laboratory, is called the Unified Data System (UDS) and is illustrated in Fig. 5.26. The UDS consists of several SCCMs interconnected via redundant buses. The UDS uses standby redundancy with spare SCCMs being provided to replace failed SCCMs. Each SCCM is re-

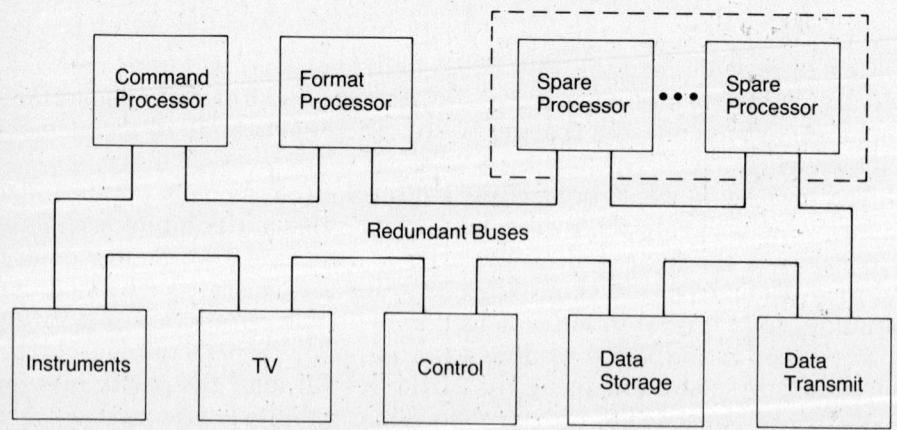

Fig. 5.26 Architecture of the United Data System (UDS). Spare processors are unpowered and are switched on-line manually via ground commands if a problem develops.

sponsible for its own fault detection using the mechanisms included within each SCCM. One SCCM is designated as a system controller and has the responsibility of periodically polling each additional SCCM to determine if a fault has been detected. If a fault is detected, the system controller implements a reconfiguration action that includes bringing on-line a spare SCCM. The spares can be either powered or unpowered depending on the requirements of the application. The SCCM that is serving as the system controller will always be provided with a powered spare. In many critical applications, two SCCMs can be run concurrently with comparisons performed between the two.

One of the most important outcomes of the FTBBC development was the definition of the basic building blocks including the SCCM. VLSI implementations of the building blocks allow a very efficient and effective set of circuits to be developed to support the design of fault-tolerant systems. The availability of such a set of building blocks should significantly improve the capabilities of the design engineer in the development of a fault-tolerant design.

5.5.2 Critical-Computation Applications

We will consider six systems that have been designed for critical-computation environments: the space shuttle [Sklaroff 1976], the Fault-Tolerant Multiprocessor (FTMP) [Hopkins, Smith, and Lala 1978], the Software Implemented Fault Tolerance (SIFT) [Wensley et al., 1978] system, the August Systems CS-3001 Control Computer [August 1986], the Honeywell Multi-Microprocessor Flight Control System (MMFCS) [White et al., 1981], and the Agusta A129 Integrated Multiplex System (IMS) [Johnson and Julich 1985]. Each system has been designed to achieve stringent reliability and fault tolerance requirements, yet each system uses different approaches to achieving those goals. In many cases, the system designed for a computation-critical application will be required to maintain correct operation after two or more faults. In addition, the system can be required to fail in a safe manner when correct operation can no longer be continued.

Space Shuttle

The space shuttle computer system is one of the most highly visible applications of fault-tolerant computing, to date. Few systems are as complex as the shuttle or have received the level of national exposure. The uniqueness of the shuttle is partly due to its being the first application of off-the-shelf, nonredundant computers in the development of a redundant, fault-tolerant system [Sklaroff 1976]. The shuttle computer system is responsible for providing flight-critical functions during the ascent, reentry, and descent of the shuttle craft.

The shuttle computer system consists of five identical, general-purpose computers, each capable of performing flight-critical or noncritical functions. During critical operations such as ascent, reentry, and landing, four of the five computers are configured into a redundant set that performs all flight-critical functions. The four computers form an NMR arrangement, where the value of N is four. The four computers receive the same input data, perform the same computations, and provide the same results. The results produced by the four computers are voted upon to ensure that error-free commands are provided to control the shuttle safely.

The shuttle computer system had to meet five basic requirements. First, each computer needed self-test features capable of detecting 96% of the faults that could occur within that computer. In addition, each of the detected faults must be reported to the crew such that the faulty computer can be precluded from performing any flight-critical functions, prior to the initiation of such functions. Because the shuttle design uses off-the-shelf, general-purpose computers, most of the fault detection mechanisms must be implemented in software that executes in each computer.

Second, the failure of the first two computers, of those in the NMR set of four computers, must be automatically identified to the crew, so that the crew can initiate an appropriate action. To the extent possible, the third computer failure must also be identified to the crew.

Third, any failed computer must be capable of automatically ceasing all its transmissions so as to prevent erroneous data from being propagated throughout the system. Although the capability to stop transmissions must be autonomous, the crew is required to enable the system to perform this autonomous operation.

Fourth, it was desired to have as much fault isolation as possible. Specifically, the failure of one computer should not result in another computer either identifying itself as failed or producing incorrect results.

Finally, the recovery from transient faults should be accomplished whenever practical to do so. The objective with regards to transient faults was to minimize the impact that they have on the overall reliability and performance of the system.

The overall architecture of the space shuttle avionics system is shown in Fig. 5.27. Each general-purpose computer consists of a central processing unit (CPU) and an input/output (I/O) processor. During operation, four of the five general-purpose computers perform identical functions of flight-control, guidance, and navigation. The fifth computer performs noncritical functions such as communications control and display processing.

The I/O processor in each general-purpose computer provides for the transmission and reception of information over serial data buses. There are twenty-eight serial data buses, of which five are used for the transfer of in-

5.5 ■ Sample Fault-Tolerant Systems   321

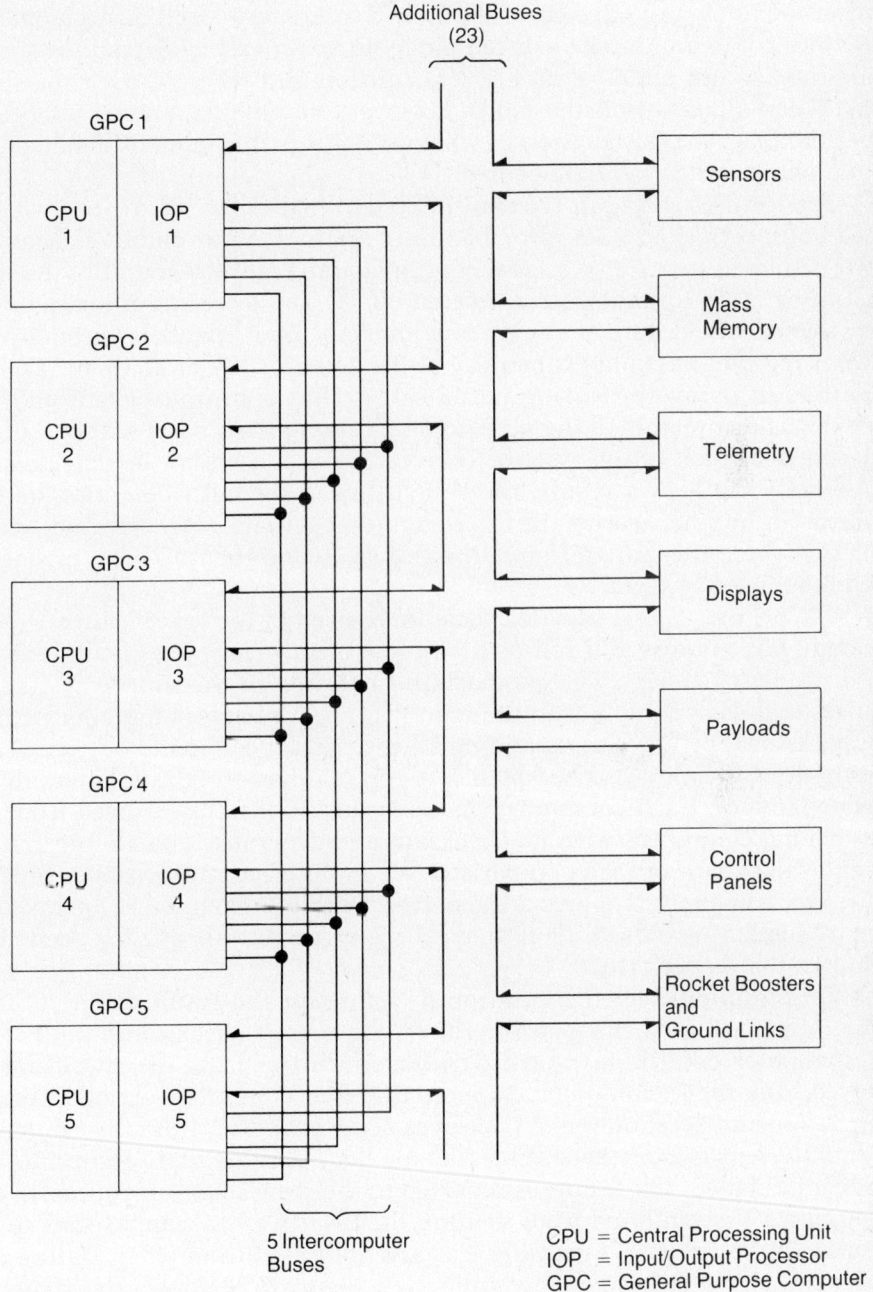

Fig. 5.27 Architecture of the Space Shuttle's avionics system.

formation among the five general-purpose computers. Each bus can operate in either command mode or listen mode. In command mode, data requests, commands, and actual data are transmitted and received over the same bus. When a bus is in listen mode, however, only the data reception occurs over the bus. The designation of whether a bus is in command mode or listen mode is under software control.

One form of voting in the shuttle system takes place at the control surface actuators. Each actuator has four distinct input channels that are force-summed to achieve a form of voting similar to that found in the flux-summing technique discussed in Chapter 3. The force-summing actuator can tolerate the failure of any three of the four input channels without compromising the functional capability of the actuator [Sklaroff 1976]. The advantage of using the voting actuators is that a computer can provide erroneous commands to the actuators on a temporary basis without affecting the operation of the system. Therefore, faults need not be detected immediately, and, as a result, the time allowed for fault detection can be increased. In other words, the force-summing actuators provide fault masking to prevent the system from failing when a computer produces erroneous commands.

The primary fault detection techniques used in the space shuttle system include built-in-test and self-test features, bus timeout tests, comparisons, and watchdog timers. The comparison operation in the shuttle system requires that each computer in the redundant set of four computers calculate a checksum word by simply adding all the critical command outputs. Each computer transmits its checksum to the remaining three computers in the redundant set. Each computer compares the checksums received from the remaining computers with its checksum and generates a single bit that indicates the result of each comparison. A computer is categorized as failed if two successive miscompares are reported. With four computers operating in the redundant set, each computer generates three bits, one for each comparison that is performed.

For simplicity, let the notation F_{ij} represent the result of the jth computer comparing its checksum to the ith computer's checksum. F_{ij} is 1 if the jth computer determines that its checksum differs from the checksum received from the ith computer. Suppose that four computers are operating in the redundant set. Computer 1 receives F_{12}, F_{13}, and F_{14} from the remaining computers. If two of those three signals have a value of 1, Computer 1 is labeled as failed. The failure is reported to the crew's instrument panel, and Computer 1 is inhibited from sending the results of its comparisons to the remaining computers. Computer 1 is also inhibited from participating as a member of the redundant computer set. Figure 5.28 shows the structure of the circuitry necessary to perform the operations of voting upon the F_{ij} signals and disabling the failed computer.

5.5 ■ Sample Fault-Tolerant Systems

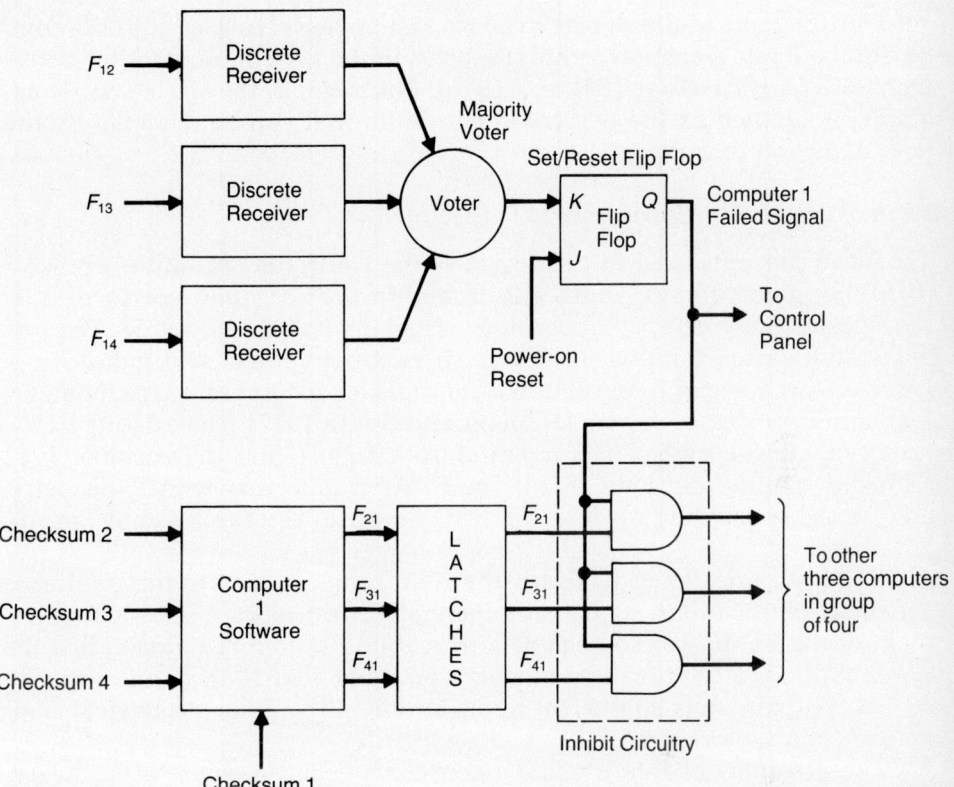

Fig. 5.28 Each computer compares its own checksum with those of the remaining three computers. Computer 1 generates F_{21}, F_{31}, and F_{41} and receives F_{12}, F_{13}, and F_{14}. A flip flop is reset if a majority of F_{12}, F_{13}, and F_{14} is 1, and computer 1 is disabled.

A watchdog timer is another important entity used to detect faults within each general-purpose computer. Each I/O processor contains a timer that must be set periodically by the central processing unit. If the timer completes its count without being reset by the processor, a timeout latch is set, and the failed computer is removed from the redundant set.

The bus timeout test is another form of a watchdog timer. Each computer has one dedicated bus and must periodically perform an operation on that bus to prevent a timer from expiring. If the timer expires, the computer is labeled as failed. The bus timeout test is useful for detecting conditions where a computer simply ceases to perform any functions. Such conditions can exist if an oscillator fails or the computer becomes caught in an infinite loop due to a software fault.

The designers of the shuttle avionics system have reported that the combination of fault detection techniques used in the system provides for better than 96% fault coverage [Sklaroff 1976]. Coverage, in the space shuttle example, is defined as the percentage of faults that can be detected by the fault detection techniques.

Fault-Tolerant Multiprocessor

The basic concepts used in the design of the Fault-Tolerant Multiprocessor (FTMP) originated in the mid-1960s in multiprocessor studies performed by the Charles Stark Draper Laboratory [Hopkins and Smith 1975]. The primary application for this work was spaceborne control and monitoring. Later research began to develop the concepts for incorporating redundancy into a multiprocessor system [Hopkins and Smith 1977]. The existing FTMP prototype was developed over a period from August 1978 to December 1982 [Hopkins, Smith, and Lala 1978]. The FTMP engineering model was delivered to the sponsoring agency, the NASA Langley Research Center, in August 1982.

FTMP was designed to fulfill the requirements of extremely reliable aerospace applications, such as commercial air transport. A generic requirement of the FTMP was to achieve a probability of failure (often called the unreliability) of 10^{-9}, or less, during a ten-hour flight. An unreliability of 10^{-9} is approximately equivalent to the unreliability of the mechanical components of a typical aircraft.

The designers of FTMP made a conscious decision to use a hardware-intensive approach in their design. It was felt that the crucial parts of the system could be most economically designed and implemented using hardware, as opposed to software. The software-implemented fault tolerance (SIFT) system, that will be described later, was a competing design to FTMP and used a software-intensive approach.

The basic architecture of FTMP is that of a conventional multiprocessor system, as shown in Fig. 5.29. The system consists of six basic elements: processing units called processor triads, system memory, a real-time clock, a control unit, input/output units, and a system bus that interconnects each of the basic elements.

The basic operation of the system is that of a conventional multiprocessor. Each processor triad can perform separate functions such that operations can be conducted in parallel. Processor triads share information through the system memory (often called the shared memory) with the transfers to and from the system memory occurring over the system bus.

Each processor triad consists of three separate processor modules, as illustrated in Fig. 5.30. A processor triad functions as if it were a single processor; that is, each processor within the triad executes the exact same software. All processors are tightly synchronized using a common, fault-

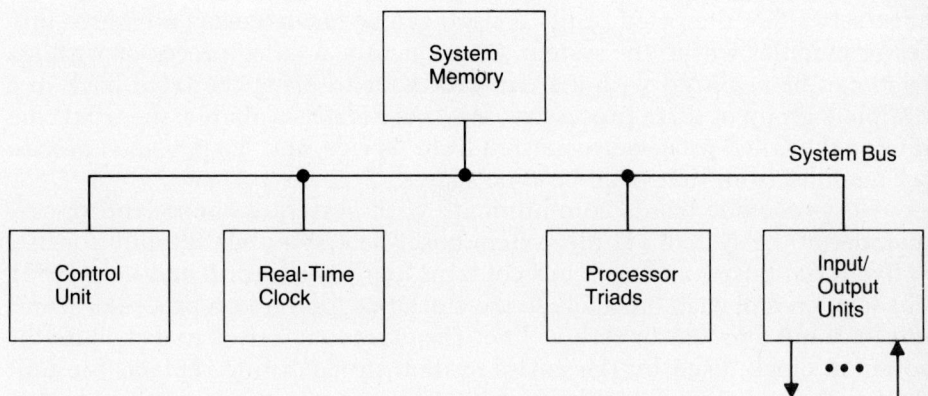

Fig. 5.29 FTMP consists of six basic elements (system memory, control unit, real-time clock, processor units, input/output units, and system bus) that are organized as a conventional multiprocessor. Each element, however, contains redundancy.

tolerant clock. The results produced by the three processors within a triad are voted upon, using a majority-voting scheme, to determine the correct result; therefore, the failure of a single processor module in a triad does not

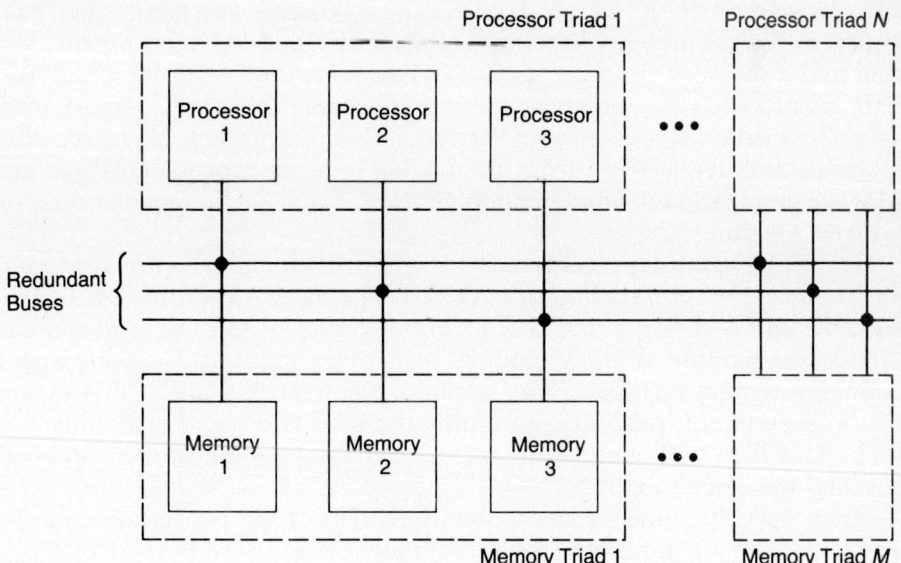

Fig. 5.30 In the FTMP system processors and memories are organized into triads such that majority voting can be used.

adversely affect the voted result. A triad can be formed using any three processor modules within the system. Consequently, a failed processor within a triad can be replaced with a spare processor to bring the triad back to a complete group of three processors. If a spare is not available, the triad containing the failed processor is retired from service, and the two good processor modules from that triad become spares.

The processor triads communicate with system memory and the remainder of the system via the system bus. The system bus actually consists of five serial buses, and each bus contains four lines [Smith and Lala 1983]. The four lines of each bus include the clock line, poll line, a processor transmit line, and a processor receive line. The clock line is used to distribute the common clock, used by the entire system, to each line replaceable unit (LRU). The poll line is used to arbitrate among the processor triads that wish to place information on the bus. A processor triad examines the poll line to determine if the bus is free and issues a special polling sequence over the poll line to obtain control of the bus. The processor transmit line transmits the processor read and write commands, and the processor receive line receives the responses to processor read commands.

Three of the five buses are configured into a bus triad with the remaining two buses being used as spares. Each processor module reads and writes information over one of the three buses of the bus triad. As a result, any unit receiving information from a processor triad actually receives three copies of that information, one copy from each processor module. The receiving unit can vote on the three received results to correct any errors that have occurred. An example of the connections provided between a processor triad and a memory triad are shown in Fig. 5.31. Data from the processors to the memory is voted on at the memory to ensure that each memory module of the triad receives a correct version of the information. Likewise, when information is transferred from the memory to the processors, three versions are obtained and voted upon to guarantee that each processor receives a correct version.

The FTMP prototype achieved the architecture described above using units called line replaceable units (LRU). The basic structure of a line replaceable unit is shown in Fig. 5.32 [Smith and Lala 1983]. The LRU contains a processor module, a clock module, a memory module, an input/output module, a remote terminal clock module, bus guardian units (BGUs), and several system control and communications registers and functions. The discussions here will concentrate on the processor modules, memory modules, and the clock modules.

Each LRU has one processor module. The three processor modules needed to form a triad can be taken from any three LRUs. In the FTMP prototype, there are ten LRUs, so as many as three processor triads can be formed with one processor module remaining as a spare. The three proces-

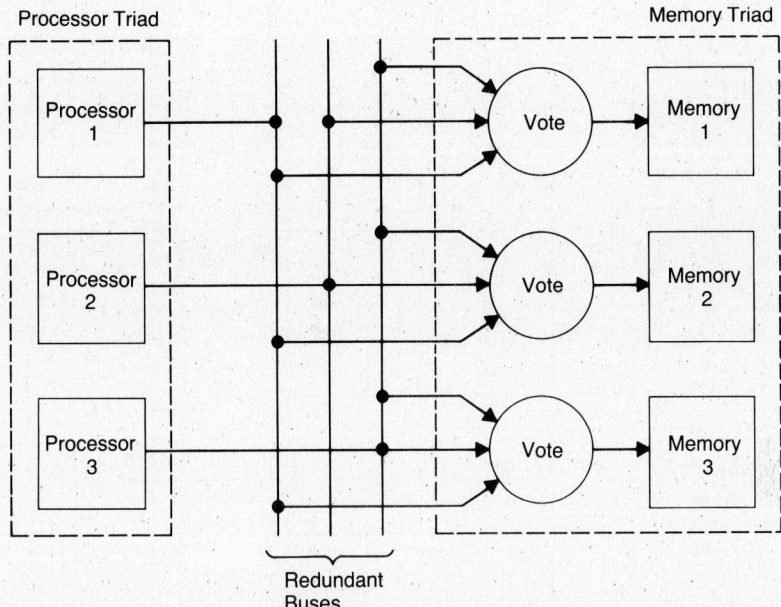

Fig. 5.31 Information transferred from the processors to memory is voted upon at the memory to ensure correct results are placed in each memory.

sor triads can perform computations in parallel, whereas the processor modules within each triad perform exactly the same computations.

Like processor modules, the memory modules from three LRUs can be combined into a single memory triad. Once again, the ten LRUs provide ten memory modules that can be configured into as many as three memory triads and one remaining spare.

All elements of the FTMP operate using a common clock. The clock generation modules of the LRUs are used to form a reliable and fault-tolerant clock generator. The clock generator modules from four of the LRUs each transmit a clock signal that is phase-locked to the other three to form a *clock quad*. The clock signals produced by the four members of the clock quad are transmitted continuously over the clock lines of four of the five buses. Clock receivers in the remaining LRUs receive three of the four clock signals and use a majority vote to derive their own clock. Therefore, one of the clock generator modules can fail without affecting the clock signal at any LRU. Once a clock generator module has failed, a spare can be provided by one of the LRUs.

The bus guardian units (BGUs) form a key component of the FTMP system. The fundamental purpose of the BGU is fault containment; each LRU

328 The Design of Practical Fault-Tolerant Systems

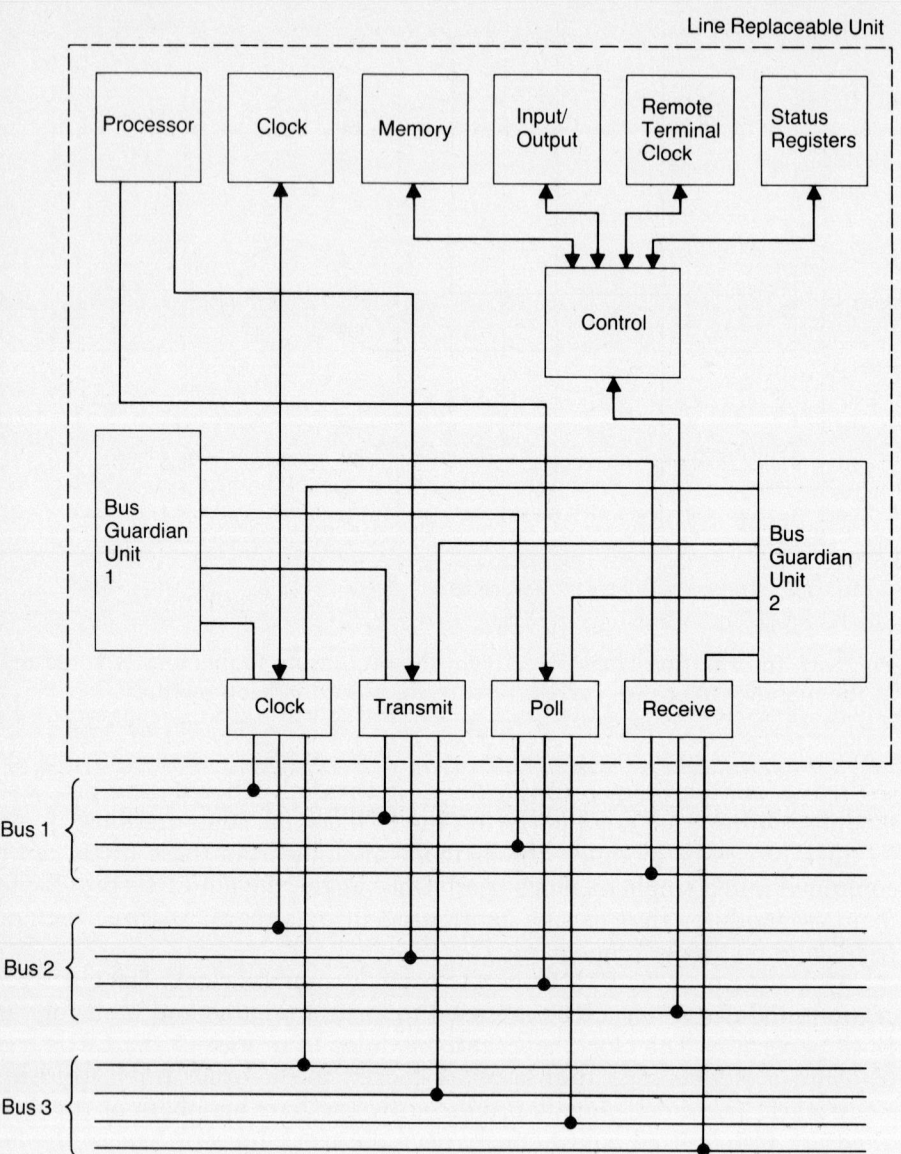

Fig. 5.32 Architecture of a line replaceable unit. Three buses are used at any one time, however, spare buses are also provided.

has two BGUs that isolate the LRU from the system in the event of a fault. Figure 5.32 shows the BGUs as part of each LRU. Each BGU has control gates, and the control gates of both BGUs within an LRU must be enabled for that LRU to place information on the system bus. In other words, either BGU can prevent the LRU from transmitting on the bus. Each BGU receives all five processor transmit lines and selects three of those five lines for its use. The BGU processes the information from the three selected processor transmit lines through the voting circuitry as well as other processing circuits.

Software Implemented Fault Tolerance

The software implemented fault tolerance (SIFT) system was designed as a competitor of FTMP [Wensley et al., 1978]. Both systems were intended for the same application of commercial, fly-by-wire, aircraft control, and both were required to achieve an unreliability of 10^{-9}, or less, during a ten-hour flight. In addition, the development of prototypes of both SIFT and FTMP was funded by the NASA Langley Research Center, and the prototypes were delivered to NASA in early 1983 [Goldberg et al., 1984].

The fundamental difference between SIFT and FTMP is that SIFT selected a software-intensive approach, whereas FTMP chose a hardware-intensive approach. The primary advantage of the SIFT technique is that off-the-shelf hardware can be used as opposed to requiring the development of new, application-dependent hardware modules. A second advantage of SIFT is flexibility since system changes can be implemented by modifying the software rather than the hardware. The disadvantage of using a software-intensive approach is that significant amounts of the system's processing capability must be used to perform the fault tolerance functions; therefore, more processors must be used or fewer applications functions performed.

The designers of the SIFT system held four viewpoints that significantly influenced the resulting design. First, they wanted to maximize the use of standard hardware. The opinion was that component reliability tends to increase as the maturity of the component production process increases; therefore, the use of standard hardware should improve the overall system reliability. Second, the design should minimize, if not eliminate, the dependency associated with shared elements. For example, if two computers share a power supply, both computers are susceptible to the failure of that power supply. Third, the system software should be written in a high-level programming language to decrease the probability of software bugs and to enhance the modularity of the design. Finally, the designers felt that the best designs were the simplest ones, so they attempted to keep the design as simple as possible. Simplicity also helps to enhance the ease with which the design is verified.

The primary features of the SIFT architecture support the four design philosophies mentioned above. First, SIFT uses conventional computers, consisting of both processor and memory, as the smallest unit of hardware.

A fault in any part of the computer results in the system being reconfigured to remove that complete computer. Second, only two elements of the SIFT architecture are shared; the intercommunications bus and the primary power supply, both of which are designed to tolerate faults. Third, the system software is indeed written in a high-level language to enhance the software reliability and enforce modularity. Finally, the system is made simple by using majority voting as the sole means of fault detection, fault containment, and fault masking.

The SIFT architecture consists of four main elements: computer modules, broadcast transmitters, broadcast receivers, and an input/output bus [Goldberg et al., 1984]. The architecture of the SIFT system is that of a multiple computer array that uses point-to-point links between each computer to provide for computer-to-computer communications. The basic architecture of the SIFT system is shown in Fig. 5.33. The important point in Fig. 5.33 is that each computer can broadcast information to all other com-

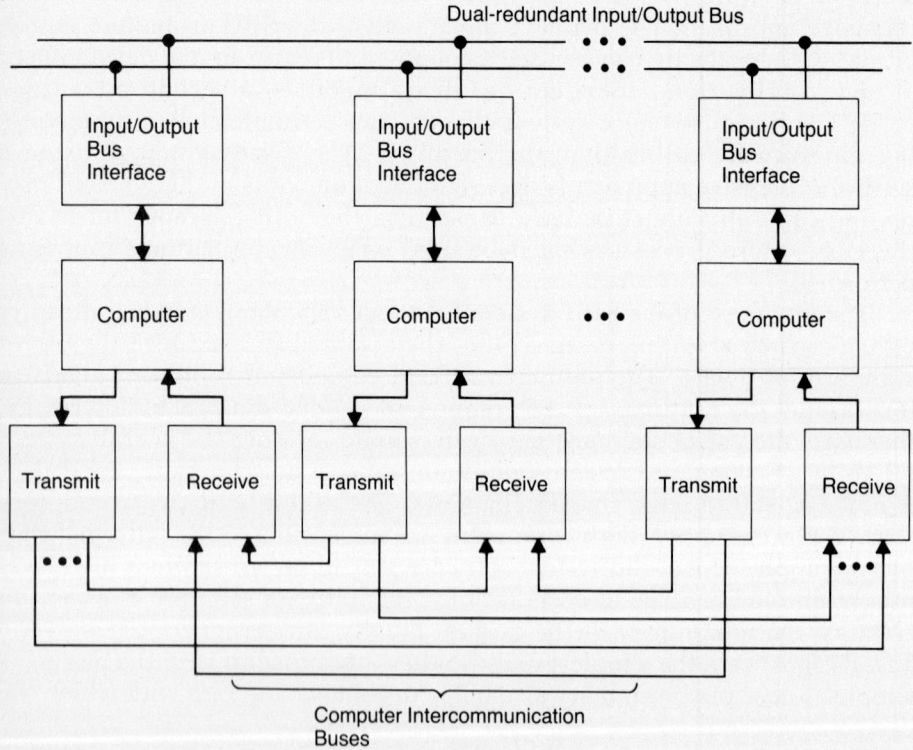

Fig. 5.33 Point-to-point interconnections are used for processor-to-processor communications in the SIFT computer system.

puters via the point-to-point links. Likewise, each computer can receive information from all the other computers. The point-to-point links between the computers are serial data buses. In addition, each computer is connected to a serial, input/output bus that is used to receive information from the aircraft's sensors and to send information to the aircraft's actuators.

Because the SIFT system is designed using a software-oriented approach, some of the most important aspects of the system are related to the system's software [Meliar-Smith and Schwartz 1982]. The SIFT system executive software provides the following functions:

1. Controls the execution of each task within the system (the designers of the SIFT system call this the *scheduling* function).
2. Performs voting to provide fault masking.
3. Detects and diagnoses faults that occur within the system.
4. Reconfigures the system to prevent the use of faulty hardware components.
5. Maintains a sufficient degree of synchronism among the computers.
6. Ensures that the necessary results of tasks are transmitted to all the computers in a timely manner.
7. Provides identical copies of all input data to the computers (this is called the *consistency* function).
8. Performs the input/output function to guarantee that sensor data is received and the actuator data is provided.

The SIFT executive software is divided into three parts: a global executive, a local executive, and a communications executive. The global executive is responsible for the complete system and is executed simultaneously by three processors using majority voting to determine the outcome of each task. The global executive performs fault diagnosis to determine which computer units have failed and reconfigures the system by allocating the tasks of a failed computer unit to some other computer. For example, if a computer module were determined to be failed, the global executive would cease to give tasks for execution to that failed module. In effect, the failed module is disconnected from the system.

A copy of the local executive executes in each computer module and controls the execution of tasks within that computer. The local executive is responsible for:

1. Running each task allocated to the computer module in which the local executive is executing.
2. Providing results to other tasks.
3. Receiving results, and voting upon those results, from other tasks.
4. Reporting errors that occur.

Errors are reported when data that is voted upon does not agree. For example, if three copies of a data item are available but only two of those copies agree, an error condition is reported.

The communications executive provides for the transfer of information between the local and the global executives. For example, any error conditions detected by the local executive must be reported to the global executive via the communications executive.

The SIFT prototype was constructed using the Bendix BDX-930 central processing unit (CPU) [Goldberg et al., 1984]. The BDX-930 is a 16-bit processor designed specifically for aerospace applications. The BDX-930 is contained on a single printed circuit board and can operate with an average operation rate of approximately 942,000 operations per second (942 KOPS).

August Systems CS-3001 Control Computer

The August Systems CS-3001 Control Computer was designed for critical control and monitoring functions in an industrial environment [August 1986]. Although the system supports a variety of applications, typical examples are chemical process control or the monitoring, and potential shutdown, of a nuclear reactor. The design also supports the maintenance process by using online diagnostics to detect, locate, and report problems.

The CS-3001 contains three primary components: the control computer module (CCM), a dataport module (DPM), and a peripheral interface module (PIM). The CCM is the heart of the control system and contains the primary processing capability of the system. The DPM provides a high-speed interface between the CCM and, if needed, other computers. Finally, the PIM provides an interface to the CCM through the DPM. Terminals or mass storage devices are examples of peripherals that can be interfaced through the PIM.

The CS-3001 achieves fault tolerance and high reliability through the use of triple modular redundancy (TMR). Figure 5.34 shows the basic architecture of the CS-3001. The CCM normally contains three processors that are arranged in a TMR voting configuration. The processors are tightly synchronized and operate on the same data using exactly the same software. An internal bus is used to provide a means of communication between the three processors. The architecture provides the flexibility to use less than triple redundancy, if the application requirements can be met with a dual or simplex system.

Incoming data arrives via the triplicated buses. Each processor can receive information from each bus and perform a majority vote to determine the value of the data to be used. Results produced by the processors are transmitted over the triplicated bus to the DPM where a single voted output is produced and provided to the appropriate interface.

5.5 ■ Sample Fault-Tolerant Systems 333

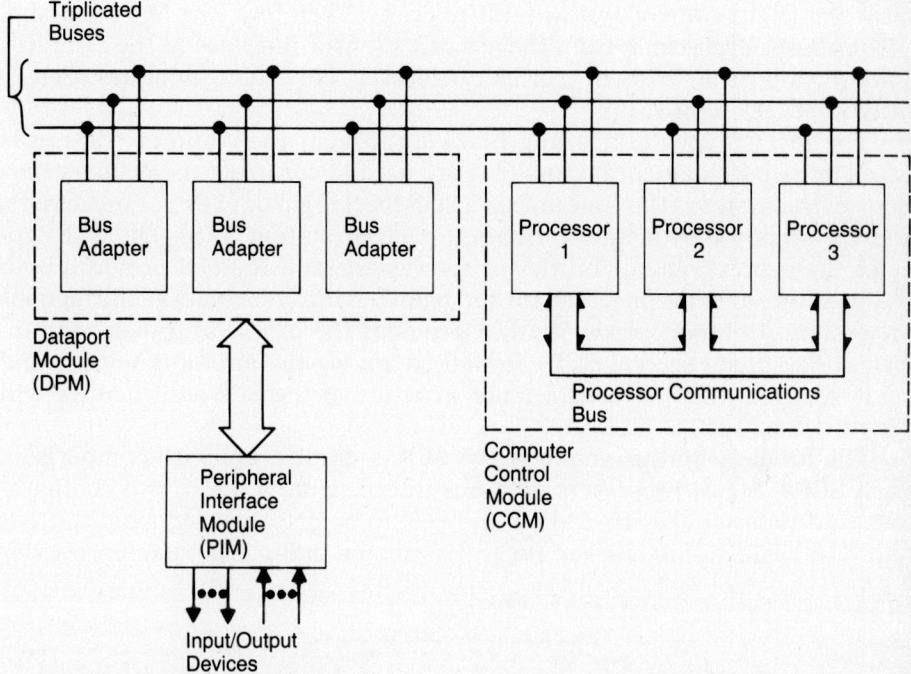

Fig. 5.34 The CS-3001 Control Computer uses triplication with voting to achieve high reliability and availability.

The processors used in the CS-3001 are the 32-bit Motorola 68020. Each processor board can contain as much as 4.5 million bytes of memory. In addition, each processor board can be configured to include a cache memory to improve performance.

Multi-Microprocessor Flight Control System

The Multi-Microprocessor Flight Control System (M^2FCS) was designed as an ultra-reliable architecture for implementing flight-control algorithms [White et al., 1981]. The design of M^2FCS was started in 1978 by Honeywell and has evolved significantly during the past few years. The primary design objectives for M^2FCS were to achieve high reliability and survivability, while maintaining an ease of testing, maintenance, and expansion.

The architects of M^2FCS chose to use the word "system" in the title of their design because they felt that they had to consider all aspects of the aircraft control system to truly develop a fault-tolerant and reliable design. In other words, simply designing a fault-tolerant computer does not make the resulting flight-control *system* fault-tolerant. Therefore, the M²FCS design encompasses everything from the aircraft's sensor interface to the aircraft's actuator interface. We will consider, however, only the techniques used in the computer architecture.

The M²FCS uses a building block approach; the complete design is based on two fundamental units that are used to construct the primary portions of the system. The first unit is a self-checking pair (SCP) of processors, and the second unit is an information transfer system (ITS). The SCP provides the processing capabilities of the system, and the ITS provides both the medium and the mechanism for transferring information throughout the system. In other words, the ITS provides the means of transferring information among several SCPs as well as among the aircraft's sensors and actuators. We will first consider the structure of the SCP and then we will examine the ITS.

The fundamental principle of the SCP is duplication with comparison. Each SCP has two processors, two bus interface units (BIU), two comparators, and two switches that allow the SCP to be disconnected from the system. The basic architecture of the SCP is shown in Fig. 5.35. Each processor

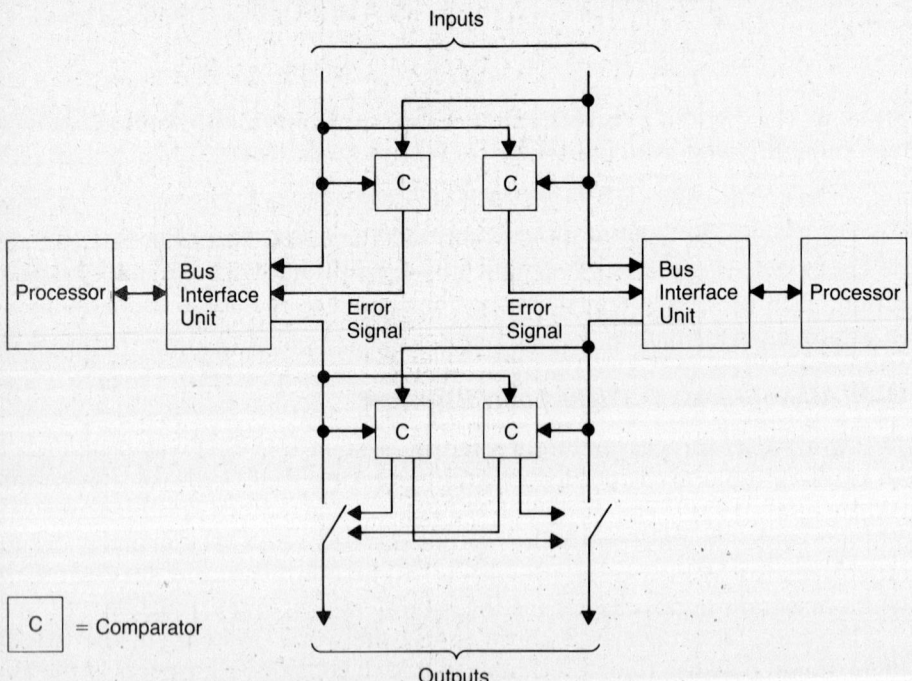

Fig. 5.35 The self-checking pair (SCP) contains two processors and two bus interface units. All inputs and outputs are compared to detect failures.

in an SCP performs exactly the same computations on exactly the same data, and the results are compared by the two comparators. If a comparator detects a disagreement, it opens the switches that connect the SCP to the bus, thereby disconnecting the malfunctioning SCP from the system. Both comparators must agree that the processors are performing correctly and are in agreement before the SCP is allowed to remain a part of the system. The comparators provide a fault-containment function by preventing erroneous data from leaving the SCP. Comparators are also provided for data coming to the SCP to detect disagreements in the data before it is processed.

The bus interface unit (BIU) provides a means for the processors to communicate with the remainder of the system. Each processor has its own BIU that is responsible for providing data to both the processor and the comparators at the appropriate times. The BIU frees the processors from considering bus transmission or protocol issues. Each processor interfaces with its BIU, not with any other element of the system. It is the responsibility of the BIU to get data from a processor to some other part of the system, and vice versa.

Although the overall concept of the SCP is simple, the implementation can be very difficult. The two primary problems are synchronization and the data comparisons. If two processors within an SCP are performing the same computations, they must be synchronized so that they complete the computations at approximately the same time. Otherwise, you might be comparing one processor's results from one time period with the other processor's results from a completely different time period. Some degree of synchronization of the processors within a SCP is required.

The data comparison is also a difficult problem, as we have seen in Chapter 3. The fundamental problem is particularly acute in real-time control systems where two signals may not agree exactly even though both data items are completely correct. For example, two analog-to-digital converters may not produce exactly the same values, even when identical signals are applied to each. Quantization errors and noise can produce values that do not agree exactly. Therefore, the comparison process can seldom be as simple as a bit-by-bit comparison.

Two possible solutions to the synchronization problem are called *tight synchronization* and *loose synchronization*. Tight synchronization implies that all processors are synchronized with a common oscillator, or clock source. In a fault-tolerant design, the common clock must have an extremely high reliability and also be capable of tolerating faults that can occur. Otherwise, the clock becomes the weak point of the system, and the system can be no more reliable, or fault tolerant, than the clock.

Honeywell's M^{2}FCS uses the loose synchronization approach. In loose synchronization, all processors need not be synchronized to the same clock.

Instead, each processor has its own clock that operates independently of any other clock. The only synchronization that is imposed upon the system is at the point where comparisons must be performed.

Loose synchronization can be accomplished with a small synchronization unit located in each SCP. The function of the synchronization unit is to provide for the comparison of consistent data. The synchronization unit can be a simple interrupt device that provides an interrupt to each BIU at specified time intervals. Upon the occurrence of each interrupt, the processors and the BIU perform specified functions. For example, upon the occurrence of the first interrupt, each processor can obtain from its BIU the data needed to perform its calculations. At the next interrupt, each processor provides to its BIU the results of the computations. The results are provided to the comparators for comparison.

The M^2FCS ensures that each processor in a SCP receives identical data by comparing the inputs to the SCP and disconnecting that SCP if the inputs are not identical. As was shown in Fig. 5.35, two comparators perform the comparison of the inputs. The results of both comparisons are provided to both processors in the SCP to ensure that each processor knows when a miscompare has occurred.

The second basic building block of the M^2FCS is the information transfer system (ITS). As mentioned earlier, the ITS provides for the transfer of information among SCPs and the other components of the complete flight-control system. The ITS contains redundant buses connected to the BIUs of the system. For example, Fig. 5.36 shows the structure of a M^2FCS with three SCPs on three different buses. Each bus is actually dual redundant. Each BIU can receive information from one of the two buses in each bus pair, but a BIU is restricted to transmit on only one bus in each bus pair.

A laboratory model of the M^2FCS has been constructed by Honeywell and used to demonstrate the basic concepts of the approach [White et al., 1981]. The laboratory model was constructed using the 16-bit Intel 8086 processor augmented with the Intel 8087 floating-point processor and the Intel 8089 input/output processor.

Agusta A129 Integrated Multiplex System

The A129 combat helicopter is a lightweight, and extremely agile helicopter designed for attack and scout missions [Johnson and Julich 1985]. The fault-tolerant computer system that performs the on-board processing is called the Integrated Multiplex System (IMS). The IMS plays a significant role in the achievement of a high reliability for the overall helicopter system.

The IMS provides a number of aircraft functions. Examples include automatic flight control, navigation, stability augmentation, weapons system

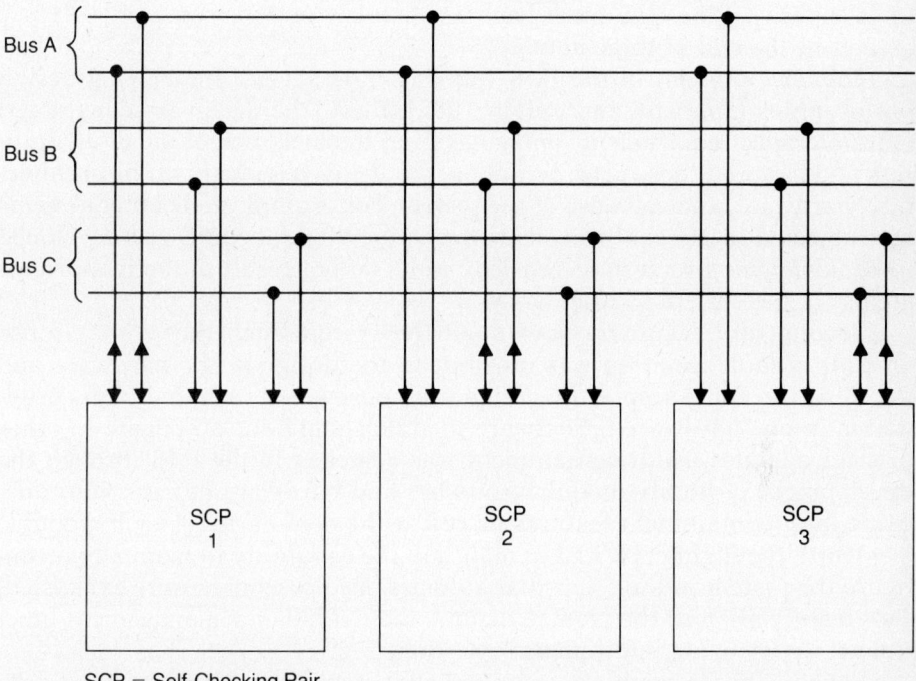

Fig. 5.36 Self-checking pairs of processors are interconnected in the M²FCS using redundant buses. Each SCP can receive data from any bus pair but transmits data on only one bus pair.

management, cockpit interfaces, redundancy management, main rotor fly-by-wire control, and various other system monitoring functions. The automatic flight control modes include attitude hold, heading hold, altitude hold, ground-speed hold, air-speed hold, vertical-speed hold, and course hold. The main rotor fly-by-wire system is an emergency backup designed to be used in the event of ballistic damage to the main rotor mechanical linkage. The IMS performs several system monitoring functions, including rotor transmission monitoring, hydraulics monitoring, vibration monitoring, engine monitoring, and fuel monitoring. Finally, the IMS provides the control of the hydraulics and fuel subsystems.

The IMS was required to be reliable enough to complete 220 three-hour missions without experiencing a mission-critical malfunction. This corresponds to a failure rate of one system failure every 660 hours. Using the exponential failure law, the required system reliability at the end of a three-hour time period can be approximated as

$$R_{\text{system}}(3 \text{ hours}) = e^{-3(1/660)} = 0.9955$$

In other words, the IMS must have a reliability of approximately 0.995 or greater at the end of three hours.

The development of the IMS was based on several underlying design philosophies [Johnson and Julich 1985]. First, the designers considered fault tolerance requirements and goals from the inception of the design process and factored those considerations into the development of the architecture, hardware, and software of the system. For example, helicopter systems are subjected to frequent transient effects; therefore, the IMS was developed to be *nonlatching* such that transient faults do not result in the inappropriate use of the system's redundancy.

Second, the designers developed the system such that a maximum amount of fault isolation was inherent in the design of the hardware and the software. For example, the design was constructed such that faults originating in one hardware or software module should not propagate to other distinct modules. Fault containment was achieved in the IMS through the development of highly modular software and hardware designs with adequate fault containment features placed at the boundaries of each module.

Third, the flight crew had to maintain the capability to manually reconfigure the system in the event that autonomous procedures were exhausted. This feature allowed the crew to assume the redundancy management functions if catastrophic, multiple failures occurred.

Fourth, the designers determined that maximum flexibility could be obtained if the majority of the fault tolerance features were software-implemented [Johnson and Julich 1985], including fault detection, fault location, fault containment, and system reconfiguration. The approach selected is similar to that adopted by the designers of the SIFT system [Wensley et al., 1978].

Fifth, the IMS required graceful degradation or passivation when unmanageable faults occurred. Graceful degradation, as defined by the designers, implied disabling all modes that could not be reliably performed. Passivation, again as defined by the designers, meant to force the critical outputs of the system to a safe state such that possible erroneous commands could not produce harmful effects.

The final design philosophy concerned the amount of redundancy to be employed. The weight limitations of the helicopter application severely restricted the redundancy that the designers could employ. Therefore, the architecture was limited to dual redundancy. The designers judged additional redundancy to be unnecessary to meet the reliability requirements and unacceptable in terms of the weight restrictions.

The architecture of the IMS, which is shown in Fig. 5.37, is dual redundant and based on a MIL-STD-1553B time-multiplexed bus [Aircraft 1978]. The system consists of two *master* units (MU1 and MU2), two *local* units (LU1 and LU2), two *remote* units (RU1 and RU2), and two 1553B data buses

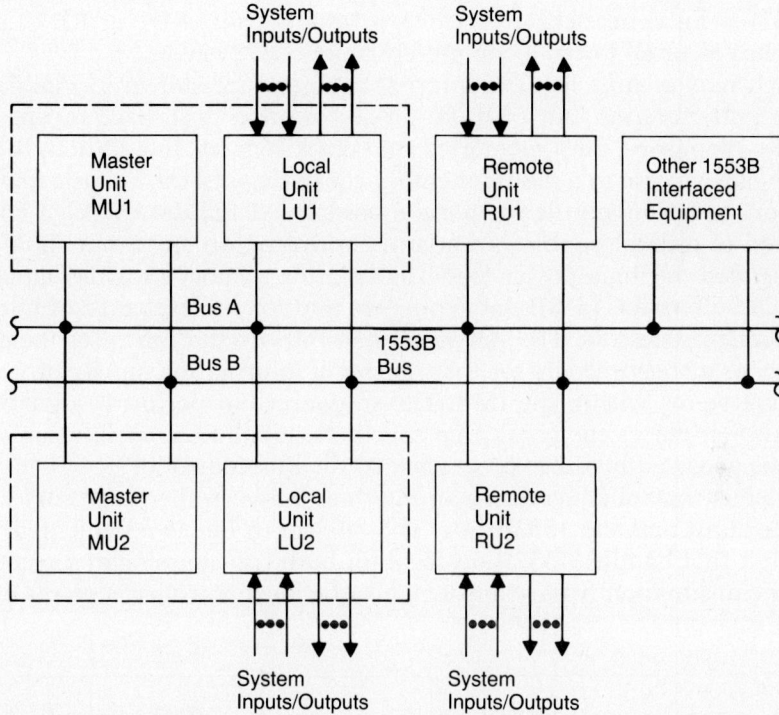

Fig. 5.37 Dual-redundant architecture of the A129 Integrated Multiplex System. (From [Johnson and Julich 1985] © 1985 IEEE)

(A and B). The master units provide dual-redundant processing capabilities, whereas the local units and the remote units form a dual-redundant set of interfaces between the master units and the external sensors and actuators. The discussions here focus on the master units and the fault detection and reconfiguration techniques used within those units. One of the unique aspects of the system is its ability to achieve high fault coverage and reliability with only dual redundancy. Most of the fault detection and reconfiguration features reside within the master units.

On power-up, a contention procedure causes one master unit to assume control of the 1553B bus. The second master unit serves as a *hot* backup that monitors all information over the 1553B and performs all computations. Equivalency checks are performed between the two master units to guarantee the validity of the critical computations. If the master unit controlling the 1553B fails, it ceases to issue commands over the 1553B, and the backup unit senses the silence and assumes control of the 1553B. All communications between the master units and the local and remote units occur over the 1553B bus. Therefore any combination of a master unit (MU1 or MU2), a local unit (LU1 or LU2), a remote unit (RU1 or RU2), and a 1553B bus (A or B) forms a completely functional system.

Each master unit has the internal architecture shown in Fig. 5.38. A master unit contains four or more processors that each have independent program memories, data memories, and clock sources. In addition, the processors have access to a global memory consisting of both volatile and nonvolatile RAM. Nonvolatile memory is used for global data storage and the retention of critical configuration and control parameters that should not be corrupted during a power loss. The volatile memory provides input/output (I/O) buffers for 1553B data transfers and for the exchange of information among processors. The software in an IMS master unit uses the global memory as a *bulletin board* for the transfer of information among processors and subsystems within the IMS. The software that performs a particular computation places the result in a specified global memory location where it can be accessed by other processors or the bus controller.

The bus controller serves as an interface between the processors within a master unit and the 1553B bus. The bus controller is a dedicated hardware unit that handles 1553B protocol for both transmission and reception. During transmission from a master unit, the bus controller accesses output

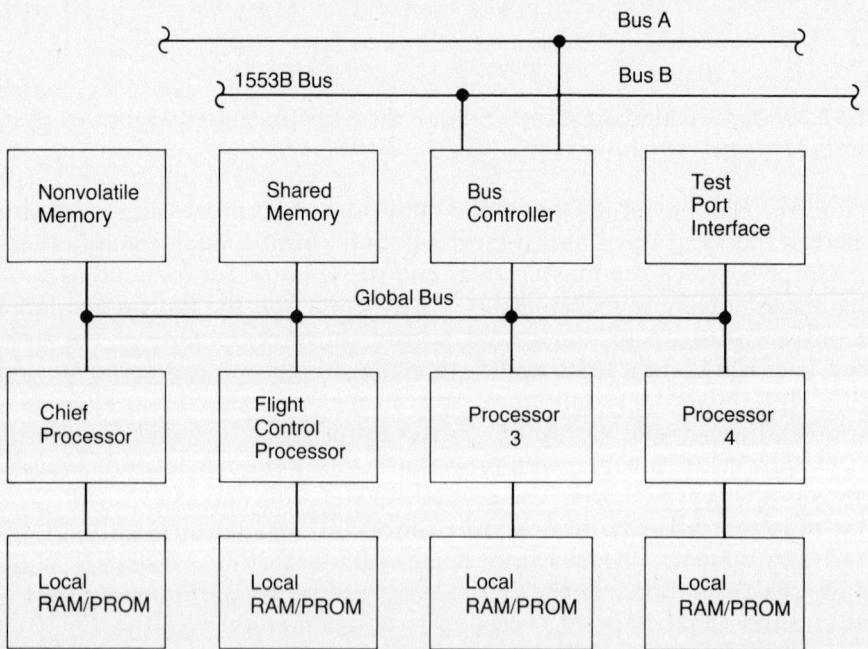

Fig. 5.38 Each master unit in the IMS is a multicomputer architecture. (From [Johnson and Julich 1985] © 1985 IEEE)

buffers in either nonvolatile memory or shared memory and converts the information into standard 1553B format. During reception over the 1553B, the bus controller converts the information from 1553B format and stores the results in a specified location in shared memory. This technique allows the processors to remain free of 1553B formatting and protocol issues; the processors need only access locations in global memory to transmit or receive information over the 1553B.

A general philosophy designed into the master units is that the processors have to continuously prove their operational capabilities to serve as bus master. This is accomplished by two flags that are incorporated into the bus controller and serve as *watchdog* timers. A master unit that is functioning properly must signal its bus controller that it is capable of performing as bus master by setting a dedicated flag. If both master units request bus mastership, a hardware contention mechanism determines which is to be master. Because hardware clears the *master request* flag at 180 Hz, a master unit must refresh the flag at that rate to maintain bus mastership.

A second control flag must also be set by the master unit's software at 180 Hz to prevent its bus controller from resetting all the processors in a master unit. A master unit's software ceases setting the flag when fatal errors such as a memory parity error are detected by the master unit's fault detection features. When a reset command is issued by a bus controller to its processors, the affected master unit goes to an offline state and initializes itself in an attempt to recover to backup status. During the initialization time, the backup master unit assumes control of the 1553B bus. This is the normal means of transferring bus control from one master unit to the other.

The fault tolerance attributes of the IMS consist of a collection of hardware and software that provides for the detection of faults and the management of the system's redundancy. Because the system was constrained to dual redundancy, the basic fault tolerance philosophies had to be carefully defined if the desired high reliability and fault tolerance were to be attained. For example, comparisons between the two master units can detect faults but cannot identify the faulty unit such that reconfiguration can occur. Consequently, comparisons cannot be relied on as a primary fault tolerance mechanism.

The basic fault tolerance philosophy was to consider the system as hierarchical in nature, as is shown in Fig. 5.39. Because the majority of the fault detection and redundancy management techniques are implemented in software, the master units form the highest level in the hierarchy. Fault detection techniques were devised to concurrently check the operation of each level in the hierarchy and to check the paths between levels to guarantee the error-free performance of the system. The results of the fault detection algorithms are then used to form a decision concerning the *health* of

342 The Design of Practical Fault-Tolerant Systems

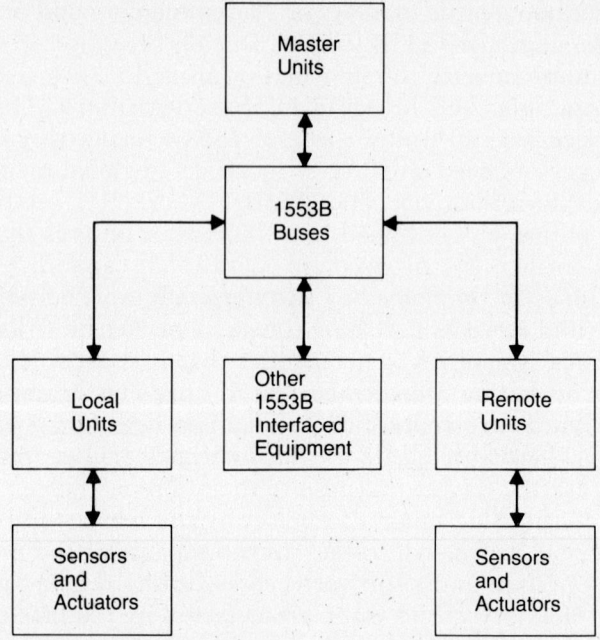

Fig. 5.39 Hierarchical organization of the fault tolerance features in the IMS. (From [Johnson and Julich 1985] © 1985 IEEE)

each element in the system and to develop a redundancy management action if any element is not functioning correctly. The possible redundancy management actions include:

1. Master unit swap-over
2. Bus swap-over
3. Local unit swap-over
4. Remote unit swap-over
5. Sensor swap-over
6. Degradation of the system
7. Passivation of the system

The basic concept of each master unit's fault detection capability is that each processor runs a series of tests to verify its own capability. Each processor uses the results of its tests to form a decision concerning its own health. The resulting decisions are transferred to the *chief* processor, which combines the information with the results of several additional master unit tests to form a decision concerning the health of the complete master unit. If the master unit is determined to be faulty, the chief processor ceases setting the master request flag for the bus controller, and a master unit swap-over occurs, provided that the second master unit is capable of assuming control of the 1553B bus. The bus controller of the faulty master unit forces a processor reset and a recovery to backup status is attempted. In both the interprocessor and the chief/bus controller interfaces, there is a *watchdog timer effect* in that the lack of a response from an element is interpreted as a failure. This means that a failed element need not explicitly recognize that a failure has occurred as long as the failure prevents the appropriate interchange from occurring.

The state diagram presented in Fig. 5.40 illustrates the fault recovery procedure for the master units. Each master unit can exist in one of three logical states: bus master (implies control of the 1553B), slave (backup mode), and offline. While serving as bus master, a master unit has complete control of the 1553B. To become bus master, a master unit must first determine that it is capable of serving as such and must be granted bus mastership by the bus controller.

If a master unit fails and that failure is detected, the affected master unit goes to the offline state and attempts to perform an initialization and restart procedure. If the failed master unit is the bus master, the backup master unit, if it is fault free, senses silence on the 1553B bus and assumes

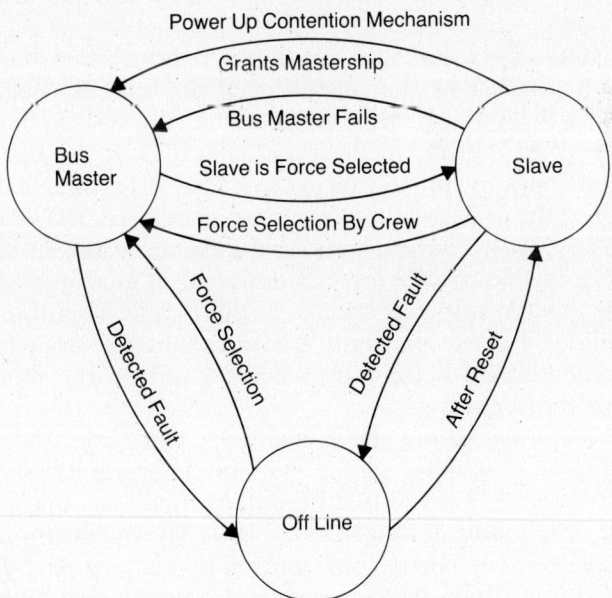

Fig. 5.40 State diagram illustrating the reconfiguration operations in the IMS. (From [Johnson and Julich 1985] © 1985 IEEE)

the bus master position. It is important to point out that a master unit is never locked into the offline state. If a fault clears, as might be the case if a transient fault occurs, the master unit can return to the backup status, fully prepared to assume the bus mastership again, if necessary. This prevents the misuse of the system's redundancy as a result of transient faults.

To be in the slave state, a master unit must possess enough capability to serve as bus master. The slave state represents a master unit that is serving as a *hot* backup and is ready to assume mastership of the 1553B if the bus master fails. The slave unit also responds to requests from the bus master for equivalency checks. Periodically the bus master will request and receive critical results from the slave unit such that comparisons with the results of the bus master can occur. The only exception to the performance of the equivalency checks is when the slave unit is known to be performing erroneously; the bus master does not attempt to compare its results with those of a known faulty slave unit. If both the bus master and the slave unit are perceived as fault free, the bus master compares the results of the two units and passivates the system if a disagreement occurs. Passivation is required because the faulty unit cannot be identified from the comparison of the two results. The units continue running their fault detection and recovery algorithms, however, and attempt to recover from this condition. This illustrates the importance of the fault detection techniques other than comparisons; the only course of action when a miscompare occurs is passivation.

The final state transition that can occur results when the crew forces one master unit to serve as the controller of the bus. The selected master unit continues to detect faults that occur within the system, but it does not relinquish control of the bus in the event that a fault occurs. Instead, the master unit continues to control the system as best it can.

The design of the system allows the master unit to exist in all combinations of states except two bus masters, two slave units, or one slave and one offline unit. The bus controller logic is designed to guarantee that one and only one of the master units has control of the 1553B. Also, if one unit is capable of serving as a slave, that unit is also capable of serving as bus master, so the system should never have a master unit in the slave state with none in the bus master state.

Before the state transitions described above can occur, the master units must detect faults that exist in the system. There are three hardware-implemented tests and seven software-implemented tests that form the core of each master unit's fault detection capability. The hardware-implemented tests consists of memory parity, bus controller memory parity, and illegal instruction detection. Upon the occurrence of a parity error, a hardware interrupt is generated, which results in the failed master unit transitioning to the offline state. The occurrence of illegal instructions results in an identical action.

The primary fault detection techniques are implemented in software. Perhaps the most important of these is the *interprocessor handshake*, which requires each processor to set a flag at, or above, a specified frequency to indicate the status of that processor. The chief processor is responsible for examining each flag to determine if the flag has been appropriately set. If any processor fails to refresh the flag, the chief processor ceases setting the master request flag, and the master unit transitions to the offline state. The chief processor clears the flags immediately after examining them.

Periodic self-test functions are executed within each processor to verify the arithmetic logic unit (ALU) of the processor and to provide additional protection against memory faults. Each processor checks its ALU by periodically executing a representative set of arithmetic and logic functions and comparing the results to the correct ones stored in PROM. Each memory is also checked on a periodic basis by writing specified patterns to the memory and subsequently reading the same location to make sure that the identical data is returned. The memory test is performed on global memory, nonvolatile memory, and local memory and provides protection against faults in the interfaces between the processors and the memory.

As in any system that contains large quantities of software, the occurrence of software faults is of major concern [Wensley et al., 1978]. As a means of detecting software faults, two tests are included. The *task overflow test* uses hardware timers to detect cases where a particular software routine is taking too much time to execute. This type of test can detect the existence of infinite loops and other software faults. In addition, the task overflow test can detect certain hardware faults; for example, a processor that simply *hangs up* will be detected. The *task execution test* is designed to verify that all tasks are executing at some time. Tasks that fail to execute can be indicative of software faults in the scheduling routine or hardware faults in the execution tables. In either case, the task execution test detects the problem.

The reliability analysis performed on the IMS has shown that the system is capable of flying more than 2400 three-hour missions without experiencing a failure that prevents the system from continuing its desired functions. The IMS was flight tested during the latter half of 1985 and early 1986.

5.5.3 High-Availability Applications

We will consider four systems that have been designed to achieve high availability; the Tandem NonStop computer system [Katzman 1977], the

Stratus/32 system [Serlin 1984], the Electronic Switching System (ESS) [Toy 1978], and the Synapse N+1 architecture [Frank and Inselberg 1984]. Each of the systems has been designed and developed for commercial applications such as banking or telephone switching. The primary intent of each design is to develop a system that achieves a high availability.

The Tandem 16 NonStop System

The Tandem 16 NonStop computer system was one of the first commercially produced systems to be designed for high availability [Katzman 1977]. The Tandem system was designed to support the transactions processing market.

The design goals of the Tandem system were similar to those of many fault-tolerant systems. First, the system was designed to provide autonomous fault detection, system reconfiguration, and system repair without interrupting the functions of the fault-free components within the system. An important attribute of the Tandem system is its ability to allow the removal of failed components and the replacement of repaired components without impairing the ability of the remaining components to function correctly. Second, it was desired to eliminate all hardware single points of failure in the system. No single, hardware fault should threaten the integrity of the data within the system. This requirement is particularly important in the transactions processing industry where large volumes of data, such as banking records, are being processed and stored. Finally, the designers of the Tandem system had as an objective the use of significant modularity in their design. The modularity should allow the system to be easily expanded without impacting the applications software. Modularity also improves the maintainability of the system via the use of module replacement schemes.

The basic architecture of the Tandem system is shown in Fig. 5.41. The system presently supports up to 16 processors, each of which consists of a 16-bit central processing unit, bus control circuitry, input/output control, and up to 512K bytes of main memory. Interprocessor communications occur over the Dynabus, which actually comprises two independent buses, called the X-bus and the Y-bus. Both the X-bus and the Y-bus each consist of a 16-bit data field and a control field. The Dynabus is designed such that the failure of the X-bus, the Y-bus, or any single component connected to either bus will not curtail interprocessor communications. Attached to the input/output buses are various peripheral devices such as disks, terminals, and tape drives. Devices on the input/output buses have multiple ports that allow each device to be accessed from more than one input/output bus, in the event of failure.

The Tandem 16 system contains a number of hardware-implemented fault detection features. Included among these are the use of checksums, parity error checking, error-correcting memories, and the use of watchdog

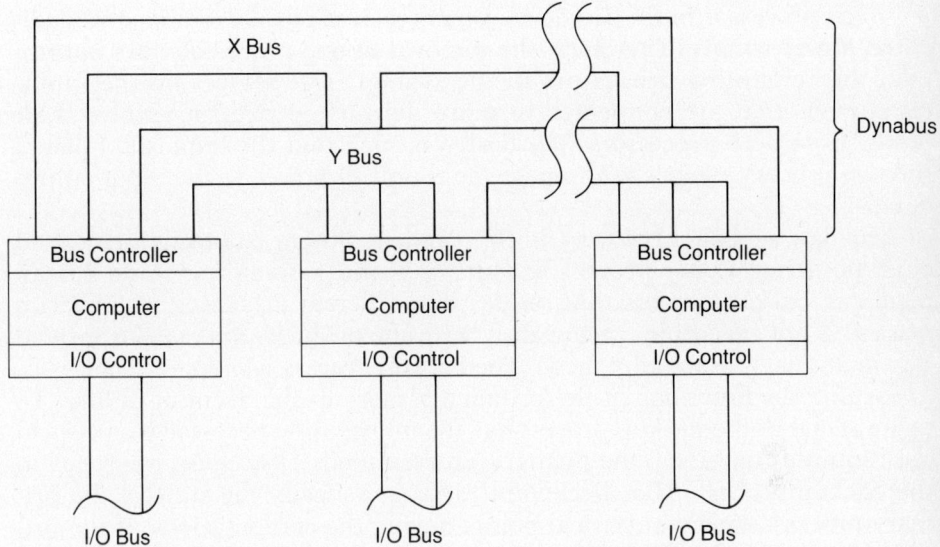

Fig. 5.41 Architecture of the Tandem Nonstop Computer System.

timers. For example, data packets transferred from processor to processor via the Dynabus are protected with checksums. A repeated transmission occurs if a checksum error is detected.

Memories within the Tandem system use the modified Hamming code that we discussed in Chapter 3 to provide correction of single-bit errors and the detection of double-bit errors. Each 16-bit word in memory has 5 Hamming check bits associated with it, and single-bit parity is taken over the total group of 16 data bits and 5 Hamming check bits. The Hamming check bits and the parity bit are generated prior to writing the word to memory and are stored with the 16 data bits in a 22-bit memory. When a word is read from memory, a single-bit error is detected by the existence of a nonzero syndrome and incorrect parity. Double-bit errors are characterized by correct parity and a nonzero syndrome. Single-bit errors are corrected, and double-bit errors are detected.

Watchdog timers play an important role in fault detection within the Tandem 16 system. Transmissions over the Dynabus must be completed within a specified time; otherwise, an error indication is given, and a recovery action is initiated. A second watchdog timer requires that each processor in the system place a special message on each bus once each second. Once every two seconds, each processor checks to make sure that it has received each message from all the remaining processors. The failure to receive a message from a processor is assumed to be indicative of that processor's failure.

Each processor in the Tandem system receives power from its own dedicated power supply. Therefore, the removal of one processor does not impact the remaining processors in the system. The devices on the input/output bus that are connected to more than one processor receive their power from both processors. The design is such that the removal of one of the two processors does not impact the source of power to the input/output device.

One key aspect of recovery in the Tandem system is a technique called checkpointing. Every process executing in the Tandem machine has an identical, backup process that resides in a different processor. The backup process is not executing concurrently with the primary process, but instead it is in an inactive standby mode. The backup process is prepared, however, to assume the functions of the primary process in the event of failure. To guarantee that the backup process has the information necessary to perform the required functions, the primary process sends *checkpoint* messages to the backup process. The checkpoint messages specify the state of the primary process at certain critical points during the computations. If the primary process fails, the backup process is started using the most recent checkpoint information as the starting point.

The first Tandem systems were introduced in 1975, and several product improvements have occurred since that time. Most of the improvements have been in terms of performance (processing capability) and extended memory addressing capabilities.

The Stratus/32 System

The Stratus/32 system [Serlin 1984] was designed to compete in the same transactions processing market as the Tandem system. However, the Stratus system used a completely different approach. Stratus employs a tightly coupled, hardware approach to fault tolerance, whereas the Tandem system is more software oriented.

The Stratus/32 system uses a variation of the concept of a pair-and-a-spare that we discussed in Chapter 3. Recall that the general pair-and-a-spare approach uses two modules whose outputs are continuously compared to achieve error detection. If a miscompare occurs, the system attempts to identify the failed module using self-test capabilities, and a spare module is used to replace the failed one. In the Stratus system, modules use the duplication with comparison approach, and each duplicated module has a spare that is an identical, duplicated module. If a miscompare occurs, the failed pair is replaced with a spare pair. Fault diagnosis routines then attempt to identify if the failed pair contains a permanent fault or a transient fault. If it is determined that a transient fault was the culprit, the pair is placed back into operation and serves as a spare pair.

5.5 ■ Sample Fault-Tolerant Systems

The overall concept of the Stratus system is illustrated in Fig. 5.42. Each subsystem is usually tightly synchronized with its spare so that reconfiguration can occur without disrupting the operation of the system. In other words, the failure of the primary module results in the spare immediately assuming the functions of the primary without any loss of service. The

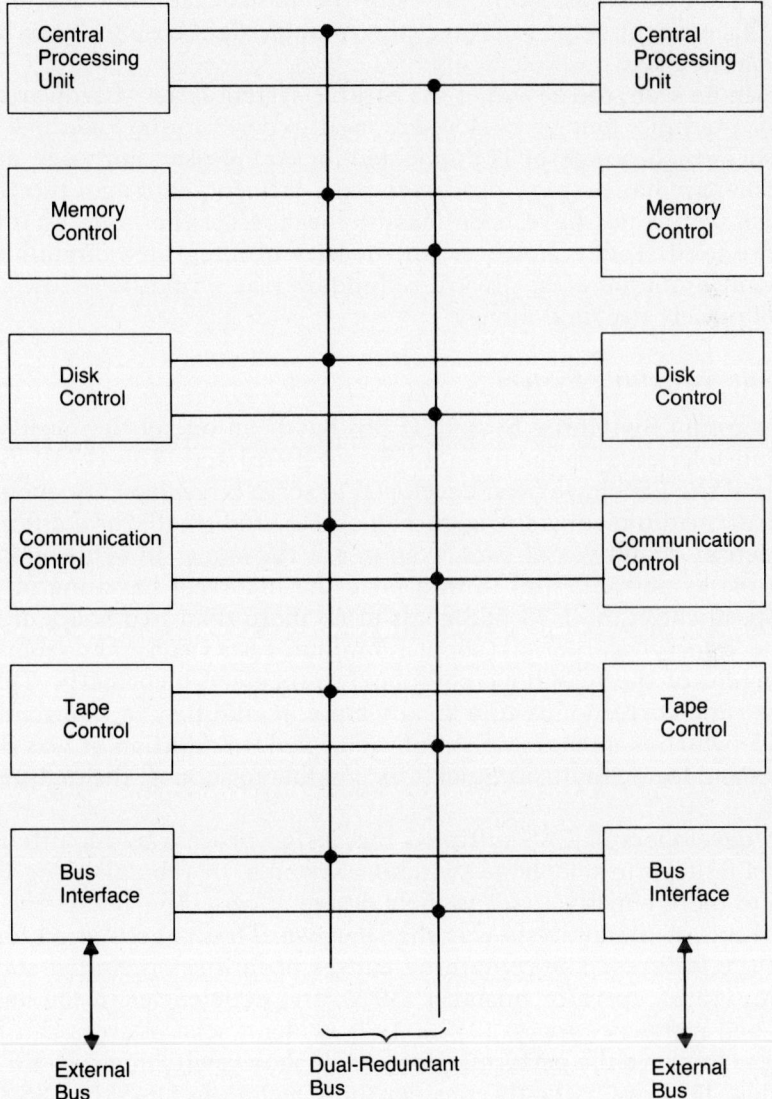

Fig. 5.42 Architecture of the Stratus Computer System. (From [Serlin 1984] © 1984 IEEE)

subsystems of the Stratus system are interconnected via a dual-redundant bus such that the failure of a bus can also be tolerated.

The concept of duplication with comparison is employed throughout the Stratus system. The disk controllers, for example, are designed with duplicated read/write control circuits; the disk controller will not perform a write to the disk unless both halves of the control circuitry are in agreement. This concept helps to protect the valuable data base that can be contained on the disk.

As can be seen, the design of the Stratus system is very hardware intensive. For example, four processors are used to perform the functions of one processor—each processor is duplicated for comparison purposes, and the processor pair has a spare processor pair. Fifteen years ago, the Stratus approach would not have been feasible because of the large quantity of hardware used. Today, however, the density of integrated circuits allows significant amounts of hardware redundancy at a relatively low cost in terms of power, size, and money.

Electronic Switching System

The Electronic Switching System (ESS) has been one of the most widely used fault-tolerant systems to date ([Storey 1976] and [Toy 1978]). Several versions of the ESS have been developed to serve large-capacity, metropolitan offices, medium-capacity, suburban offices, and small, rural offices. The fundamental objectives of each system are the same: to achieve as little down-time as possible and to minimize the incorrect handling of phone calls. Specifically, the ESS had goals of no more than two hours of down-time over a forty-year period (about 1.5 minutes per year) and no more than 0.01 percent of the telephone calls being processed incorrectly. The goal of down-time corresponds to a steady-state availability of approximately 0.999995, which is an extremely ambitious goal. In addition, it was desired to detect and locate faults as quickly as possible to support the maintenance process.

The developers of ESS initiated the design process by identifying the causes of failures in telephone switching systems, thereby allowing the designers to more effectively focus their design efforts [Toy 1978]. The interesting result of the analysis was that hardware faults accounted for only 20% of the failures. The remaining causes of failures included software problems (15%), operator mistakes (30%), and deficiencies in the fault detection and recovery process (35%). As is evident, several areas can be addressed to improve the performance of a telephone switching system. In the discussions here, we will only consider those techniques used in ESS to prevent failures resulting from hardware faults.

One of the more recent versions of the ESS uses the ESS 3A processor and is intended for applications in rural areas. The ESS 3A processor was designed to be reliable, flexible, and inexpensive. We will discuss the general architecture of the system as well as some of the error detection techniques that were used. The first ESS 3A system was placed into operation in 1975.

The basic technique used in the 3A processor, as well as all of the ESSs, is duplication. As shown in Fig. 5.43, the 3A processor has all of the processor functions duplicated. A fault in any element of a processor results in that processor being switched out of operation. In other words, reconfiguration occurs at the processor level as opposed to within a processor.

One unique feature of the ESS 3A, however, is that the duplicated processors are not compared to achieve fault detection. Instead, each processor is designed to perform its own fault detection. In other words, the 3A processor is designed to be self-checking. The primary advantage of self-checking is that the processors no longer must operate in synchronization.

The ESS 3A processor uses standby sparing. One processor will be the online unit while the other is in a halt state (standby state) until a fault is detected. Once a fault is detected in the online processor, the standby processor assumes the responsibilities of processing calls. Faults are identified using the self-checking features that are designed as an integral part of each processor.

One interesting self-checking technique used in the 3A processor is the error-detection coding in the microprogram storage. The microprogram

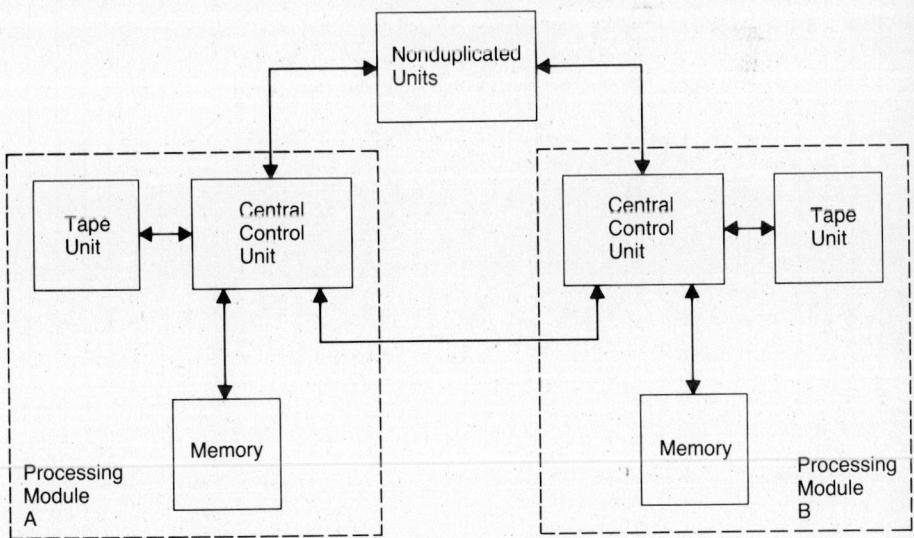

Fig. 5.43 Processing modules are duplicated in the ESS 3A. Comparisons and self-tests are used to detect failures.

352   The Design of Practical Fault-Tolerant Systems

storage is simply a read-only memory that contains address and control information. Faults in either the memory or the interface between the memory and the remainder of the processor will corrupt the operation of the system and must be detected as quickly as possible.

Two different codes are used to provide error detection for the address and control information in the microprogram store. A single parity bit is used to detect errors in the address, whereas a 4-of-8 code is used on the control. The detection of multiple, adjacent bit errors is accomplished by interleaving the address and control bits, as shown in Fig. 5.44. Any single-bit errors in the control field are detected using the 4-of-8 code. Single-bit errors in the address field are detected by the parity. Because of the interleaving, any two, adjacent bits that are erroneous are also detected because one error impacts the single-bit parity and the other impacts the 4-of-8 code.

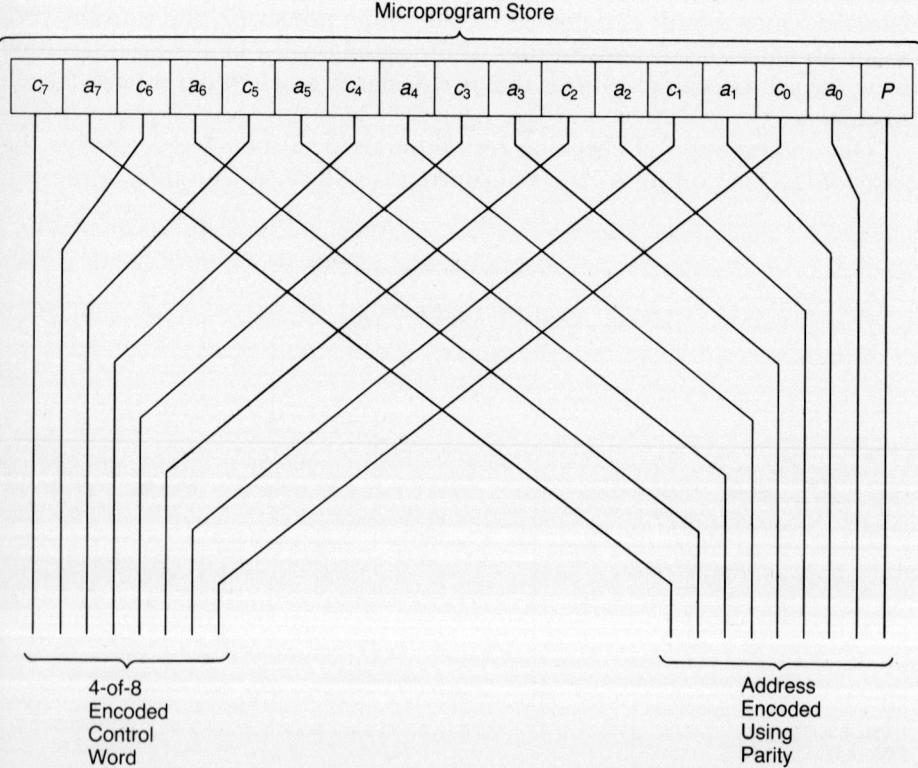

Fig. 5.44 The address is encoded using parity while the control is encoded using a 4-of-8 code. Bits from the address are interleaved with bits from the control word to allow detection of adjacent bit errors. (From [Toy 1978] © 1978 IEEE)

Error checking of the registers and the main memory contained within the 3A processor is accomplished using a *multiple-chips-per-bit* parity approach. Each 16-bit register and each 16-bit word of memory is constructed using eight, 2-bit *slices*. An extra 2-bit slice is provided to store two parity bits P_1 and P_2. P_1 is determined by taking parity over a group of bits that contains one bit from each 2-bit slice. The remaining bit in each 2-bit slice is incorporated into a group that determines P_2. The basic concept of this approach is illustrated in Fig. 5.45. A single parity checker is placed on the internal data bus of the processor and used to check all data transfers that occur via that bus. As we have seen previously in Chapter 3, the primary advantage of the multiple-chips-per-bit approach is that whole-chip failures affect only one bit per parity group and can, therefore, be detected.

Various versions of the ESS have been in operation since 1965. By 1977 over 1000 of the first-generation ESSs had been installed and were providing telephone service to over 15 million subscribers. In 1977, those systems had an average outage time of approximately five minutes per year.

The Synapse N+1 Architecture

Similar to the Tandem and Stratus systems, the Synapse N+1 computer system is targeted for the transactions processing industry, including banking, airline reservations systems, and information retrieval systems such as those found in libraries [Frank and Iselberg 1984]. The Synapse N+1 system derives its name from the fact that only one additional copy of each resource (processor, memory, bus, and so on) must be provided to achieve a certain level of performance if some component of the system fails. This is contrasted to other architectures that require duplication (or more) to accomplish error detection and successful reconfiguration.

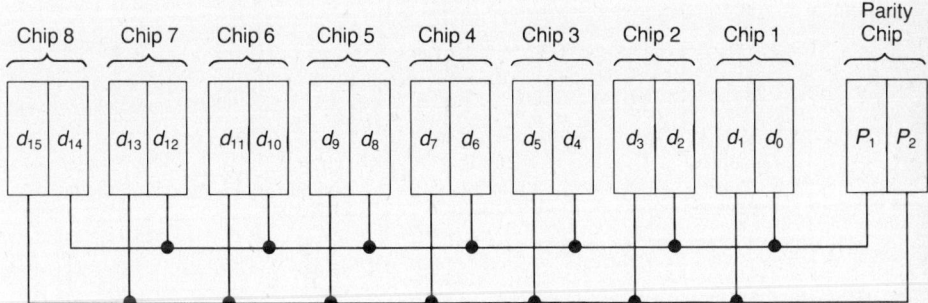

Fig. 5.45 In the ESS, memory is constructed using nine 2-bit chips. One bit from each chip is used to form the parity bit P_1. The second bit from each chip is used to form parity bit P_2. The complete failure of one chip affects no more than one bit in each of the two parity groups. (From [Toy 1978] © 1978 IEEE)

The Synapse N+1 system uses a shared memory architecture, similar to that found in the Agusta A129 system, as shown in Fig. 5.46. The system consists of general-purpose processors (GPPs), input/output processors (IOPs), shared memory, and a bus that allows processors to deliver and receive data via the shared memory. All of the operating system software and the applications programs reside in the shared memory and can be executed by any of the GPPs. Consequently, a failed GPP can be replaced with any other available GPP. The Synapse N+1 architecture supports up to a total of 28 processors. The total number of processors includes both GPPs and IOPs.

The shared memory provides a means of communicating among processors as well as providing storage for all operating system and applications software. The shared memory of the Synapse system can be expanded in 1-megabyte increments up to a total of 16 megabytes. A set of up to four memory control units can be used with each providing job queues to handle memory requests and responses. The GPPs each contain a cache memory that allows the use of the shared bus to be minimized.

The processors within the Synapse system are referred to as *self-dispatching*. In other words, the processors assign tasks to themselves by

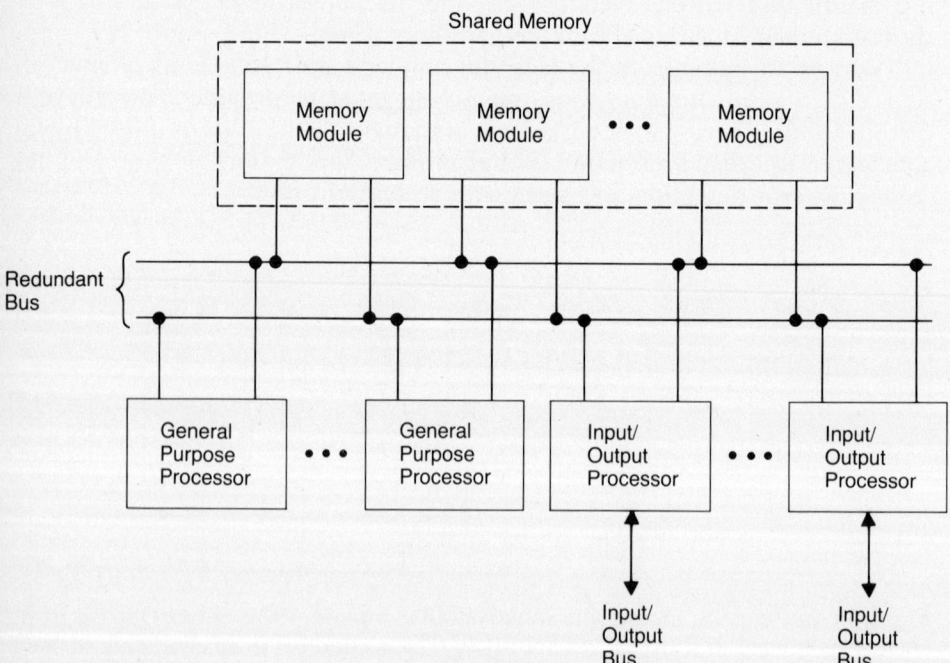

Fig. 5.46 Synapse N + 1 system architecture.

looking at a task queue in shared memory. When a processor is idle, that processor examines the task queue and selects the next task to perform. The software for the task resides in shared memory and is available to any processor in the system. The GPPs and IOPs can schedule tasks for each other by making an entry in the task queue. For example, suppose that one task (for example, Task A) requires information from another task (for example, Task B) before it can be completed. The processor performing Task A can enter Task B into the job queue so that the appropriate computations are performed.

The Synapse architecture relies on self-test features of the processors to achieve fault detection and fault tolerance. If a processor determines that it is faulty, that processor ceases assigning itself tasks from the task queue. The effect is that other processors will start performing the tasks no longer performed by the failed processor. Provided that there are enough processors available to ensure that all tasks are performed on time, the system will continue to function properly. The Synapse approach relies on the self-dispatching feature of the processors in the system and the self-testing capability of each processor to achieve the fault tolerance attributes of the system.

Summary

This chapter has described and demonstrated the design process that is typically undertaken in the development of a fault-tolerant system. In addition, thirteen practical systems have been described in enough detail to allow understanding the basic techniques used to achieve fault tolerance in each system. It is hoped that the information contained in this chapter will be sufficient to stimulate ideas that enable the development of new fault-tolerant designs. The following list summarizes the important concepts and terms developed in this chapter.

Conceptual Design Review—a detailed check of the basic concepts of candidate designs.

Conceptual Phase—the portion of the design process where candidate designs are conceived.

Design Phase—the portion of the design process where the actual hardware and software design is conducted.

Design Rules—guidelines by which a design is performed. The strict adherence to design rules often minimizes design mistakes.

Detailed Design Review—a detailed check of the hardware and software designs prior to implementation.

Final Review—a last check of the performance of a prototype design and the results of the system analysis.

Partitioning—the process of dividing a system into manageable pieces, or subsystems.

Problem Definition—the process of clearly and completely defining the design problem to be addressed.

Requirement—an attribute demanded of a design.

Requirements Design Review—a detailed check of a system's derived requirements.

Requirements Phase—the portion of the design process where requirements are determined.

Specification—a detailed plan for a design that enables the design to meet requirements.

Specifications Design Review—a detailed check of the specifications developed for the system hardware and software.

Specifications Phase—the portion of the design process where specifications are developed.

System Integration—the process of combining hardware and software into an operational system.

Test Phase—the portion of the design process where the validity of the design is verified.

Tradeoffs—the process of compromise in a design. In other words, one attribute is partially sacrificed so that another attribute can be improved.

References

1. Aircraft Internal Time Division Command/Response Multiple Data Bus, Military Standard MIL-STD-1553B, September 1978.
2. August Systems Data Sheet, *The CS-3001 Control Computer System Description*, August Systems, 18277 SW Boones Ferry Road, Tigard, Ore., 97224-7673, 1986.
3. Avizienis, A., G.C. Gilley, F.P. Mathur, D.A. Rennels, J.A. Rohr, and D.K. Rubin. "The STAR (Self-Testing And Repairing) computer: An investigation of the theory and practice of fault-tolerant computer design," *IEEE Transactions on Computers*, Vol. C-20, No. 11, November 1971, pp. 1312–1321.
4. Breuer, M.A., and A.D. Friedman. *Diagnosis and Reliable Design of Digital Systems*, Computer Science Press, Inc., Potomac, Md., 1976.
5. Burchby, D.B., L.W. Kern, and W.A. Sturm. "Specification of the Fault-Tolerant Spaceborne Computer (FTSC)," *Proceedings of the 1976 International Symposium on Fault-Tolerant Computing*, Pittsburgh, Pa., June 1976, pp. 129–133.

6. Frank, S., and A. Inselberg. "Synapse tightly coupled multiprocessors: A new approach to solve old problems," *AFIPS Conference Proceedings: 1984 National Computer Conference*, Las Vegas, Nev., July 1984, pp. 41–50.
7. Goldberg, J., W.H. Kautz, P.M. Melliar-Smith, M.W. Green, K.N. Levitt, R.L. Schwartz, and C.B. Weinstock. "Development and analysis of the software implemented fault-tolerance (SIFT) computer," NASA Contractor Report #172146, SRI International, Menlo Park, Calif., February 1984.
8. Hopkins, Jr., A.L., and T.B. Smith, III. "The architectural elements of a symmetric fault-tolerant multiprocessor," *IEEE Transactions on Computers*, Vol. C-24, No. 5, May 1975, pp. 498–505.
9. Hopkins, Jr., A.L., and T.B. Smith, III. "OSIRIS—A distributed fault-tolerant control system," *Proceedings of the 14th IEEE Computer Society International Conference (COMPCON '77)*, San Francisco, March 1977.
10. Hopkins, Jr., A.L., T.B. Smith, III, and J.H. Lala. "FTMP—A highly reliable fault-tolerant multiprocessor for aircraft," *Proceedings of the IEEE*, Vol. 66, No. 10, October 1978, pp. 1221–1239.
11. Johnson, B.W., and P.M. Julich. "Reliability analysis of the A129 integrated multiplex system," *Proceedings of the National Aerospace and Electronics Conference (NAECON)*, Dayton, Ohio, May 1984, pp. 1229–1236.
12. Johnson, B.W., and P.M. Julich. "Fault tolerant computer system for the A129 helicopter," *IEEE Transactions on Aerospace and Electronic Systems*, Vol. AES-21, No. 2, March 1985, pp. 220–229.
13. Johnson, B.W. "A course on the design of reliable digital systems," *IEEE Transactions on Education*, Vol. E-30, No. 1, February 1987, pp. 27–36.
14. Katzman, J.A. "System architecture for NonStop computing," *Proceedings of the 14th IEEE Computer Society International Conference (COMPCON '77)*, San Francisco, March 1977, pp. 77–80.
15. Meliar-Smith, P.M., and R.L. Schwartz. "Formal specification and mechanical verification of SIFT: A fault-tolerant flight control system," Technical Report CSL-133, SRI International, Menlo Park, Calif., January 1982.
16. Rennels, D.A., A. Avizienis, and M. Ercegovac. "A study of standard building blocks for the design of fault-tolerant distributed computer systems," *Proceedings of the 1978 International Symposium on Fault-Tolerant Computing*, Toulouse, France, June 1978, pp. 144–149.
17. Serlin, O. "Fault-tolerant systems in commercial applications," *Computer*, Vol. 17, No. 8, August 1984, pp. 19–30.
18. Sklaroff, J.R. "Redundancy management technique for space shuttle computers," *IBM Journal of Research and Development*, Vol. 20, No. 1, January 1976, pp. 20–28.
19. Smith, T.B., and J.H. Lala. "Development and evaluation of a fault-tolerant multiprocessor (FTMP) computer: FTMP principles of operation," NASA Contractor Report #166071, The Charles Stark Draper Laboratory, Inc., Cambridge, Mass., May 1983.

20. Smith, T.B., and J.H. Lala. "Development and evaluation of a fault-tolerant multiprocessor (FTMP) computer: FTMP executive summary," NASA Contractor Report #172286, The Charles Stark Draper Laboratory, Inc., Cambridge, Mass., February 1984.
21. Stiffler, J.J. "The SERF fault-tolerant computer. Part 1: Conceptual design," *Proceedings of the 1973 International Symposium on Fault-Tolerant Computing*, Palo Alto, Calif., June 1973, pp. 23–26.
22. Storey, T.F. "Design of a microprogram control for a processor in an electronic switching system," *Bell System Technical Journal*, Vol. 55, No. 2, February 1976, pp. 183–232.
23. Toy, W.N. "Fault-tolerant design of local ESS processor," *Proceedings of the IEEE*, Vol. 66, No. 10, October 1978, pp. 1126–1145.
24. Wensley, J.H., L. Lamport, J. Goldberg, M.W. Green, K.N. Levitt, P.M. Melliar-Smith, R.E. Shostak, and C.B. Weinstock. "SIFT: Design and analysis of a fault-tolerant computer for aircraft control," *Proceedings of the IEEE*, Vol. 66, No. 10, October 1978, pp. 1240–1255.
25. White, J.A., G.L. Hartmann, R.E. Pope, and J.W. Hunger. "A multi-microprocessor flight control system," Honeywell Systems and Research Center, Minneapolis, Minn., Final Report, Contract No. AFWAL-TR-81-3044, May 1981.

Additional Reading

The past twenty-five years have witnessed a tremendous amount of research and development in the fault-tolerant computing area. It has been impossible to encapsulate all of the important work in the contents of this one chapter or, for that matter, in the contents of this one book. Consequently, it is extremely important to examine the material placed at the end of each chapter in the additional reading sections, if more information is needed or desired.

Several journals frequently publish special issues on fault-tolerant computing. For examples, see the August 1984 issue of *Computer*, the December 1984 issue of *IEEE Micro*, the April 1986 issue of the *IEEE Transactions on Computers*, and the October 1978 issue of the *Proceedings of the IEEE*.

Ayache, J, J. Courtiat, and M. Diaz. "REBUS: A fault-tolerant distributed system for industrial real-time control," *IEEE Transactions on Computers*, Vol. C-31, No. 7, July 1982.

Boone, L.A., H.L. Liebergot, and R.M. Sedmak. "Availability, reliability, and maintainability aspects of the Sperry Univac 1100/60," *Proceedings of the 10th Annual International Symposium on Fault-Tolerant Computing*, Kyoto, Japan, October 1980.

Bosch, J.A., and W.J. Kuehl. "Reconfigurable redundancy management for aircraft flight control," *Journal of Aircraft*, Vol. 14, No. 10, October 1977.

Chen, Y., and T. Chen. "DFT: Distributed fault tolerance—analysis and design," *Proceedings of the 15th Annual International Symposium on Fault-Tolerant Computing*, Ann Arbor, Mich., June 1985.

Choi, Y.H., and M. Malek. "A fault-tolerant FFT processor," *Proceedings of the 15th Annual International Symposium on Fault-Tolerant Computing*, Ann Arbor, Mich., June 1985.

Cooper, A.E., and W.T. Chow. "Development of on-board space computer systems," *IBM Journal of Research and Development*, Vol. 20, No. 1, January 1976.

Deswarte, Y., J.L. Rosseboeuf, P. Cohen, N. Gargir, and J. Lerouge. "A fault-tolerant multimicroprocessor architecture for SARGOS," *Proceedings of the 11th Annual International Symposium on Fault-Tolerant Computing*, Portland, Me., June 1981.

Deswarte, Y., K. Alami, and O. Tedaldi. "Realization, validation, and operation of a fault-tolerant multiprocessor: ARMURE," *Proceedings of the 16th Annual International Symposium on Fault-Tolerant Computing*, Vienna, Austria, July 1986.

Fernandez, M. "A state-of-the-art fault tolerant computer," *Proceedings of the 3rd Digital Avionics Systems Conference*, Fort Worth, Tex., November 1979.

Forman, P., and K. Moses. "SIFT: Multiprocessor architecture for software implemented fault tolerance flight control and avionics computers," *Proceedings of the 3rd Digital Avionics Systems Conference*, Fort Worth, Tex., November 1979.

Gupta, R., A. Zorat, and I.V. Ramakrishnan. "A fault-tolerant multipipeline architecture," *Proceedings of the 16th Annual International Symposium on Fault-Tolerant Computing*, Vienna, Austria, July 1986.

Hopkins, Jr., A.L. "A fault-tolerant information processing concept for space vehicles," *IEEE Transactions on Computers*, Vol. C-20, No. 11, November 1971.

Hopkins, Jr., A.L., and T.B. Smith, III. "The architectural elements of a symmetric fault-tolerant multiprocessor," *IEEE Transactions on Computers*, Vol. C-24, No. 5, May 1975.

Hopkins, Jr., A.L., and T.B. Smith, III. "OSIRIS: A distributed fault-tolerant control system," *Proceedings of the 14th Computer Society International Conference*, San Francisco, February 1977.

Hsaio, M.Y., W.C. Carter, J.W. Thomas, and W.R. Stringfellow. "Reliability, availability, and serviceability of IBM computer systems: A quarter century of progress," *IBM Journal of Research and Development*, Vol. 25, No. 5, September 1981.

Ihara, H., K. Fukuoka, Y. Kubo, and S. Yokota. "Fault-tolerant computer system with three symmetric computers," *Proceedings of the IEEE*, Vol. 66, No. 10, October 1978.

Ihara, H., and K. Mori. "Autonomous decentralized computer control systems," *Computer*, Vol. 17, No. 8, August 1984.

Johnson, D. "The Intel 432: A VLSI architecture for fault-tolerant computer systems," *Computer*, Vol. 17, No. 8, August 1984.

Katsuki, D., E.S. Elsam, W.F. Mann, E.S. Roberts, J.G. Robinson, F.S. Skowronski, and E.W. Wolf. "Pluribus: An operational fault-tolerant multiprocessor," *Proceedings of the IEEE*, Vol. 66, No. 10, October 1978.

Kopetz, H., and W. Merker. "The architecture of MARS," *Proceedings of the 15th Annual International Symposium on Fault-Tolerant Computing*, Ann Arbor, Mich., June 1985.

Krol, T. "The '(4,2)-Concept' fault-tolerant computer," *Proceedings of the 12th Annual International Symposium on Fault-Tolerant Computing*, Santa Monica, Calif., June 1982.

Lala, J.H. "A Byzantine resilient fault-tolerant computer for nuclear power plant applications," *Proceedings of the 16th Annual International Symposium on Fault-Tolerant Computing*, Vienna, Austria, July 1986.

Lala, P.K. *Fault-Tolerant and Fault-Testable Hardware Design*, Prentice-Hall International, New York, 1985.

Manuel, T. "New architecture cuts redundancies in fail-safe processing," *Electronics*, August 25, 1982.

Pradhan, D.K., ed., *Fault-Tolerant Computing: Theory and Techniques*, Prentice-Hall Publishing Company, New York, 1986.

Scherson, I.D., and M. Tachnai. "A fault-tolerant microprocessor system and its application to fuel management instrumentation," *Proceedings of Melecon '81*, Tel-Aviv, Israel, May 1981.

Schmid, H., J. Lam, R. Naro, and K. Weir. "Critical issues in the design of a reconfigurable control computer," *Proceedings of the 14th Annual International Symposium on Fault-Tolerant Computing*, Kissimmee, Fla., June 1984.

Schoenman, R.L., and J.E. Templeman. "Airborne advanced reconfigurable computer system," *Proceedings of EASCON '76*, Washington, D.C., September 1976.

Siewiorek, D.P., and R.S. Swarz. *The Theory and Practice of Reliable System Design*, Digital Press, Bedford, Mass., 1982.

Siewiorek, D.P., V. Kini, H. Mashburn, S.R. McConnel, and M.M. Tsao. "A case study of C.mmp, Cm*, and C.vmp: Part I—Experiences with fault tolerance in multiprocessor systems," *Proceedings of the IEEE*, Vol. 66, No. 10, October 1978.

Siewiorek, D.P. "Architecture of fault-tolerant computers," *Computer*, Vol. 17, No. 8, August 1984.

Smith, III, T.B. "High performance fault-tolerant real-time computer architecture," *Proceedings of the 16th Annual International Symposium on Fault-Tolerant Computing*, Vienna, Austria, July 1986.

Smith, III, T.B. "Fault-tolerant processor concepts and operation," *Proceedings of the 14th Annual International Symposium on Fault-Tolerant Computing*, Kissimmee, Fla., June 1984.

Zorpette, G. "Computers that are 'never' down," *IEEE Spectrum*, Vol. 22, No. 4, April 1985.

Projects

5.1. The Reliable Aircraft Corporation (RAC) is currently designing the next generation military tactical fighter and requests help in developing candidate flight-control architectures. RAC is still in the predesign phase of the development, so design philosophies that substantiate candidate architectures are extremely important. Although specific hardware has not yet been defined, the long-range goals should be to reduce cost, weight, size, and power consumption while maximizing reliability and safety.

The architecture that must be developed is required to implement the flight control laws for the tactical fighter. The primary requirements placed on the architecture are the achievement of a reliability of 0.9_7 for a three-hour mission and the capability to tolerate any two faults. A block diagram of the nonredundant flight-control system is shown in Fig. 5.47. As part of this project you are to develop a fully redundant, fault-tolerant, and reliable version of the flight-control architecture. The written results should include design philosophies, block diagrams, a reliability analysis, complete descriptions of the fault detection capabilities, and proposed implementations (again at the block diagram level). A quantitative and qualitative evaluation of the approach should be presented. You are not expected to complete the design of your system but instead present a top-level description and evaluation of the approach. Do not become involved in the specifics of the circuits but remain at a block diagram level.

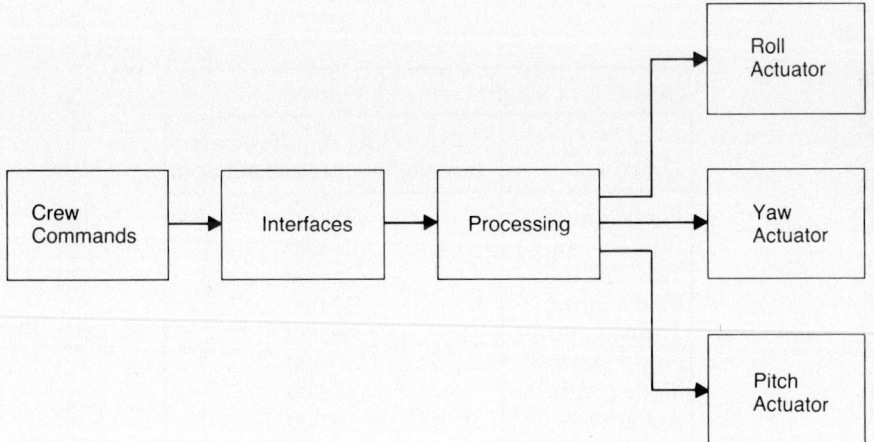

Fig. 5.47 Architecture of the nonredundant flight-control system.

The functions that the flight-control processors must implement are presented in Table 5.12 where the processing requirements of each task are also detailed. The design should provide approximately 50% spare processing resources to allow for future growth.

The tactical fighter is being developed for the Air Force, and they have requested that we consider two candidate approaches that they have been investigating. You should critique these two approaches as part of your development process. The architecture that you develop into a solution can be based on one of the two suggested by the Air Force or one of your own. You should provide the most detail on the approach that you ultimately select, but you need to justify your selection.

The first candidate approach is called the *triple-duplex* method and is illustrated in Fig. 5.48. Its fundamental characteristic is the use of triple modular redundancy (TMR) where each individual module is constructed using the *duplication with comparison* technique. When the comparison process detects a fault, the affected module is removed from the voting process. The voter is implemented using a flux-summing approach.

The second candidate approach is called the *quad redundancy* method and is illustrated in Fig. 5.49. Its basic feature is the use of four modules that exchange data and vote independently (in software) to form four *voted* results. The four results are flux-summed at the actuator drives. The quad redundancy method depends on the exchange of information among processors to determine which processors have failed.

The analysis of the above two approaches should include the following:
- Implementation approaches and problems
- Reliability comparisons
- Weight, size, and power estimates
- Ease of testing comparisons

TABLE 5.12 Flight-control functions

Functions	Processing requirements instructions per second (ips)
Pitch control	26,000
Roll control	26,000
Yaw control	26,000
Flap control	20,000
Thrust control	40,000
Power control	40,000
Mode control	35,000
Executive	10,000
Total	223,000

Projects 363

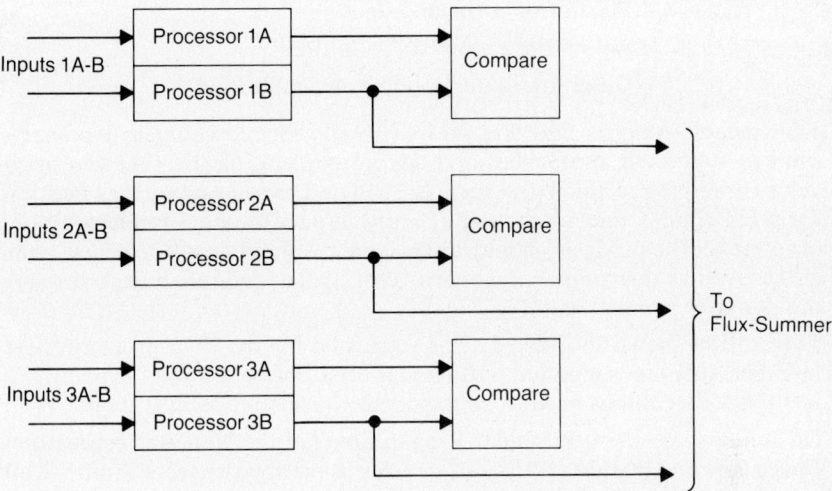

Fig. 5.48 Triple-duplex architecture for flight-control system.

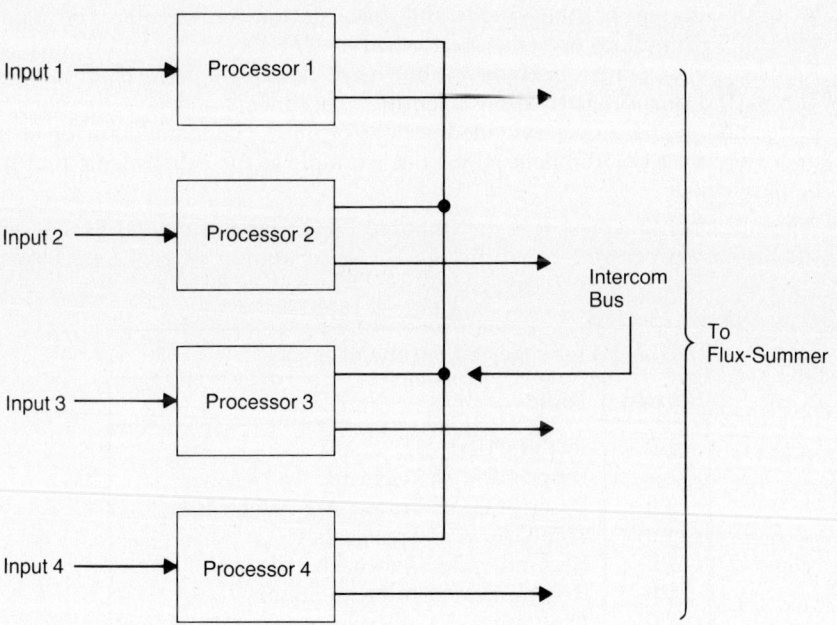

Fig. 5.49 Quad redundancy approach for the flight-control system.

- Flexibility of the architecture
- Amount of redundancy required
- Other attributes you feel are important

The tradeoff analysis that you present should form a comparative analysis of the two suggested approaches and any unique approaches that you might develop. Emerging technologies such as VLSI and techniques such as multiprocessor architectures that might significantly impact the final implementation are of interest. The analysis should discuss the availability of these new technologies as well as their impact (primarily cost, speed, and reliability) on the resulting design.

The outline shown in Table 5.13 is suggested for the final project report. Remember, this is a suggested outline and need not be followed explicitly. If you feel that other things need to be included, you should include them.

5.2. On January 25, 1984, President Reagan directed the National Aeronautics and Space Administration (NASA) to develop a manned space station within the next decade. The present concept for the space station is that it will contain a permanently manned station, a human-tended platform, and a number of free-flying, but visible (from the station) satellites. Several constellations of the satellites will be placed in different orbits close to the station itself and used in experiments that are controlled from the station, or the platform.

The space station has two primary purposes. The first purpose is to allow space exploration to be controlled and monitored, by humans, from the low-level orbit of the station. In other words, the space station could replace, or augment, some of the functions presently performed by NASA's earth stations (such as the Johnson Space Center in Houston) but more effectively performed from space. The second purpose is to allow scientific experiments and investigations to be performed in space over extended periods of time. The manufacturing of items in a weightless environment is just one example of the experiments that might be performed.

The subject of this project is the Onboard Electronic System (OES) that will be the primary processing capability for the space station. The OES performs two

TABLE 5.13 Suggested outline of report

Section	Topic
1.0	Introduction
2.0	Description of Suggested Approaches
3.0	Description of Unique Approach (if any)
4.0	Comparison of Approaches
5.0	Recommended Approach
6.0	Recommended Implementation
7.0	Evaluation of the Recommended Approach
8.0	Conclusions

basic functions: (1) controlling the transfer of vehicles, such as the space shuttle and satellites, from one orbit to another, and (2) docking the shuttle with the space station. To complete the two basic functions, the OES must perform guidance and navigation, communications control, database management, flight control during rendezvous and docking, and telemetry control. Because of the criticality of the functions of the OES, fault tolerance and high reliability are mandatory.

A block diagram of the overall OES is shown in Fig. 5.50. The system is a hierarchical network of processing elements with one large network (called the *system network*) that interconnects a number of smaller networks (called the *module networks*). Each of the module networks provides for the interconnection of several computers that perform functions such as the processing of images from cameras, controlling the earth-to-space and space-to-earth communications, docking control, and navigation and guidance of the station, the shuttle, and the associated satellites. The computers within a module network communicate with each other via the network. Computers on one of the module networks can communicate with those on another via the system network.

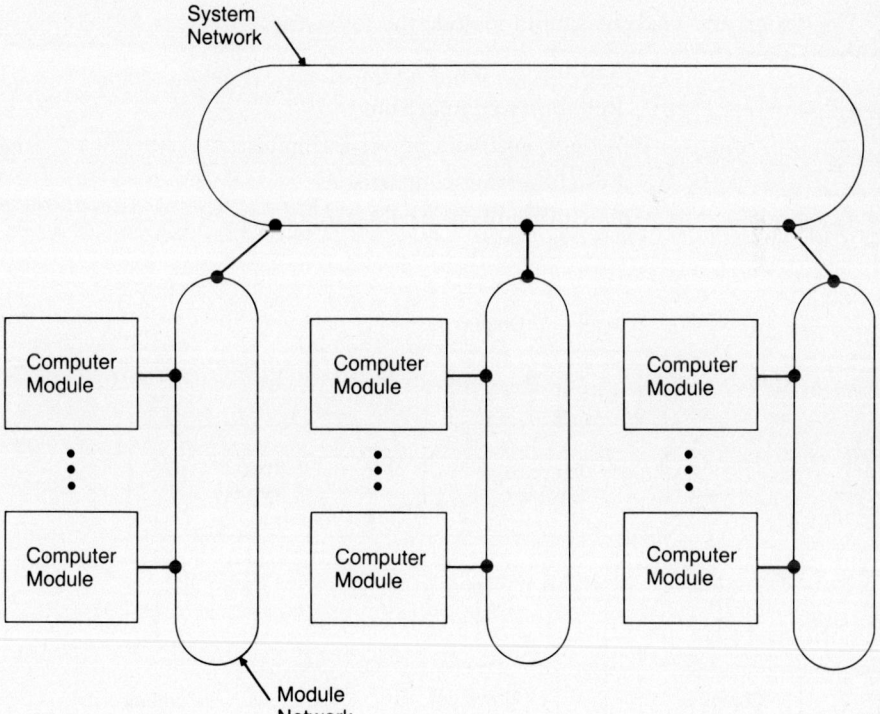

Fig. 5.50 Architecture of the Onboard Electronic System (OES) to be examined in Project 5.2.

Of concern in this project is the module network responsible for flight and vehicle control functions. The primary issues to be addressed are the design of a computer module and an interconnection network to provide for the transfer of information among the computer modules. Each computer module must be capable of detecting its own faults, and the interconnection network must be capable of tolerating the failure of any two computer modules. The network must consist of six computer modules and must achieve a reliability of at least 0.9_75 after five hours. The design should show block diagrams of at least two candidates for both the computer module and the interconnection network, as well as a complete analysis of both candidates. For your benefit, an example architecture of a computer module is shown in Fig. 5.51. The computer module is a multiprocessor system.

The functions that must be performed by each computer module and the processing requirements are presented in Table 5.14. Each computer module executes the functions in Table 5.14, and at least four of the six computer modules are required for the flight and vehicle control functions to be performed correctly. The design should provide approximately 50% spare processing resources to allow for future growth. Assume that you have available a processor with a throughput of approximately 1 million ips for use in the computer modules.

The design and analysis should include the following:

- Approaches and problems
- Reliability comparisons
- Weight, size, and power estimates
- Ease of testing comparisons
- Flexibility of the architecture

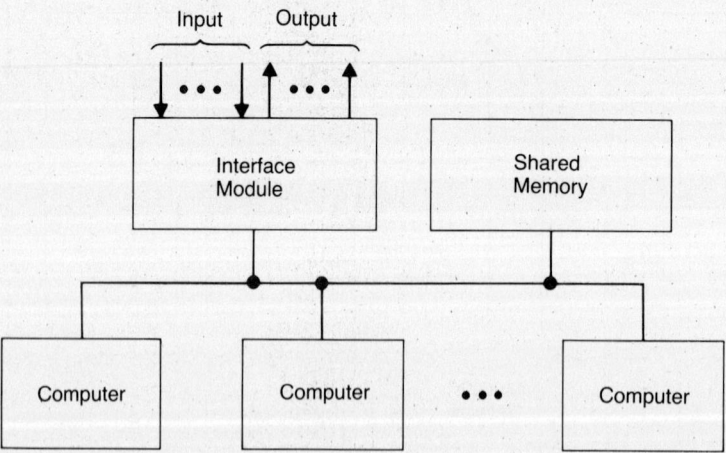

Fig. 5.51 Example architecture of a computer module for use in the OES.

TABLE 5.14 Flight and vehicle control functions

Functions	Processing requirements instructions per second (ips)
Guidance	300,000
Navigation	300,000
Database management	300,000
Docking control	200,000
Orbital transfer	400,000
Telemetry	400,000
Crew interface	200,000
Executive	100,000
Total	1,500,000

- Amount of redundancy required
- Other attributes you feel are important

The tradeoff analysis that you present should form a comparative analysis of at least two candidate approaches and any other novel approaches that you might develop. Emerging technologies such as VLSI and techniques such as multiprocessor architectures that might impact significantly the final implementation are of interest. The analysis should discuss the availability of these new technologies as well as their impact (primarily cost, speed, and reliability) on the resulting design.

The outline shown in Table 5.15 is suggested for your report. Remember, this is a suggested outline and need not be followed explicitly. If you feel that other things need to be included, you should include them.

5.3. The purpose of this project is to design a fault-tolerant, microprocessor-based controller for an electric wheelchair. The basic structure of the electric wheelchair system is shown in Fig. 5.52. The system includes command sensors for

TABLE 5.15 Suggested outline of report

Section	Topic
1.0	Introduction
2.0	Problem Description
3.0	Description of Suggested Approaches
4.0	Comparison of Approaches
5.0	Recommended Approach
6.0	Further Evaluation of the Recommended Approach
7.0	Conclusions

368 The Design of Practical Fault-Tolerant Systems

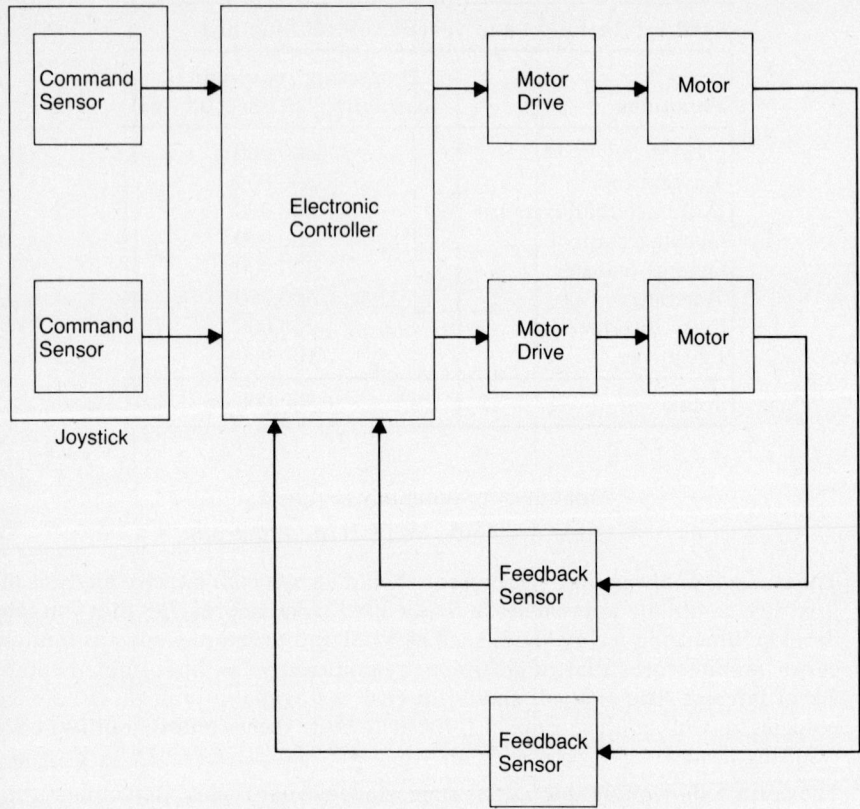

Fig. 5.52 Basic structure of the electric wheelchair control system.

the user of the chair to specify the desired direction and speed of travel. Examples of command sensors include joysticks, sip-and-puff devices (the user controls direction by blowing into or sucking on a tube while the velocity of the air flow controls the speed), and voice recognition units.

The most important feedback sensor is one used to measure the velocity of the wheelchair for feedback control purposes. Each motor of the wheelchair has a feedback controller that uses velocity information from the feedback sensor and a reference signal from the command sensor to generate the desired command to the motor. The feedback algorithm used is typically a simple proportional, integral, and derivative controller. An electrical current measurement device is also provided as a feedback sensor to allow monitoring the capacity of the batteries that power the chair.

The motor-drive circuitry consists of a set of power transistors for each of the two wheelchair motors. The power transistors are arranged in a *bridge* configu-

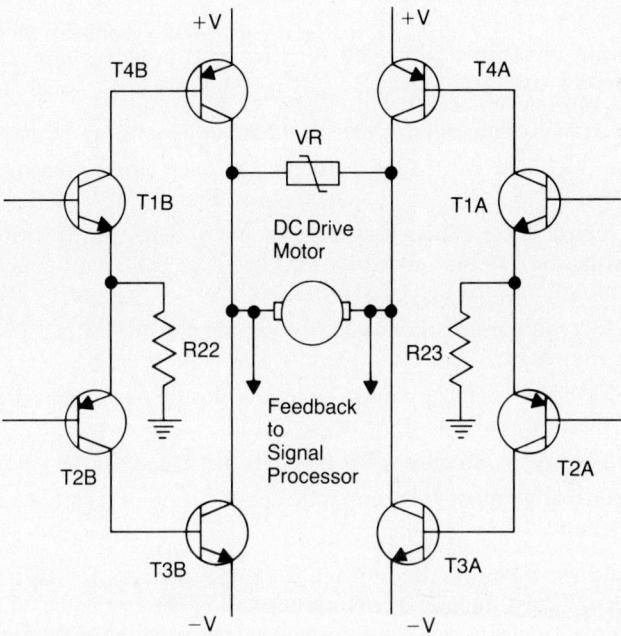

Fig. 5.53 Transistor implementation of the motor-drive circuitry.

ration as shown in Fig. 5.53. The circuit of Fig. 5.53 has four inputs that allow complete control of the speed and direction of rotation of the motor. For example, if transistor T2B is turned on, T3B is also on, and a current path from one terminal of the motor to the negative power supply is created. If, at the same time, T4A and T1A are turned on, a path for current flow will be created from the positive power supply to the negative power supply through the dc drive motor. The direction of current flow through the motor, and consequently the direction of rotation of the motor, can be changed by turning on T1B, T4B, T2A, and T3A instead of T2B, T3B, T1A, and T4A. The speed of the motor is determined by the degree to which the transistors are turned on. If each transistor is saturated, for example, the motor is running at its maximum speed.

The heart of the wheelchair system is the electronic controller that contains all of the analog-to-digital conversion, digital-to-analog conversion, and the processing capability. The microprocessor in the electronic controller samples the command signals from the user, samples the feedback signals, calculates a command for the motors, and applies the resulting command to the motor drive circuitry. Clearly, one of the critical design issues is to guarantee the correct operation of the electronic controller. If the controller fails in such a way that erroneous commands are sent to the motor, the user of the wheelchair can be physically harmed.

The objective of this project is to perform a complete design and analysis of a fault-tolerant electronic controller for the electric wheelchair. The specific requirements for the system are:

1. The electronic controller must have a steady-state availability of 0.99.
2. The safety of the controller at the end of one year of operation must be no less than 0.998.
3. The electronic controller must be capable of detecting 98% of its own internal faults and 95% of all faults that occur in other components of the system (motors, motor drives, sensors, etc.).
4. The electronic controller must be capable of tolerating 95% of the first faults that occur.
5. The electronic controller must be able to locate to a *replaceable module* 95% of all the first faults that occur.
6. The electronic controller must be operated from a battery power source.
7. The controller must be compact enough to fit on the back of an electric wheelchair.

You should go through the complete design process for this system. Your project report must include descriptions of all of the tradeoffs, analyses, and design decisions made, as well as complete schematic diagrams and software listings for the system. Imagine that you are a design engineer in charge of the design, and your project report must be the document that completely defines the electronic controller.

5.4. The purpose of this project is to design and analyze a fault-tolerant, microprocessor-based controller that might be useful in a number of control applications. To make the project more general, we will not consider a specific application, but instead will concentrate on the processing system that would form the heart of most any real-time control system. Develop your project results as if you were writing a proposal to a specific funding agency or company. In other words, you are trying to convince the funding agency that your approach is the best!

In this project you will start with a basic controller that currently *does not* incorporate fault tolerance, and you will create a design that *does* incorporate fault tolerance. At the same time, you will use the fault tolerance to meet very specific reliability, safety, and availability goals. All the requirements should be met while attempting to minimize power consumption, size, weight, and cost.

The architecture of the present nonredundant controller is shown in Fig. 5.54. The nonredundant system uses a single central processing unit (CPU) to provide the necessary computational power for the controller. In addition, a separate input/output (I/O) controller is available to supply commands to the device under control. The CPU has its own memory for program storage and temporary data storage. There is also a memory that is shared by the CPU and the I/O controller. Data from the CPU to the I/O controller is first written, by the CPU, into the

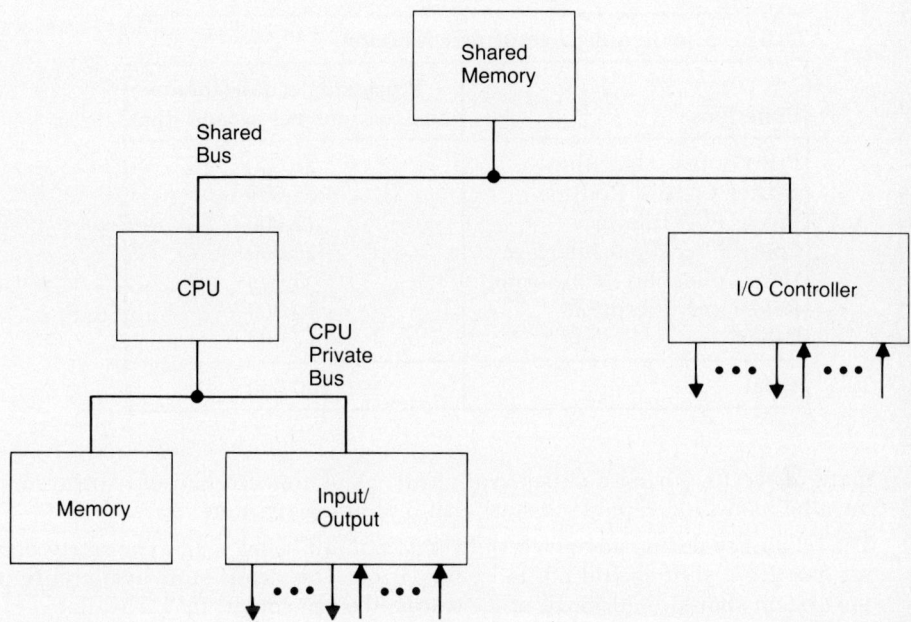

Fig. 5.54 Architecture of the nonredundant controller.

shared memory where the I/O controller can then access the information. Likewise, data from the I/O controller to the CPU is first written, by the I/O controller, into shared memory where the CPU can then obtain the data. As is easily seen, the failure of any element (CPU, I/O controller, shared memory, or the bus) will result in the failure of the complete system because there is no redundancy.

The system is typically used in cases where control algorithms such as the proportional-integral-derivative (PID) type are employed. Table 5.16 lists the programs typically used and the processing requirements for each. The *applications* processing requirements of the system will remain unchanged in the new, fault-tolerant design.

The three primary requirements of the system are reliability, availability, and safety. Before specifying the requirements, it is important to completely understand how the system is typically used. The system is usually run in *shifts*. A shift is ten hours long, and the time between shifts is no less than two hours. During a shift, the system is performing some control action that must be performed reliably and safely. For example, the system can control the creation of a chemical that requires ten hours to make. In between shifts, the system is not being used and could be repaired, if necessary. Reliability and safety are the

TABLE 5.16 Required controller functions

Functions	Processing requirements instructions per second (ips)
PID Control Algorithm	26,000
Sensor Control Routine	15,000
Process I/O Interface	12,000
Control Terminal Interface	20,000
Data Collection (Monitoring)	5,000
Real-Time Executive	10,000
Other	12,000
Total	100,000

main concerns during a shift! Availability is of concern because the system must be ready and capable to start a shift at any given time.

The reliability during any given shift must not fall below 0.9_45. The safety during any given shift should not fall below 0.9_65. The steady-state availability of the system should be 0.99. In other words, the system should be available for use at the beginning of a shift 99% of the time. In addition to reliability, safety, and availability requirements, the system (when fault free) should have a 50% reserve of processing capability to allow for the possible addition of processing tasks. In other words, the system requires 100,000 ips (from Table 5.16) to perform all the basic functions, so the total processing capability available should be at least 150,000 ips. Note that these numbers do not include any self-test or reconfiguration functions!

As part of this project you must develop a minimum of two candidate, fault-tolerant architectures and analyze those approaches to select the best overall technique. Your results should include:

1. Block diagrams of the candidate approaches
2. Reliability, safety, and availability models for use in analyzing the candidates
3. Results of analysis, including a comparison of the new fault-tolerant approach with the nonredundant controller
4. Selection of a preferred approach based on the analysis results and any other tradeoffs you consider important
5. More detailed block diagrams of the selected system's hardware and software
6. Recommendations for any further work that must be performed to actually realize the system

To allow you to make some realistic tradeoffs, you should use the modules presented in Table 5.17 in creating your candidate systems. Each processor is capable of executing 100,000 ips.

TABLE 5.17 Standard modules to use in fault-tolerant controller

Module	Power (watts)	Weight (pounds)	Cost (dollars)	Failure rate (per 10^6 hours)	Card slots
Processor with Memory	21.0	0.8	595.00	85.0	1.0
Shared Memory	35.0	2.5	295.00	42.0	1.0
I/O Controller	33.0	1.5	495.00	95.0	1.0
Power Supply	20.0	11.5	500.00	20.0	2.0
8-card Chassis	—	7.5	200.00	—	—
12-card Chassis	—	12.0	300.00	—	—
14-card Chassis	—	16.0	400.00	—	—

The project report should be a complete description of the work performed. You should be sure to cover each of the six different areas specifically defined in the work requirements. A suggested outline for the report is presented in Table 5.18.

TABLE 5.18 Suggested outline of project report

Section	Topic
1.0	Introduction
2.0	Description of Candidate Approaches
3.0	Analysis and Comparison of Approaches
4.0	Recommended Approach and Justification
5.0	Hardware Description
6.0	Software Description
7.0	Any Additional Considerations
8.0	Conclusions

6

Fault-Tolerant Design of VLSI Circuits and Systems

6.1 Introduction
6.2 VLSI Technology
6.3 Failure Modes in VLSI Technology
6.4 Distribution of Faults in VLSI Technology
6.5 Opportunities Presented by VLSI
6.6 Problems Presented by VLSI
6.7 Redundancy Techniques in a VLSI Design Environment
Summary
References
Additional Reading
Problems

6.1 Introduction

The advent of very large scale integration (VLSI) has had a tremendous impact on the design of fault-tolerant circuits and systems. The increasing density, decreasing power consumption, and decreasing costs of integrated circuits (ICs), due in part to VLSI, have made it possible and practical to implement the redundancy approaches used in fault-tolerant computing. Only a few years ago, a *single* computer occupied too much space and consumed too much power for use in a commercial aircraft, so designers could not even consider adding *redundant* computers in any but the most critical of applications. Today, microcomputers and other digital systems are compact enough to allow many of the redundancy techniques presented in Chapter 3 to be efficiently and effectively implemented. In fact, today's technology often allows multiple computers to be placed on a single IC so that fault detection or fault tolerance is attained within the IC itself.

Along with the advantages of VLSI have come a number of disadvantages. For example, failure modes are different in VLSI than in the earlier small scale integration (SSI), medium scale integration (MSI), and large scale integration (LSI). Whereas in SSI and MSI, the pins and the package of the IC dominated the failure rate, devices developed using LSI and VLSI are experiencing more and more faults internal to the IC itself. In addition, single-fault assumptions are inadequate in VLSI because the smaller feature sizes imply that even tiny imperfections can corrupt a significant portion of the IC chip area. Finally, the complexity of VLSI is resulting in a significant increase in design mistakes, thereby increasing the probability of common-mode failures where two or more identical devices fail in exactly the same way at exactly the same time. For example, if duplication with comparison is used as a means of fault detection, and the duplicated modules experience a common-mode failure, the comparison process will not be able to detect the problem because both modules will be affected identically.

The advantages and disadvantages mentioned in the previous paragraphs are just a few of those associated with designing fault-tolerant systems using VLSI. The advent of VLSI has had perhaps the single most important impact on fault-tolerant computing of any technological accomplishment. Many people feel, for example, that no new fault tolerance techniques will be developed in the future. Instead, we will devote our efforts to developing practical and efficient methods of implementing redundancy techniques in VLSI. Although the opportunities of combining fault tolerance and VLSI seem almost unlimited, there are also significant problems, as we have alluded to in the previous paragraph.

The purpose of this chapter is to study the many aspects of designing fault-tolerant systems in a VLSI environment. First, we provide an overview of VLSI technology and a brief history of the development of VLSI. Second, we consider the unique ways in which VLSI-based devices and systems fail. As previously mentioned, the failure modes of VLSI differ from those of SSI, MSI, and LSI. Also, the fault models that we used in MSI and LSI are somewhat limited when applied to VLSI technology. Third, we expound upon the opportunities and problems presented by VLSI technology. We consider in detail the importance of design mistakes, common-mode failures, and transient faults in VLSI. Finally, we examine the techniques available to implement redundancy using VLSI and the problems associated with these techniques.

6.2 VLSI Technology

Before beginning the discussions of the design of fault-tolerant systems using VLSI technology, it is important to clearly understand the basic concepts of VLSI. We do not intend to delve into semiconductor device theory or fabrication techniques; however, we do want to understand many of the commonly used VLSI terms. It is also important to understand some of the history behind the progression from early SSI to VLSI technology.

The progress in IC technology is evident in our daily lives. For example, the calculators that once cost three hundred dollars, or more, and performed only the four basic functions of addition, subtraction, multiplication, and division are now often given away with subscriptions to popular magazines. More powerful calculators capable of performing complex arithmetic and trigonometric operations and possessing programming capabilities now sell for tens of dollars rather than hundreds of dollars. In examining today's calculators we find that they are smaller, perform more functions, and are less expensive than the mainframe computers of ten to thirty years ago. As a dramatic example, consider the ENIAC computer (the first electronic computer) that was developed in 1946 and performed only the most basic of functions [Muroga 1982]. The ENIAC consumed 50,000 watts, occupied 4000 ft^3, cost \$480,000, and weighed 30 tons. The HP-16C is one of several examples of *handheld* calculators with essentially the same computational capabilities as the ENIAC computer. Similar developments are apparent in video games, electronic toys, home appliances, personal computers, cameras, watches, television sets, automobile electronics, and numerous other products.

Integrated circuit (IC) simply implies that more than one component is placed on a single piece of semiconductor material, and all of the components and associated wiring are an integral part of the IC and cannot be physically separated [Muroga 1982]. An IC is in contrast to *discrete devices*, where each component is packaged separately and all wiring is external to the component itself. For example, single transistors, diodes, resistors, or capacitors could be packaged as discrete devices and then interconnected to form a simple circuit. An IC, however, might contain a number of transistors, diodes, and other elements on a single piece of semiconductor material and placed in a single package. The actual semiconductor material in an IC is called a *chip* or *die* and can have an area of several hundred square mils. The chip is usually placed in ceramic or plastic packages that contain metal pins to provide electrical connections.

The number of components placed on an IC is representative of the level of integration [Muroga 1982] and [Hodges and Jackson 1988]. When IC technology was first introduced only one or two gates could be placed on a single chip. Today, however, it is possible to place more than 100,000 gates on a single chip. **Small-Scale Integration (SSI)** refers to an IC that contains fewer than 10 logic gates or, equivalently, fewer than 30 transistors. **Medium-Scale Integration (MSI)** refers to ICs containing between 10 and 100 logic gates (30 to 300 transistors). **Large-Scale Integration (LSI)** implies that an IC contains 100 to 10,000 logic gates (300 to 30,000 transistors). Finally, **Very-Large-Scale Integration (VLSI)** represents ICs containing more than 10,000 logic gates (more than 30,000 transistors). The precise number of logic gates used in the definitions of SSI, MSI, LSI, and VLSI are not universally accepted. However, it does appear that most people accept the 10,000 gate level as the beginning of VLSI because ICs with fewer than 10,000 gates can be manufactured with the same technology, whereas those containing more than 10,000 gates typically require new technology.

There are a number of advantages to using ICs rather than discrete devices, and many of the advantages are extremely important in the design of fault-tolerant circuits and systems. For example, it is usually less expensive to design with ICs as opposed to discrete devices. Cost savings occur not only in the actual construction of the system but also in the design process. For example, it is much easier to design using off-the-shelf ICs such as adders, multipliers, and so forth than to create each of these functions using discrete devices. Also, systems developed using ICs are usually faster than those that use discrete components because many of the signal propagation delays are on the chip where wire lengths are much shorter.

Perhaps the three most important advantages of using IC technology in the design of fault-tolerant systems are:

1. Reduced power consumption
2. Increased reliability
3. Decreased size and weight

First, when the majority of the connections in a system are made on a single chip, the parasitic capacitances are reduced, and the power required to drive signals thoughout the system is also reduced. Consequently, overall power consumption is minimized. Second, placing components on a single chip reduces the probability of loose connections, broken wires, and poor solder joints, therefore decreasing the overall failure rate. Finally, integration simply decreases size and, as a result, physical weight and volume. Many of the critical applications that require fault tolerance also mandate minimum weight and volume.

6.3 Failure Modes in VLSI Technology

The term **failure mode** is used to describe the underlying cause of failure in an IC. Using the terminology from Chapter 2, a failure mode is, therefore, equivalent to a fault. In the subsequent sections the term "failure mode" will be used frequently to imply a fault.

As might be expected, it would be extremely difficult to list all the possible causes of failure in an integrated circuit simply because the list would be extremely long. In addition, the list of faults would undoubtedly be different for bipolar devices than for metal-oxide semiconductor (MOS) ones. Finally, we would expect that changes in technology would produce subsequent changes in the distribution of faults among the various categories; for example, the common faults in SSI technology may be insignificant in LSI and VLSI technology. In this section, we want to discuss the common categories of faults that occur in integrated circuits and investigate the changes that we have seen in the distribution of faults as we have progressed from SSI to MSI and to even higher levels of integration. Note that the list of faults is in no way exhaustive; it is simply intended to illustrate the types of faults that can occur.

At the highest level, faults in ICs can be divided into two primary categories: (1) those resulting from the manufacturing process and (2) those resulting from wear-out or other phenomena in the field [Peattie et al. 1974]. During the manufacturing process, scratches, dust, or other foreign particles often corrupt a device. Likewise, the improper packaging of the device or the bonding of the leads to the actual semiconductor material can result in a failed device. While a device is in operation, electromagnetic fields can wear down an insulator and cause a short to occur. Similarly, moisture can become trapped in the IC and produce metal corrosion that leads to broken lines or degraded conductivity of a particular device.

We will consider seven categories when examining fault distributions: metal systems, diffusion, foreign material, oxide, package and bonding, chip mounting, and misapplication. We will discuss briefly the meaning of each fault category and examine the percentage of faults that falls within each of the categories.

6.3.1 Metal Systems

Metal systems, often called metalization, are the thin layers of metal (usually aluminum) that are placed on an integrated circuit to interconnect the various devices ([Muroga 1982] and [Peattie et al. 1974]). As circuit complexities and densities have increased, it has become necessary to provide several layers of metalization such that a larger number of interconnections can be made. Multiple layers of metal are separated by insulating material with contacts to lower layers made by etching away parts of the insulating material and allowing metal to flow from one layer to the next.

Intuitively, we could name a number of ways in which the metal can fail. For example, it is certainly possible that any particular strip of metal can break, resulting in an open connection on the IC. Likewise, two or more adjacent metal lines can become erroneously connected producing a short circuit on the IC. Finally, the metal can corrode, resulting in a degradation of the conductivity of the material.

The three most important failure modes in the metalization are: electromigration, open metal at an oxide step, and metal corrosion or degradation.

Electromigration is the movement of mass within a conductor when sufficiently large currents are passed through the conductor. In other words, currents in a conductor can eventually cause the conductor to break, creating an open circuit. In many cases, electromigration can be minimized, or even eliminated, through the careful selection of metals, such as gold or certain types of alloys, and the adherence to specific design rules [Peattie et al. 1974].

An open metal at an oxide step occurs when a layer of metal is placed over an uneven or improperly cut layer of oxide. In many cases, the uneven oxide can cause the metal to break or be poorly connected. The result can be an open circuit in the IC. Finally, metal corrosion is a typical cause of faults in ICs. Aluminum is a common material used to form the metal interconnections, and aluminum very readily oxidizes. As a result, interconnections can be weak, and the conductivity of the material can be dramatically affected.

6.3.2 Diffusion

Diffusion is the process by which certain types of impurities are spread throughout regions of a particular semiconductor device to form, in the case of an MOS device, the source and the drain ([Muroga 1982] and [Peattie et al. 1974]). Diffusion is usually performed by placing the wafers in a furnace at temperatures of approximately 1000° C and passing mixtures of gases through the furnace. Exposed areas of the silicon substrate will be implanted with impurities, thereby creating the desired device regions (source, drain, and so on) and junctions.

The diffusion process can introduce faults into an integrated circuit in numerous ways. For example, improper amounts of the impurities can be deposited in certain areas, thus affecting the characteristics of the device. For example, the diffusion process can affect the resistivity of a junction and cause a nonuniform distribution of current throughout a conductor [Peattie et al. 1974]. Likewise, the impurities can be implanted in the wrong areas of the substrate. Because of the nature of the diffusion process, one can easily envision that the decreasing geometries associated with LSI and VLSI can increase the number of faults that arise from the diffusion process.

Diffusion is rapidly being replaced with a process known as ion implantation, which achieves the same results as diffusion using a completely different method. In ion implantation the dopant ions are injected into the silicon by accelerating the ions in an electric field. The primary advantages of ion implantation are that the doping levels can be easily controlled, uniformity can be achieved, and low temperatures are used. The primary limitations of ion implantation are the high cost of the equipment needed to perform the process and the high voltages needed to accelerate the particles to be implanted.

Regardless of whether diffusion or ion implantation is used, the same types of problems can occur in the device and result in faults or potential faults. Ion implantation, however, can reduce the number of faults because of the improved ability to control the accuracy of the process.

6.3.3 Foreign Material

Perhaps one of the most difficult problems in the manufacture of semiconductor devices is controlling dust, moisture, or other foreign materials during the manufacturing process. Foreign particles can enter the process through a variety of means including the handling of the wafers or dust contained in the air of the processing facility. Semiconductor device manufacturers spend enormous sums of money to guarantee the cleanliness of the processing facility and the appropriate handling of the materials by the employees of the facility.

As we progress from SSI to VLSI, we might expect the percentage of faults resulting from foreign particles to increase because of the decreasing geometries of the device. Smaller geometries mean that foreign particles can be similar in size to the separation between lines or between transistors on the integrated circuit. Consequently, the particles have a higher probability of contacting a line, or a device, and affecting the operation.

6.3.4 Oxide

Oxide films are used throughout the manufacturing process for semiconductor devices. The oxide acts as a barrier to dopants during the diffusion process and as an insulating material upon which metal can be deposited and patterned into the appropriate interconnections ([Muroga 1982] and [Peattie et al. 1974]). During the early stages of the manufacturing process, the entire silicon wafer is covered with a film of silicon dioxide. Later stages in the process remove portions of the oxide layers so that impurities can be implanted in the appropriate places during the diffusion process. Thicker layers of oxide are then deposited onto the device to provide an insulating material between the silicon substrate and the subsequent layers of metal that will form the chip's interconnections.

Numerous faults can be introduced by imperfections in the oxide or the process by which the oxide is applied to, or removed from, the silicon substrate. For example, impurities in the oxide material can affect the insulating capabilities of the material and, eventually, the electrical characteristics of the device. Similarly, the thickness of the oxide can affect its ability to isolate metal from the substrate or layers of metal from each other. Also, variations in the oxide thickness can produce breaks in any metal lines placed on top of the oxide, thereby producing an open circuit. Finally, if the oxide is inappropriately removed in certain areas, metal connections can exist where they should not.

6.3.5 Package and Bonding

Bonding is the process of connecting wire leads from the semiconductor device to the pins of the package that houses the device. The package is typically one of four common types: metal can, ceramic, plastic, or glass [Peattie et al. 1974]. The most common packages used today to house integrated circuits are ceramic and plastic. Metal cans were used in the days of discrete devices to hold single transistors or diodes. Glass packages seem to be used primarily to house diodes.

There are clearly a number of ways in which mechanical problems can arise in the package or the bonding and introduce faults. For example, one of the primary functions of a good package is to protect the semiconductor device from moisture and other contaminants. As previously discussed, moisture on the semiconductor device can result in corrosion of the metal interconnects or the bonds, thereby creating an open line or changing the conductivity of the line. A secondary function of the package is to protect the device from mechanical shock and vibration that can lead to physical damage of the device.

Problems that can occur due to the bonding include wire breakage, corrosion, or the degradation of the device due to the wire's poor conductivity. Bonding is a very difficult process because of the small size of the wires and the small size of the points on the semiconductor device to which the wires must be connected.

6.3.6 Mounting

Prior to the packaging and bonding process, the semiconductor device is mounted on a metallic or ceramic substrate to protect the device from mechanical shock and to allow the packaging and bonding processes to be more easily performed [Peattie et al. 1974]. Problems can occur in the mount because of expansion of the mount material or undesirable thermal characteristics of the mount. Expansion of the mount material, for example, can result in wire breakage or other physical damage to the semiconductor device.

6.3.7 Misapplication

In performing fault analysis on ICs, scientists and engineers examine failed parts that have been returned from the field and attempt to determine why the particular part failed. In many cases, the IC has failed because of one of the failure modes discussed thus far in this chapter. However, it is not unusual for the failure to be the result of misuse of the IC. For example, the device may have been subjected to unreasonable temperatures, overstress conditions, or may have been improperly wired in a circuit. Misapplication, in many cases, can be a function of the cost of the IC. For example, misapplication can dominate the failure causes when the IC is an inexpensive SSI circuit. However, when the IC is a very expensive VLSI circuit, misapplications often occur less frequently.

Note that misapplication is a legitimate, and extremely important, cause of faults. Recall from Chapter 2 that faults can be caused by implementation mistakes, design mistakes, random component defects, and external disturbances. Misapplication is an example of either an implementation mistake or a design mistake.

6.4 Distribution of Faults in VLSI Technology

Now that we have identified some of the major causes of faults in ICs, we will examine the percentage of total faults that is the result of each individual cause. We will examine this *fault distribution* as a function of the level of integration (SSI, MSI, and so on). Unfortunately, very little data is available on advanced VLSI devices because such devices are just now beginning to be manufactured and used in large quantities. However, by examining the data for SSI, MSI, and LSI, we can see a number of trends that clearly indicate the dominant sources of faults in higher levels of integration.

Table 6.1 shows the fault distribution percentages for bipolar ICs as a function of the technology [Peattie et al. 1974]. Several interesting points are apparent in Table 6.1. First, the percentage of faults resulting from misapplication has been steadily decreasing as a function of the integration level. In SSI technology, misapplication accounted for 35% of the faults,

TABLE 6.1 Fault distribution percentages for bipolar ICS

Failure mode	SSI (%)	MSI (%)	LSI (%)
Metal systems	9.5	17.5	27.0
Diffusion	8.0	12.0	24.5
Foreign material	4.0	11.0	12.0
Oxide	17.5	20.0	13.5
Package and Bonding	13.5	7.0	4.0
Chip mounting	5.5	3.0	1.5
Misapplication	35.0	16.0	5.5
Miscellaneous	7.0	13.5	12.0

whereas in LSI technology misapplication accounts for slightly more than 5% of the faults. Several factors can explain the marked decrease in the percentage of misapplications. For example, many of the systems designed using LSI technology are simpler because they involve fewer components and less wiring. Consequently, design engineers and technicians are less likely to make mistakes because of the inherent simplicity of the circuits and systems. In addition, the cost of misapplication has been significantly increasing because of the cost of many of today's complex ICs. An SSI circuit that contains two or three gates can cost only a few cents, whereas a microprocessor with several thousands of gates can cost several dollars. More complex functions or custom-designed devices can cost hundreds or even thousands of dollars. As a result of cost, designers have an increased pressure to ensure that the ICs are used properly and are not damaged due to misapplication.

A second interesting feature apparent in Table 6.1 is the significant decrease in the percentage of faults due to packaging and bonding. Over 13% of all faults in SSI technology were the result of packaging or bonding problems, but the same problems cause only about 4% of the faults in LSI technology. The decrease in packaging and bonding problems indicates an improvement in the quality of the packaging and bonding techniques and perhaps an increase in the faults that are occurring on the IC itself.

Finally, note in Table 6.1 the significant increase in the percentage of faults associated with the actual semiconductor material of the IC. Suppose we consider metalization, diffusion, foreign material, and oxide problems as *internal* faults associated with the semiconductor material itself. Also, suppose we consider packaging, bonding, mounting, and misapplication as faults resulting from *external* factors. Table 6.1 shows that, for SSI technology, *internal* factors account for approximately 39% of all faults, whereas *external* factors account for approximately 54%. However, in LSI technology, the *internal* factors account for over 75% of the faults, whereas the *external* factors account for approximately 11% of the faults. Consequently, we have seen a significant change in the primary source of faults as we have gone from SSI to LSI technology.

Table 6.2 shows the fault distribution for LSI MOS devices [Peattie et al. 1974]. Although the failure modes are categorized a little differently in Table 6.2, it is still easy to see that the dominant cause of faults is due to the *internal* factors rather than the *external* factors. In fact, oxide problems account for approximately 33% of all faults in MOS LSI devices. Faults due to the package, bonding, and mounting account for only 10% of the total faults in MOS devices.

In summary, the data for ICs indicates that more and more faults are resulting from factors unrelated to the packaging, mounting, bonding, or wiring of the IC. Several important points result from knowledge of the change in the fault distribution. First, in the days of SSI and MSI, it did not make

TABLE 6.2 Fault distribution percentages for LSI MOS ICS

Failure mode	Percentage of faults LSI MOS technology
Oxide	33
Electrical overstress	15
Electrical	13
Metallization and particles	3
Package and wire	5
Bond and chip mount	5
Photolithographic	5
Other	21

sense to incorporate redundancy into the ICs because the most common way for the device to fail was in the bonding or the packaging. Consequently, internal redundancy would have a very limited impact (if any) on the overall reliability of the device or the system. In addition, the SSI technology did not provide the capability to include any redundant devices on the chip, primarily because of space limitations. In LSI and subsequently VLSI, the packaging, bonding, and wiring are becoming less of a factor, so it is possible to improve the reliability of an IC by incorporating redundancy into the IC itself. Also, the use of LSI and VLSI technology provides the capability to incorporate redundant circuitry in many applications.

6.5 Opportunities Presented by VLSI

In this section, we want to consider in more detail the advantages and opportunities offered by VLSI capabilities for the design of fault-tolerant circuits and systems. In general, VLSI provides one simple capability: the ability to put more circuitry in a smaller, more reliable, and, in many cases, less expensive package. For the designer of fault-tolerant circuits and systems, the ability to have more circuitry implies that many of the approaches that were previously not cost effective can now be used. For example, duplicated processors can now be placed on a single chip, whereas previously they required multiple boards or even multiple cabinets. Likewise, triple modular redundancy, quad redundancy, or even higher levels of redundancy can now be practical because of the size, power, and cost savings attributable to VLSI technology.

An extremely important benefit of VLSI is that it is now possible to provide fault detection and fault tolerance within the IC itself as opposed to only providing such capabilities at the board or the system level. A resulting advantage is that the faults are either detected or tolerated as close as possible to the site of their occurrence. Consequently, the propagation of erroneous information throughout a system is minimized. The containment of faults in such a manner is often crucial to the successful operation of fault-tolerant system. Faults whose effects are allowed to propagate throughout a system can be devastating to the system.

As is evident from many of the fault tolerance techniques and designs studied thus far in this text, the design of fault-tolerant systems has previously been very much an *ad hoc* process. In many applications, for example, numerous approaches can be employed to solve a particular problem and meet a specific set of requirements. As a result, there is often very little consistency from one design to the next. Also, there is very little opportunity to create design rules to govern the design process because there is no well-defined design method that can be employed.

An advantage created by VLSI is the possibility of formalizing the design process for fault-tolerant systems so that an acceptable set of design rules and practices can be created. One such formalization is the use of standard, VLSI-based building blocks. For example, suppose that we had a processor with built-in fault detection capability. Since fault detection is the cornerstone of many fault tolerance techniques, our design can be significantly simplified because our processor supports fault detection. In our *library* of building blocks, we could also have devices such as memories, input/output units, bus interface units, analog-to-digital and digital-to-analog converters, and other modules necessary to put together many different types of processing and control systems. If we could properly define the set of building blocks and create a methodology by which the building blocks could be used, the design process for fault-tolerant systems could be formalized and significantly improved.

A final potential advantage of VLSI is the possibility of using redundancy to improve the yield of VLSI circuits. In many cases, the yield of complex ICs is less than 10%. Because of the cost of manufacturing a device, low yields result in extremely expensive devices and systems that use those devices. In many cases, it might be possible to continue to use the IC if the defective modules within the circuit could somehow be replaced or repaired. Because VLSI allows additional circuitry to be included on an IC, some, or all, modules might be duplicated so that defective modules can be replaced with spares, and the IC made usable. In very regular structures such as memories or processing arrays, it is possible to use a small number of spare elements, each of which is capable of replacing a single faulty element.

The redundancy techniques presented in this chapter focus on the opportunities that VLSI technology provides. Specifically, we examine the use of redundancy to enhance yield, the use of significant amounts of redundancy in certain fault tolerance techniques, redundancy strategies in certain regular structures such as array processors, and the use of standard VLSI building blocks in the design of fault-tolerant systems.

6.6 Problems Presented by VLSI

The basic problem associated with VLSI is that the failure modes are now different. We have already seen, for example, that as the level of integration has increased from SSI to LSI, the common faults have moved from the pins and the package to the semiconductor material. In addition, the increased complexity of the design has increased the probability of design mistakes. Finally, the lower operating voltages of integrated circuits (many VLSI devices use less than 5.0 volts) has decreased the noise margin of the devices and increased the frequency of transient faults.

In this section, we consider common-mode failures, design mistakes, and the increasing occurrence of transient faults in VLSI circuits. Each of these topics presents unique problems not encountered when designing using technologies other than VLSI.

6.6.1 Common-Mode Failures

In general, a **common-mode failure** occurs when two or more identical modules are affected by faults in exactly the same way at exactly the same time [Tamir and Sequin 1984]. The problems with such failures are numerous. For example, if two modules in a triple modular redundancy system experience a common-mode failure, the two faulty modules could force the output of the majority voter to become erroneous, even though a majority of the modules (the two faulty ones) would agree. Similarly, a common-mode failure in a duplication with comparison scheme would go undetected because the duplicated modules would produce identical, erroneous results.

There are numerous causes of common-mode failures. For example, if two modules are operated from the same power supply, a failure of the supply or a fluctuation in its output voltage can equally affect each module. Similarly, the effects of electromagnetic radiation or lightning can affect several identical modules in exactly the same manner. Finally, design mistakes can result in replicated modules producing the same erroneous results at exactly the same time. For example, suppose that a complex module is designed and then duplicated to create a duplication with comparison system. A latent design mistake in the module could appear in both modules at the same time because both modules are performing exactly the same operations at, in many cases, exactly the same times.

A VLSI design environment increases the probability of common-mode failures for several reasons. First, VLSI devices are extremely complex; therefore, the possibility of design mistakes is increased significantly. Many modern ICs contain hundreds of thousands of logic gates and often require the efforts of tens of designers to complete. The expectation that the design is never going to contain latent design mistakes is often unrealistic.

Second, identical modules can be located very close to one another on a single chip or wafer. Consequently, stuck-type faults can easily affect both modules. For example, it is very reasonable that two lines, one from each of two identical modules, could become physically stuck at the same logical value; the result could be a common-mode failure.

Finally, the small feature sizes and resulting low operating voltages of VLSI devices are increasing the likelihood that external disturbances will impact the operation of the device. For example, an IC can be subjected to radiation or lightning that equally affects all identical modules contained on the IC.

It is important at this point to formally define the common-mode failure [Tamir and Sequin 1984]. Suppose that we have two modules implementing functions denoted as A and B. Under fault-free conditions, A and B are identical and produce outputs equal to those of the design-specified function that we will call Z. For example, under fault-free circumstances, any input combination I produces $A(I) = B(I) = Z(I)$. In the discussions here we will assume that we are comparing the outputs of A and B for the purpose of fault detection.

Suppose that we have a list F of all possible, single faults that can occur in either A or B. We will denote the functions performed when affected by some fault f from F as A^f and B^f. A number of conditions can occur. First, both functions can be affected by exactly the same fault f at exactly the same time, in which case we have $A^f = B^f$. If we are simply comparing the outputs of the two functions, the occurrence of the fault will go undetected. If there exists some input combination, I_s, such that $A^f(I_s) = B^f(I_s) \neq Z(I_s)$, the fault will have produced an error, and the error will go undetected.

There can also be cases where different faults affect the functions, but because of the structure of the functions, the outputs remain identical. For example, suppose that fault f affects A, whereas q affects B. There could exist certain input combinations I that would result in $A^f(I) = B^q(I) \neq Z(I)$. In other words, the outputs are identical, but erroneous.

Common-mode failures are said to occur when there exists at least one input combination for which the outputs of the two functions are erroneous, *and* the outputs are identical for all possible input combinations. Using the notation that we have presented, two faults f and q are said to produce common-mode failures if, and only if, there exists some input combination I_s for which $A^f(I_s) \neq Z(I_s)$, $B^q(I_s) \neq Z(I_s)$, and for all possible input combinations I, $A^f(I) = B^q(I)$. In other words, common-mode failures occur when two functions have experienced faults, the faulty functions respond to all inputs in exactly the same way, and there is at least one input combination that causes the two functions to respond erroneously.

From the definition of common-mode failures, several interesting points arise. First, it is possible for two faults to produce common-mode failures but be physically different faults. The only restriction is that the faults cause the two functions to respond in exactly the same way for all possible input combinations, and for one of the inputs to result in erroneous outputs from the functions. Second, common-mode failures will go undetected in traditional implementations of voting and comparison techniques. Because common-mode failures are realistic ways for circuits to fail, they must be accounted for in the design process.

6.6.2 Increased Design Mistakes

Recall from Chapter 2 that there are four primary causes of faults:

1. Implementation mistakes
2. Design mistakes
3. External disturbances
4. Random component defects

The fault tolerance approaches considered thus far have ignored the category of design mistakes and have assumed that such causes of faults are handled via fault avoidance techniques. When designs are relatively simple, fault avoidance is easy to accomplish and can be extremely effective in preventing design mistakes. However, with VLSI technology, designs are no longer simple. Single ICs can have hundreds of thousands of gates, and there can be hundreds of ICs within a given system. Clearly, the likelihood of design mistakes is increased when designing in a VLSI environment.

The problem of design mistakes in VLSI is similar, in many respects, to the problem of latent bugs in software. Because of the large number of paths through many software routines, it is not practical, from cost and time viewpoints, to exhaustively verify the correctness of each possible path. Yet, the failure to exhaustively verify the correctness of the routines creates the possibility that design mistakes go undetected until certain operating conditions occur in the field, at which point it may be too late to correctly handle the problem.

In the design of VLSI-based systems, the design tools (both hardware and software) we use are often comparable in complexity to the system being designed. Consequently, we not only have to worry about design mistakes occurring because of the limitations of the human designers, but we must also be concerned with design mistakes that are a result of problems in the design tools.

A possible solution to the problem of design mistakes is the use of *design diversity*, in which multiple designs of the same module, IC, or system are created by independent design teams [Avizienis 1982]. The concept is that design mistakes, if they occur, should be different in the various designs because of the independence between the designers. An obvious disadvantage of design diversity is the cost associated with providing multiple design teams and tools. Also, each team must perform the design from a common set of design requirements, so now there must be some approach to guarantee the validity of the requirements that are developed.

6.6.3 Increased Susceptibility to External Disturbances

The dimensions of VLSI devices have been steadily decreasing from approximately 10 microns in 1970 to less than 0.5 microns in 1986 [Muroga 1982]. In the circuits with a larger separation between adjacent lines, certain particles such as α particles had very little effect because the separation between lines was much larger than the size of the particle. In the smaller, present-day circuits, the particles are close in size to the separation between lines, so the effect of such particles on the operation of the circuit is increasing ([Aichelmann 1984] and [Bossen 1980]).

A second factor is the lower operating voltages of VLSI circuits. In 1970, a typical IC operated with power supply voltages of +12 volts, −12 volts, and ground, whereas today the typical power supply voltages are +5 volts and ground. In some VLSI devices the difference between the two operating voltages may be 3.5 volts or less. The lower operating voltages have decreased the noise margins of the circuits, so the circuit is more sensitive to variations in the supply voltages or external disturbances. Consequently, transient faults are an increasing problem in VLSI-based systems.

6.7 Redundancy Techniques in a VLSI Design Environment

We will consider the following four primary categories dealing with the use of redundancy in a VLSI environment:

1. Duplication with complementary logic
2. Self-checking circuits
3. Reconfigurable arrays
4. Yield enhancement

The use of complementary logic is a technique developed to overcome the problem of common-mode failures. Complementary logic has been applied in a number of different situations and has proven to be very effective ([Sedmak and Liebergot 1980] and [Tamir and Sequin 1984]). Self-checking circuits were developed initially to solve the "Who checks the checker?" problem. For example, in duplication with comparison, the comparator is a weak link whose failure can result in the system either erroneously indicating that a fault has occurred or ignoring the occurrence of a legitimate fault. Reconfigurable arrays are becoming more popular as the ability to put large numbers of processors on a single chip or wafer continues to increase. Redundancy is often added to processing arrays to achieve real-time fault tolerance or to improve the probability of initially obtaining an operational array. Finally, redundancy has been used in many cases to improve the yield of VLSI devices by placing spare elements on the IC; memories are excellent examples of where such techniques have been employed. In the following sections, we consider each of the four categories mentioned above.

6.7.1 Duplication with Complementary Logic

In Chapter 3 we studied in detail the concept of fault detection using duplication with comparison. Recall that the technique simply duplicates a given module and compares the outputs of the resulting two modules. As long as the comparator is working properly, the failure of any one of the two modules is detected. The problems with duplication with comparison are: (1) the comparator is subject to failure, and (2) the approach relies on the assumption that only one of the duplicated modules will fail at a given time. As we have already seen, the possibility of common-mode failure implies that we cannot safely assume that only one of the two modules will fail. Consequently, we need to modify the design of the duplication with comparison scheme to ensure that the effect of common-mode failures is minimized.

Several approaches can be used to help alleviate the problem of common-mode failures in duplication with comparison. One technique focuses on minimizing the possibility of identical design mistakes appearing in both modules by requiring that two separate design teams independently develop the two modules [Avizienis 1982]. The hope is that any design mistakes that occur will not be identical because the designers have been working independently. The problem with such an approach is that the expense is often intolerable. The company must provide twice as many designers and design resources. Also, the procedure does not address other causes of common-mode failures such as those occurring during fabrication, packaging, or as a result of external disturbances during normal operation.

A second approach relies completely on fault avoidance techniques. Designs are checked and double checked, and the production process is closely monitored to attempt to ensure that the potential causes of common-mode failures are eliminated. The difficulty here is that fault avoidance techniques are seldom 100% effective, and once again there is no consideration for problems that arise during the normal operation of the circuit or system.

One technique useful in overcoming many of the difficulties with common-mode failures is the use of complementary logic. In complementary logic, one module is designed using positive logic while the other module uses negative logic [Roth 1975]. As a reminder, positive logic simply implies that the higher of the two voltages used in a logic circuit represents a logic 1, whereas the lower of the two voltages represents a logic 0. Negative logic implies that the higher of the two voltages represents the logic 0, whereas the lower of the two voltages represents the logic 1. In most logic systems, +5 volts and 0 volts (ground) are the two voltages used. In this case, positive logic uses +5 volts for logic 1 and 0 volts for logic 0, and negative logic uses 0 volts for logic 1 and +5 volts for logic 0.

If we know the function realized by a circuit using positive logic, it is very easy to determine the function realized by the same circuit using negative logic because of the concept of duality. Every combinational Boolean function has a corresponding dual function. If we let X be a vector consisting of n input bits given by $X = (x_1, x_2, \ldots, x_n)$, the function $f(X)$ has a dual function $f_d(X)$, defined by $f_d(X') = f'(X)$ [Kohavi 1978]. In other words, if we apply X to the function f and then apply X' to the function f_d, and the functions f and f_d are duals, the resulting outputs will be complementary. A function, f, is said to be self-dual if for a given input, X, $f(X') = f'(X)$.

Recall from Boolean algebra that the dual of a Boolean function can be formed by replacing AND operations with OR operations, OR operations with AND operations, 1s with 0s, and 0s with 1s. The variables and complement operations are not changed. As an example, consider the function

$$f(x_1, x_2, x_3) = x_1 x_2' + x_3$$

The dual of the function f is given by

$$f_d(x_1, x_2, x_3) = (x_1 + x_2') x_3$$

We can also see from the definition of the dual function that we can find f_d for a given f by first complementing f and then replacing each variable with its complement. For example, if

$$f(x_1, x_2, x_3) = x_1 x_2' + x_3$$

then

$$f'(x_1, x_2, x_3) = (x_1' + x_2) x_3'$$

and finally

$$f_d(x_1, x_2, x_3) = f'(x_1', x_2', x_3') = (x_1 + x_2') x_3$$

Complementary logic can be used to implement a duplication with comparison approach to fault detection. Rather than use exact replicas of each module and directly compare the outputs of those modules, in **duplication with complementary logic,** the modules are designed as duals of each other so that one module operates using positive logic and the other using negative logic. The normal input is applied to one of the two modules, and the complement of the input is applied to the other. If both modules are functioning properly, the outputs will be complementary.

There are three primary advantages of using complementary logic in duplication with comparison approaches. First, the use of dual implementations forces the use of separate masks to create the two modules because the modules are different. Consequently, the possibility of common-mode failures resulting from design mistakes or mask problems is reduced. Second, the voltage transitions on corresponding lines in the two modules are in opposite directions so that the possibility of faults that are sensitive to voltage transitions producing identical effects is reduced. Finally, the corresponding lines in the two modules are always at different voltage levels, so a short between two such lines always results in one of the two lines having an erroneous value and the other line having the correct value. Consequently, the fault can be detected.

As an example, consider the design of a simple circuit that realizes the function

$$f(x_1, x_2, x_3) = x_1 x_2 x_3' + x_1' x_2' x_3$$

using duplication with comparison and the concept of complementary logic. Figure 6.1 shows the logic diagram of the circuit that implements the desired function. The dual of the function f is given by

$$f_d(x_1, x_2, x_3) = (x_1 + x_2 + x_3')(x_1' + x_2' + x_3)$$

Figure 6.2 shows the logic diagram of the circuit that implements the dual function f_d.

The original function and its dual are now operated in parallel using complementary input combinations. The outputs, in the fault-free case, will

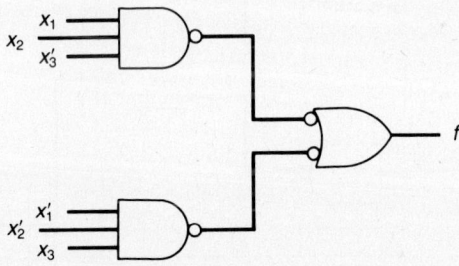

Fig. 6.1 Example implementation of $f(x_1, x_2, x_3) = x_1 x_2 x_3' + x_1' x_2' x_3$.

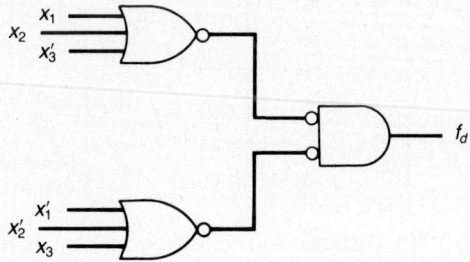

Fig. 6.2 Implementation of the dual of the function shown in Fig. 6.1.

also be complements and can be compared to achieve fault detection. The circuit diagram of the system is shown in Fig. 6.3. Logic values on corresponding lines in the two modules are complementary.

Although the use of complementary logic has many promising features, there are also many problems. First, the conversion process from positive logic to negative logic is not always easy, particularly if the functions are complex. Also, the impact on the design of the VLSI device can be extremely severe because the topology of the circuit and its layout will be changed because one module is the dual implementation of the other. For example, it is possible to design NOR gates with a large number of inputs while NAND gates are restricted to no more than three or four inputs [Tamir and Sequin 1984]. One of the modules is built around NAND gates while its dual is built around NOR gates, so the two modules can be quite different.

6.7.2 Self-Checking Logic

The concept of **self-checking logic** has increased in popularity because of the traditional "checking the checker" problem. In many designs that use

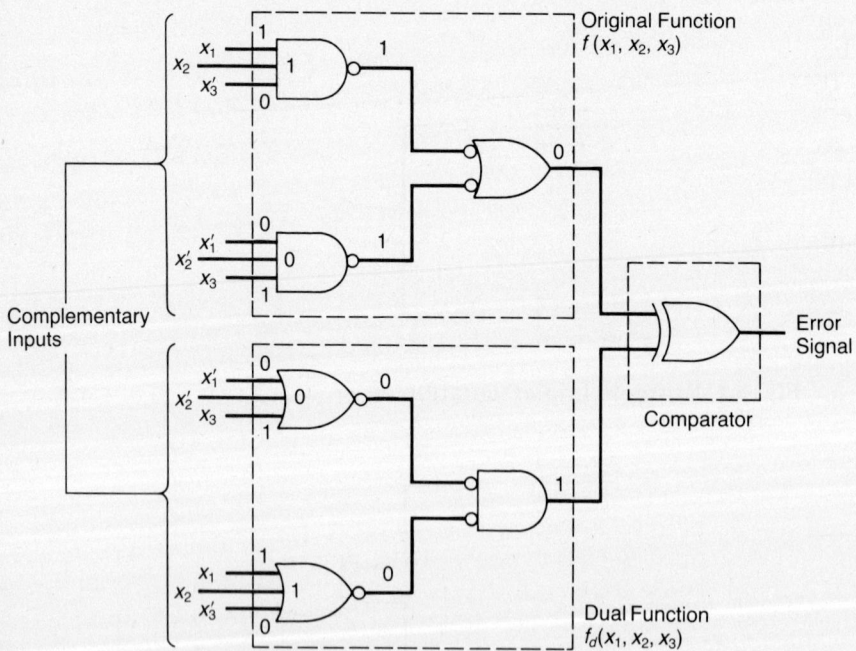

Fig. 6.3 Duplication with comparison using complementary logic. An example input pattern and the resulting logic values on each line are shown. Note the complementary values on corresponding lines.

coding schemes or duplication with comparison, it is necessary to compare the outputs of two modules or to verify that the output is a valid code word. The basic problem with such techniques, as we have seen, is the reliance of the approaches on the correct operation of comparators or code checkers. If the code checker fails, for example, the system can indicate that an error exists when in fact one does not, or the system can fail to detect a legitimate error that occurs. In many applications either condition is unacceptable. One possible solution is to design comparators and code checkers that are capable of detecting their own faults. Consequently, the concept of self-checking logic has been developed. Before beginning the discussions of self-checking logic, we must first introduce several important terms that are crucial to the understanding of self-checking technology.

In general, a circuit is said to be self-checking if it has the ability to automatically detect the existence of a fault without the need for any externally applied stimulus [Lala 1985]. In other words, a self-checking circuit determines if it contains a fault during the normal course of its operations. Self-checking logic is typically designed using coding techniques similar to those discussed under information redundancy in Chapter 3. The basic idea is to design a circuit that, when fault free and presented a valid input code word, will produce the correct output code word. If a fault exists, however, the circuit should produce an invalid output code word so that the existence of the fault can be detected.

To formalize the concept of self-checking logic, we will define fault secure, self-testing, and totally self-checking. In each definition, note that we are considering circuits designed to accept code words on their input lines and produce code words on their output lines.

A circuit is said to be **fault secure** if any single fault within the circuit results in that circuit either producing the correct code word or producing a noncode word, for any valid input code word [Lala 1985]. In other words, a circuit is fault secure if the fault either has no effect on the output or the output is affected such that it becomes an invalid code word. A circuit would not be fault secure, for example, if a fault resulted in the output becoming incorrect but still a valid code word.

A circuit is said to be **self-testing** if there exists at least one valid input code word that will produce an invalid output code word when a single fault is present in the circuit [Lala 1985]. In other words, a circuit is self-testing if each single fault is detectable since a fault that resulted in valid output code words for each possible input code word would be undetectable.

Finally, a circuit is said to be **totally self-checking** if it is both fault secure and self-testing [Lala 1985]. The fault secure property guarantees that the circuit will either produce the correct code word output or an invalid code word output when any single fault occurs. The self-testing property guarantees that there is at least one input code word that will produce an invalid code word output from the circuit when a fault is present. In summary, a circuit is totally self-checking if all single faults are detectable by at least one valid code word input, and when a given input combination does not detect the fault, the output is the correct code word output.

Figure 6.4 illustrates the definition of the properties of a totally self-checking circuit. The primary inputs of the circuit are encoded to produce the set of valid input code words that is a subset of the total set of inputs. For example, a 4-bit binary number has 16 possible values of which only 8 are legal, odd-parity code words. Similarly, the total set of output values is partitioned into valid output code words and is further partitioned into the correct output code words. During normal operation, a fault-free circuit accepts a valid input code word and produces the correct output code word. A fault secure circuit accepts a valid input code word, and, when a fault is present, produces either the correct output code word or a noncode word. A self-testing circuit produces correct code word outputs, valid but incorrect code word outputs, or completely invalid code word outputs. However, for any single fault that can be present, you are guaranteed that at least one valid input code word will result in the output being an invalid code word. Finally, a totally self-checking circuit always produces either the correct code word at the output or an invalid code word. Also, at least one valid input code word will result in an invalid output code word when any single fault is present.

The general structure of a totally self-checking (TSC) circuit is shown in Fig. 6.5. During normal operation, coded inputs are applied to the func-

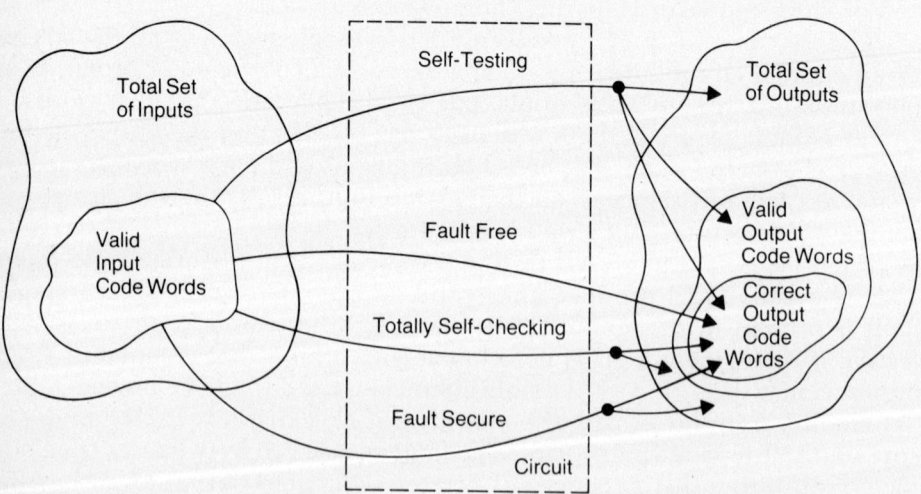

Fig. 6.4 Illustration of the concepts of self-testing, fault secure, and totally self-checking.

6.7 ■ Redundancy Techniques in a VLSI Design Environment

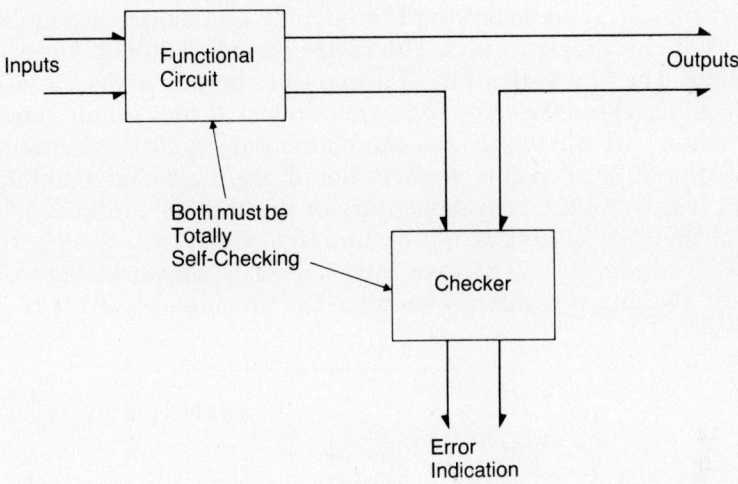

Fig. 6.5 Basic structure of a totally self-checking circuit.

tional circuit and coded outputs are produced at the circuit's output. A checker verifies that the outputs are indeed valid code words and provides an error indication if they are not. To provide a truly TSC design, both the functional circuit and the checker must possess the TSC property. Perhaps the key to the correct operation of the circuit is the TSC checker, so we will first consider its operation and design.

The function of the checker is to determine if the output of the functional circuit is a valid code word or not. In addition, the checker must indicate if any faults have occurred in the checker itself. To accomplish both tasks, the output of the checker is encoded to produce a coded error signal. Rather than have a single-bit output that provides a "faulty" or "not faulty" indication, the output consists of two bits that are: (1) complementary if the input to the checker is a valid code word *and* the checker is fault-free, or (2) noncomplementary if the input to the checker is not a valid code word *or* the checker contains a fault. One obvious reason for using two checker outputs is to overcome the problem of the checker output becoming *stuck* at either the logic 0 or the logic 1 value.

The checker must possess the **code disjoint** property [Lala 1985]. Code disjoint implies that when the checker is fault free, valid code words on the checker's input lines must be mapped into valid error codes (the checker's outputs are complementary) on the checker's output lines. Likewise, invalid code words on the checker's input lines must be mapped into invalid error code words (the checker's outputs are not complementary) on the checker's outputs.

398 Fault-Tolerant Design of VLSI Circuits and Systems

Unfortunately, there is no simple, systematic method to use in the design of totally self-checking checkers. The most common TSC checker is the **two-rail checker.** The block diagram of a two-rail checker is shown in Fig. 6.6. The two-rail checker is used to compare two words that should normally be complementary. If the words are complementary *and* the checker itself is fault free, the outputs of the checker should also be complementary. If the two input words are not complementary *or* the checker contains a fault, the outputs of the checker should not be complementary.

A simple design of a TSC two-rail checker is shown in Fig. 6.7 where each of the two input words is two bits. The first input word is (x_0, x_1), and

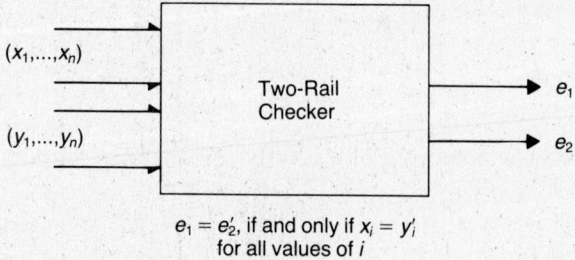

$e_1 = e_2'$, if and only if $x_i = y_i'$
for all values of i

Fig. 6.6 Basic block diagram of the two-rail checker.

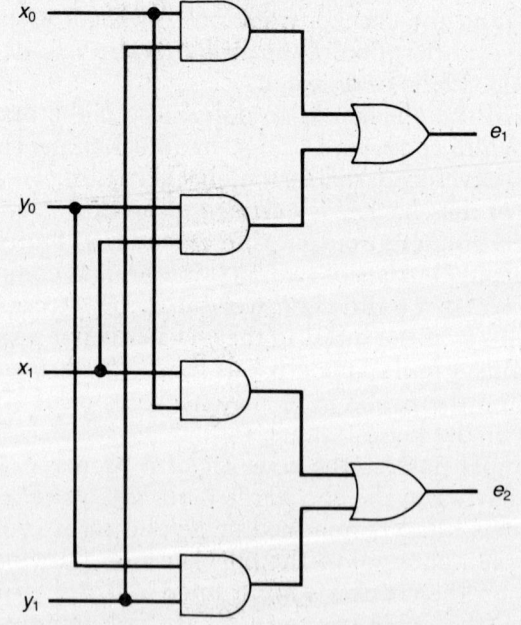

Fig. 6.7 A simple, 2-bit, totally self-checking, two-rail checker.

the second input word is (y_0, y_1). Valid code words on the inputs will have $x_0 = y'_0$ and $x_1 = y'_1$. It is easy to verify from the logic of the circuit that

$$e_1 = x_0 y_1 + y_0 x_1$$
$$e_2 = x_0 x_1 + y_0 y_1$$

Provided the checker is fault free and $x_0 = y'_0$ and $x_1 = y'_1$, the outputs of the TSC two-rail checker reduce to

$$e_1 = x_0 x'_1 + x'_0 x_1 = x_0 \oplus x_1$$
$$e_2 = x_0 x_1 + x'_0 x'_1 = (x_0 \oplus x_1)'$$

and e_1 and e_2 are always complementary.

Now consider the cases where the checker is fault free, but the inputs are not complementary. In the first case where $x_0 = y_0$ and $x_1 = y'_1$, the outputs of the checker become

$$e_1 = x_0 x'_1 + x_0 x_1$$
$$e_2 = x_0 x_1 + x_0 x'_1$$

which are identical for all possible values of x_0 and x_1. In the second case, where $x_0 = y'_0$ and $x_1 = y_1$, the outputs of the checker become

$$e_1 = x_0 x_1 + x'_0 x_1$$
$$e_2 = x_0 x_1 + x'_0 x_1$$

which are also identical for all possible values of x_0 and x_1. In the final case, where $x_0 = y_0$ and $x_1 = y_1$, the outputs of the checker become

$$e_1 = x_0 x_1 + x_0 x_1 = x_0 x_1$$
$$e_2 = x_0 x_1 + x_0 x_1 = x_0 x_1$$

which are identical. In summary, the fault-free checker will produce complementary error signals e_1 and e_2 if the inputs are complementary; otherwise, the error signals will be noncomplementary.

Now that we have verified that the checker of Fig. 6.7 is indeed a code checker, we must verify that it provides a TSC operation. We can accomplish this goal by verifying that the circuit has both the fault secure and the self-testing properties.

Referring to Fig. 6.7, we can see that the error signals e_1 and e_2 are generated with two physically separate logic circuits. As a result, we know that any single fault on any line other than a primary input can affect no more than one of the two outputs. Because the error signals are defined as complementary, we know that any fault that affects only one of the two signals will cause the signals to be noncomplementary. Consequently, a fault either does not impact the error signals or it forces the error signals to no longer remain complementary, in which case the error will be detected. Faults on the primary inputs to the checker simply appear to the checker as invalid code words and result in noncomplementary outputs from the checker. As a result, the two-rail checker can be seen to possess the fault secure property required for it to be TSC.

To verify the self-testing property of the two-rail checker, it is sufficient to note that the checker is a nonredundant circuit. A combinational circuit is said to be **nonredundant** if for every line k within the circuit, the value on line k directly affects the output for at least one input combination [Kohavi 1978]. In other words, an erroneous value on line k will corrupt at least one output value, and the fault will be detectable, if the circuit is nonredundant. From inspection of the circuit in Fig. 6.7, we can see that it is nonredundant. As a result of the two-rail checker being both fault secure and self-testing, we know that it is also TSC.

We could easily question the usefulness of a TSC two-rail checker with only two bits in each of the two input words, like that shown in Fig. 6.7. However, it is possible to create TSC two-rail checkers with a larger number of input bits using the circuit of Fig. 6.7 as a basic building block. Consequently, the circuit in Fig. 6.7 is useful for more than simply demonstrating the basic concepts of TSC checkers.

Suppose that we wish to design a TSC two-rail checker to compare two words with eight bits each, and the two words should normally be complementary. Figure 6.8 shows the structure of such a checker when designed using the circuit of Fig. 6.7 as the basic building block. The notation e_i^j is used in Fig. 6.8 to represent the ith error signal from the jth checker. e_1 and e_2 represent the primary error signals from the complete checking process. Note that the four checkers in the first level of the hierarchy each compare two bits from the 8-bit operands and each produce two error signals. Checkers in the second and third levels of the hierarchy verify that the error signals from the checkers in the first level are indeed complementary; if they are not, an error condition is propagated to the primary error signals. If no fault exists, the primary error signals will be complementary as long as the 8-bit input operands are complementary. If the two 8-bit inputs are not complementary, the outputs of one of the checkers in the first level of the hierarchy will also not be complementary, and this fact will propagate to the primary outputs of the checker. If a fault exists in any one of the 2-bit checkers, its outputs will be noncomplementary for at least one input combination, and the faulty condition will be detected.

In general, a TSC checker for n-bit operands can be designed using the technique illustrated in Fig. 6.8. Under ideal circumstances where $n = 2^b$, the checker will have b levels and will require $2^b - 1$ checker modules to construct [Lala 1985].

A natural feature of the two-rail checker to question is the requirement that the two input operands be complements in the error-free and fault-free

6.7 ■ Redundancy Techniques in a VLSI Design Environment

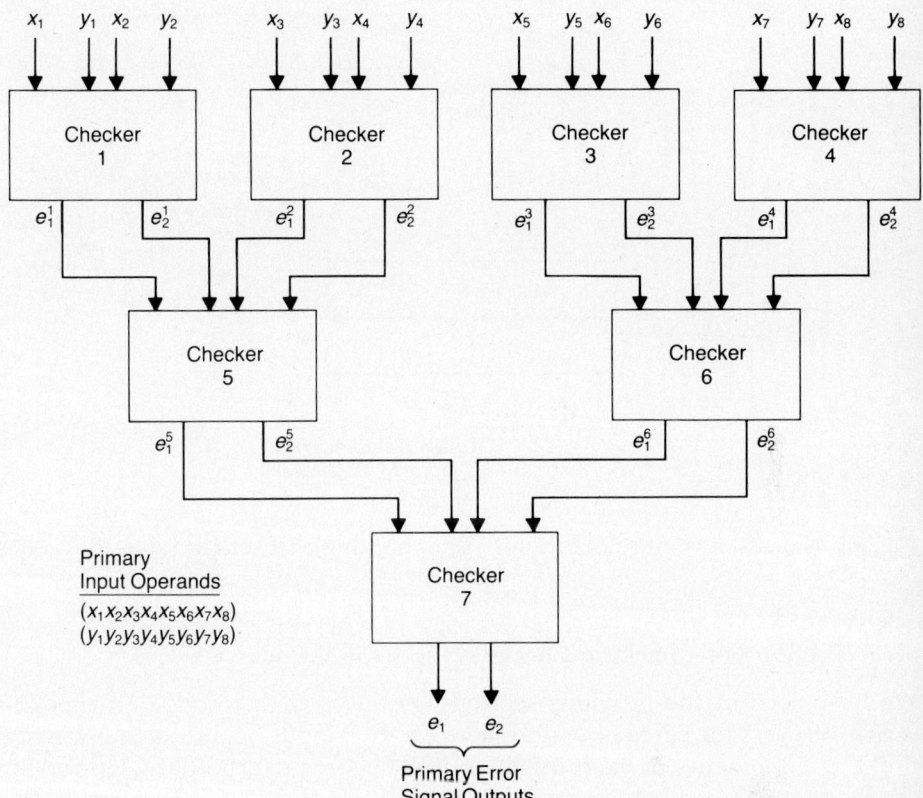

Fig. 6.8 Construction of an 8-bit, totally self-checking checker using 2-bit totally self-checking checkers as building blocks.

case. If you consider parity codes, arithmetic codes, or simply duplication with comparison, it is not the normal condition that two words to be compared are complements. However, it is easy to structure the problem of performing any comparison or code-checking process as that of comparing two words to determine if they are complementary. Figure 6.9 shows the general structure of a TSC checker for any type of separable coding scheme. The checking process consists of regenerating the code from the received information, complementing the regenerated code bits, and then using a two-rail checker to compare the result with the original code bits. The coding scheme used could be parity, arithmetic codes of some type, duplication, or any of a number of different techniques. The complement of the regenerated code bits can be created by simply inverting them. Consequently, TSC checkers can be designed, using the approach presented here, in a variety of applications and using a number of different coding schemes.

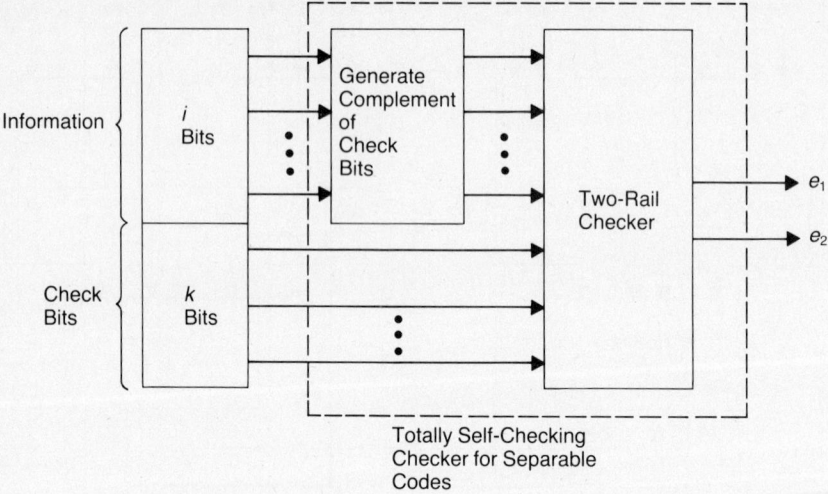

Fig. 6.9 General structure of a totally self-checking checker for separable codes.

6.7.3 Totally Self-Checking Checkers for M-of-N Codes

We have seen in the previous sections how to design totally self-checking (TSC) checkers for cases in which there are two input operands that are normally complements of each other. A TSC checker can be designed for any separable coding scheme using the techniques that we have already presented. In this section we want to consider TSC checkers specifically for M-of-N codes. Recall that an M-of-N code creates code words that have exactly M 1s and $(N - M)$ 0s out of a total of N bits. In general, the M-of-N codes are not separable; however, if N is selected as $2M$, the code can be designed as a separable code (M-of-$2M$ code). The function of the checker in the M-of-N code is to verify that the code word has exactly M 1s.

As we saw in the case of the two-rail checkers described earlier, the fault secure property necessary to be TSC is achieved by having two independent subcircuits that produce the two error signals at the output of the checker. Independence of this type can easily be achieved if the code is an M-of-$2M$ code because one subcircuit can operate on M of the total of $2M$ bits while the other subcircuit operates on the remaining M bits. Once again the self-testing property can be attained by guaranteeing that the circuit is not redundant. In the developments presented here we will only consider checkers for M-of-$2M$ codes.

Suppose that we have an M-of-$2M$ code word, $(x_1, x_2, \ldots, x_M, x_{M+1}, \ldots, x_{2M-1}, x_{2M})$, and we wish to design a TSC code checker [Lala 1985]. First, we partition the $2M$ bits of the code word into two disjoint subsets as

$$A = (x_1, x_2, \ldots, x_M)$$
$$B = (x_{M+1}, x_{M+2}, \ldots, x_{2M})$$

Now define k_A as the number of 1s contained in subset A and k_B as the number of 1s contained in subset B. The problem is to design a TSC checker that will produce two complementary outputs if $k_A + k_B$ is exactly M and will produce identical outputs if $k_A + k_B$ is greater than or less than M. Also, the checker must be fault secure and self-testing. We will first present the design approach and then use an example to illustrate the approach.

The two outputs for the TSC checker for an M-of-$2M$ code are [Lala 1985]

$$z_1 = \sum_{i=1}^{M} T(k_A \geq i) \cdot T(k_B \geq M - i)$$

$$z_2 = \sum_{j=0}^{M} T(k_A \geq j) \cdot T(k_B \geq M - j)$$

where i is always an odd number and j is always an even number. The summation represents the logical OR operation, and the "dot" represents the logical AND operation. The T functions are threshold functions that assume a value of 1 if and only if their arguments are satisfied. For example, the function $T(k_A \geq i)$ is 1 if and only if $k_A \geq i$; otherwise, the function assumes a value of 0.

Suppose we consider an example of the design of a TSC checker for a 2-of-4 code. Let the input code word be represented by the vector (x_1, x_2, x_3, x_4). In this example, the value of M is 2. Since the variable i is all odd numbers between 1 and M, i only assumes the value of 1. Likewise, j is all even numbers between 0 and M, so j assumes the values of 0 and 2. The outputs of the checker are defined by

$$z_1 = \sum_{i=1}^{2} T(k_A \geq i) \cdot T(k_B \geq 2 - i)$$

$$z_2 = \sum_{j=0}^{2} T(k_A \geq j) \cdot T(k_B \geq 2 - j)$$

where i is odd and j is even. In other terms,

$$z_1 = T(k_A \geq 1) \cdot T(k_B \geq 1)$$
$$z_2 = T(k_A \geq 0) \cdot T(k_B \geq 2) + T(k_A \geq 2) \cdot T(k_B \geq 0)$$

Logic functions can now be written for z_1 and z_2 as

$$z_1 = (x_1 + x_2) \cdot (x_3 + x_4)$$
$$z_2 = 1 \cdot (x_3 x_4) + (x_1 x_2) \cdot 1 = x_1 x_2 + x_3 x_4$$

A closer examination of the functions z_1 and z_2 allows us to better understand the operation of the checker under fault-free circumstances. Table 6.3 shows the various values that k_A and k_B can have and the resulting values of z_1 and z_2 for the 2-of-4 checker being designed in this example. As can be seen in Table 6.3, the values of z_1 and z_2 will always be complementary when valid code words are presented to the circuit. Likewise, the values of z_1 and z_2 will not be complementary when any possible invalid code word is presented to the circuit.

The resulting TSC checker for the 2-of-4 code is shown in Fig. 6.10. The circuit of Fig. 6.10 has the fault secure property because z_1 and z_2 are generated with separate logic circuits. Consequently, any fault within the circuit that is not on a primary input will affect one and only one of the two outputs. Faults on the primary inputs will appear as invalid codewords being applied to the circuit and will result in noncomplementary outputs. Finally, the circuit of Fig. 6.10 is self-testing because it is nonredundant. Because the circuit is both self-testing and fault secure it is TSC.

6.7.4 Reconfigurable Array Structures

Perhaps one of the most promising areas of research in the fault-tolerant computing field is the design and analysis of array structures for highly parallel and high-speed processing. The goal of a parallel processing system is to exploit the fact that the individual operations required in a given calculation do not necessarily have to be performed sequentially. As a simplistic example, suppose that a system must sample eight temperatures, convert each temperature to degrees Celsius, and display each temperature on one of eight separate display units. One design approach for this simple problem would be sequential in nature and would use a single processor that samples the temperatures in sequence, performs the conversion of each

TABLE 6.3 Values for the Z_1 and Z_2 functions as a function of k_A and k_B

	k_A	k_B	Z_1	Z_2
Valid code words	0	2	0	1
	1	1	1	0
	2	0	0	1
Invalid code words	0	1	0	0
	0	0	0	0
	1	0	0	0
	1	2	1	1
	2	1	1	1
	2	2	1	1

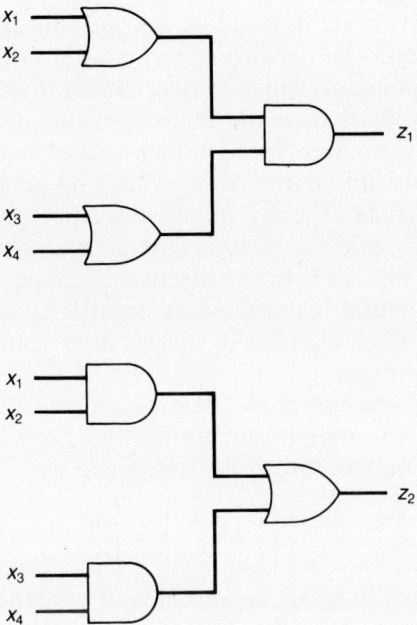

Fig. 6.10 Totally self-checking checker for a 2-of-4 code.

temperature in sequence, and provides the results to the appropriate display unit, again in sequence. If, however, you wanted to sample the temperatures as often as possible, you might consider a structure whereby eight processors are used, and each temperature is sampled, processed, and displayed in parallel. Theoretically, the parallel approach would be capable of sampling each temperature eight times more frequently than the sequential technique.

Perhaps a better example is that of matrix multiplication. Suppose that we wish to multiply two matrices, **A** and **B**, each having three rows and three columns, to obtain a result matrix **C** that also has three rows and three columns. We will let the notation a_{ij}, b_{ij}, and c_{ij} represent the element in the ith row and jth column of matrices **A**, **B**, and **C**, respectively. Elements c_{11} and c_{12}, for example, can be written as

$$c_{11} = a_{11}b_{11} + a_{12}b_{21} + a_{13}b_{31}$$
$$c_{12} = a_{11}b_{12} + a_{12}b_{22} + a_{13}b_{32}$$

which requires three multiplications and two additions to calculate each of the two elements.

Parallelism can be incorporated into the matrix multiplication example in several ways. First, the three products required in creating each result element could be calculated in parallel. Second, the calculation of each element of the result could be performed in parallel because the calculations are completely independent. Since matrix multiplication is a common calculation in many applications, a number of parallel architectures have been developed to support the high-speed calculation of matrix products.

A very useful parallel structure is called the array processor [Hwang 1984]. Several variations of array processors exist including simple linear arrays, systolic arrays, and the near-neighbor mesh connection. Each structure is illustrated in Fig. 6.11. In the discussions here, we focus on the near-neighbor mesh connection because of its popularity and usefulness. In the near-neighbor mesh, each element of the network communicates with each of its immediate neighbors.

As an example of the use of the near-neighbor interconnection scheme, consider once again the matrix multiplication problem. Suppose that we wish to perform the matrix multiplication given by

$$\begin{bmatrix} a_{11} & a_{12} \\ a_{21} & a_{22} \end{bmatrix} \begin{bmatrix} b_{11} & b_{12} \\ b_{21} & b_{22} \end{bmatrix} = \begin{bmatrix} c_{11} & c_{12} \\ c_{21} & c_{22} \end{bmatrix}$$

Figure 6.12 illustrates how an array using the near-neighbor connection technique can be used to calculate the matrix product. Each element of the array is capable of performing a multiply and accumulate operation during one clock cycle. The complete calculation of the matrix product requires a total of four clock cycles. The operations performed during each clock cycle are described below. Note that the operation could be performed much faster if all the input data was made available simultaneously to every element in the array. In such cases, each element of the result matrix could be calculated completely in parallel.

Clock Cycle 1

- Element $(1,1)$ calculates the product term $a_{11}b_{11}$.
- Element $(1,1)$ passes a_{11} and b_{11} to elements $(1,2)$ and $(2,1)$, respectively.

Clock Cycle 2

- Element $(1,1)$ calculates the product term $a_{12}b_{21}$ and adds it to $a_{11}b_{11}$.
- Element $(1,2)$ calculates the product term $a_{11}b_{12}$.
- Element $(2,1)$ calculates the product term $a_{21}b_{11}$.
- Element $(1,1)$ passes a_{12} and b_{21} to elements $(1,2)$ and $(2,1)$, respectively.
- Element $(1,2)$ passes b_{12} to element $(2,2)$.
- Element $(2,1)$ passes a_{21} to element $(2,2)$.

Clock Cycle 3

- Element $(1,2)$ calculates the product term $a_{12}b_{22}$ and adds it to $a_{11}b_{12}$.

6.7 ■ Redundancy Techniques in a VLSI Design Environment

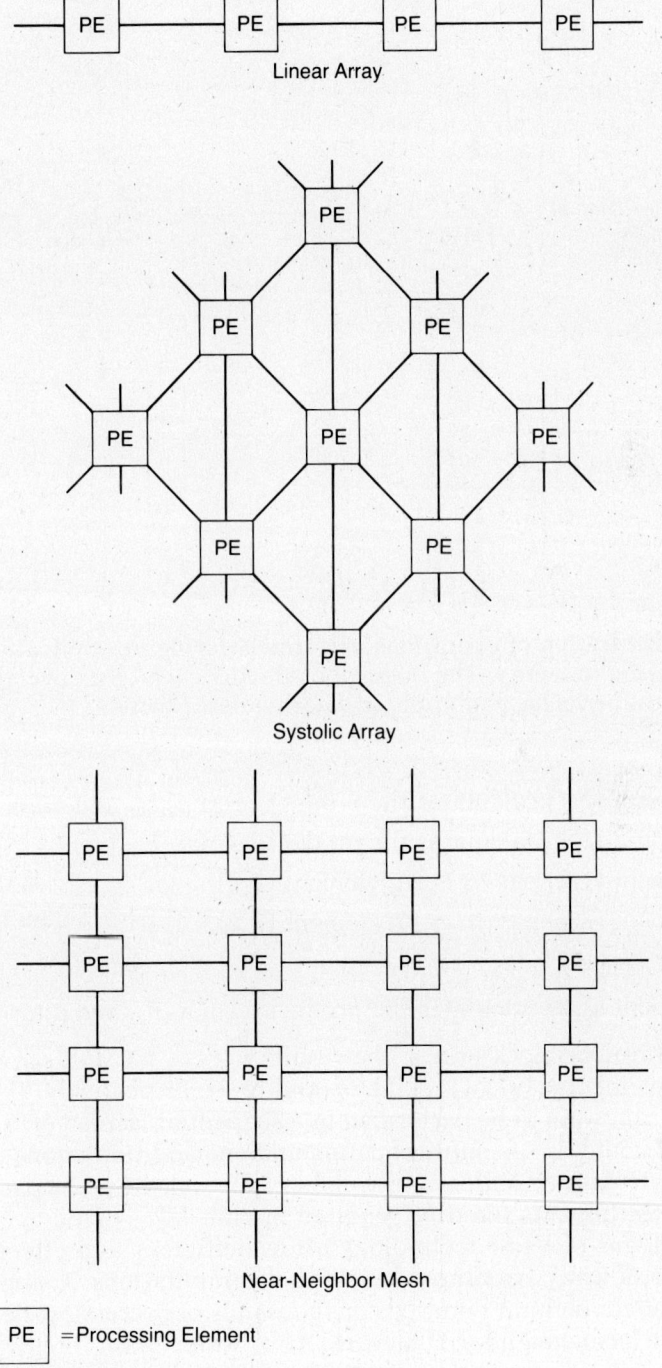

Fig. 6.11 Three variations of structures for array processors.

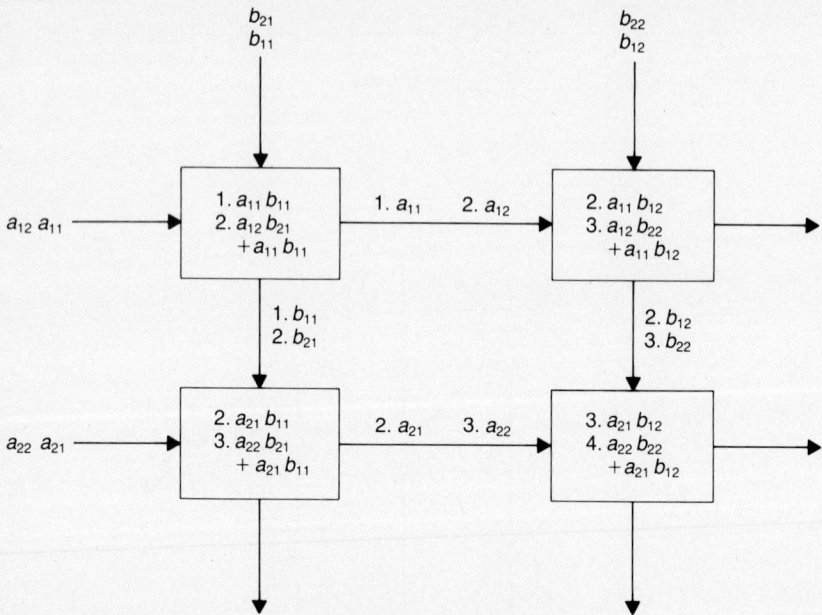

Fig. 6.12 Illustration of operations performed during multiplication of 2-by-2 matrices using an array. The operations during each clock cycle are shown. Each element performs a multiply and accumulate function.

- Element $(2,1)$ calculates the product term $a_{22}b_{21}$ and adds it to $a_{21}b_{11}$.
- Element $(2,2)$ calculates the product term $a_{21}b_{12}$.
- Element $(1,2)$ passes b_{22} to element $(2,2)$.
- Element $(2,1)$ passes a_{22} to element $(2,2)$.

Clock Cycle 4

- Element $(2,2)$ calculates the product term $a_{22}b_{22}$ and adds it to $a_{21}b_{12}$.

At the end of four clock cycles, the results c_{11}, c_{12}, c_{21}, and c_{22} will be contained in elements $(1,1)$, $(1,2)$, $(2,1)$, and $(2,2)$, respectively. If the matrix multiplication were to be performed in a sequential fashion on a single processor that could do one multiplication and one addition in one clock cycle, the total calculation would require eight clock cycles. So, the use of the parallel architecture cuts the time required in half.

The advent of VLSI technology allows efficient and effective implementations of array structures for parallel computations. A single chip, for example, might contain hundreds or thousands of processing elements connected in a near-neighbor structure. Also, wafer-scale integration (WSI) might allow hundreds or thousands of chips to be interconnected [Hwang 1984]. The potential for such designs seems almost unlimited.

Two fundamental problems, however, must be solved via fault tolerance techniques. First, a wafer or chip manufactured with thousands of processing elements will likely contain failed elements as soon as it comes off the production line. If the design depends on all the elements being operational to be useful, you may never obtain a useful device. So, the design must be performed such that the faulty elements within the array can be bypassed and the fault-free elements interconnected to achieve a functional array. Second, many applications will require that the array be capable of handling element failures that occur during the normal operation of the array. In some applications, the array will be allowed to shut down to perform a reconfiguration, whereas in other cases the array must continue its normal processing during the reconfiguration process.

Three specific types of reconfiguration can be identified: (1) **fabrication-time reconfiguration** that is performed immediately after manufacturing to produce an operational processing array, (2) **compile-time reconfiguration** that is performed before each use of the array, but not while the array is performing its normal operations, and (3) **real-time reconfiguration** that is performed while the array is in operation and continues to provide uninterrupted performance of its normal operations [Kung et al. 1987]. The most difficult reconfiguration to perform is real-time reconfiguration, and the easiest may very well be fabrication-time reconfiguration.

Fabrication-time reconfiguration consists of a unique action determined immediately after the array is manufactured; the reconfiguration action is usually irreversible. After the array is fabricated, external test procedures determine which of the processing elements are failed, and a reconfiguration algorithm attempts to find an interconnection pattern that will allow the remaining fault-free elements to be interconnected to create a fully functional array. Once an interconnection pattern is established, the required connections are made using procedures such as fusible links. The primary objective of fabrication-time reconfiguration is to increase the yield of the production process by *repairing* circuits that would normally not be usable.

Compile-time reconfiguration is used to provide for quick repair of the array prior to each individual use. If the array fails during operation, it is shut down based on a fault detection algorithm indicating that some fault has occurred. A fault location and reconfiguration algorithm then determines which particular processing element, or elements, has failed and attempts to reconfigure the array to remove the failed element. The key attribute of compile-time reconfiguration is that no time constraints are placed on the repair time. Clearly, you want to repair the array as quickly as possible, but you are not attempting to perform the repair without discontinuing the normal operations of the array. One primary objective of compile-time reconfiguration is to allow an operational array to be determined prior to each use so as to improve the availability of the processing array.

Real-time reconfiguration is used to provide for fault tolerance during the normal operation of the processing array. If an element fails, the effect of the failure is either masked, or the failure is immediately recognized, located, and a reconfiguration is implemented. In some situations both masking and reconfiguration are performed. The ideal objective is to prevent the array from generating any erroneous results. However, in some applications, such as real-time control systems, it may be acceptable to generate erroneous results for a very brief time period, provided that the problems can be quickly corrected. For example, in the control of certain dynamic systems, the response of the system under control is slow enough that brief periods of erroneous control commands are not noticeable. The primary objective of real-time reconfiguration is to improve the reliability and safety of the processing array by continuing normal operations even after the occurrence of certain failures.

In the discussions here, we will examine each of the three types of reconfiguration techniques and investigate the problems associated with each. We will look at the reconfiguration techniques and examine their ability to handle various types of failures.

Fabrication-Time and Compile-Time Reconfiguration

Fabrication-time and compile-time reconfiguration have many similar attributes. For example, both applications have the luxury of not requiring that reconfiguration be performed without interrupting the normal operations of the system. In addition, the length of time required to perform the reconfiguration, although extremely important to the system's availability, is not restricted. The key differences between fabrication-time and compile-time reconfiguration are the time at which the reconfiguration is performed and the permanence of the reconfiguration decisions. As the name implies, fabrication-time reconfiguration is performed immediately after the fabrication of the array during the production process. The reconfiguration is permanent, so the array cannot be reconfigured again at a later time. Compile-time reconfiguration can be performed in the field after the array has been operational for some period of time. The compile-time reconfiguration might use the same techniques as the fabrication-time approach, but the reconfiguration decisions are not permanent. Consequently, a system can be reconfigured numerous times using the compile-time reconfiguration philosophy.

Fabrication-time reconfiguration has become very popular with companies in an attempt to significantly improve the yield of VLSI devices, particularly those devices having regular structures such as arrays. A good example is the memory chip where the storage cells are organized in an array format. However, the fabrication-time reconfiguration algorithms that have been developed are applicable to devices where the individual elements are memories, processors, or any other function.

Arrays can be designed using a number of methods to support fabrication-time and compile-time reconfiguration. We will consider only three primary approaches:
1. The use of a single spare row or column and the *rippling* replacement strategy [Negrini, Sami, and Stefanelli 1986]
2. The use of both a spare row and a spare column and the *fault stealing* technique [Negrini, Sami, and Stefanelli 1986]
3. The use of multiple spare rows and columns and the *repair-most* algorithm [Kuo and Fuchs 1987].

The Rippling Replacement Strategy

The simplest redundancy approach in an array structure is to add a single column to the basic array. The elements in the additional column serve as spares to replace faulty elements in the original array. The algorithm used to configure the array is based on the concept of "rippling," which we discussed in Chapter 5. In the rippling replacement strategy, a faulty element is eliminated by simply replacing it with its neighbor [Negrini, Sami and Stefanelli 1986]. The functions of the element used to replace the faulty one must then be transferred to the next neighboring cell. This process continues until the element from the spare column is used to replace the last element in the row.

As an example of the rippling replacement strategy, consider the array shown in Fig. 6.13. The original 4×4 array shown in Fig. 6.13 has been augmented with a spare column of elements. Each element is labeled using the notation (i,j) to represent the element in the ith row and the jth column. If element $(2,3)$ is faulty, it is replaced by element $(2,4)$. The functions of element $(2,4)$ are then picked up by the spare element $(2,5)$. The result is a completely operational, 4×4 array with the faulty element $(2,3)$ fully bypassed.

In general, the rippling algorithm augments an $N \times N$ array with one spare column, so that the resulting array is $N \times (N + 1)$. A faulty element (i,j) is then replaced with element $(i,j + 1)$, which is in turn replaced with element $(i,j + 2)$. The process continues until element (i,N) is replaced with the spare element $(i,N + 1)$. This process continues until all the rows have been properly configured. The name of the rippling technique is derived from the fact that the reconfiguration process "ripples" from the faulty element to its neighbor and eventually to the spare element in the spare column.

The rippling algorithm can accommodate multiple faulty elements as long as no two faulty elements are contained in the same row. For example, Fig. 6.14 illustrates a reconfigured, 4×4 array that contains a faulty element in each row. If more than one faulty element exists within a given row, the rippling algorithm is unable to replace all the faulty elements in that particular row, and the array cannot be repaired.

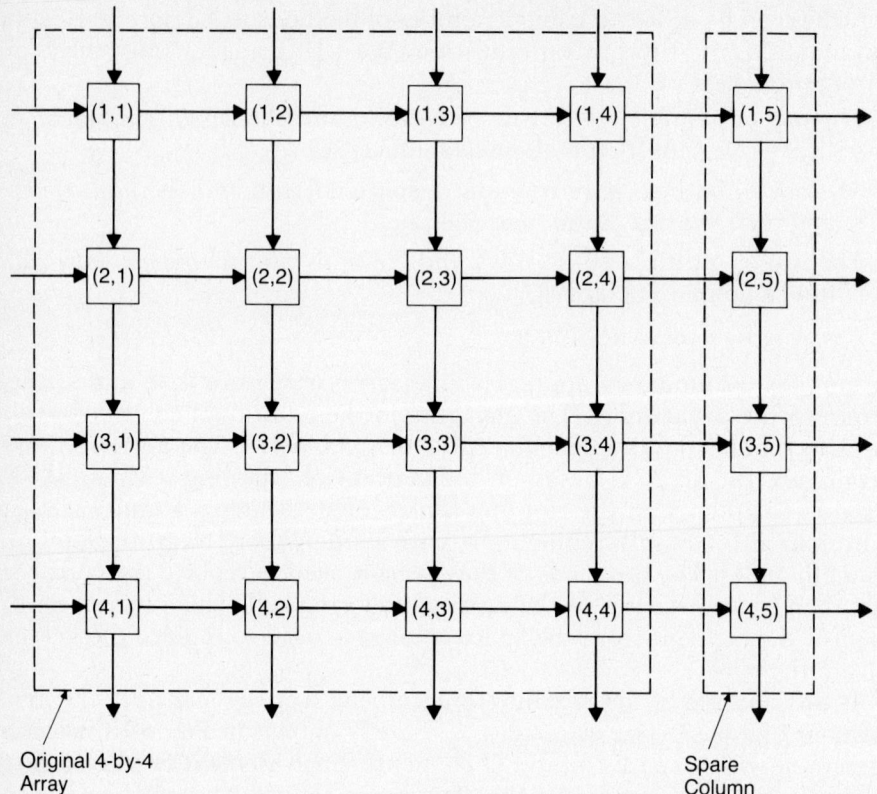

Fig. 6.13 Sample 4-by-4 array augmented with one spare column.

Each element contained in an array designed using the rippling replacement strategy must support the reconfiguration process. Specifically, we know that element (i,j) must receive information from elements $(i, j-1)$ and $(i-1, j)$ during the normal, fault-free operation of the array. In addition, element (i,j) must provide information to elements $(i, j+1)$ and $(i+1, j)$ during normal, fault-free operation. However, to support the reconfiguration process, each element (i,j) must be able to receive information from elements $(i, j-2)$, $(i-1, j-1)$, and $(i-1, j+1)$ and provide information to elements $(i, j+2)$, $(i+1, j-1)$, and $(i+1, j+1)$. A typical element structure capable of supporting the rippling replacement technique is shown in Fig. 6.15. Each element's inputs and outputs are connected to multiplexer circuits, which can be configured to properly route the data. In fabrication-time reconfiguration, the multiplexers would be permanently set to route data in a particular direction once the configuration of the array was estab-

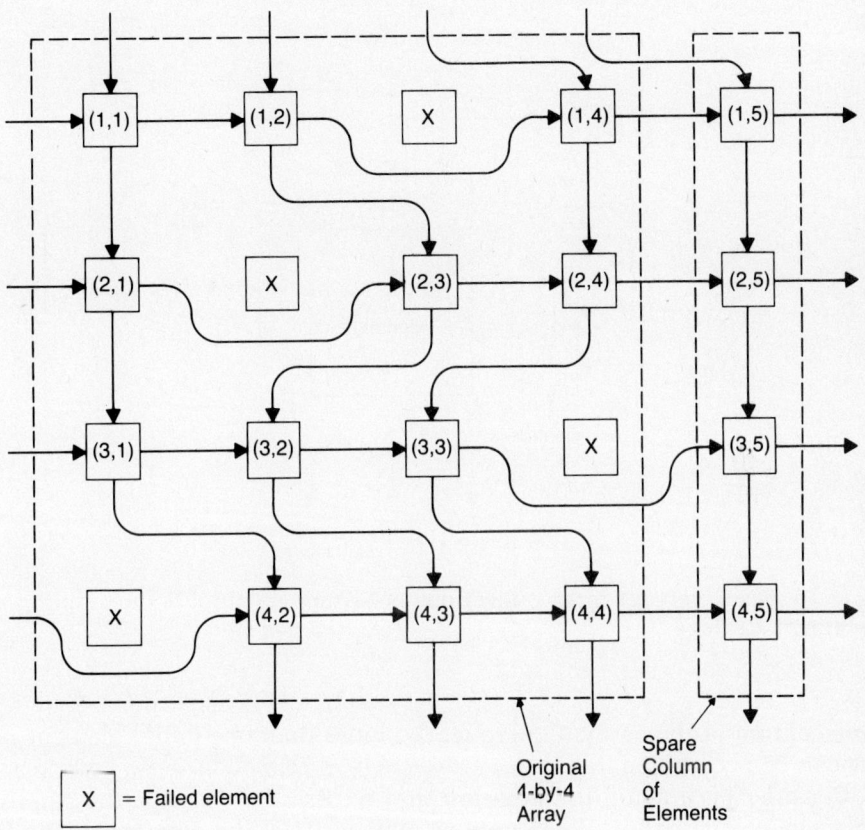

Fig. 6.14 Example of an array reconfigured using the rippling replacement strategy. The resulting array is 4-by-4.

lished. In compile-time reconfiguration, the multiplexers might be changed periodically depending on the failures that occur at particular times.

The Fault Stealing Replacement Strategy

The major difficulty with the rippling reconfiguration technique is its inability to accommodate more than one faulty element in any one row of a given array. The fault stealing replacement strategy overcomes the multiple fault problem by efficiently using both a spare column and a spare row of elements [Negrini, Sami and Stefanelli 1986]. In the fault stealing approach, an $N \times N$ array is augmented with one spare column and one spare row to produce an $(N + 1) \times (N + 1)$ array containing $2N + 1$ spare elements. During the reconfiguration process, a faulty element can be replaced by its neighbor within the same row of the array or its neighbor within the

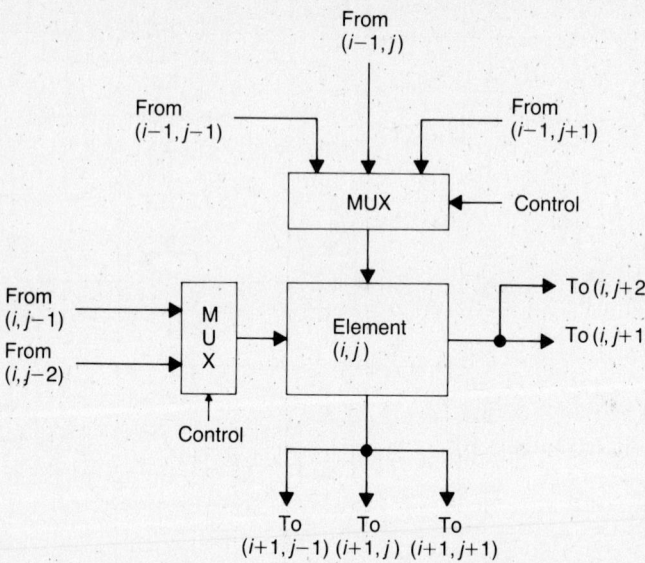

Fig. 6.15 Basic element interconnections necessary to support the rippling replacement strategy.

same column of the array. Consequently, more than one faulty element per row can be accommodated.

Consider once again the notation that we used to describe the rippling replacement algorithm. We denote as element (i,j) the element in the ith row and the jth column of the array. The element in the upper left-hand corner of the array is considered as element $(1,1)$. Figure 6.16 shows a 4×4 array that has been augmented with one row and one column of spare elements to create a 5×5 array containing nine spare elements.

The fault stealing reconfiguration algorithm begins at element $(1,1)$ and scans the rows of the array from top to bottom searching for faulty elements. If a particular row contains only one faulty element, the rippling reconfiguration philosophy is employed to reconfigure the row. In other words, a faulty element (i,j) is replaced with element $(i,j + 1)$. In turn, element $(i,j + 1)$ is replaced with element $(i,j + 2)$ and so on until element (i,N) is replaced with element $(i,N + 1)$. As long as there are no rows with more than one faulty element, the fault stealing algorithm is identical to the rippling algorithm.

If a row contains more than one faulty element, the rightmost faulty element is replaced using the rippling strategy. Additional faulty elements within the row are replaced with the elements immediately beneath them in the array. For example, suppose that row i contains the two faulty ele-

6.7 ■ Redundancy Techniques in a VLSI Design Environment

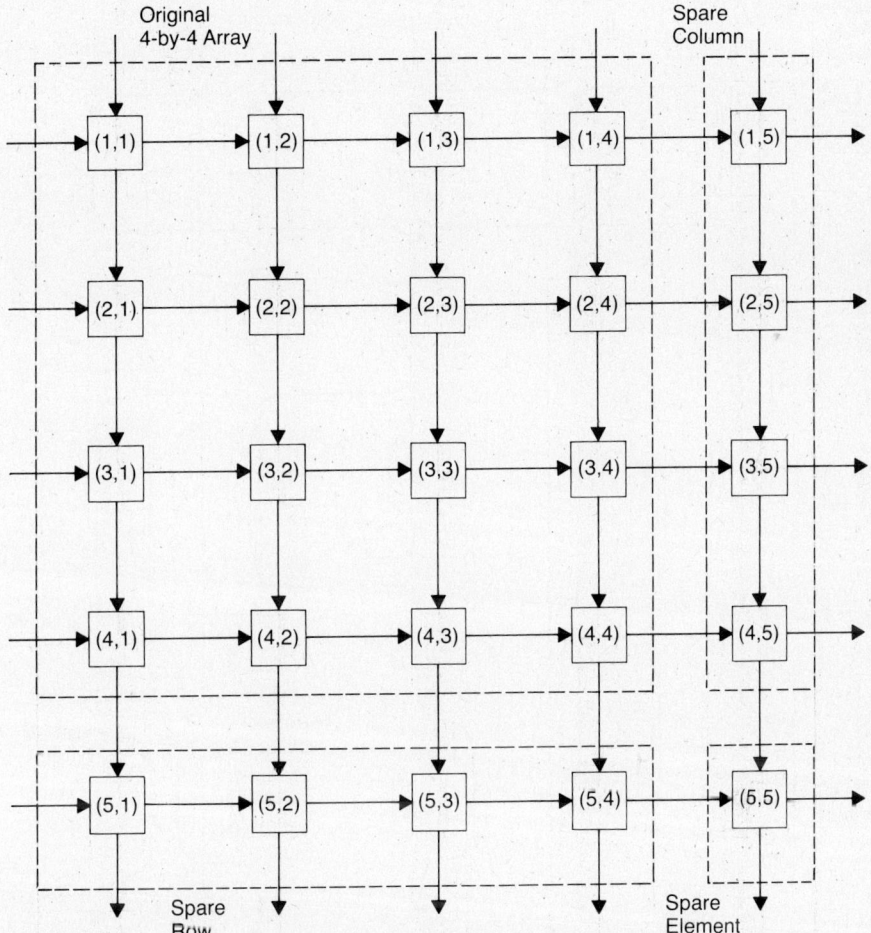

Fig. 6.16 4-by-4 array augmented with spares to support the fault stealing replacement strategy.

ments (i,j) and (i,k), where $k > j$. Element (i,k) will be replaced by element $(i,k+1)$ using the rippling strategy, whereas element (i,j) will be replaced by element $(i+1,j)$. In essence an element has been "stolen" from row $i+1$ to replace a faulty element in row i; thus the name fault stealing. When a given row is examined, any elements that have been stolen by the previous row must be considered unavailable and replaced as if they were faulty using the fault stealing strategy.

An example reconfiguration using the fault stealing strategy is shown in Fig. 6.17. Row 1 contains only a single faulty element which is bypassed us-

416 Fault-Tolerant Design of VLSI Circuits and Systems

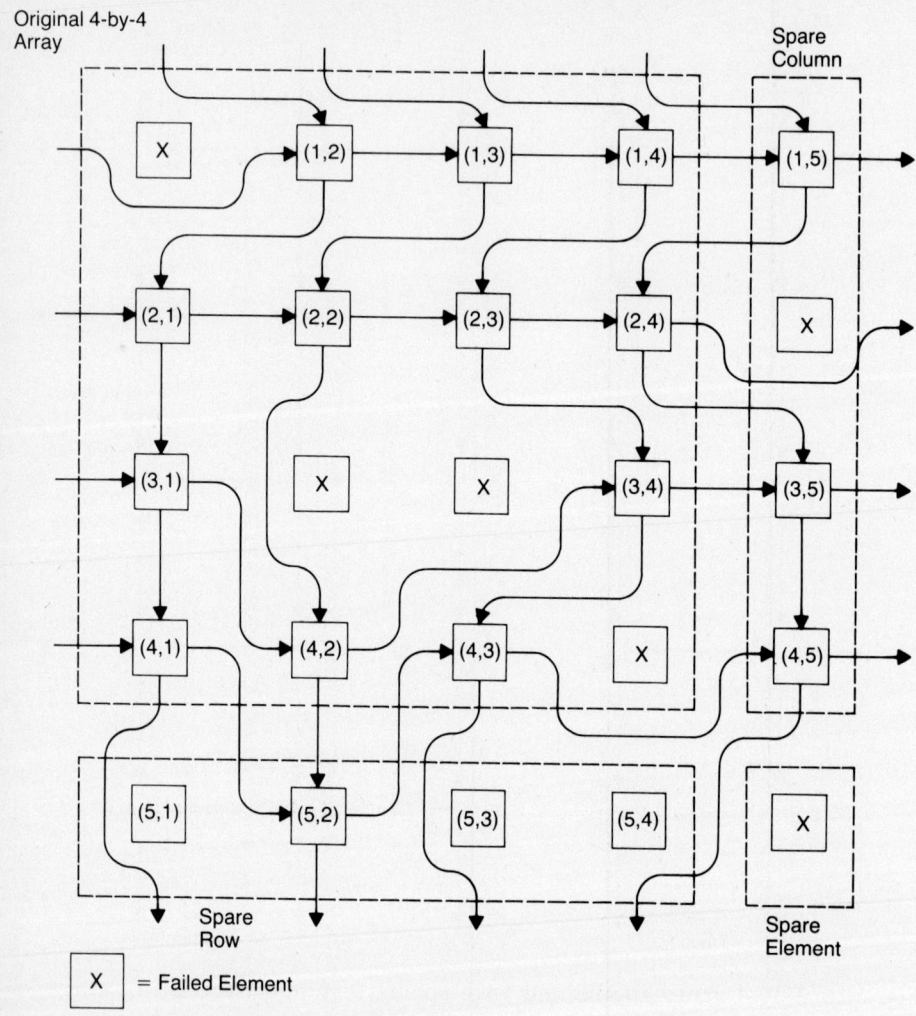

Fig. 6.17 Example of an array reconfigured using the fault stealing strategy. The resulting array is 4-by-4.

ing the rippling replacement algorithm. Note, however, that in row 3, the adjacent elements (3, 2) and (3, 3) are both faulty. Using the philosophy of the fault stealing approach, element (3, 3) is replaced with element (3, 4), and element (3, 4) is then replaced with element (3, 5). Element (3, 2) is replaced with element (4, 2), and element (4, 2) must then be replaced with element (4, 3). The remaining reconfigurations shown in Fig. 6.17 follow the same procedure illustrated in this example.

The fault stealing algorithm determines an operational array unless it reaches a point where an element that needs to be stolen is faulty. Several scenarios will cause such a condition to exist. For example, if row i contains faulty elements in columns j and k, where $k > j$, and row $i + 1$ contains a faulty element in column j, the fault stealing technique will not be able to bypass all the faulty elements. In this case, element (i,k) is replaced with element $(i,k + 1)$, and element (i,j) must then be replaced with element $(i + 1,j)$. However, element $(i + 1,j)$ is faulty, so the algorithm fails.

The Repair-Most Replacement Strategy

The repair-most replacement strategy requires the availability of multiple spare columns and rows within the array. The fundamental idea behind the repair-most strategy is that complete rows or columns of the original array are replaced with spare rows or columns [Kuo and Fuchs 1987]. This is in contrast to the rippling and fault stealing techniques where individual faulty cells are replaced. To best understand the repair-most algorithm, we consider the repair problem graphically.

Suppose that we have an array with M rows and K columns, and for the purposes of redundancy, we have augmented the $M \times K$ array with S_r spare rows and S_c spare columns. Also suppose that the original array has several faulty elements that are to be repaired using the spare rows and columns. Once again, we use the notation (i,j) to represent the element found at the intersection of the ith row and the jth column.

The repair problem can be described graphically using what is called a bipartite graph [Kuo and Fuchs 1987]. In general, a graph can be formally described using a set N of nodes and a set E of edges. The elements of E are pairs of nodes that are connected in the graph. In a bipartite graph, the set of nodes is divided into two subsets, N_1 and N_2, where each edge of the graph has one node in N_1 and one node in N_2.

We can represent the array repair problem by designating one node of a graph for each column and one node for each row. For example, node R_1 represents row 1, whereas C_1 represents column 1. A connection is made within the graph between nodes R_i and C_j if the element (i,j) is faulty. So, if we have a 4×4 array with faulty elements $(1,3)$, $(2,2)$, $(2,4)$, $(3,1)$, $(3,3)$, $(4,2)$, and $(4,3)$, the graph will be as shown in Fig. 6.18. Note, for example, that nodes 1 and 3 are connected to illustrate that element $(1,3)$ is faulty.

The graphical representation allows us to visualize the repair problem. In essence, our problem is to replace rows and columns with spare rows and columns such that all connections between nodes within the graph are eliminated. For example, if we replace row 3 in the example of Fig. 6.18 with a spare row, the faulty elements $(3,1)$ and $(3,3)$ will be repaired. From the graphical viewpoint, the replacement of row 3 eliminates the connection between nodes R_3 and C_1 and the connection between nodes R_3 and C_3. How-

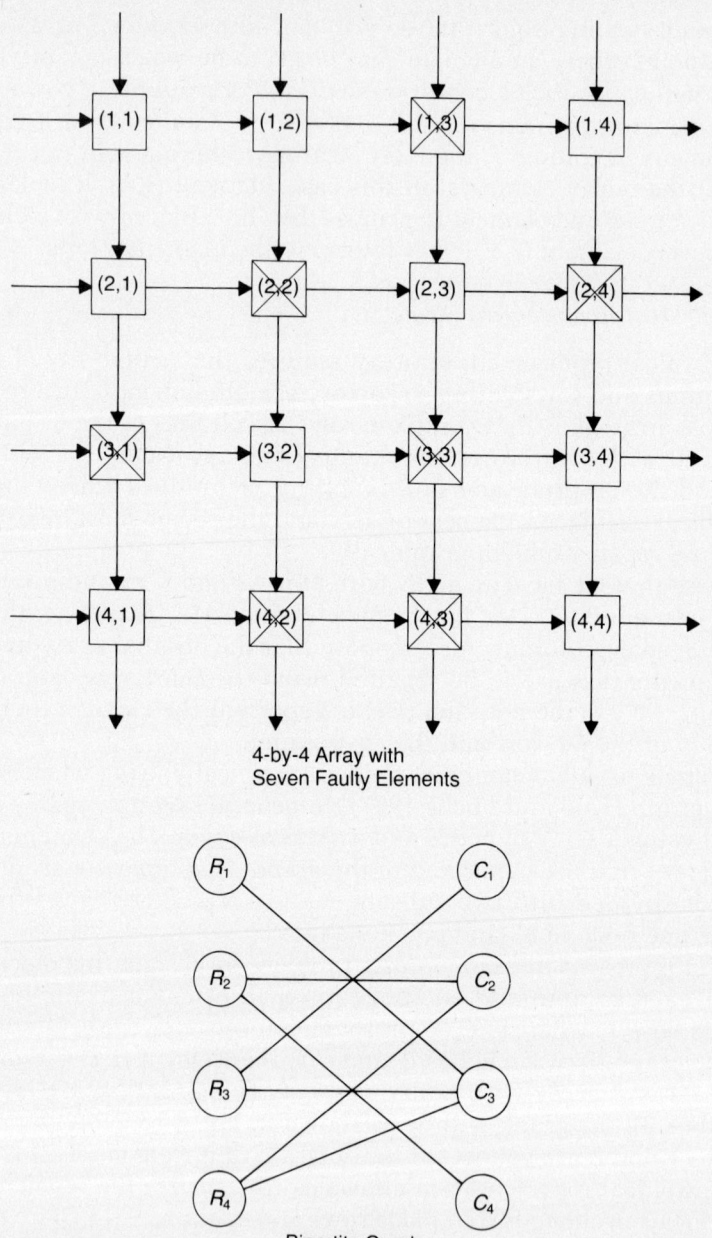

Fig. 6.18 A sample array, with faulty elements, and its corresponding bipartite graph.

6.7 ■ Redundancy Techniques in a VLSI Design Environment

ever, we could replace column 3 with a spare column and repair the faulty elements (1, 3), (3, 3), and (4, 3). In this simple example, the best initial step would be to replace the third column of the array because it replaces the largest number of faulty cells. Herein lies the basic concept of the repair-most algorithm.

The repair-most algorithm chooses for replacement the node with the largest number of connections to other nodes. In other words, the repair-most strategy first replaces the row or column with the largest number of faulty elements. Once a row or column has been replaced, the node corresponding to that row or column and all of its connections are removed from the graph. The process continues until either all faulty cells are replaced or the spare rows or columns are exhausted. The algorithm fails if all spare rows and columns are used and faulty cells still remain.

Consider as an example, the 4×4 array with one spare column and one spare row and four faulty cells, as shown in Fig. 6.19. The corresponding graph for this problem is also shown in Fig. 6.19. In the repair-most strategy, the graph is searched to find the node with the largest number of connections to other nodes. In Fig. 6.19, node C_3 has the largest number of connections, so column 3 would be replaced first. The graph is then reduced to simply nodes R_3 and C_1. In this simple example, R_3 would be replaced next because all the spare columns have been used. The solution in this simple example is to replace column 3 and then replace row 3 with spare columns and rows, respectively, to yield an operational 4×4 array.

The repair-most replacement strategy is very straightforward and simple to implement, but it has several disadvantages. The primary disadvantage is that it can fail to find a solution for repairing a given array, even though the array can be repaired. For example, suppose we want to repair the 7×7 array shown in Fig. 6.20 using the three spare rows and three spare columns [Kuo and Fuchs 1987]. Figure 6.20 also shows the bipartite graph for this problem. The repair-most algorithm would first select row 1 and column 7 for replacement. The reduced graph would then contain six rows and six columns, none of which share any edges. Consequently, either six additional spare rows, six additional spare columns, or any other combination of spare rows and columns totaling six would be required to complete the repair of the array. Since only four spare rows and columns remain, the algorithm will fail. However, a solution to the repair of the array in Fig. 6.20 does exist. For example, if we replace rows 5, 6, and 7, and then replace columns 1, 2, and 3, the array can be repaired using the available three spare rows and three spare columns.

A second disadvantage of the repair-most algorithm is that it does not always use the spare rows and columns in the most efficient manner. For example, the solution found by the repair-most algorithm can use all avail-

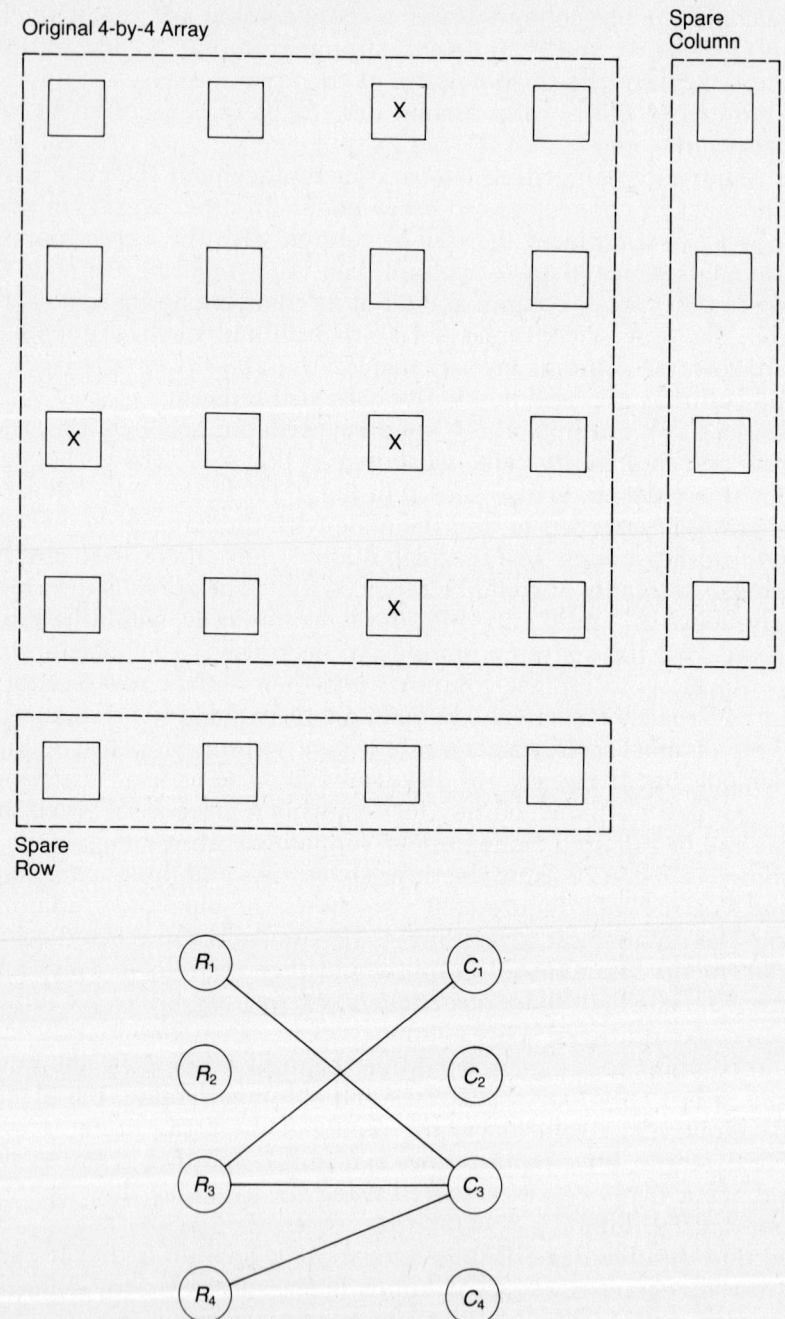

Fig. 6.19 A sample 4-by-4 array, with four faulty elements, and its corresponding graph. The repair-most algorithm would first replace column 3 and then row 3 to yield an operational array.

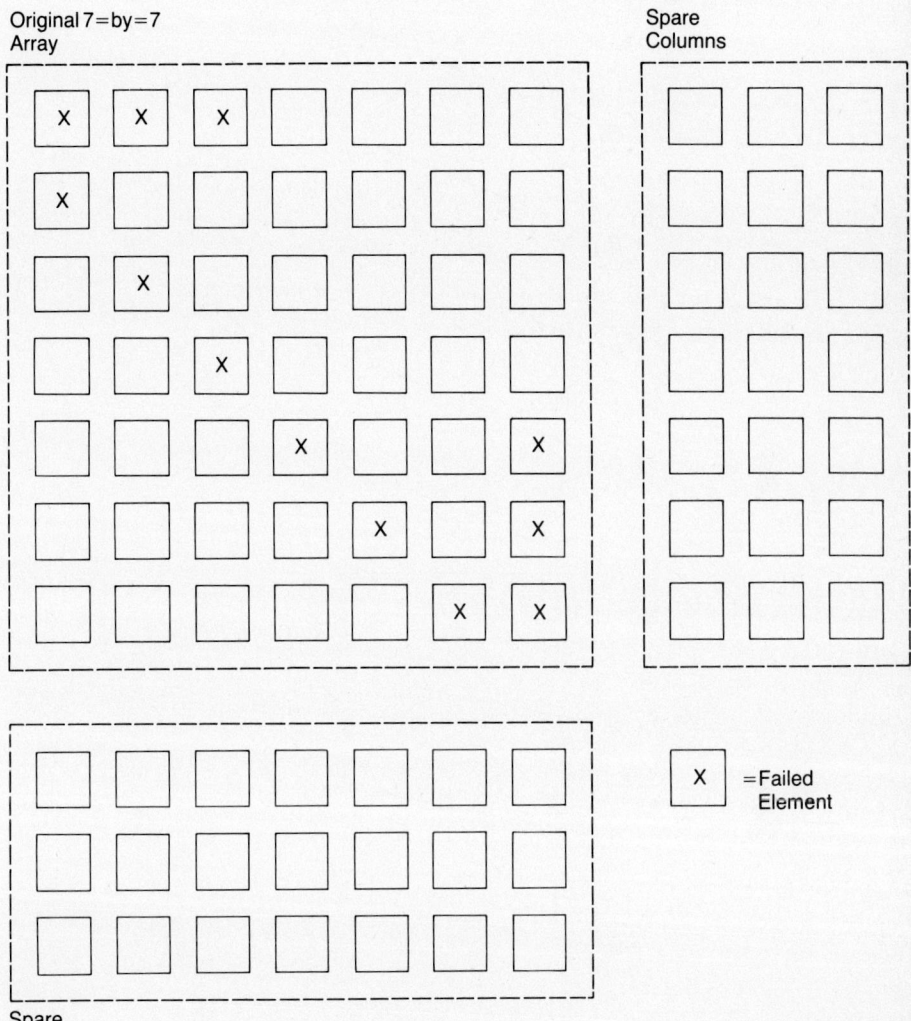

Fig. 6.20 A sample array for which the repair-most reconfiguration technique fails to find a repair solution. The bipartite graph for the array is also shown.

able spare rows and columns when a solution exists that uses fewer spare rows and columns. In many cases, it is this inefficiency that causes the algorithm to fail to find a reconfiguration solution for a given, faulty array.

Real-Time Reconfiguration

Real-time reconfiguration is the most difficult of all reconfigurations because it must be performed while maintaining the array in an operational

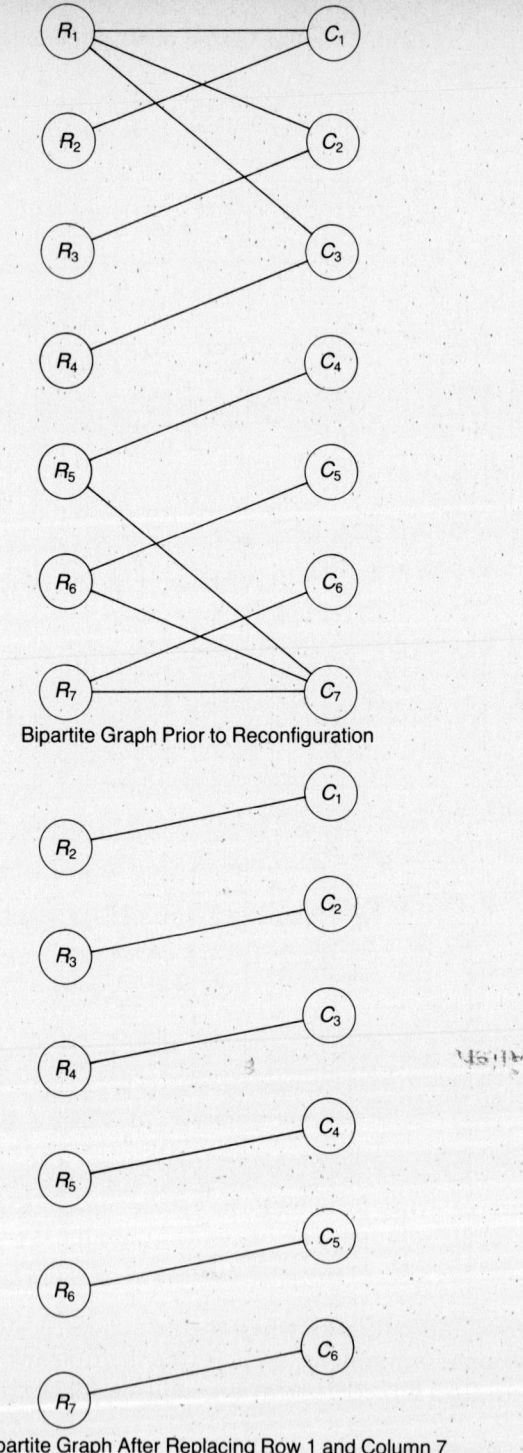

Bipartite Graph Prior to Reconfiguration

Bipartite Graph After Replacing Row 1 and Column 7

Fig. 6.20 (continued)

state. In many applications, particularly those found in life- or equipment-critical situations, the time available to reconfigure an array before system malfunctions begin to occur is extremely short. In fact, in many applications, the time available for reconfiguration is zero; if an erroneous result is created, a malfunction is immediately present. In systems that do not allow time for reconfiguration, we must use passive or hybrid redundancy techniques to prevent system malfunctions. In situations where some limited time is available for fault detection and reconfiguration, active redundancy approaches may be employed.

We will examine two primary techniques that could be useful in real-time reconfiguration. The first technique is based on the successive elimination of rows and/or columns once a faulty element is detected. The approaches include Successive Row Elimination (SRE), Successive Column Elimination (SCE), and Alternate Row and Column Elimination (ARCE) [Fortes and Raghavendra 1984]. The second technique is based on the correction of erroneous information using algorithm-based fault tolerance approaches [Abraham et al. 1987].

Successive Row Elimination Reconfiguration

The basic idea of Successive Row Elimination (SRE) is to eliminate an entire row of an array once a faulty element within that row is detected [Fortes and Raghavendra 1984]. The array, after a row is eliminated, would simply be reduced to an array containing one less row. In many applications, the concept of SRE is practical because the problem being solved can be mapped onto arrays of varying size. For example, the only difference between solving a particular problem on an $N \times N$ array and an $(N - 1) \times N$ array might be the speed of execution, in some applications. If the value of N for the $N \times N$ array is selected such that the performance of the array is acceptable with only $N - 2$ rows, two rows may be eliminated and the array will still perform acceptably.

The elimination of rows using the SRE technique is accomplished using programmable switches that may be set to an appropriate state to allow the bypassing of a complete row of the array. As an example, consider the 4×4 array shown in Fig. 6.21. Figure 6.21 also shows each possible state of the programmable switches. Programmable switches and alternate interconnections are placed between each processing element in each column. Note from Fig. 6.21 that, in general, an $N \times N$ array requires $(N + 1)N$ programmable switches. If any processing element fails in row 3, for example, the switches will be set to completely bypass the third row of the array. The array of Fig. 6.21 then becomes a 3×4 array.

The switches shown in Fig. 6.21 are two-input, two-output devices capable of directing incoming data to either of two destinations, based on the value of a control signal C_i. The normal switch mode occurs when C_i is 1

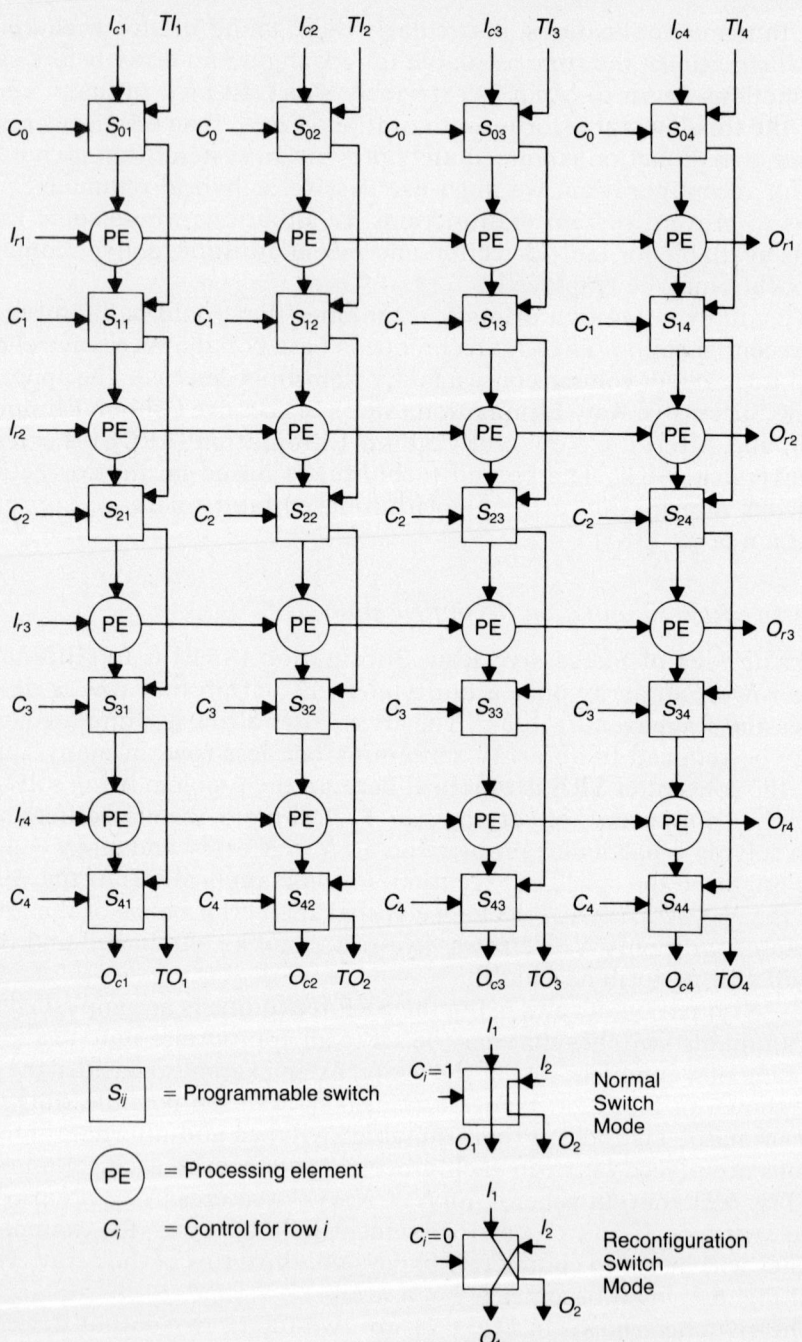

Fig. 6.21 Structure of a 4-by-4 array using successive row elimination reconfiguration.

and input I_1 is connected to output O_1, while input I_2 is connected to output O_2. When the switch is in reconfiguration mode ($C_i = 0$), input I_1 is connected to output O_2, and input I_2 is connected to output O_1.

Figure 6.22 shows an array reconfigured via the bypassing of the third row, which contains a single failed processing element. Each switch preced-

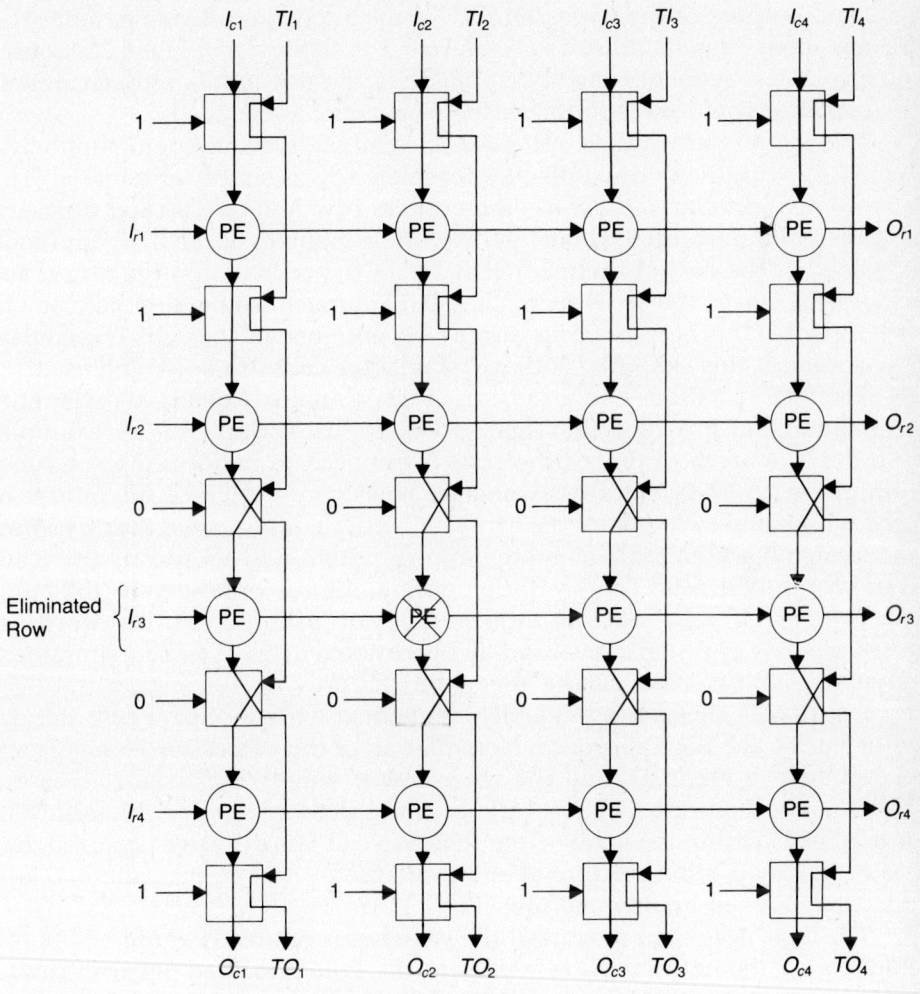

Fig. 6.22 Reconfigured array that bypasses row 3. The resulting operational array is 3-by-4.

ing row 3 has been placed in reconfiguration mode so that the bypass is implemented. Note the flow of information through the array; for example, the inputs to each column, I_{c1}, I_{c2}, I_{c3}, and I_{c4}, pass through rows 1 and 2 before bypassing row 3 and entering row 4. A 3×4 array is obtained by providing inputs to each column and using rows 1, 2, and 4 of the array.

Each column of the array in Fig. 6.22 also has a test input TI_i and a test output TO_i. The test inputs may be easily used to provide access to the inputs and outputs of the bypassed row of the array. For example, note that the test inputs bypass all rows except row 3 in the array of Fig. 6.22. Consequently, a test sequence may be applied via the test inputs to determine if the bypassed row (row 3 in this case) is operational or failed.

The primary advantage of the SRE technique is its inherent simplicity. If any cell within any row fails, the complete row is simply eliminated. The operations necessary to bypass a complete row are very straightforward and easily implemented. Unfortunately, the simplicity of the SRE approach is gained at the cost of the inefficient use of the cells within the array. For example, a 100×100 array using the SRE approach would eliminate an entire row of 100 processing cells simply because one of the cells was faulty. The result, in this example, is that 99 fault-free cells are not being used.

The SRE technique can also be used when an array of a given size must be maintained, provided that spare rows are used. Suppose, for example, that a 4×4 array of processing cells is required to perform a given function. An array that contains two spare rows could tolerate the failure of cells within any two rows of the array and still maintain a 4×4 array. Also, this design could tolerate as many failures within a given row as there are cells within that row. The limitation on the number of failures is the number of rows affected, not the total number of failures. In other words, if there are only two spare rows and an operational array is to be maintained, no more than two rows can be affected by failures.

Figure 6.23 shows a 3×3 array augmented with one spare row and designed using the SRE approach. Note that all of the processing elements are operational in Fig. 6.23, and the spare row is simply bypassed to create a 3×3 array. Figure 6.24 shows the reconfigured array after the failure of several cells within the array. Note that two cell failures have occurred, but a 3×3 array is still operational and available for use because the two failures are confined to only one row.

The logic necessary to control the switches is relatively simple. Suppose that we use the notation S_{ij} to represent the switch located between rows i and $i + 1$ and in column j. Switch S_{0j} is located in the jth column between the primary inputs and the first row of cells. If row i experiences a failure, the purpose of the switches is to direct the information around row i. This is accomplished by routing data that normally goes from row $i - 1$ to row i

6.7 ■ Redundancy Techniques in a VLSI Design Environment

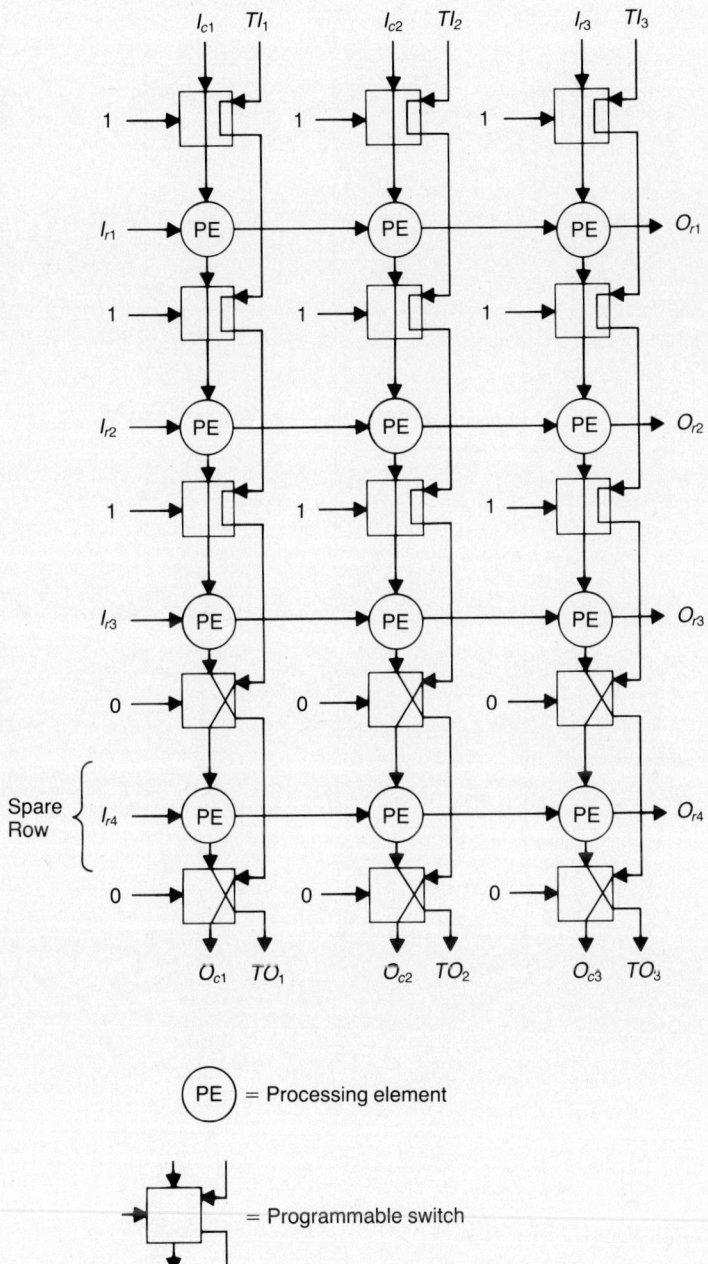

Fig. 6.23 A 3-by-3 array augmented with one spare row using the SRE technique.

428   Fault-Tolerant Design of VLSI Circuits and Systems

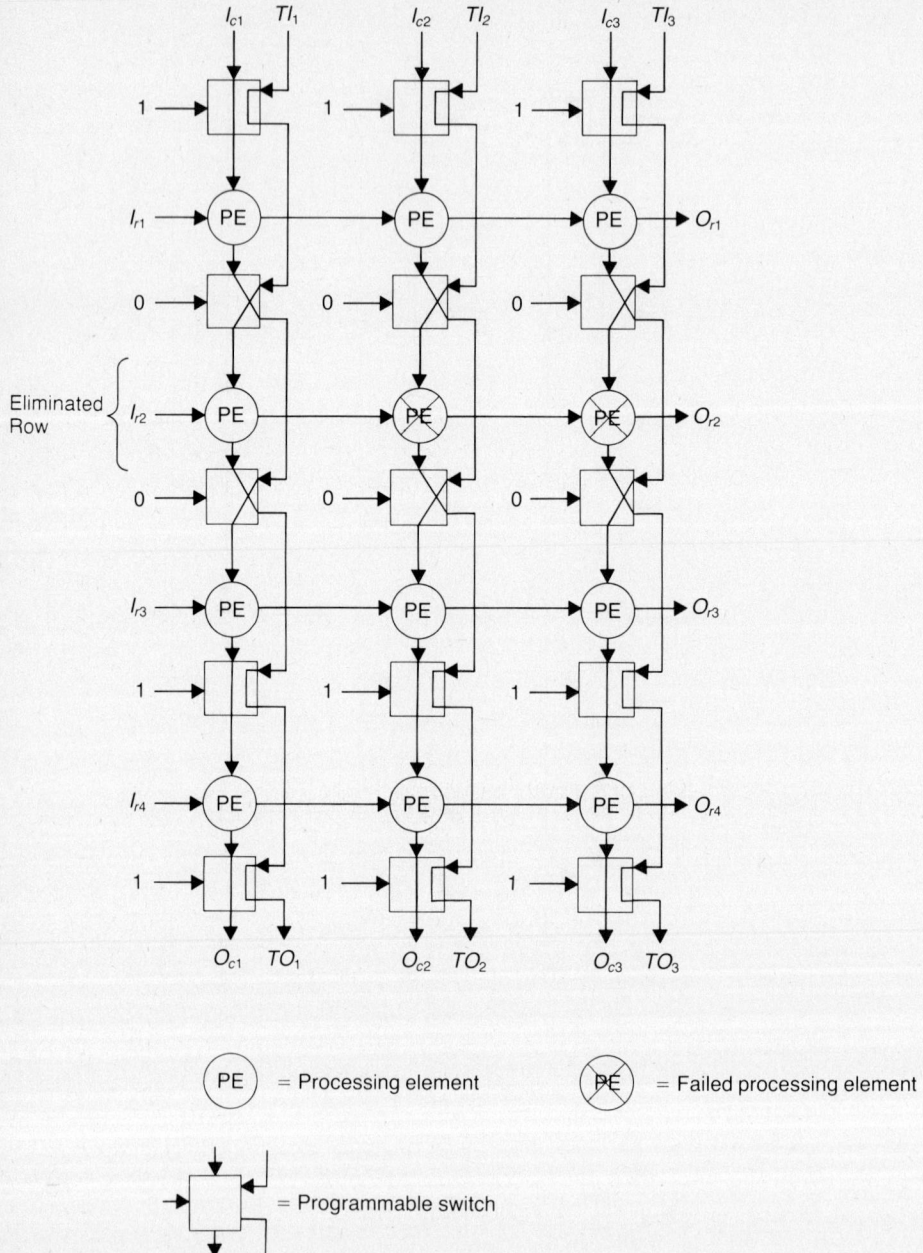

Fig. 6.24 An operational 3-by-3 array obtained by eliminating a faulty row.

so that it goes from row $i - 1$ to row $i + 1$. The switches S_{ij} and $S_{(i-1)j}$ must be set to reconfiguration mode (for all possible values of j), and the remaining switches must be set to normal mode (for all possible values of j). Figures 6.22 and 6.23 clearly illustrate the different modes required for the various switches for an example reconfiguration.

Successive Column Elimination Reconfiguration

The concept of Successive Column Elimination (SCE) is identical to that of Successive Row Elimination (SRE) except that columns rather than rows are eliminated when failures occur [Fortes and Raghavendra 1984]. Because SRE and SCE are essentially identical, we will not devote a significant amount of time and space to SCE. The choice between SRE and SCE might be made based on the number of processing elements contained within each row and column. For example, if an array has 100 rows and 25 columns, the addition of a spare row and the use of SRE would be more cost effective because fewer additional processing elements are required to add an extra row as opposed to an extra column. However, if the array has 50 rows and 100 columns, the addition of spare columns would be more cost effective. As an example of SCE, Fig. 6.25 illustrates a 3×3 array augmented with one spare column. Figure 6.26 shows the same array reconfigured using SCE after the failure of two processing elements in column 1.

Alternate Row and Column Elimination Reconfiguration

As previously mentioned, the use of Alternate Row and Column Elimination (ARCE) can improve the ability of an array to tolerate multiple, failed processing elements. The disadvantage of the ARCE approach is the increase in the hardware overhead required to implement the reconfiguration. The fundamental idea behind ARCE is to use both the elimination of rows and columns, in an alternating fashion [Fortes and Raghavendra 1984]. In other words, if the previous failure was tolerated using a row elimination, then the present failure should be handled using a column elimination, and vice versa. The primary advantage of the ARCE approach is that certain multiple failures can be tolerated using the simple switch structures originally presented in Fig. 6.21. The ARCE approach uses both the switches found in SRE and those found in SCE to achieve the reconfiguration capability. In addition, both spare rows and spare columns must be provided, if the original size of the array is to be maintained after the occurrence of failures.

An example of the ARCE technique is presented in Figs. 6.27 and 6.28 where a 3×3 array, augmented with a spare column and a spare row, is shown. Figure 6.27 shows the original array prior to the occurrence of any

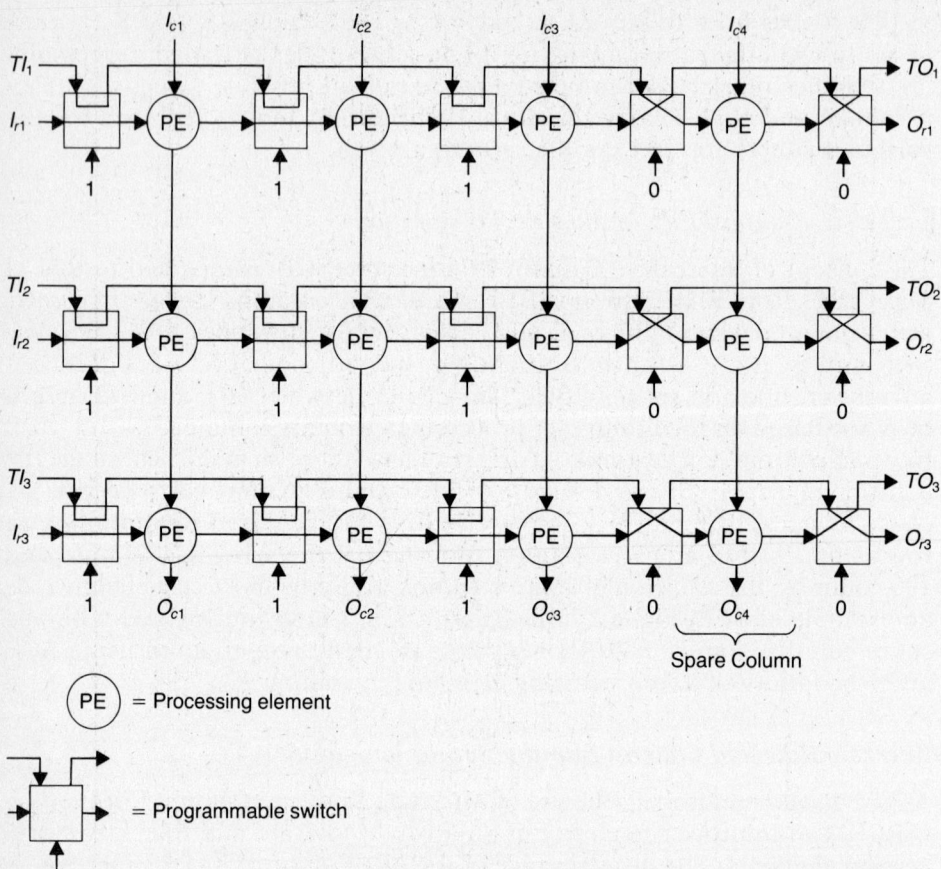

Fig. 6.25 A 3-by-3 array augmented with one spare column using the SCE reconfiguration technique. Note that the spare column is bypassed during fault-free operation.

failures. The important points of Fig. 6.27 are the availability of a spare column and a spare row and the inclusion of switches between both the columns and the rows. Figure 6.28 shows the reconfigured array after the occurrence of two failures. The reconfiguration has been handled by eliminating a row and column of the array and utilizing the spare row and spare column.

Algorithm-Based Fault Tolerance

Algorithm-based fault tolerance is a new concept that attempts to develop fault detection and fault tolerance techniques specific to the particular algorithm being employed [Abraham et al. 1987]. The majority of the published

6.7 ■ Redundancy Techniques in a VLSI Design Environment

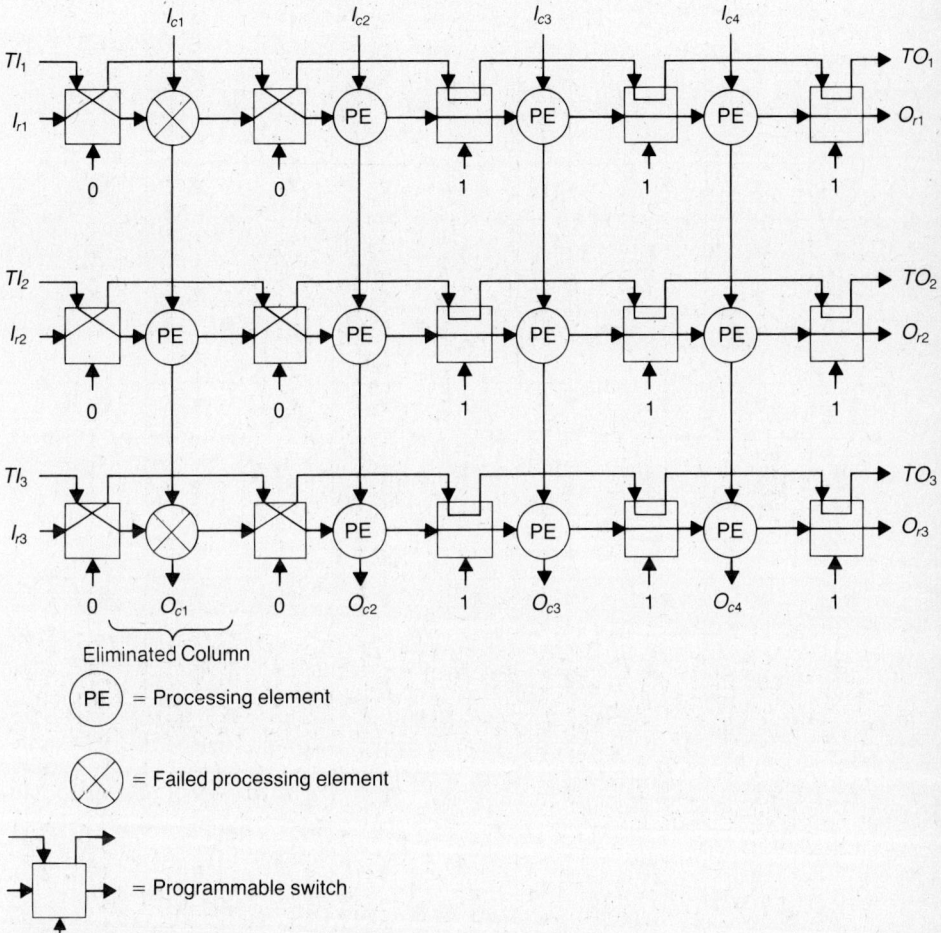

Fig. 6.26 An operational 3-by-3 array obtained by eliminating a faulty column.

results on algorithm-based fault tolerance have dealt with the problem of matrix multiplication. Techniques have been presented that allow errors in computations to be detected, and in some cases corrected, when matrix multiplication is performed. This section examines the basics of such techniques. The first approach provides only error detection. Next, we examine an approach that allows both error detection and correction.

Recall from our previous discussions in this chapter that it is relatively easy to perform matrix multiplication on a mesh-type array processing system. In fact, we have examined, in detail, how the matrix multiplication is performed, the structure of the array necessary to implement the multi-

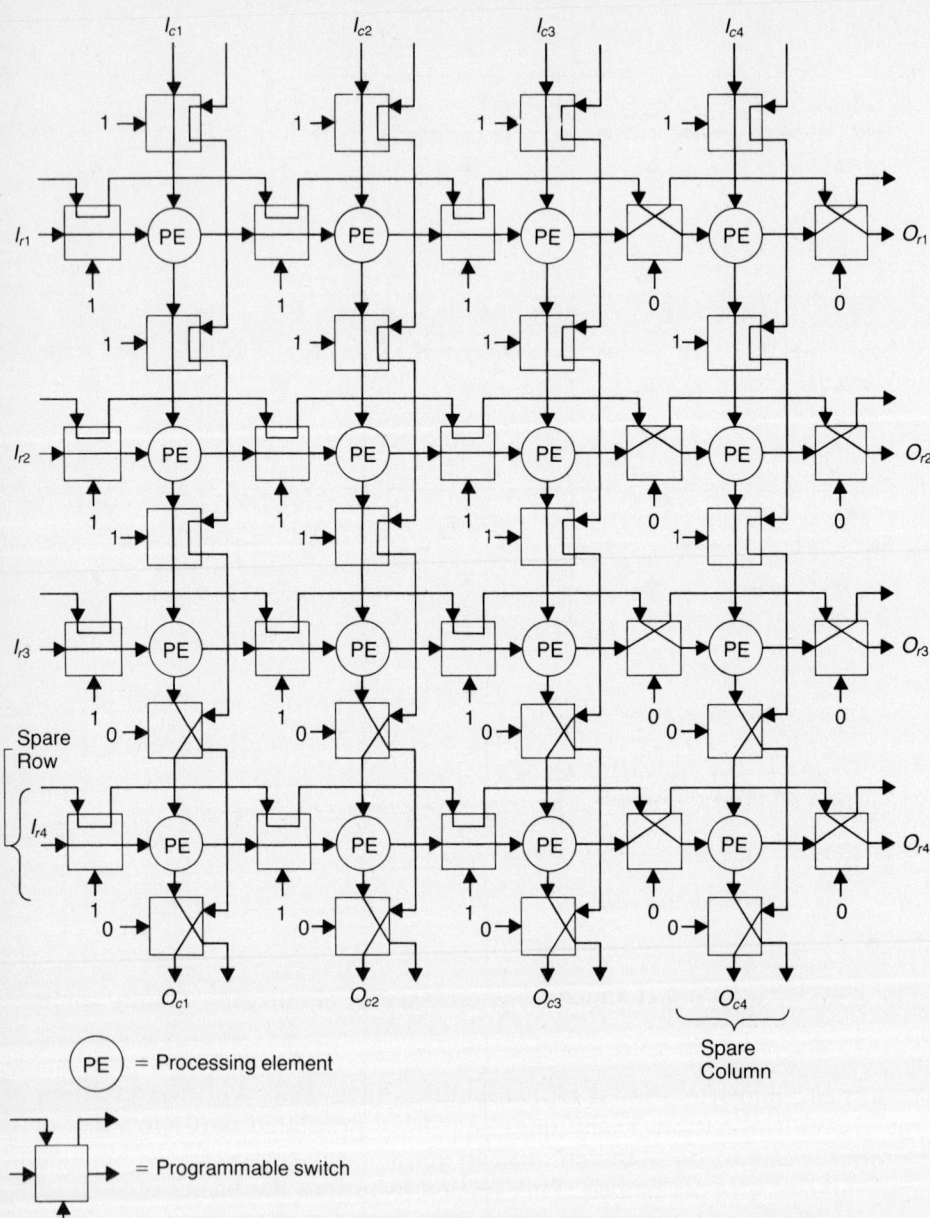

Fig. 6.27 An array augmented with both spare row and spare column elements. Note that spares are bypassed here to create a 3-by-3 operational array.

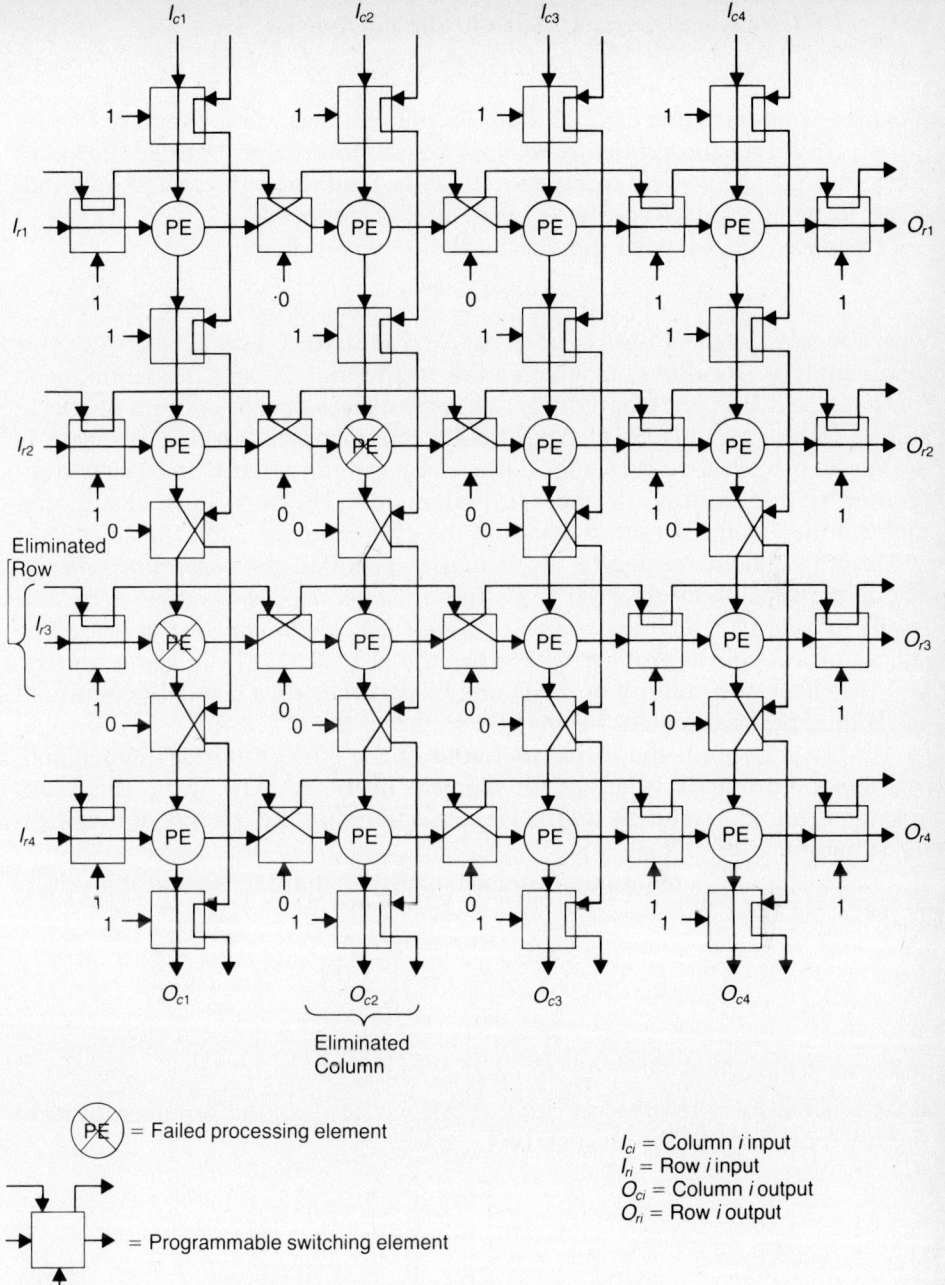

Fig. 6.28 A reconfigured array that maintains an operational 3-by-3 configuration after the occurrence of two failures, each of which is in a unique row and column.

plication, and the time savings that can be realized when compared to sequential matrix multiplication. Now we examine a technique that uses conventional checksums to encode the data contained in each of two matrices to be multiplied so that error detection can occur.

Suppose that we wish to perform the matrix multiplication

$$\mathbf{AB} = \mathbf{C}$$

where $\mathbf{A}$ is an $M \times N$ matrix, $\mathbf{B}$ is an $N \times M$ matrix, and $\mathbf{C}$ is the $M \times M$ result matrix. To allow error detection in the matrix multiplication, each matrix $\mathbf{A}$ and $\mathbf{B}$ is augmented with an appropriate row or column of checksums [Abraham et al. 1987]. Specifically, the $\mathbf{A}$ matrix, in this example, is augmented with a row of checksum elements, and the $\mathbf{B}$ matrix is augmented with a column of checksum elements. The row of checksum elements added to the $\mathbf{A}$ matrix is called the *checksum row*, and the column of checksum elements added to the $\mathbf{B}$ matrix is called the *checksum column*. The $\mathbf{A}$ matrix is now referred to as the *row checksum encoded matrix*, and the $\mathbf{B}$ matrix is referred to as the *column checksum encoded matrix*. The checksum row of the $\mathbf{A}$ matrix becomes the $(M + 1)$th row of the $\mathbf{A}$ matrix, and the checksum column of the $\mathbf{B}$ matrix becomes the $(M + 1)$th column of the $\mathbf{B}$ matrix. The element in the ith column of the checksum row is formed by simply adding all the elements found in the ith column of the original, unaugmented matrix, whereas the element in the ith row of the checksum column is found by adding all the elements found in the ith row of the original, unaugmented matrix.

As an example, suppose that we have the 2×2 matrices, $\mathbf{A}$ and $\mathbf{B}$ given by

$$\mathbf{A} = \begin{bmatrix} a_{11} & a_{12} \\ a_{21} & a_{22} \end{bmatrix}$$

$$\mathbf{B} = \begin{bmatrix} b_{11} & b_{12} \\ b_{21} & b_{22} \end{bmatrix}$$

If we wish to form the matrix product $\mathbf{AB}$ we first use the $\mathbf{A}$ matrix to form the row encoded checksum matrix $\mathbf{A}_{\text{rcs}}$ given by

$$\mathbf{A}_{\text{rcs}} = \begin{bmatrix} a_{11} & a_{12} \\ a_{21} & a_{22} \\ a_{31} & a_{32} \end{bmatrix}$$

where

$$a_{31} = a_{11} + a_{21}$$
$$a_{32} = a_{12} + a_{22}$$

In other words, the elements a_{31} and a_{32} are simply the checksums of columns 1 and 2, respectively. Next, we use the $\mathbf{B}$ matrix to form the column encoded checksum matrix $\mathbf{B}_{\text{ccs}}$ given by

6.7 ■ Redundancy Techniques in a VLSI Design Environment

$$\mathbf{B}_{ccs} = \begin{bmatrix} b_{11} & b_{12} & b_{13} \\ b_{21} & b_{22} & b_{23} \end{bmatrix}$$

where

$$b_{13} = b_{11} + b_{12}$$
$$b_{23} = b_{21} + b_{22}$$

In other words, the elements b_{13} and b_{23} are simply the checksums of rows 1 and 2, respectively.

Performing the multiplication

$$\mathbf{A}_{rcs}\mathbf{B}_{ccs} = \mathbf{C}_{cs}$$

yields

$$\mathbf{C}_{cs} = \begin{bmatrix} a_{11} & a_{12} \\ a_{21} & a_{22} \\ a_{31} & a_{32} \end{bmatrix} \begin{bmatrix} b_{11} & b_{12} & b_{13} \\ b_{21} & b_{22} & b_{23} \end{bmatrix}$$

$$= \begin{bmatrix} (a_{11}b_{11} + a_{12}b_{21}) & (a_{11}b_{12} + a_{12}b_{22}) & (a_{11}b_{13} + a_{12}b_{23}) \\ (a_{21}b_{11} + a_{22}b_{21}) & (a_{21}b_{12} + a_{22}b_{22}) & (a_{21}b_{13} + a_{22}b_{23}) \\ (a_{31}b_{11} + a_{32}b_{21}) & (a_{31}b_{12} + a_{32}b_{22}) & (a_{31}b_{13} + a_{32}b_{23}) \end{bmatrix}$$

Note that the elements of the result matrix $\mathbf{C}_{cs}$ contain the expected result of the multiplication (elements c_{11}, c_{12}, c_{21}, and c_{22}) as well as an additional row and column. Further examination of the additional row and column will reveal that they are the checksum row and column of the result matrix. Specifically,

$c_{13} = (a_{11}b_{13} + a_{12}b_{23})$
$\quad = (a_{11}b_{11} + a_{12}b_{21}) + (a_{11}b_{12} + a_{12}b_{22})$
$\quad = c_{11} + c_{12}$

$c_{23} = (a_{21}b_{13} + a_{22}b_{23})$
$\quad = (a_{21}b_{11} + a_{22}b_{21}) + (a_{21}b_{12} + a_{22}b_{22})$
$\quad = c_{21} + c_{22}$

$c_{31} = (a_{31}b_{11} + a_{32}b_{21})$
$\quad = (a_{11}b_{11} + a_{12}b_{21}) + (a_{21}b_{11} + a_{22}b_{21})$
$\quad = c_{11} + c_{21}$

$c_{32} = (a_{31}b_{12} + a_{32}b_{22})$
$\quad = (a_{11}b_{12} + a_{12}b_{22}) + (a_{21}b_{12} + a_{22}b_{22})$
$\quad = c_{12} + c_{22}$

$c_{33} = (a_{31}b_{13} + a_{32}b_{23})$
$\quad = (a_{11}b_{11} + a_{12}b_{21}) + (a_{11}b_{12} + a_{12}b_{22}) + (a_{21}b_{11} + a_{22}b_{21}) + (a_{21}b_{12} + a_{22}b_{22})$
$\quad = c_{13} + c_{23}$
$\quad = c_{31} + c_{32}$

Note from the above equations that element c_{13} is the checksum of row 1 for the result matrix, c_{23} is the checksum of row 2, c_{31} is the checksum of column 1, c_{32} is the checksum of column 2, and c_{33} is the checksum of both column 3 and row 3.

To obtain a better understanding of the matrix checksum approach, consider the following specific numerical example. Suppose that we have the 2 × 2 matrices **A** and **B**, and we wish to calculate the matrix product **AB** = **C** to obtain the 2 × 2 result matrix **C**. In addition, suppose that the matrices **A** and **B** are given by

$$\mathbf{A} = \begin{bmatrix} 8 & 3 \\ 5 & 4 \end{bmatrix}$$

$$\mathbf{B} = \begin{bmatrix} 2 & 4 \\ 6 & 7 \end{bmatrix}$$

First, we use **A** to calculate the row checksum matrix, $\mathbf{A}_{rcs}$, given by

$$\mathbf{A}_{rcs} = \begin{bmatrix} 8 & 3 \\ 5 & 4 \\ 13 & 7 \end{bmatrix}$$

Note that the original information from matrix **A** is contained in the first two rows of $\mathbf{A}_{rcs}$, and the third row is the checksum elements.

Next, we use **B** to calculate the column checksum matrix, $\mathbf{B}_{ccs}$ given by

$$\mathbf{B}_{ccs} = \begin{bmatrix} 2 & 4 & 6 \\ 6 & 7 & 13 \end{bmatrix}$$

Note again that the original information from matrix **B** is contained in the first two columns of $\mathbf{B}_{ccs}$, and the third column is the checksum elements. Calculating the product $\mathbf{A}_{rcs}\mathbf{B}_{ccs}$ yields

$$\begin{bmatrix} 8 & 3 \\ 5 & 4 \\ 13 & 7 \end{bmatrix} \begin{bmatrix} 2 & 4 & 6 \\ 6 & 7 & 13 \end{bmatrix} = \begin{bmatrix} 34 & 53 & 87 \\ 34 & 48 & 82 \\ 68 & 101 & 169 \end{bmatrix}$$

As you can see from examining the result matrix from the preceding calculation, the result matrix for the product **AB** is contained in elements (1,1), (1,2), (2,1), and (2,2). Element (1,3) is the checksum of row 1, element (2,3) is the checksum of row 2, element (3,1) is the checksum of column 1, element (3,2) is the checksum of column 2, and element (3,3) is the checksum of both row 3 and column 3. If any of the checksums are incorrect, an error has occurred during some step of the calculation.

To better understand the capabilities of matrix checksum technique for the detection of errors, consider the hardware implementation of a simple matrix multiplier that uses the approach. Suppose that we wish to multiply

6.7 ■ Redundancy Techniques in a VLSI Design Environment

a pair of 2 × 2 matrices using the matrix checksum technique to provide error detection. Recall from our earlier discussions in this chapter that matrix multiplication can be performed with a mesh array where each element of the array is a multiplier with an accumulator. Figure 6.29 shows a numerical example of a 3 × 3 mesh array performing the matrix multiplication

$$\begin{bmatrix} 8 & 3 \\ 5 & 4 \\ 13 & 7 \end{bmatrix} \begin{bmatrix} 2 & 4 & 6 \\ 6 & 7 & 13 \end{bmatrix} = \begin{bmatrix} 34 & 53 & 87 \\ 34 & 48 & 82 \\ 68 & 101 & 169 \end{bmatrix}$$

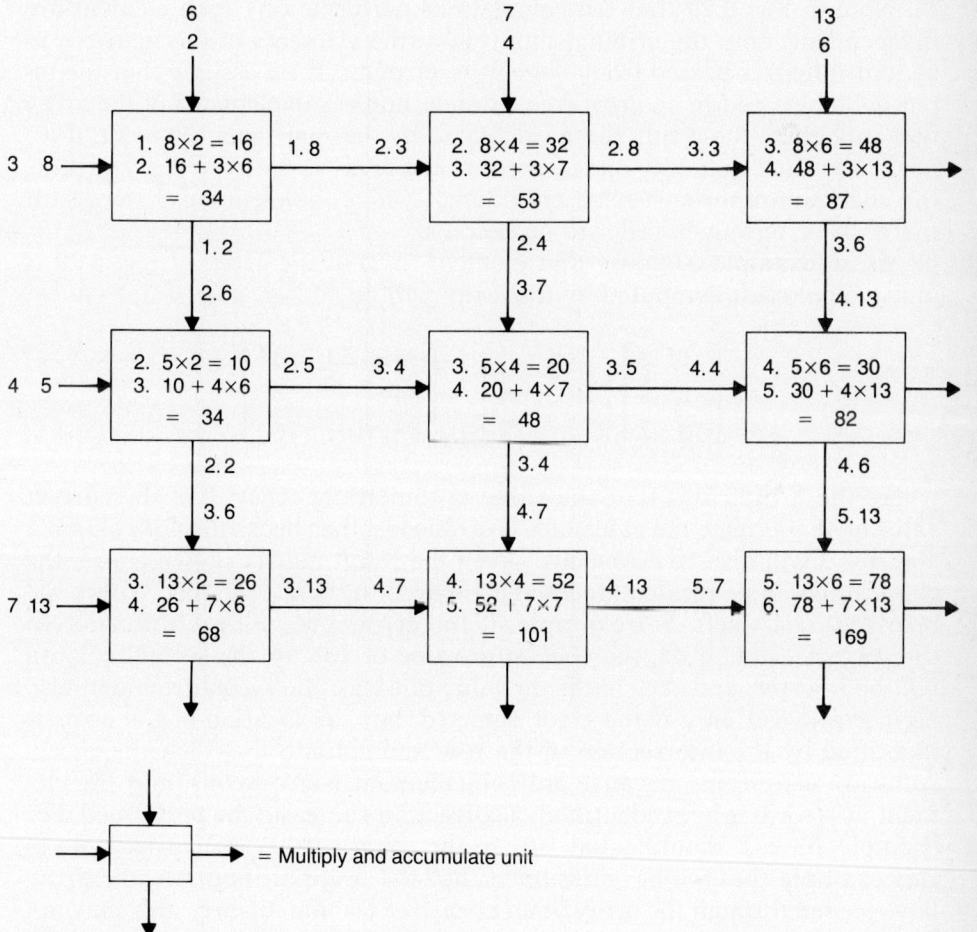

Fig. 6.29 Multiplication of a row checksum encoded matrix and a column checksum encoded matrix.

The information written inside each element of the array shows the calculations performed by that element at each clock step. For example, element (1,1) multiplies 8 times 2 on the first clock step and stores the result, which in this case is 16. On the second clock step, element (1,1) will multiply 3 times 6 and add the result to 16 to obtain element (1,1) of the result matrix, which in this example is 34. Element (2,2), on the other hand, must wait until the third clock step to perform any calculations, at which time it creates the product of 4 and 5. On the fourth clock step, element (2,2) will multiply 4 times 7 to obtain 28, and add 28 to 20 to obtain 48, which is element (2,2) of the result matrix. The process continues until all elements of the result matrix are computed.

Note in Fig. 6.29 that the calculations performed by each element are independent; only the original input data (the elements of the matrices to be multiplied) is passed from element to element. If we assume that the input data is passed in an error-free manner, failures in elements of the array will only affect the result being calculated by that particular element. If we further assume that only one element of the array is faulty at a given time, the checksum approach need only detect when one element of the result matrix is erroneous in order to be effective.

As an example, suppose that element (2,2) of the array of Fig. 6.29 is faulty. The result computed by the array will be

$$\begin{bmatrix} 8 & 3 \\ 5 & 4 \\ 13 & 7 \end{bmatrix} \begin{bmatrix} 2 & 4 & 6 \\ 6 & 7 & 13 \end{bmatrix} = \begin{bmatrix} 34 & 53 & 87 \\ 34 & X & 82 \\ 68 & 101 & 169 \end{bmatrix}$$

where the X indicates that the value is something other than the correct value of 48. Because the value of X is erroneous, the checksums for column 2 and row 2 will also be erroneous. Given the result matrix shown above, the checksums can be recalculated and compared to those contained within the result matrix itself. For example, if the erroneous value X differs from the correct value of 48, the checksum value of 101 for the second column will be in error, and the checksum value of 82 for the second row will also be in error. Not only is the error detected, but the location of the error is identified by the intersection of the row and column in which the checksums are erroneous. Because only one element is erroneous, and the element in error has been identified, a correction can easily be performed. For example, $53 + X$ should equal 101, so the correct value of X is 48. Note in this example that we have assumed that the original inputs to the array have passed through the array in an error-free fashion. Clearly, this may not be an accurate assumption, and techniques will have to be used to protect against errors in the input data that affect all calculations using that data. For example, a simple parity technique might be applied to the input data to guarantee its correctness before it is used at each element.

6.7.5 Redundancy to Enhance Yield of VLSI Circuits

An important application of redundancy techniques in a VLSI design environment is to improve the yield of integrated circuits. For our purposes, **yield** is defined as the number of fabricated devices that work correctly divided by the total number of fabricated devices ([Bernard 1978] and [Siewiorek and Swarz 1982]). For example, if a company fabricates 50,000 processor ICs and finds that 5000 of those processors perform their functions correctly, the yield is 5000/50,000, or 0.1. Expressed as a percentage, the yield in this example is 10%.

In many VLSI applications, it is not unusual to experience yields on the order of 10%, or perhaps even less. Consequently, the cost of manufacturing a circuit can become prohibitive since ten ICs must be produced to obtain one that works. A number of investigators have examined approaches that attempt to use redundant elements to replace failed ones so that a faulty circuit can be repaired and made usable [Mangir and Avizienis 1982]. For example, it is quite common in the semiconductor memory industry to include spare rows or columns of storage elements such that a faulty row or column can be replaced using the fabrication-time reconfiguration techniques described in the previous section for array-type structures. If used solely for yield improvement, the reconfiguration is performed immediately after fabrication and is irreversible.

The purpose of this section is to examine the improvement that can be obtained in circuit yield by using redundancy. The specific redundancy techniques that may be used for yield improvement include many of the approaches already considered in the previous sections. For example, the most common application of yield improvement techniques is memory ICs, and the most common redundancy technique is the addition of redundant rows or columns that can be used to replace faulty rows or columns.

Several yield models have been developed in the past, and the majority are based on a similar set of assumptions ([Bernard 1978] and [Siewiorek and Swarz 1982]). First, all failures on the IC are assumed to be the result of what are called *spot defects*. Spot defects are localized to a given area within an IC and are assumed to affect no more than one module within that IC. The specific size of the module is not important here; the important point is that it is assumed that the defect is not large enough to span more than one module. However, a module may have more than one spot defect. If we assume, for example, that spot defects affect no more than one module, we can use duplicate modules to provide redundancy, and a single defect cannot render both modules faulty. Spot defects are in contrast to *area defects*, which affect complete sections of a chip or wafer. Specifically, an area defect might result in the failure of several modules within the same chip or several chips on the same wafer.

The second major assumption in yield modeling is that any single spot defect will result in the chip being inoperative unless some type of redundancy is included. In other words, all defects are considered to be fatal defects in nonredundant chips. In practice, a defect might result in performance degradations that do not render the chip completely useless. The yield models, however, assume that any single defect in a nonredundant chip results in the chip being completely inoperative.

The final major assumption in yield modeling is that spot defects are randomly distributed, in a physical sense, throughout a chip and a wafer. In other words, the number of defects contained in any one area of a device is a random variable.

The simplest yield model assumes that the probability p of a chip being free of defects, and consequently being completely operational, is [Price 1970]

$$p = e^{-DA}$$

where A is the area of the chip and D is the defect density of the chip. The defect density D is the expected number of defects per unit area. Although its effectiveness in modeling the details of a complex fabrication process has been questioned, this exponential model of yield has been considered as a good approximation for comparing candidate chip designs.

Several interesting points can be examined using the exponential yield model. For example, clearly it is important to keep chips small, due to the exponential relationship between area and the probability of a given chip being operational. If $A = n/D$, then

$$p = e^{-Dn/D} = e^{-n}$$

Consequently, the probability p of having an operational device decreases exponentially as a function of n. The conclusion of this analysis is that, for a given defect density, the exponential expression allows us to determine the size of the chip that produces a given yield. If you want yield to be approximately 0.32, for example, the chip area should be no more than $1/D$.

In yield modeling, we are interested in determining the yield of a given fabrication run or collection of runs. The preceding exponential equation simply tells us the probability of a specific chip with a given area and defect density being completely operational. In calculating the yield of a fabrication run, we have many chips of the same area, but possibly having different defect densities. To account for varying defect densities, we must consider the distribution function $f(D)$ of the defect density D. Given $f(D)$, we can calculate the yield Y, as

$$Y = \int_0^\infty e^{-DA} f(D) \, dD$$

Many different models of the distribution function $f(D)$ have been developed and investigated. Recall from probability theory that the integral of the distribution function over all possible values of D must be 1. Figure 6.30 shows several distribution functions that have been investigated in the past.

6.7 ■ Redundancy Techniques in a VLSI Design Environment

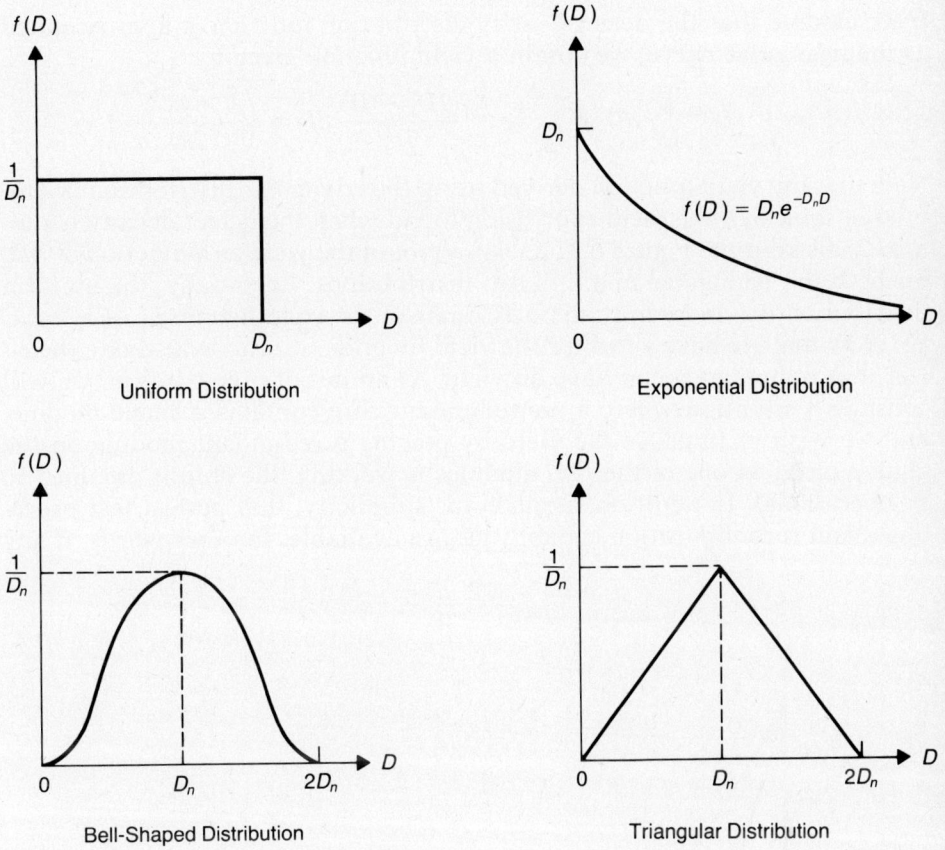

Fig. 6.30 Examples of defect distributions commonly used in yield analysis.

Suppose that we consider $f(D)$ to be uniform, having a value of $1/D_n$ for all values of D between 0 and D_n. The resulting yield function in such a situation would be

$$Y = \int_0^{D_n} e^{-DA} \frac{1}{D_n} dD = \frac{1}{D_n} \int_0^{D_n} e^{-DA} dD = \frac{1}{AD_n}[1 - e^{-D_n A}]$$

More complex distribution functions for the defect density allow more accurate yield models to be developed. For example, the triangular distribution shown in Fig. 6.30 is often used to approximate the bell-shaped curve also shown in Fig. 6.30. The triangular distribution function is given by

$$f(D) = \begin{cases} \dfrac{D}{D_n^2} & \text{if } 0 \leq D \leq D_n \\ \dfrac{(2D_n - D)}{D_n^2} & \text{if } D_n \leq D \leq 2D_n \end{cases}$$

If we assume that the defect density distribution function is approximated as the triangular curve, we obtain a yield function given by

$$Y = \int_0^{D_n} e^{-DA} \frac{D}{D_n^2} dD + \int_{D_n}^{2D_n} e^{-DA} \frac{(2D_n - D)}{D_n^2} dD = \left(\frac{1 - e^{-D_n A}}{AD_n} \right)^2$$

Note that the yield function derived using the triangular distribution is simply the square of the yield expression found when the defect density is uniformly distributed. Figure 6.31 shows a plot of the yield as a function of AD_n for both the triangular and uniform distributions. Essentially, the uniform distribution is a more optimistic estimate of the yield function.

Now that we have a feel for the yield function, we can investigate the effect that redundancy can have on yield. As an initial investigation, we will consider a situation where a nonredundant chip contains a single module, and we wish to improve the yield by placing a redundant module on the chip. As long as one of the two modules is working, the chip is assumed to be operational. In addition, assume, for simplicity, that perfect test procedures and reconfiguration capabilities are available. In other words, if any

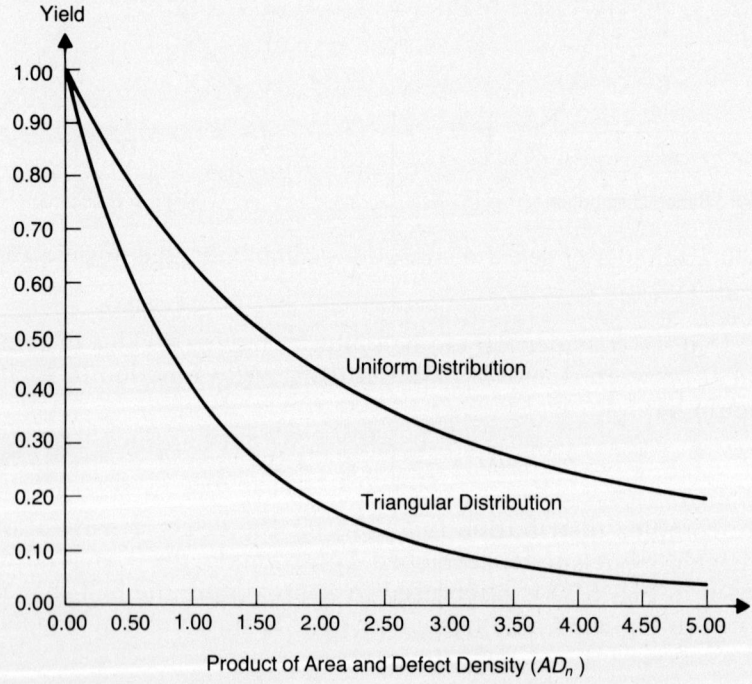

Fig. 6.31 Comparison of uniform versus triangular distributions for the defect density.

6.7 ■ Redundancy Techniques in a VLSI Design Environment

module is faulty, we will be able to identify the faulty module and, provided the redundant module is operational, correctly reconfigure the chip. The yield obtained in the nonredundant case and that obtained when the redundancy is included will be compared.

The redundant chip is shown in Fig. 6.32 where it is seen that two identical modules are provided, and a single multiplexer is used to select the outputs from one of the two modules. The chip is operational if, and only if, one of the two modules works and the multiplexer is defect free. We assume that the area of each module is A, and the area of the multiplexer is some fraction of A, say, αA. Given the exponential yield model developed above, the probability that module 1 is free of defects is equivalent to the probability that module 2 is free of defects and is given by

$$p_1 = p_2 = e^{-DA}$$

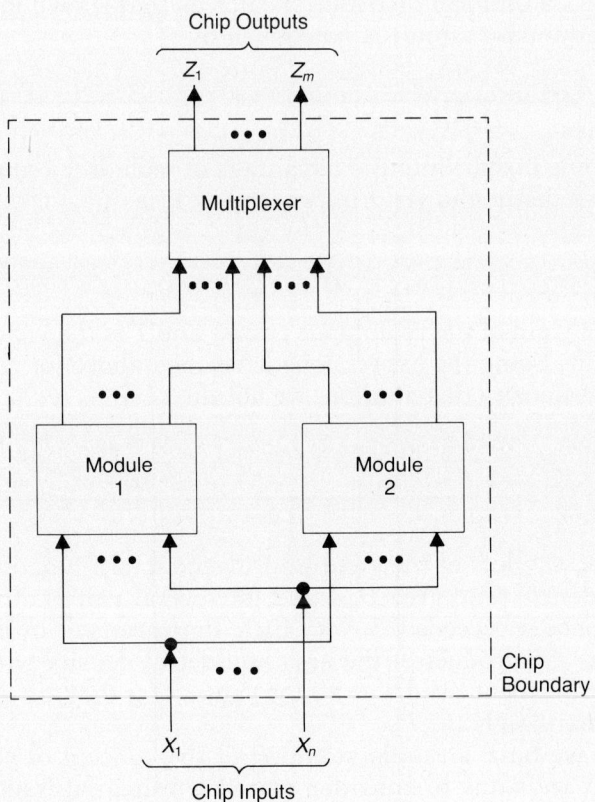

Fig. 6.32 Simple chip that includes redundancy for yield enhancement.

where p_1 and p_2 are the probabilities that modules 1 and 2 are free of defects, respectively. Once again, D is the defect density. The probability that the multiplexer is free of defects is given by

$$p_m = e^{-D\alpha A}$$

The probability that the chip will be completely operational is the probability that the multiplexer is operational and one of the two modules is operational. This probability may be written as

$$p_w = p_m (p_1 + p_2 - p_1 p_2)$$

where p_w is the probability that the chip operates properly. Using the expressions available for p_1, p_2, and p_m, we can write p_w as

$$p_w = e^{-\alpha AD}(e^{-AD} + e^{-AD} - e^{-2AD}) = e^{-\alpha AD}(2e^{-AD} - e^{-2AD})$$

Now suppose that we wish to calculate the yield of a manufacturing process for the circuit of Fig. 6.32. For simplicity, assume that the defect density D obeys a uniform distribution such as that shown in Fig. 6.30. The yield of the redundant circuit is now given by

$$Y_r = \frac{1}{D_n} \int_0^{D_n} p_w \, dD = \frac{2}{D_n A(1+\alpha)}[1 - e^{-D_n A(1+\alpha)}] - \frac{1}{D_n A(2+\alpha)}[1 - e^{-D_n A(2+\alpha)}]$$

To gain some insight into the advantage of including redundancy to improve yield, we define the yield improvement YI as

$$YI = \frac{Y_r}{Y}$$

where Y_r is the yield of the redundant circuit and Y is the yield of the nonredundant circuit. Using the expressions developed above for the yield with a uniform defect density distribution, we obtain

$$YI = \frac{Y_r}{Y} = \frac{\frac{2}{(1+\alpha)}[1 - e^{-D_n A(1+\alpha)}] - \frac{1}{(2+\alpha)}[1 - e^{-D_n A(2+\alpha)}]}{[1 - e^{-D_n A}]}$$

Figure 6.33 shows plots of the yield improvement as a function of the product of defect density and area $D_n A$ and for several values of the parameter α. Note that once α exceeds 0.5, very little improvement in the yield is obtained. In fact, the product of the area and defect density will determine if any yield improvement results at all for values of α that are greater than or equal to approximately 0.5.

Now that we have an understanding of the concept of yield and yield improvement, we want to consider more complicated issues and circuit structures. The first important issue is coverage. To use the redundancy provided on chip for yield improvement, we must first be capable of detect-

6.7 ■ Redundancy Techniques in a VLSI Design Environment

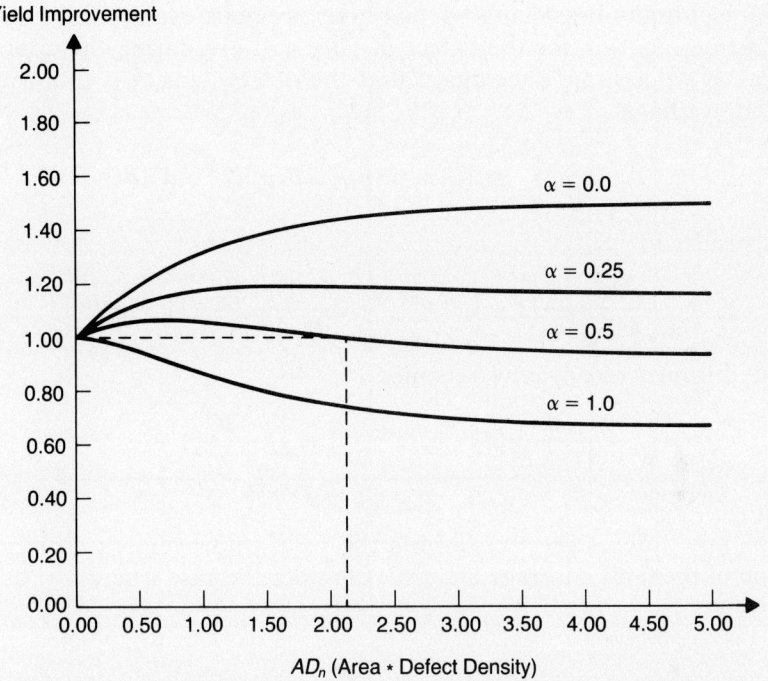

Fig. 6.33 Yield improvement as a function of AD_n and alpha.

ing and locating the failed element *and* successfully reconfiguring the circuit. As we have seen in many instances, it may not be possible to provide 100% coverage of the element failures or provide perfect reconfiguration techniques. Consequently, an accurate model of yield must account for the coverage of the fault detection and reconfiguration process.

Consider once again the example of Fig. 6.32, which has two identical modules and some multiplexing circuitry. Accounting for fault coverage, the probability that a properly operating circuit can be obtained is given by

$$p_w = p_m[C(p_1 + p_2) - p_1 p_2 (2C - 1)]$$

Note that for perfect coverage ($C = 1$), the above equation reduces to

$$p_w = p_m[(p_1 + p_2) - p_1 p_2]$$

which implies that either module working properly will result in an operational circuit, as long as the multiplexer is operational. Also note that for zero coverage the equation reduces to

$$p_w = p_m(p_1 p_2)$$

which implies that both modules must work for the circuit to work.

The yield may be calculated, as in our previous examples, by using the distribution function for the defect density and evaluating the appropriate integral. If we assume once again that the defect density is uniformly distributed, we have

$$Y_r = \frac{1}{D_n} \int_0^{D_n} p_m[C(p_1 + p_2) - p_1 p_2(2C - 1)] dD$$

which results in

$$Y_r = \frac{2C}{D_n A(1 + \alpha)}[1 - e^{-D_n A(1+\alpha)}] + \frac{1 - 2C}{D_n A(2 + \alpha)}[1 - e^{-D_n A(2+\alpha)}]$$

The yield improvement now becomes

$$YI = \frac{Y_r}{Y} = \frac{\dfrac{2C}{(1 + \alpha)}[1 - e^{-D_n A(1+\alpha)}] + \dfrac{1 - 2C}{(2 + \alpha)}[1 - e^{-D_n A(2+\alpha)}]}{[1 - e^{-D_n A}]}$$

It is now interesting to examine the yield improvement obtained as a function of the fault coverage factor C. Consider the case where $\alpha = 0.25$. Fig-

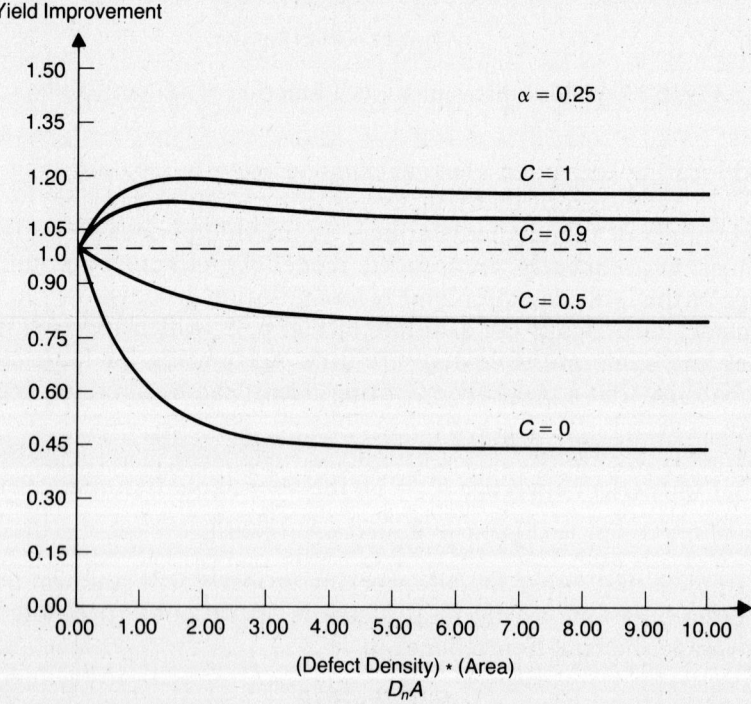

Fig. 6.34 Yield improvement as a function of $D_n A$ and coverage for $\alpha = 0.25$.

6.7 ■ Redundancy Techniques in a VLSI Design Environment

ure 6.34 shows a plot of the yield improvement as a function of both the product of defect density and area ($D_n A$) and the fault coverage factor C. Note that the coverage determines whether there is any yield improvement at all! Figures 6.35 and 6.36 show the impact of coverage and the factor α more explicitly. Both curves show the variation of yield improvement as a function of coverage and α for a fixed value of $D_n A$. Figure 6.35 shows yield improvement plotted versus coverage for several values of α. Figure 6.36 shows yield improvement plotted versus α for several values of coverage. Figure 6.35 clearly shows that a coverage factor of more than 0.5 is required to achieve any yield improvement, even in the case where $\alpha = 0$. For a more reasonable α of 0.25, a coverage of 0.7 or greater is necessary to obtain any yield improvement. Finally, Fig. 6.36 shows that perfect coverage allows α to increase to as high as approximately 0.65 before the yield improvement ceases. For a coverage of 0.9, α may be as high as approximately 0.55 before the yield improvement disappears.

Now we want to examine a more complicated and realistic design to assess the impact of redundancy to improve the yield. Suppose we consider an array of elements, as shown in Fig. 6.37. Figure 6.37 might represent an array of memory elements, processing elements, or other functional devices.

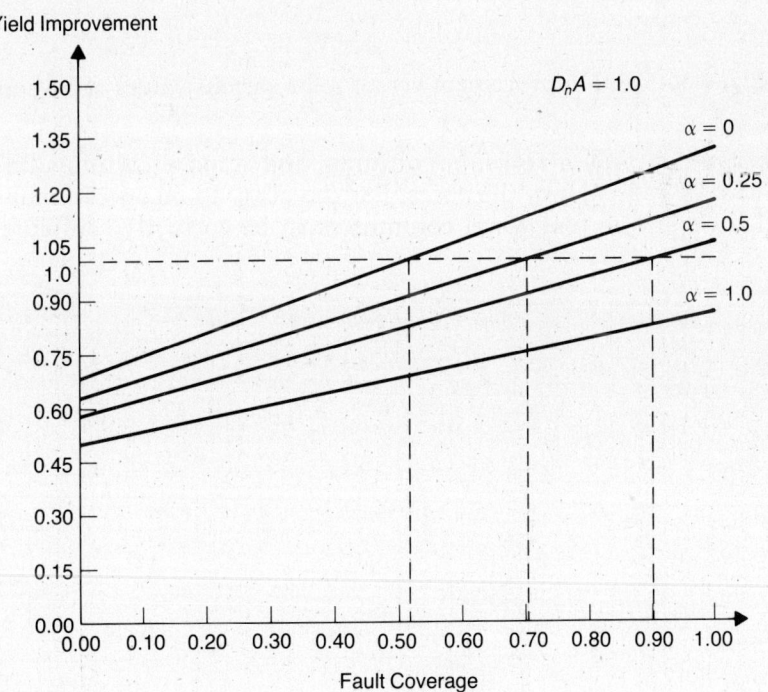

Fig. 6.35 Yield improvement versus coverage for various values of α.

448 **Fault-Tolerant Design of VLSI Circuits and Systems**

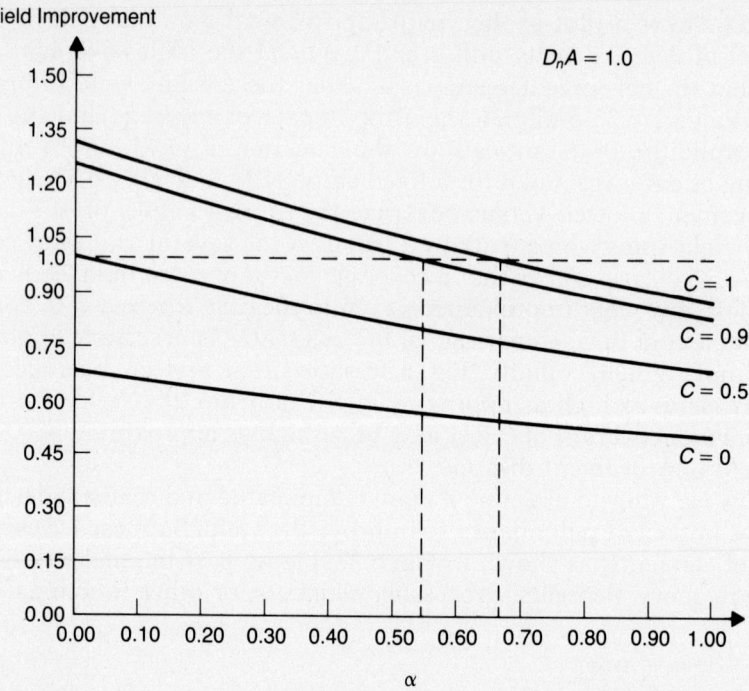

Fig. 6.36 Yield improvement versus α for various values of coverage.

The array contains n rows, m columns, and s spare columns have been provided for redundancy purposes. For the array to be operational, n functional rows and m functional columns must be correctly configured as an

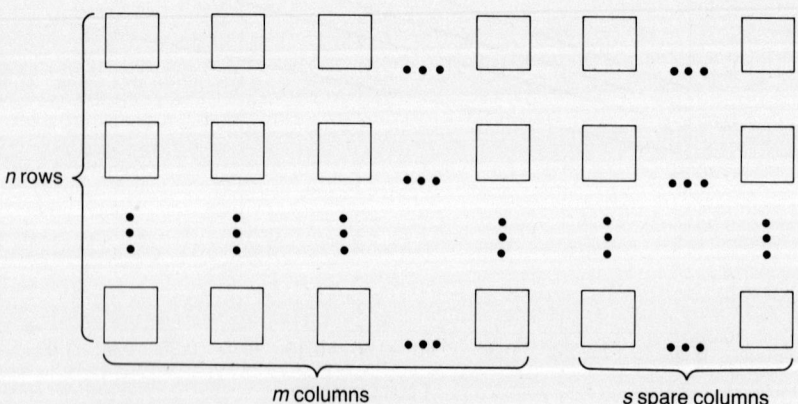

Fig. 6.37 An n-by-m array structure with s spare columns.

6.7 ■ Redundancy Techniques in a VLSI Design Environment

operational system. The objective is to replace columns containing failed elements with the spare columns so that an operational array can be constructed. We assume that each element has an area A, and the defect density will once again be represented as D. Also, we assume for simplicity that the defect density varies according to a uniform distribution with magnitude $1/D_n$ between 0 and D_n.

The probability that any one element will be defect free is

$$p_e = e^{-DA}$$

whereas the probability that a complete column of n elements will be defect free is

$$p_c = e^{-nDA}$$

Note that the probability of a complete column being defect free, p_c, is simply the probability that all n elements within that column will be defect free.

Finally, we must calculate the probability that m of the $m + s$ columns in the array will be free of defects. We know from Chapter 4 that the probability of m of the $m + s$ columns being defect free may be written as

$$p_o = \sum_{i=0}^{s} \binom{m+s}{i} p_c^{m+s-i}(1 - p_c)^i$$

where

$$\binom{m+s}{i} = \frac{(m+s)!}{(m+s-i)!\, i!}$$

p_o is the probability that there are sufficient elements to make the array completely operational. Because each column failure must be identified and successful reconfiguration performed, we must include both a coverage factor C and a control circuit. If we assume that the control circuit has an area of αA, the probability of the control circuitry being defect free is

$$p_m = e^{-\alpha DA}$$

Finally, the probability that a completely functional array is produced is given by

$$p_a = p_m \left[\sum_{i=0}^{s} \binom{m+s}{i} p_c^{m+s-i}(1 - p_c)^i C^i \right]$$

For simplicity, we will examine a 4 × 4 array containing two spare columns. The probability of obtaining a completely functional array in this case may be written as

$$p_a = p_m \left[\binom{6}{0} p_c^6 + \binom{6}{1} p_c^5 (1 - p_c) C + \binom{6}{2} p_c^4 (1 - p_c)^2 C^2 \right]$$

which reduces to
$$p_a = p_m[(1 + 15C^2 - 6C)p_c^6 + (6C - 30C^2)p_c^5 + (15C^2)p_c^4]$$
For convenience, we will rewrite the above expression as
$$p_a = p_m[C_1 p_c^6 + C_2 p_c^5 + C_3 p_c^4]$$
where
$$C_1 = 1 + 15C^2 - 6C$$
$$C_2 = 6C - 30C^2$$
$$C_3 = 15C^2$$

Using
$$p_c = e^{-4DA}$$
and
$$p_m = e^{-\alpha DA}$$
allows us to write
$$p_a = C_1 e^{-(24+\alpha)DA} + C_2 e^{-(20+\alpha)DA} + C_3 e^{-(16+\alpha)DA}$$
The yield of the redundant circuit can be calculated as
$$Y_r = \frac{1}{D_n} \int_0^{D_n} p_a \, dD$$
which becomes
$$Y_r = \frac{C_1}{D_n(24+\alpha)A}[1 - e^{-(24+\alpha)D_n A}] + \frac{C_2}{D_n(20+\alpha)A}[1 - e^{-(20+\alpha)D_n A}]$$
$$+ \frac{C_3}{D_n(16+\alpha)A}[1 - e^{-(16+\alpha)D_n A}]$$

For comparison purposes, the yield of the nonredundant circuit can be shown to be
$$Y = \frac{1}{D_n} \int_0^{D_n} e^{-16DA} \, dD = \frac{1}{16 D_n A}[1 - e^{-16 D_n A}]$$
The yield improvement can finally be written as
$$YI = \frac{Y_r}{Y}$$
$$= \frac{\frac{C_1}{(24+\alpha)}[1 - e^{-(24+\alpha)D_n A}] + \frac{C_2}{(20+\alpha)}[1 - e^{-(20+\alpha)D_n A}] + \frac{C_3}{(16+\alpha)}[1 - e^{-(16+\alpha)D_n A}]}{\frac{1}{16}[1 - e^{-16 D_n A}]}$$

Figure 6.38 shows the improvement in yield that can be obtained as a function of both the fault coverage factor C and the control circuit complexity factor α. Note that the product αA is the area of the complete control circuitry, where A is the area of one element of the array. Consequently, an α of 4.0 in this example implies that the control circuitry occupies the same area as one column of the array. Based on the information provided in Fig. 6.38, approximately a 35% improvement in yield can be obtained in this example when two spare columns are included, and the coverage factor is 0.8 and $\alpha = 4.0$.

Summary

This chapter has presented an introduction to the extremely vital topic of designing fault-tolerant systems in a VLSI environment, using the advantages of VLSI and accounting for its disadvantages. The following is a summary of the key terms and concepts introduced in this chapter.

Algorithm-Based Fault Tolerance—fault detection and fault tolerance techniques incorporated into a particular application algorithm.

Code Disjoint—under fault-free conditions valid code words on the input map into valid code words on the output.

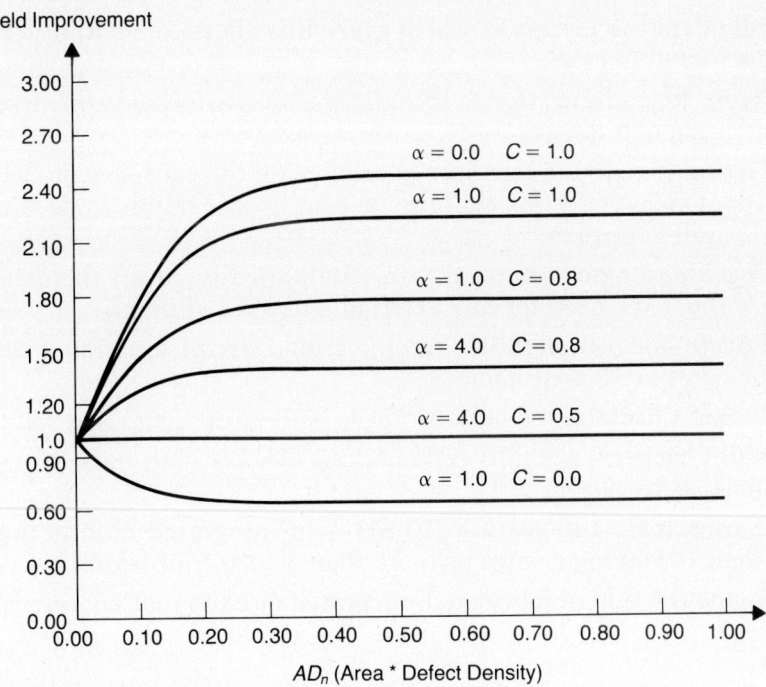

Fig. 6.38 Yield improvement versus AD_n for various values of α and c.

Common-Mode Failure—a failure occurring in two or more identical modules at exactly the same time and having exactly the same effect.

Compile-Time Reconfiguration—reconfiguration performed immediately prior to performing a desired operation.

Discrete Device—a device in which each component is individually packaged, for example, a transistor, resistor, or capacitor.

Duplication with Complementary Logic—duplication with comparison with one module designed using positive logic and the other using negative logic.

Fabrication-Time Reconfiguration—reconfiguration performed immediately after a device is fabricated.

Failure Mode—the underlying cause of failure in an integrated circuit.

Fault Secure—any single fault results in the circuit producing either the correct code word output or a non-code word, for any valid code word input.

Integrated Circuit (IC)—a single piece of semiconductor material containing more than one component and all associated wiring.

Large-Scale Integration (LSI)—an integrated circuit containing between 100 and 10,000 logic gates or between 300 and 30,000 transistors.

Medium-Scale Integration (MSI)—an integrated circuit containing between 10 and 100 logic gates or between 30 and 300 transistors.

Nonredundant—a circuit in which every line affects the output for at least one input combination.

Real-Time Reconfiguration—reconfiguration performed concurrent with the operation of a device.

Self-Testing—a circuit attribute guaranteeing the existence of at least one valid input code word that will produce an invalid output code word when a single fault is present.

Self-Checking Logic—an ability to automatically detect the existence of faults without the need for any externally applied stimulus.

Small-Scale Integration (SSI)—an integrated circuit containing fewer than 10 logic gates or 30 transistors.

Totally Self-Checking—a circuit that is both fault secure and self-testing.

Two-Rail Checker—a totally self-checking checker with complementary error signals as its output.

Very-Large-Scale Integration (VLSI)—an integrated circuit containing more than 10,000 logic gates or more than 30,000 transistors.

Yield—the fraction of fabricated integrated circuits that operate correctly.

References

1. Abraham, J.A., P. Banerjee, C-Y Chen, W.K. Fuchs, S-Y Kuo, and A.L.N. Reddy "Fault tolerance techniques for systolic arrays," *Computer*, Vol. 20, No. 7, July 1987, pp. 65–74.
2. Aichelmann, Jr., F.J. "Fault-tolerant design techniques for semiconductor memory applications," *IBM Journal of Research and Development*, Vol. 28, No. 2, March 1984, pp. 177–183.
3. Avizienis, A. "Design diversity—The challenge of the eighties," *Proceedings of the 12th Annual International Symposium on Fault-Tolerant Computing*, June 22–24, 1982, Santa Monica, Calif., pp. 44–45.
4. Bernard, J. "The IC yield problem: A tentative analysis for MOS/SOS circuits," *IEEE Transactions on Electron Devices*, Vol. ED-25, No. 8, August 1978, pp. 939–944.
5. Bossen, D.C. "A system solution to the memory soft error problem," *IBM Journal of Research and Development*, Vol. 24, No. 3, May 1980, pp. 390–397.
6. Fortes, J.A.B., and C.S. Raghavendra. "Dynamically reconfigurable fault-tolerant array processors," *Proceedings of the Fourteenth International Conference on Fault-Tolerant Computing*, Kissimmee, Fla., June 22–24, 1984, pp. 386–392.
7. Hodges, D.A., and H.G. Jackson, Analysis and design of digital integrated circuits, McGraw Hill, Inc., New York, 1988.
8. Huang, K.H., and J.A. Abraham. "Algorithm-based fault tolerance for matrix operations," *IEEE Transactions on Computers*, Vol. C-33, No. 6, June 1984, pp. 518–528.
9. Kohavi, Z. *Switching and Finite Automata Theory*, McGraw-Hill Publishing Company, New York, 1978.
10. Kung, S.Y., S.C. Lo, S.N. Jean, and J.N. Hwang. "Wavefront array processors—concept to implementation," *Computer*, Vol. 20, No. 7, July 1987, pp. 18–33.
11. Kuo, S-Y., and W.K. Fuchs. "Efficient spare allocation in reconfigurable arrays," *IEEE Design and Test of Computers*, Vol. 4, No. 1, February 1987, pp. 24–31.
12. Lala, P.K. *Fault-Tolerant and Fault-Testable Hardware Design*, Prentice-Hall International, London, 1985.
13. Mangir, T.E., and A. Avizienis. "Fault-tolerant design for VLSI: Effect of interconnect requirements on yield improvement of VLSI designs," *IEEE Transactions on Computers*, Vol. C-31, No. 7, July 1982, pp. 609–616.
14. Muroga, S. *VLSI System Design*, John Wiley and Sons, New York, 1982.
15. Negrini, R., M. Sami, and R. Stefanelli. "Fault tolerance techniques for array structures used in supercomputing," *Computer*, Vol. 19, No. 2, February 1986, pp. 78–87.
16. Peattie, C.G., J.D. Adams, S.J. Carrell, T.D. George, and M.H. Valek. "Elements of semiconductor device reliability," *Proceedings of the IEEE*, Vol. 62, No. 2,

February 1974, pp. 149–168.
17. Price, J.E. "A new look at yield of integrated circuits," *Proceedings of the IEEE*, Vol. 57, No. 8, August 1970, pp. 1290–1291.
18. Roth, Jr, C.H. *Fundamentals of Logic Design*, West Publishing Company, St. Paul, Minn., 1975.
19. Sedmak, R.M., and H.L. Liebergot. "Fault tolerance of a general purpose computer implemented by very large scale integration," *IEEE Transactions on Computers*, Vol. C-29, No. 6, June 1980, pp. 492–500.
20. Siewiorek, D.P., and R.S. Swarz. *The Theory and Practice of Reliable System Design*, Digital Press, Bedford, Mass., 1982.
21. Strapper, C.H. "Defect density distributions for LSI yield calculations," *IEEE Transactions on Electron Devices*, Vol. ED-20, 1973.
22. Strapper, C.H. "Yield modeling and process monitoring," *IBM Journal of Research and Development*, Vol. 20, pp. 228–234, 1976.
23. Strapper, C.H., et al., "Yield model for productivity optimization of product," *IBM Journal of Research and Development*, Vol. 23, May 1980, pp. 298–409.
24. Tamir, Y., and C.H. Sequin. "Reducing common mode failures in duplicate modules," *Proceedings of the 1984 IEEE International Conference on Computer Design*, Port Chester, N.Y., October 8–11, 1984, pp. 302–307.

Additional Reading

The following references provide additional reading material on the topics, as well as related ones, discussed in this chapter. These references cover a broad selection of material for the reader interested in further pursuing fault-tolerant design in a VLSI environment.

Arai, E. "VLSI fine technology and its problems," *Proceedings of the Thirteenth Conference on Solid-State Devices*, Tokyo, Japan, August 26-27, 1981, pp. 43–49.

Baille, G., L. Bergher, B. Courtois, J. Laurent, and C. Merac du Rubat. "Testing for failure analysis: New tools and new test methods," *Proceedings of the 13th Annual Symposium on Fault-Tolerant Computing*, Milan, Italy, June 28-30, 1983, pp. 266–269.

Barto, R.L. "The impact of reliability and performance requirements on VLSI design (spaceborne processor)," *Proceedings of the IEEE International Conference on Computer Design: VLSI in Computers*, Port Chester, N.Y., October 31-November 3, 1983, pp. 109–112.

Bindels, J.F.M., J.D. Chlipala, F.H. Fischer, T.F. Mantz, R.G. Nelson, and R.T. Smith. "Cost effective yield improvement in fault-tolerant VLSI memory," *Proceedings of the 1981 IEEE International Solid-State Circuits Conference*, New York, February 18-20, 1981, p. 82.

Blackley, W. S., M. A. Jack, and J. R. Jordan. "High speed VLSI digital correlator with architecture for yield enhancement and fault tolerance," *Proceedings of the Second International Conference on the Impact of High Speed and VLSI Technology on Communications Systems*, London, England, November 30-December 1, 1983, pp. 62–66.

Butner, S. E. "A constructive approach to fault tolerance in VLSI-based systems," *Proceedings of the 1981 International Conference on Parallel Processing*, Columbus, Ohio, August 25-28, 1981, pp. 264–265.

Chung, F. R. K., F. T. Leighton, and A. L. Rosenberg. "DIOGENES: A methodology for designing fault-tolerant processor arrays," *Proceedings of the 13th Annual Symposium on Fault-Tolerant Computing*, Milan, Italy, June 28-30, 1983, pp. 26–32.

Ciminiera, L., and A. Sierra. "A fault-tolerant connecting network for multiprocessor systems," *Proceedings of the 1982 International Conference on Parallel Processing*, Bellaire, Mich., August 24-27, 1982, pp. 113–122.

Disparte, C. P. "A design approach for an electronic engine controller self-checking microprocessor," *Proceedings of the Seventh EUROMICRO Symposium on Microprocessing and Microprogramming*, Paris, France, September 8-10, 1981, pp. 243–247.

Falavarjani, K. M., and D. K. Pradhan. "Fault diagnosis of parallel processor interconnection networks," *Proceedings of the Eleventh International Symposium on Fault-Tolerant Computing*, Portland, Me., June 24-26, 1981, pp. 209–212.

Frank, G., N. Kanopoulos, and R. Preister. "Yield improvement of wafer-scale integrated systolic structures via redundancy," *Proceedings of the 16th Annual IEEE Electronics and Aerospace Systems Conference and Exposition*, Washington, D.C., September 19-21, 1983, pp. 317–322.

Grosspietsch, K. E., J. Kaiser, and E. Nett. "A concept for test and reconfiguration of a fault-tolerant VLSI processor system," *Proceedings of the 7th Annual Symposium on Computer Architecture*, La Baule, France, May 6-8, 1980, pp. 37–43.

Fussell, D., and P. Varman. "Fault-tolerant wafer-scale architecture for VLSI," *Proceedings of the 9th Annual Symposium on Computer Architecture*, Austin, Tex., April 26-29, 1982, pp. 190–198.

Johnson, D., D. Budde, D. Carson, and C. Peterson. "Intel IAPX 432 — VLSI building blocks for a fault-tolerant computer," *Proceedings of the 1983 National Computer Conference*, Anaheim, Calif., May 16-19, 1983, pp. 531–537.

Koren, I. "A reconfigurable and fault-tolerant VLSI multiprocessor array," *Proceedings of the 8th Symposium on Computer Architecture*, Minneapolis, Minn., May 12-14, 1981, pp. 425–442.

Koren, I., and M. A. Breuer. "On area and yield considerations for fault-tolerant VLSI processor arrays," *IEEE Transactions on Computers*, Vol. C-33, No. 1, January 1984, pp. 21–27.

Kuhn, R. H. "Interstitial fault tolerance—A technique for making systolic arrays fault tolerant," *Proceedings of the 16th Hawaii International Conference on System Sciences*, Honolulu, Hawaii, January 5-7, 1983, pp. 215–224.

Kung, H. T., and M. S. Lam "Fault tolerance and two-level pipelining in VLSI systolic arrays," *Proceedings of the Conference on Advanced Research in VLSI*, Cambridge, Mass., January 23-25, 1984, pp. 74–83.

Lala, P. K. "An on-chip fault-tolerant design scheme," *Computer Design*, Vol. 21, No. 8, August 1982, pp. 143–146.

Lombardi, F., and V. R. Obac. "Software implemented fault tolerance: A methodology," *Microelectronics and Reliability*, Vol. 22, No. 4, 1982, pp. 873–886.

Lowry, M. R., and A. Miller. "General purpose very large scale integration (VLSI) chip for computer vision with fault-tolerant hardware," *Proceedings of SPIE—The International Society of Optical Engineers*, Vol. 281, April 1981, pp. 342–345.

McCanny, J. V., and J. C. McWhirter. "Yield enhancement of bit-level systolic array chips using fault-tolerant techniques," *Electronic Letters*, Vol. 19, No. 14, July 7, 1983, pp. 525–527.

Moore, W. R., and M. J. Day. "Yield enhancement of a large systolic array," *Microelectronics and Reliability*, Vol. 24, No. 3, 1984, pp. 511–526.

Murray, A. F., P. B. Denyer, and D. Renshaw. "Self-testing in bit serial VLSI parts: High coverage at low cost," *Proceedings of the International Test Conference*, Philadelphia, Pennsylvania, October 18-20, 1983, pp. 260–268.

Papp, Z. "A failure rate model for memory devices," *Proceedings of the IEEE International Conference on Computer Design: VLSI in Computers*, Port Chester, New York, October 31-November 3, 1983, pp. 177–180.

Pradhan, D. K. "Fault-tolerant architectures for multiprocessors and VLSI systems," *Proceedings of the 13th Annual Symposium on Fault-Tolerant Computing*, Milan, Italy, June 28-30, 1983, pp. 436–441.

Raghavendra, C. S., and T. E. Mangir. "On the VLSI implementation of fault-tolerant architectures," *Proceedings of the IEEE International Conference on Computer Design: VLSI in Computers*, Port Chester, N.Y., October 31-November 3, 1983, pp. 744–747.

Reeves, A. P., and M. J. B. Duff. "Fault tolerance in highly parallel mesh connected processors," *Proceedings of the Workshop on Multicomputers and Image Processing—Computing Structures for Image Processing*, Abingdon, England, May 1982, pp. 77–94.

Sami, M., and R. Stefanelli. "Reconfigurable architectures for VLSI processing arrays," *Proceedings of the 1983 National Computer Conference*, Anaheim, Calif., May 16–19, 1983, pp. 565–577.

Sood, A. K., F. M. El-Turky, and S. Nanayakkara. "In chip fault tolerant interconnection networks for integrated circuits," *Proceedings of IEEE SOUTHEASTCON '82*, Destin, Florida, April 4-7, 1982, pp. 403–406.

Takefuji, Y., Y. Adachi, and H. Aiso. "A novel approach to fault-tolerant sequential circuits," *Proceedings of the IEEE International Conference on Circuits and Computers*, New York, September 28-October 1, 1982, pp. 206–209.

Takefuji, Y., Y. Adachi, and H. Aiso. "A novel approach to fault-tolerant logic and yield enhancement," *Proceedings of the Twenty-fifth IEEE Computer Society International Conference (COMPCON 82)*, Washington, D.C., September 20-23, 1982, pp. 56–64.

Takefuji, Y., and M. Ikeda. "A novel approach to fault-tolerant logic," *Journal of Information Processing*, Vol. 3, No. 3, 1980, pp. 119–126.

Tamir, Y., and C.H. Sequin. "Self-checking building blocks for fault-tolerant multicomputers," *Proceedings of the IEEE International Conference on Computer Design: VLSI in Computers*, Port Chester, N.Y., October 31-November 3, 1983, pp. 561–564.

Tanner, R.M. "Fault-tolerant 256K memory designs," *IEEE Transactions on Computers*, Vol. C-33, No. 4, April 1984, pp. 314–322.

Tsao, M.M., A.W. Wilson, R.C. McGarity, T. Chia-Jeng, and D.P. Siewiorek. "C.Fast: A fault tolerant and self testing microprocessor," *Proceedings of the CMU Conference on VLSI Systems and Computations*, Pittsburgh, Pa., October 19-21, 1981, pp. 357–366.

White, Jr., J.B. "Fault-tolerant memory system architecture for radiation induced errors," *IEEE Transactions on Aerospace and Electronic Systems*, Vol. 18, No. 1, January 1982, pp. 39–47.

Woei, L., and W. Chuan-Lin. "Design of a 2 × 2 fault-tolerant switching element," *Proceedings of the 9th Annual Symposium on Computer Architecture*, Austin, Tex., April 26-29, 1982, pp. 181–189.

Problems

6.1 Using duplication with comparison and complementary logic, design a 2-bit full-adder. Recall that the 2-bit full-adder accepts as inputs two 2-bit binary words and a carry-in bit while producing a 2-bit sum and a carry-out bit as outputs. Show the complete logic diagram of the resulting circuit, including the comparator, using only NAND and NOR gates. Assume that the comparator is always fault free. For the case where the input words are 01 and 11 and the carry-in is 1, show the logical value of each line within the circuit.

6.2 The circuit shown in Fig. 6.39 is a simple, 4-bit parity checking circuit. The output of the circuit will be 0 if the 4-bit input has even parity; otherwise, the output of the circuit will be 1. Determine if this circuit is either fault secure or self-testing for single stuck-at-0 and stuck-at-1 faults. Thoroughly explain why the circuit is, or is not, fault secure or self-testing.

6.3 Design a totally self-checking two-rail comparator that may be used in the duplication with comparison circuit of problem 6.1. Show the complete logic

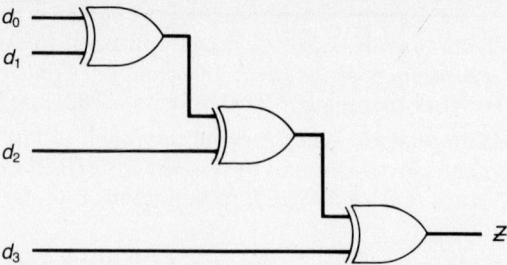

Fig. 6.39 Four-bit parity checking circuit.

diagram of the resulting circuit. Having completed the design of the comparator, use it to create a totally self-checking 2-bit full-adder using the results of Problem 6.1 as a starting point. Once again show the complete logic diagram of the circuit.

6.4 Design a totally self-checking 3-of-6 code checker using only two-input NAND gates. Show the complete logic diagram of the circuit. Illustrate the operation of the circuit using an invalid code word input and determining the output of the code checker.

6.5 Design a totally self-checking (TSC) comparator that can compare two 8-bit binary words. Construct the comparator circuit using a 2-bit totally self-checking comparator as the basic building block. Show the block diagram of the complete circuit. The block diagram should clearly show the interconnections between the 2-bit, fundamental building blocks, but it is not necessary to show gate-level implementations. How many 2-bit TSC comparators are required to complete the 8-bit TSC comparator?

6.6 Design an array that is capable of multiplying two 3 × 3 matrices and producing a 3 × 3 result matrix. Assume that you have available a basic processing element that can multiply and accumulate. In other words, the basic element can multiply two numbers and add the resulting product to a previously obtained value. How many such processing elements are required to construct the array requested in this problem? Show the sequence of events and intermediate calculations produced while multiplying the two matrices:

$$\mathbf{A} = \begin{bmatrix} 3 & 2 & 1 \\ 1 & 3 & 6 \\ 4 & 5 & 2 \end{bmatrix}$$

$$\mathbf{B} = \begin{bmatrix} 1 & 6 & 3 \\ 5 & 2 & 1 \\ 2 & 4 & 3 \end{bmatrix}$$

If each element can perform one multiply and accumulate operation per clock cycle, how many clock cycles are required to complete the multiplication of the two 3 × 3 matrices?

6.7 The array shown in Fig. 6.40 is a 6 × 5 array augmented with one spare column and possessing a number of faulty elements. Each faulty element is shaded to illustrate the fact that it is not usable. Show the interconnections that will create an operational 6 × 5 array using the rippling replacement strategy. Modify the array of Fig. 6.40, by making one additional element faulty, such that the

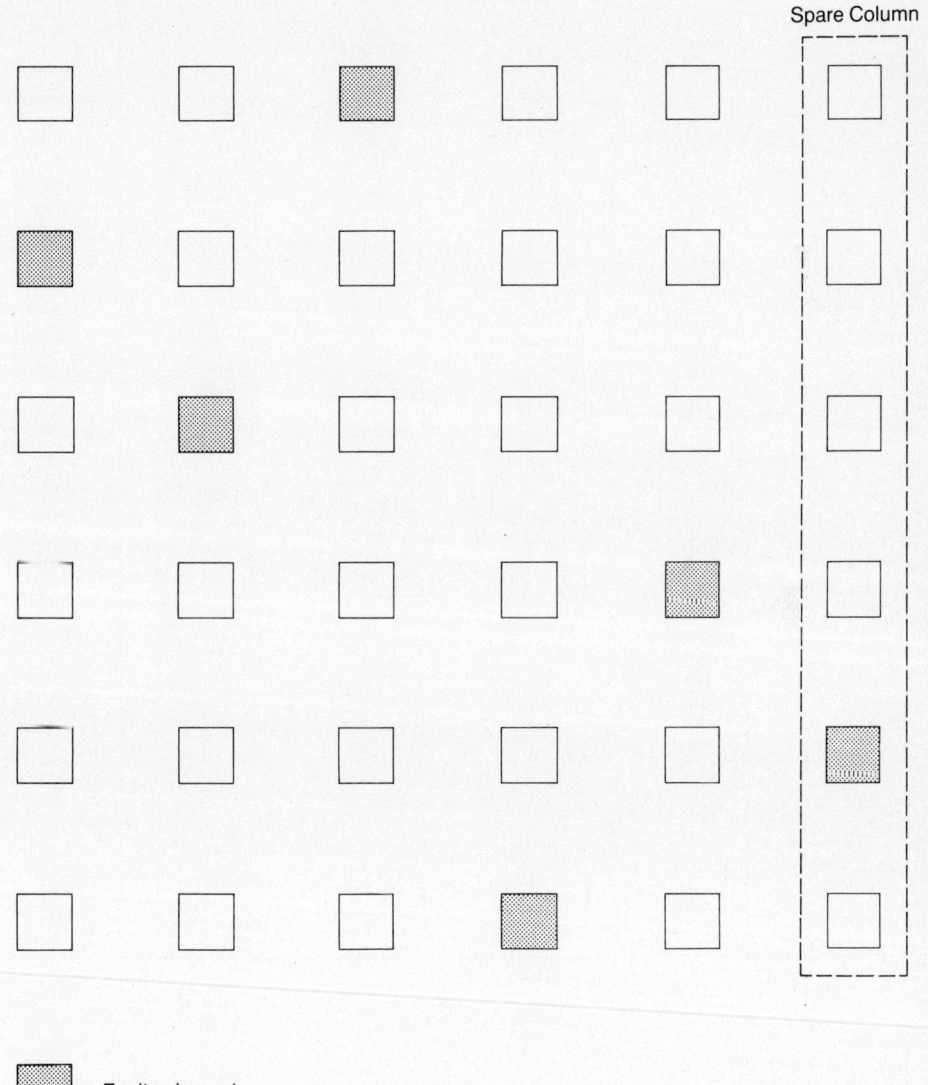

= Faulty element

Fig. 6.40 Faulty array to be reconfigured using the rippling replacement strategy.

rippling replacement strategy will not be successful in finding an interconnection pattern.

6.8 The array shown in Fig. 6.41 is a 5 × 5 array augmented with one spare row and one spare column and possessing a number of faulty elements. Each faulty

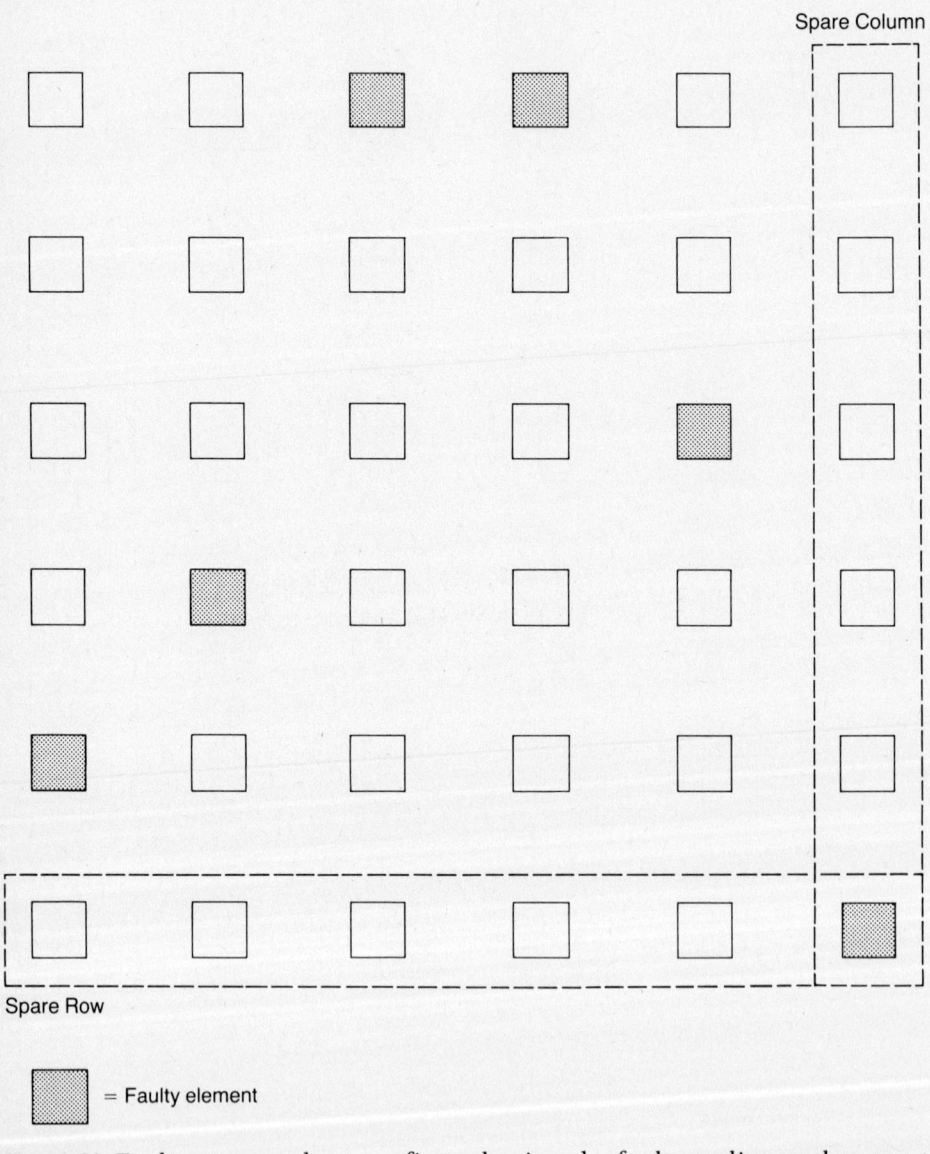

Fig. 6.41 Faulty array to be reconfigured using the fault stealing replacement strategy.

element is shaded to illustrate the fact that it is not usable. Show the interconnections that will create an operational 5 × 5 array using the fault stealing replacement strategy. Modify the array of Fig. 6.41 by making one additional element faulty, such that the fault stealing replacement strategy will not be successful in finding an interconnection pattern.

6.9 Demonstrate the use of checksums in matrix multiplication to detect, locate, and correct an error by multiplying the following two matrices. You must first augment each matrix to include the necessary checksum elements. Show the augmented matrices, the resulting matrix product, and explain how the erroneous value is located and corrected. Assume that element (2,2) of the matrix product is calculated incorrectly when demonstrating the technique.

$$A = \begin{bmatrix} 3 & 2 & 1 & 2 \\ 1 & 3 & 6 & 1 \\ 4 & 5 & 2 & 3 \\ 7 & 8 & 1 & 0 \end{bmatrix}$$

$$B = \begin{bmatrix} 1 & 6 & 3 & 2 \\ 5 & 2 & 1 & 2 \\ 2 & 4 & 3 & 1 \\ 3 & 5 & 2 & 7 \end{bmatrix}$$

6.10 Suppose that you have a device consisting of a 2 × 2 array of processing elements. Each element has an area of A, and the defect density is given by D. You may assume that the probability of any given element being defect free obeys an exponential relationship given by $p = e^{-DA}$. Derive an expression for the yield if the defect density distribution function is approximated using the triangular distribution function. Plot the yield versus the defect density for an element area of 2 cm².

6.11 Suppose that you have a single chip with area A, and the probability that the chip is defect free is given by $p = e^{-DA}$, where the defect density is given by D. It has been proposed that the density distribution function be approximated by

$$\text{PDF}(D) = \frac{e^{-D/D_0}}{D_0}$$

Derive an expression for the yield that can be expected from this process.

6.12 You are designing a memory chip that must have 10 columns and 10 rows to operate properly, and you have been given the assignment of investigating the improvement in yield that can be obtained by including redundant columns in the design. The important questions, as you well know, concern the number of redundant columns that should be used, the fault coverage required to make the approach effective, and the amount of area that can be devoted to the control circuitry. Assume that:

1. Each element of the memory chip has an area A
2. The fault coverage is represented by C

3. The defect density is represented by D and obeys a uniform distribution.
4. The area required for the control circuitry is $0.1N_r A$, where N_r is the number of redundant columns used, and A is the area of each element.
5. Each element has an exponential probability of being defect free ($p = e^{-DA}$).
6. The process has a maximum defect density of $D_n = 5.4$ defects per cm^2, and each element has an area of $A = 0.05$ cm^2.

Develop a yield analysis, including the necessary computer programs, that allows you to calculate the yield as a function of the number of spare columns and the fault coverage. If the fault coverage is perfect ($C = 1.0$), determine and plot the yield improvement as a function of the number of spare columns added.

Remember that the yield improvement is defined as the yield of the redundant circuit divided by the yield of the non-redundant circuit. Does the yield improvement continue to increase as the number of spare columns increases, or does the yield improvement reach a peak, at which point the improvement begins to decline? If the yield improvement reaches a peak, at what number of spare columns does the peak occur? If the number of spare columns is set at five, determine and plot the yield improvement as a function of the fault coverage factor. Is there a fault coverage at which the yield improvement falls below 1.0? If so, what is that fault coverage value?

7

Testing

7.1 Introduction
7.2 Fault Testing
7.3 Test Pattern Generation
7.4 Random Testing
7.5 Signature Analysis
7.6 Design for Testability
7.7 Testability Analysis
   Summary
   References
   Additional Reading
   Problems

7.1 Introduction

The testing of a digital system, or a single integrated circuit, is extremely vital to the ultimate goal of achieving high reliability, availability, safety, maintainability, fault tolerance, or some other design requirement. If it is impossible to guarantee that a system is functioning correctly when first placed into operation, it will also be impossible to predict the characteristics of the performance of that system. For example, a triple-modular redundancy system in which a single module has failed, but that failure has not been detected by the user of the system, will lack the fault tolerance and reliability that the user expects. Likewise, a company that sells integrated circuits will develop a less than desirable reputation if its products are consistently returned because of manufacturing defects that went undetected until the user placed the device into operation. As a result, it is very important that testing techniques be available and understood.

The testing process attempts to answer two very simple questions:

1. Does the device or system work?
2. Does the device or system possess its complete capability?

The first question is clearly necessary to ensure the correctness of a system before beginning to use that system. The second question can be somewhat confusing. One might argue that if a system works, it must surely have its full capability; however, this argument is particularly not true in redundant systems. A redundant system can perform its functions perfectly but have lost its capability to tolerate a fault. The lack of fault tolerance capability must be known prior to using the system. For example, the pilot of a jet aircraft would certainly like to know that one of the on-board computers has failed before starting a flight, even if the redundancy in the system has allowed correct operation to continue uninterrupted.

The critical questions in the test process can be answered by performing a variety of tests in one of several ways. Test techniques can be performed using two major approaches: **built-in test** and **external test** [Buehler and Sievers 1982]. External test techniques typically require that the device under test (DUT) be removed from its operational environment and subjected to various tests using equipment that is external to the device itself. Built-in test techniques, on the other hand, are usually incorporated into the design of the device such that testing can be performed without the need for external test equipment. Built-in test techniques can be divided into either concurrent or nonconcurrent approaches. **Concurrent test techniques** allow a device to be tested while it continues to perform its intended functions. Many of the fault detection techniques described in Chapter 3 can be used to provide concurrent, built-in test. Examples include the various coding schemes as well as duplication with comparison and other redundancy approaches. **Nonconcurrent test techniques** require that the functions of the DUT be halted before the test process can begin.

Regardless of whether the test capability is external or built in, there are three major types of tests that one would like to conduct [Breuer and Friedman 1976]. The first type of test is called a **functional test.** Functional testing attempts to verify that the DUT possesses the functional characteristics that it was intended to have. For example, functional testing would verify that a combinational circuit behaves as its truth table says it should, and a sequential circuit functions according to its state transition and output tables. The primary purpose of functional testing is to determine if design mistakes exist within the device. In other words, functional testing is testing for faults due to design mistakes.

The second type of test is called a **fault test.** Fault testing attempts to determine if faults exist within the DUT. Also, many fault tests attempt to not only detect the existence of a fault, but also to locate that fault. When considering redundant circuits, it is particularly important to understand the differences between functional and fault testing. For example, a circuit that is designed using triple modular redundancy might pass a functional test because of the fault masking properties of the system, but it could still possess a fault.

The third type of test is called a **parametric**, or **parameter test.** The purpose of the parametric test is to verify that certain parameters of the DUT are within the required ranges. An example of an important parameter is the switching time of a logic gate. A parametric test might measure switching time to confirm that it is acceptable.

Each of the three tests—functional, fault, and parametric—can be performed in either a **static** or a **dynamic** manner. In a static test, you are interested in only the steady-state, or dc, characteristics of the device's response. In other words, you would apply an input to the device and wait until the outputs have stabilized before examining the values of the outputs; in essence, the time response of the outputs is of no interest in a static test. In a dynamic test, on the other hand, the time response is of paramount importance and would be used to evaluate the DUT. A subset of the dynamic testing is the **at-speed testing** where a device is tested at, or above, its normal operating speed.

Figure 7.1 illustrates the taxonomy of testing that has been developed in these first few paragraphs. In summary, tests can be categorized according to the *type* of test performed and the *mechanism* used to perform the test. There are three major types of tests: functional, fault, and parametric. Each

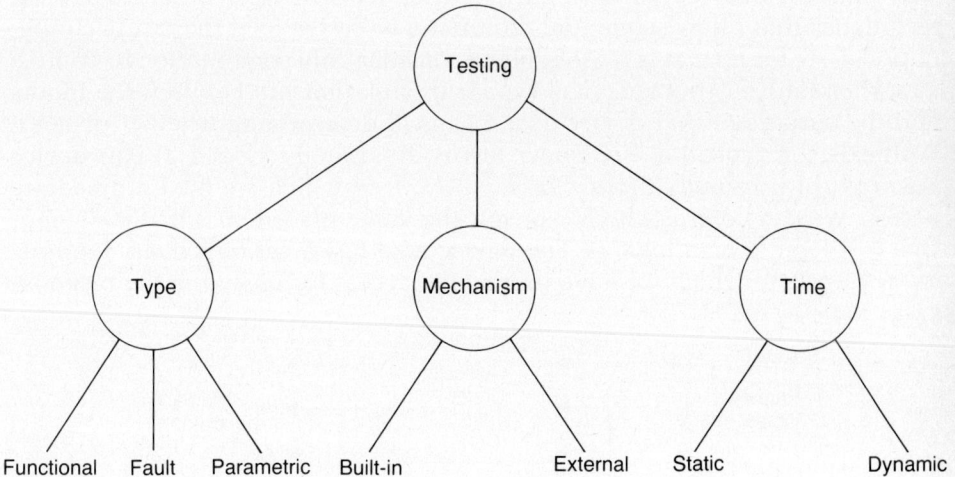

Fig. 7.1 A taxonomy of testing.

type of test can be performed in either a static or a dynamic manner. The mechanism used to perform a given test can be either external or built-in.

The primary subject of this chapter is the static fault test. We are interested in techniques that can be used to determine if faults exist within digital circuits. In addition, we will examine several ways to design digital circuits such that they are easy to test. Such approaches are referred to as **design for testability** techniques [Williams and Parker 1983]. Finally, we will discuss several methods that can be used to quantify the ease with which a digital circuit can be tested without having to actually design a test for that circuit. Such techniques are called **testability analysis** methods [McCluskey 1986].

7.2 Fault Testing

As previously discussed, the purpose of **fault testing** is to determine whether or not a fault exists in the circuit. The first question that naturally arises concerns the types of faults that you wish to detect. The test procedure will be different if you are attempting to detect stuck-open faults as opposed to stuck-at-1 or stuck-at-0 faults. For the work in this chapter, we will concentrate solely on the stuck-at-1 and stuck-at-0 faults because of the popularity and effectiveness of this fault model.

Before presenting the procedures that are available for deriving fault tests, it is important to become familiar with the terminology and assumptions. First, we will assume that all faults obey the stuck-fault model and that each circuit, or device under test, contains no more than one fault. This is the single, stuck-fault model that was described in Chapter 2. Second, we will consider only combinational circuits. We will later describe several techniques that allow sequential circuits to be tested as if they were combinational, so, for now, it is reasonable to consider only combinational circuits.

When fault testing a device, we use the relationship between the inputs and the outputs of that device as a means of determining whether or not a fault exists. A typical device under test is illustrated in Fig. 7.2. This device has a set of n primary inputs, $(x_1, x_2, \ldots, x_n)$, to which we have immediate access. We can control whether or not the value placed on a primary input line is a logic 1 or a logic 0. The device also has a set of primary outputs, $(z_1, z_2, \ldots, z_k)$, to which we have immediate access for observing the response

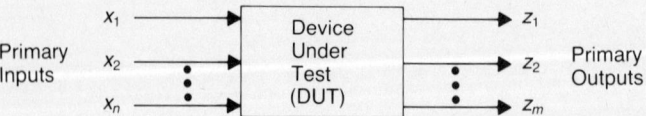

Fig. 7.2 Definition of the primary inputs and outputs of the device under test.

of the circuit. The primary inputs and primary outputs can be the pins on an integrated circuit, for example.

The terms **test, test vector, test pattern,** or **fault detection test** are used interchangeably to denote a specific, primary input pattern $(x_1, x_2, \ldots, x_n)_t$ that, when applied to the fault-free circuit, yields a primary output pattern that differs from that observed when $(x_1, x_2, \ldots, x_n)_t$ is applied to the circuit and a specific fault is present. When an input pattern causes this type of response from a circuit for a given fault, the input pattern is said to *detect* the fault. Any fault detected by a test pattern is said to be *covered* by that test pattern. The terms **experiment, fault detection experiment,** or **test experiment** are used interchangeably to denote a collection of test patterns that, once each pattern is applied to the circuit and the responses recorded, can be used to draw conclusions on whether or not a circuit contains a fault.

The process of fault testing a device requires that we first develop a fault detection experiment. This process is commonly called the **test pattern generation** process. The difficulty of generating test patterns is easily seen. If a circuit has n possible sites of faults, that circuit can have $2n$ different, single, stuck-type faults. If a separate test pattern is required to detect each possible fault, the number of test patterns can become prohibitively large as the size of the circuit increases. The objective then is to find test patterns that cover more than one fault, so that the size of the fault detection experiment can be reduced. This is the fundamental purpose of test pattern generation techniques.

For a test pattern to detect some fault within a circuit, the test pattern must accomplish two things. First, if we wish to detect a line being stuck-at-1, we must select a primary input pattern that attempts to force that line to 0. If the line is stuck-at-1 and it is suppose to be 1 anyway, we will not notice any difference in the output values. In other words, the fault will not have produced an error. Likewise, if we wish to determine if a line is stuck-at-0, we must attempt to force that line to 1. This concept is called **controllability**. The term **1-controllability** is used to denote the ability to control a line to a logic 1 value by specifying the primary inputs of the circuit. The term **0-controllability** is used to denote the ability to control a line to a logic 0 value by specifying the primary inputs of the circuit.

The second function that must be achieved by a test pattern is to propagate the effect of the fault to the primary outputs. For example, consider the two-input AND gate shown in Fig. 7.3. If we wish to detect a stuck-type fault on line A by looking at the output of the AND gate, line B should not be set to a 0 value. If line B is 0, the output of the gate is independent of the value on line A, and the fault cannot be detected. If line B is 1, however, the output of the gate will be the value that appears on line A. If this value is not as expected, the fault can be detected. This concept is called **observability** and is simply the ability to observe the value on a line internal to a circuit by looking at the primary outputs of that circuit.

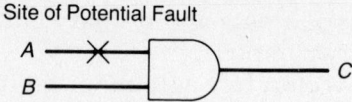

Fig. 7.3 Illustration of observability for a simple AND gate.

In summary, a test pattern for a given fault must be able to control the line on which the suspected fault occurs *and* must allow observing that line by looking at the primary outputs.

7.3 Test Pattern Generation

We will consider in this chapter several basic techniques for test pattern generation, including the fault table [Kohavi 1978], Boolean differences [Sellers, Hsiao and Bearnson 1968], literal propositions [Poage 1963], fault simulation [Goel and Moorby 1984], and the D-algorithm [Roth 1966]. Of these techniques, the D-algorithm is certainly the most widely used today because of its relative efficiency on large combinational circuits when compared to the other approaches.

The fault table technique is a *tabular* method that uses a combinational circuit's truth table to aid in the generation of test patterns. Both the Boolean difference and literal proposition methods are *algebraic* techniques that are quite good for small circuits but that lose their appeal as the size of the circuit increases. Finally, the D-algorithm is one of a class of techniques that uses **path sensitization.** The basic idea of path sensitization methods is to first select a path from the site of a potential fault to a primary output. Next, you proceed through the circuit from the site of the potential fault to the primary output, specifying the values, on lines along the path, that are required to propagate the signal value on the faulty line to a primary output. This process of propagating a signal through a circuit is called the *forward drive*. Finally, the primary inputs are determined that are necessary to produce all the signal values specified during the forward drive. This process is called the *backward trace*.

7.3.1 Fault Tables

Two types of fault tables are of interest in our discussions: the **fault detection table** and the **fault location table.** The fault detection table is used to determine test patterns for fault detection, whereas the fault location table can be used to determine test patterns that not only detect the existence of a fault, but also locate the fault.

A fault detection table is formed from the truth tables of both a fault-free circuit and a circuit that contains a fault. As an example, consider the half-adder circuit shown in Fig. 7.4. The half-adder has three inputs, A, B, and C, and a single output, Z. The output is the binary sum of the three input bits. The truth table for the fault-free circuit is also shown in Fig. 7.4. If the half-adder circuit were to contain a fault, the response of the circuit to input combinations could possibly be altered. For example, the truth table of the half-adder that contains a stuck-at-1 fault on line D is shown in Fig. 7.5. As you can see, the output of the circuit with the stuck-at-1 fault on line D differs from that of the fault-free circuit for the input combinations, 000, 001, 110, and 111. Each of these four primary input combinations can detect the stuck-at-1 fault on line D because each forces the circuit to yield an erroneous output. Likewise, note that the input patterns 010, 011, 100, and 101 are not test patterns for line D stuck-at-1 because the circuit produces correct results for those input combinations, even when the fault is present.

A fault detection table is a tabular method of identifying which input combinations detect a certain fault. To create the fault detection table, you first determine the output of the circuit for each of the single faults that can occur. Continuing with the half-adder as an example, we see that there are five possible lines that can be faulty, and each line can contain one of two possible faults. Therefore, ten single faults could occur. Table 7.1 shows the outputs of the half-adder when each of the ten faults is present in the circuit. The output of the circuit when no fault is present is also shown. The notation L_0 is used to denote the condition of some line L being stuck-at-0, and L_1 denotes the condition of some line L being stuck-at-1. The notation

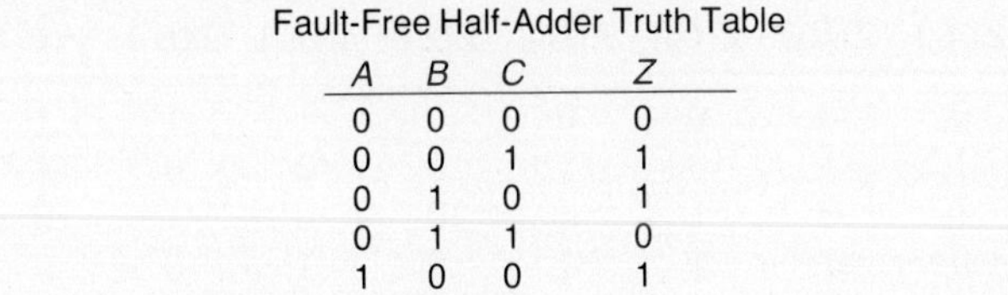

Fault-Free Half-Adder Circuit

Fault-Free Half-Adder Truth Table

A	B	C	Z
0	0	0	0
0	0	1	1
0	1	0	1
0	1	1	0
1	0	0	1
1	0	1	0
1	1	0	0
1	1	1	1

Fig. 7.4 A fault-free half-adder circuit and its corresponding truth table.

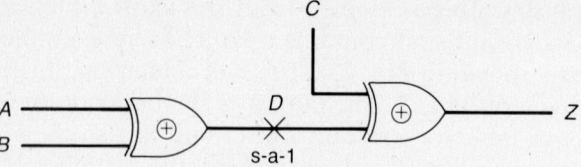

Faulty Half-Adder Circuit

Half-Adder Truth Table With D s-a-1

A	B	C	Z
0	0	0	1
0	0	1	0
0	1	0	1
0	1	1	0
1	0	0	1
1	0	1	0
1	1	0	1
1	1	1	0

Fig. 7.5 A half-adder circuit and corresponding truth table when line D is stuck-at-1.

$Z(L_0)$ represents the output when the fault L_0 is present in the circuit, and $Z(L_1)$ represents the output when the fault L_1 is present in the circuit.

The truth table of Table 7.1 can be used to construct the fault detection table. The basic purpose of the fault detection table is to identify the input patterns that detect specific faults. We already know that an input pattern will detect a fault if the output of the fault-free circuit for that particular input pattern differs from the output of the faulty circuit for that same input pattern. The fault detection table, therefore, must possess one row for each primary input combination and one column for each possible fault. The en-

TABLE 7.1 Truth table showing the effect of each single fault on the output of the half-adder

A	B	C	Z	$Z(A_0)$	$Z(B_0)$	$Z(C_0)$	$Z(D_0)$	$Z(Z_0)$	$Z(A_1)$	$Z(B_1)$	$Z(C_1)$	$Z(D_1)$	$Z(Z_1)$
0	0	0	0	0	0	0	0	0	1	1	1	1	1
0	0	1	1	1	1	0	1	0	0	0	1	0	1
0	1	0	1	1	0	1	0	0	0	1	0	1	1
0	1	1	0	0	1	1	1	0	1	0	0	0	1
1	0	0	1	0	1	1	0	0	1	0	0	1	1
1	0	1	0	1	0	1	1	0	0	1	0	0	1
1	1	0	0	1	1	0	0	0	0	0	1	1	1
1	1	1	1	0	0	0	1	0	1	1	1	0	1

try in the fault detection table at the intersection of a row and a column will be 1 if the input combination corresponding to that row detects the fault corresponding to that column. Otherwise, the entry will be 0.

To form each column of the fault detection table, you need only compare the output of the fault-free circuit to that of the circuit with a particular fault. If a difference appears in some row, a 1 is placed in that same row of the fault detection table. A method that facilitates this process is to simply form the EXCLUSIVE-OR of the fault-free output column with the output column when a given fault is present. The result creates the column in the fault detection table for that particular fault.

As an example, Table 7.2 shows the fault detection table for the half-adder circuit. Again, the appearance of a 1 at the intersection of a row and a column implies that the input combination corresponding to that row detects the fault that corresponds to that column. In other words, the input combination corresponding to the row produces differing outputs in the fault-free case and the case where the particular fault is present.

Once the fault detection table has been created, we know which input patterns will detect which faults. The objective then is to use that information to derive a fault detection experiment. We would like to develop the fault detection experiment such that all possible single faults are detected, but, at the same time, the number of test patterns used is as low as possible. This is simply a *covering* problem similar to that found when we minimize a switching function. We want to select a set of test vectors that detects all possible single faults that can occur but uses as few input combinations as possible. A fault detection experiment is said to *cover* a fault detection table if and only if each fault that appears in the fault detection table is detected by at least one of the test patterns that appears in the fault detection experiment.

If we are only interested in fault detection, we can simplify the process of covering a fault detection table by first reducing the fault detection table. First, we say that a test pattern TP_i is *covered* by another test pattern TP_j if

TABLE 7.2 Fault detection table for the half-adder circuit

A	B	C	A_0	B_0	C_0	D_0	Z_0	A_1	B_1	C_1	D_1	Z_1
0	0	0	0	0	0	0	0	1	1	1	1	1
0	0	1	0	0	1	0	1	1	1	0	1	0
0	1	0	0	1	0	1	1	1	0	1	0	0
0	1	1	0	1	1	1	0	1	0	0	0	1
1	0	0	1	0	0	1	1	0	1	1	0	0
1	0	1	1	0	1	1	0	0	1	0	0	1
1	1	0	1	0	0	0	0	0	0	1	1	1
1	1	1	1	1	1	0	1	0	0	0	1	0

and only if every fault that is detected by TP_i is also detected by TP_j. At a minimum, TP_j can detect any fault that can be detected by TP_i, so it is unnecessary to have both TP_i and TP_j. A fault detection table can be reduced by removing any row that corresponds to a test vector that is covered by any other test vector. This reduces the total number of test vectors that we have to select from for our fault detection experiment.

The number of columns in a fault detection table can also be reduced. A fault f_i is said to be *covered* by another fault f_j if and only if the set of test patterns that detects f_j is a subset of the set of test patterns that detects f_i. This means that any test pattern that is selected to detect f_j will also detect f_i because the set of test patterns that detects f_j is a subset of the test patterns that detect f_i. The column corresponding to a fault that is covered by some other fault can be removed from the fault detection table because selecting a test pattern for the covering fault automatically provides detection of the covered fault.

Once a fault detection table has been reduced, the process of covering the table is begun by first identifying the **essential test patterns.** A test pattern is *essential* if there exists at least one fault that is detected by no other test pattern. An essential test pattern must be included in any fault detection experiment that covers the fault detection table. The essential test patterns are easily spotted in the fault detection table by locating the columns in the table that have only a single 1 entry. The row in which the single 1 entry occurs corresponds to an essential test pattern. The faults detected by the essential test patterns are checked off, and nonessential test patterns are selected until the remaining faults are covered.

As an illustration of the creation, reduction, and covering of a fault detection table, we will develop a fault detection experiment for the simple combinational circuit shown in Fig. 7.6. Figure 7.6 also shows the truth

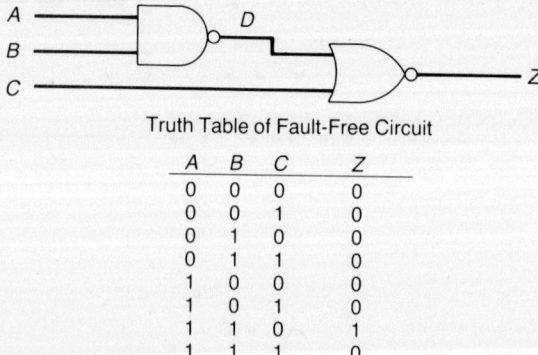

Fig. 7.6 A simple combinational circuit and its truth table for use in illustrating the reduction of a fault detection table.

TABLE 7.3 Truth table showing the response of the fault-free and faulty circuits for the circuit of Fig. 7.6

A	B	C	Z	$Z(A_0)$	$Z(B_0)$	$Z(C_0)$	$Z(D_0)$	$Z(Z_0)$	$Z(A_1)$	$Z(B_1)$	$Z(C_1)$	$Z(D_1)$	$Z(Z_1)$
0	0	0	0	0	0	0	1	0	0	0	0	0	1
0	0	1	0	0	0	0	0	0	0	0	0	0	1
0	1	0	0	0	0	0	1	0	1	0	0	0	1
0	1	1	0	0	0	0	0	0	0	0	0	0	1
1	0	0	0	0	0	0	1	0	0	1	0	0	1
1	0	1	0	0	0	0	0	0	0	0	0	0	1
1	1	0	1	0	0	1	1	0	1	1	0	0	1
1	1	1	0	0	0	1	0	0	0	0	0	0	1

table of the fault-free circuit. This circuit has three inputs, a single output, and five distinct lines that can contain faults.

As we have discussed, the first step in the process of developing the fault detection table is to determine the output of the circuit when each of the ten possible single faults is present in the circuit. Table 7.3 shows the response of the circuit when each possible fault is present. The fault detection table is then created by forming the EXCLUSIVE-OR of each output column in Table 7.3 with the output of the fault-free circuit. The result is the unreduced fault detection table shown in Table 7.4.

Several interesting features about the fault detection table are shown in Table 7.4. First, note that each of the ten possible faults is detectable. The second feature that we notice is that columns A_0, B_0, Z_0, C_1, and D_1 are equivalent, so, by definition, each of these five columns covers the remaining four. Therefore, four of the five columns can be removed to reduce the fault detection table. The third item that we notice is that the test pattern 010 covers the test patterns 000, 001, 011, and 101, so rows 000, 001, 011, and 101 can be removed from the fault detection table to provide further reduction. Next, we see that column D_0 is covered by A_1 and B_1 and can be re-

TABLE 7.4 Fault detection table for the circuit of Fig. 7.6

A	B	C	A_0	B_0	C_0	D_0	Z_0	A_1	B_1	C_1	D_1	Z_1
0	0	0	0	0	0	1	0	0	0	0	0	1
0	0	1	0	0	0	0	0	0	0	0	0	1
0	1	0	0	0	0	1	0	1	0	0	0	1
0	1	1	0	0	0	0	0	0	0	0	0	1
1	0	0	0	0	0	1	0	0	1	0	0	1
1	0	1	0	0	0	0	0	0	0	0	0	1
1	1	0	1	1	0	0	1	0	0	1	1	0
1	1	1	0	0	1	0	0	0	0	0	0	1

TABLE 7.5 The reduced fault detection table for the circuit of Fig. 7.6

A	B	C	F	C_0	A_1	B_1
0	1	0	0	0	1	0
1	0	0	0	0	0	1
1	1	0	1	0	0	0
1	1	1	0	1	0	0

F represents the equivalent faults $A_0, B_0, Z_0, C_1,$ and D_1.

moved. Finally, column Z_1 is covered by C_0, A_1, and B_1 and can also be removed. The resulting reduced fault detection table is shown in Table 7.5.

Each of the four remaining test patterns is an essential test pattern and must be included in the fault detection experiment. The fault detection experiment consists of the test patterns 010, 100, 110, and 111. To verify that each of the ten possible faults is detected by at least one of these test patterns, examine the unreduced fault detection table shown in Table 7.4. From Table 7.4, we see that the test pattern 010 detects the faults D_0, A_1, and Z_1. The test pattern 100 detects D_0, B_1, and Z_1. The test pattern 110 detects $A_0, B_0, Z_0, C_1,$ and D_1. Finally, the test pattern 111 detects C_0 and Z_1. As you can see, each fault is detected by at least one of the four test patterns.

The process of performing the fault detection experiment requires that each of the test patterns found during test pattern generation be applied to the circuit and the responses of the circuit recorded. Once all the responses are determined, they are compared to those expected from the circuit. Any discrepancies are indicative of a fault in the circuit. A *fault dictionary*, a convenient aid in performing a fault detection experiment, shows the response of the circuit to each test pattern under fault-free circumstances and when each possible fault is present. The fault dictionary for the present example is shown in Table 7.6.

In many instances, it is necessary to not only detect the existence of a fault, but also to determine the specific location of the fault. The process of locating a fault is called fault location, and the fault location table is a use-

TABLE 7.6 Fault dictionary for the circuit of Fig. 7.6.

A	B	C	Z	Z(F)	$Z(C_0)$	$Z(D_0)$	$Z(A_1)$	$Z(B_1)$	$Z(Z_1)$
0	1	0	0	0	0	1	1	0	1
1	0	0	0	0	0	1	0	1	1
1	1	0	1	0	1	1	1	1	1
1	1	1	0	0	1	0	0	0	1

F represents the equivalent faults $A_0, B_0, Z_0, C_1,$ and D_1.

ful aid to determine test patterns that achieve fault location. The fault location table is based on the premise that to determine which fault is in a circuit we must be able to distinguish each fault from all the others by the response of the circuit to specific input patterns. Once again, we use the single, stuck-fault model, and we attempt to locate which line of the circuit is stuck at either a 1 or a 0.

The fault location table contains all the rows and columns of the unreduced fault detection table. Columns (or rows) that are covered by other columns (or rows) cannot be removed from the fault detection table once we begin to consider fault location because they often provide information that allows us to distinguish between two faults. For example, recall that if test pattern A covers test pattern B, the faults detected by B are a subset of the faults detected by A. For fault detection, B is unnecessary because A detects any fault that B can detect. For fault location, however, A and B can be used to locate the fault; if A detects the fault and B does not, we then know that the fault is one of those detected by A but not detected by B. This can identify the fault specifically (if A detects only one more fault than B) or at least considerably narrow the list. Columns (or rows) that are *equivalent*, however, can be combined into a single column (or row) because no information is lost.

In addition to the rows and columns of the fault detection table, the fault location table contains one column for each pair of nonequivalent faults. Each new column is created by forming the EXCLUSIVE-OR of all possible pairs of columns of the unreduced fault detection table. For example, suppose that L_0 and M_1 are two columns of the unreduced fault detection table. In addition to containing columns L_0 and M_1, the fault location table contains a new column L_0M_1. The new column is created by performing the EXCLUSIVE-OR of columns L_0 and M_1. The resulting column has a 1 in each row if the test pattern corresponding to that row detects one of the two faults, but not the other. Any test pattern that detects one of two faults but not the other can be used to distinguish between those two faults.

As an example of the fault location table, consider again the circuit of Fig. 7.6. The unreduced fault detection table for this circuit is shown in Table 7.7. Table 7.7 has not been reduced, however, equivalent columns (or rows) have been combined into a single column (or row). As you can see, there are six unique columns in the fault detection table of Table 7.7. Since the fault location table contains the six columns of Table 7.7 and one additional column for each possible pair of those six columns, the fault location table will contain a total of 21 columns. Each of the 15 new columns is obtained by forming the EXCLUSIVE-OR of each pair of columns. The resulting fault location table is shown in Table 7.8.

A **fault location experiment** is a set of test vectors that covers the fault location table. The concept of covering the fault location table is identical to that of covering a fault detection table; a set of test patterns must be se-

TABLE 7.7 The unreduced fault detection table for the circuit of Fig. 7.6.

A	B	C	F	C_0	D_0	A_1	B_1	Z_1
0	0	0	0	0	1	0	0	1
0	0	1	0	0	0	0	0	1
0	1	0	0	0	1	1	0	1
1	0	0	0	0	1	0	1	1
1	1	0	1	0	0	0	0	0
1	1	1	0	1	0	0	0	1

F represents the equivalent faults A_0, B_0, Z_0, C_1, and D_1.

lected that is capable of detecting each fault and distinguishing between any two faults. The first step in covering the fault location table is to identify the essential tests. The essential tests are those tests that detect a fault, or distinguish between a pair of faults, that no other test pattern detects or distinguishes. In the fault location table of Table 7.8, we see that the input patterns 010, 100, 110, and 111 are essential test patterns because they are the only patterns that detect A_0, C_0, A_1, and B_1, respectively. A test pattern covers a fault if that test pattern detects that fault. Likewise, a test pattern covers a fault pair if that test pattern distinguishes those two faults. As you can see, the fault location table of Table 7.8 is completely covered by the essential tests. Therefore, the fault location experiment contains the test patterns 010, 100, 110, and 111.

The fault dictionary can be used to help locate faults. Recall that the fault dictionary is nothing more than the response of the circuit to the test patterns under fault-free conditions and when the specific faults are present. The fault dictionary for the circuit of Fig. 7.6 is shown in Table 7.6. As you can see from Table 7.6, the four responses of the circuit to the four test patterns are unique for each of the six possible faults that can exist and the fault-free case. Therefore, you can determine from the responses exactly which of the six faults is present in the circuit, if indeed a fault is present. For example, if we apply the test patterns 010, 100, 110, and 111 and obtain output responses of 0, 0, 1, and 0, respectively, we know that the circuit is fault free. However, if the output responses are 1, 1, 1, and 0, respectively, we can identify the fault as D_0 because that is the only fault that yields that particular output response.

The fault dictionary of Table 7.6 clearly illustrates the minimum number of test patterns that must be used to uniquely locate distinguishable faults within a circuit. Suppose that a circuit has k possible distinguishable faults that can occur. To be able to detect and distinguish the k faults, the test patterns must be capable of producing $k + 1$ unique output combinations in response to the set of n test patterns. The $k + 1$ unique output com-

7.3 ■ Test Pattern Generation

TABLE 7.8 Fault location table for the circuit of Fig. 7.6

A	B	C	F	C_0	D_0	A_1	B_1	Z_1	FC_0	FD_0	FA_1	FB_1
0	0	0	0	0	1	0	0	1	0	1	0	0
0	0	1	0	0	0	0	0	1	0	0	0	0
*0	1	0	0	0	1	1	0	1	0	1	1	0
*1	0	0	0	0	1	0	1	1	0	1	0	1
*1	1	0	1	0	0	0	0	0	1	1	1	1
*1	1	1	0	1	0	0	0	1	1	0	0	0

* = Essential test patterns

FZ_1	C_0D_0	C_0A_1	C_0B_1	C_0Z_1	D_0A_1	D_0B_1	D_0Z_1	A_1B_1	A_1Z_1	B_1Z_1
1	1	0	0	1	1	1	0	0	1	1
1	0	0	0	1	0	0	1	0	1	1
1	1	1	0	1	0	1	0	1	0	1
1	1	0	1	1	1	0	0	1	1	0
1	0	0	0	0	0	0	0	0	0	0
1	1	1	1	0	0	0	1	0	1	1

binations include one unique combination for each possible fault and one unique combination for the fault-free case. If there are n test patterns, there are 2^n possible combinations of outputs that can result from the application of all n test patterns. To locate the k faults, we must have

$$2^n \geq k + 1$$

to allow the n test patterns to produce at least $k + 1$ unique responses. This represents an absolute lower bound on the number of test patterns required for fault location. Certainly, the structure of the circuit can mandate that even more test patterns be used to achieve fault location.

7.3.2 Adaptive Experiments

The test patterns developed in the previous example form a preset fault location experiment. In a **preset experiment,** all the test patterns of the experiment are applied to the circuit and the responses recorded. Once all test patterns are applied, the responses are analyzed to determine if a fault exists and, if so, which fault exists. In many cases, the use of the preset test philosophy requires the application of more test patterns than necessary to locate a specific fault. For example, if we examine the fault dictionary of Table 7.6, we see that if the response of the circuit to the test patterns 010 and 100 is 1 and 0, respectively, we know immediately that the fault is A_1

because no other fault results in that particular response from the circuit for those two test patterns. Likewise, if the response to 010 and 100 is 0 and 1, respectively, the fault is B_1. What we see here is that, in many cases, it is unnecessary to apply the complete test set to detect and locate the fault, provided that we examine the response of the circuit following each test pattern. This type of test experiment is known as an **adaptive experiment.**

The underlying concept of the adaptive experiment is that we use the response of the circuit following the application of each test pattern to determine the next course of action. The appropriate action can be to terminate the test process because we know immediately which fault is contained within the circuit or to select the next test pattern based on the response of the circuit to the previous test pattern.

A convenient tool for describing an adaptive experiment is a directed graph known as a tree. The tree contains a collection of nodes that each represent a single test pattern. At the top of the tree is the *root node* that represents the first test pattern that will be applied to the circuit during the experiment. At the bottom of the tree are the *leaves* that represent the faults that can exist within the circuit. Also, there is one leaf to represent the case where the circuit is fault free. With the exception of the leaves, which are terminating points, each node has two *branches* emanating from it. Each branch represents one of the two possible responses, pass or fail, that a circuit can have to the test pattern corresponding to the given node. If the circuit passes (meaning that the correct output is obtained for that test pattern) some test, T_i, one of the two branches directs the test process to test pattern T_j. If the circuit fails T_i, the other branch directs the test process to some other test pattern, T_k. A tree in which each node has no more than two emanating branches is called a *binary tree*.

As an example of an adaptive experiment, return to the simple circuit of Fig. 7.6. The unreduced (although equivalent faults have been combined) fault detection table for the circuit of Fig. 7.6 is shown in Table 7.7. An adaptive experiment for this circuit is shown in Fig. 7.7. The first test pattern to be applied to the circuit is 010. If the circuit fails this test pattern, we immediately know that one of the three faults, D_0, A_1, and Z_1 is contained within the circuit because these are the only three faults detected by the test pattern 010. If the circuit passes this test pattern, the circuit is either fault free or contains one of the three faults that are not detected by 010. If the circuit passes the test pattern 010, the next test pattern to be applied is 110. Prior to the application of the test pattern 110, we know that the circuit is either fault free or contains one of the faults F, C_0, or B_1 because of the results of test pattern 010. If the circuit fails the test 110, the fault is immediately identified as F because none of the other possible faults is detected by 110. The test procedure continues in this manner until the circuit is identified as fault free or a specific fault is located.

7.3 ■ Test Pattern Generation 479

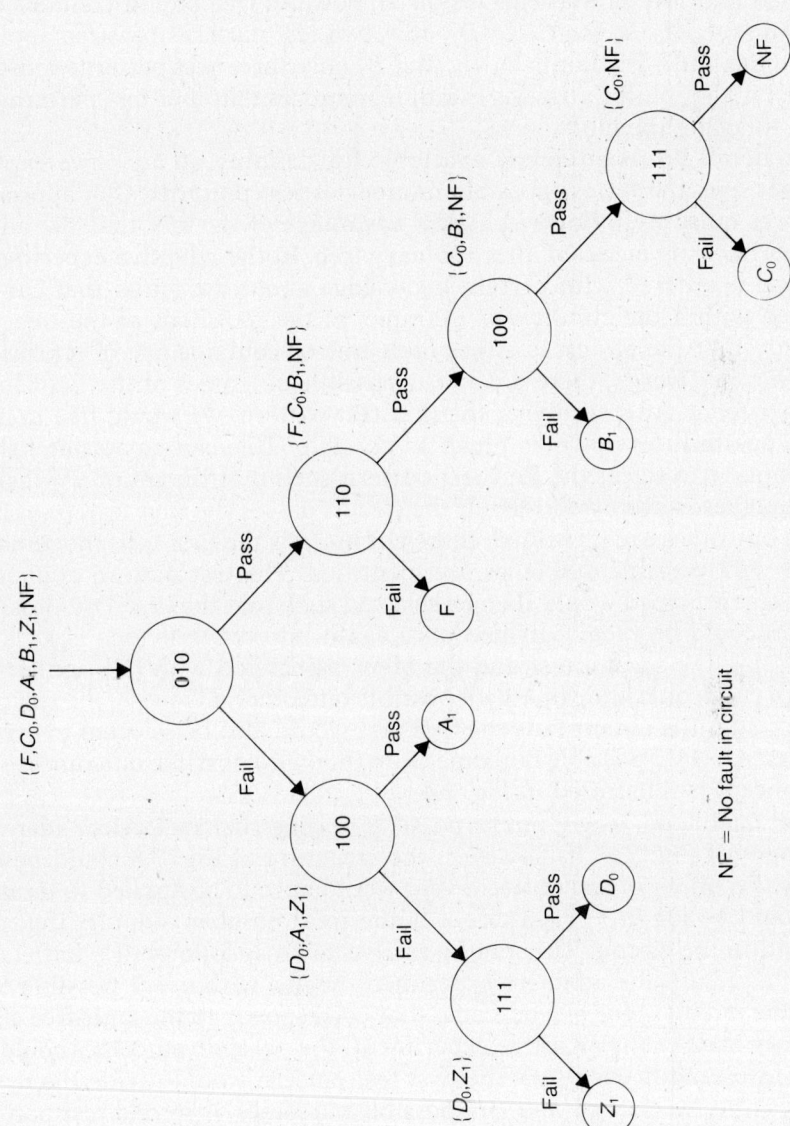

Fig. 7.7 An adaptive experiment for the circuit of Fig. 7.6.

The interesting feature of the adaptive experiment is that no more than four test patterns need ever be applied to make a decision on the correctness of the circuit. If the circuit is indeed fault free, four test patterns must be applied to arrive at that conclusion. If, however, the circuit contains one of the faults A_1, A_0, B_0, Z_0, C_1, or D_1, only two test patterns must be applied to locate the fault. For faults Z_1, D_0, and B_1 only three test patterns must be applied. Finally, fault C_0, in this example, requires that four test patterns be applied to locate the fault.

In general, no technique is available for defining an adaptive experiment that uses the fewest possible number of test patterns. One approach that works quite well, however, takes advantage of the fact that the adaptive experiment is modeled after a binary tree. In the adaptive experiment, each node is entered with certain knowledge about the faults that can be contained within the circuit. For example, at the root node of the tree, we know only that the circuit is either fault free or contains one of n possible faults. In other words, there are $n + 1$ possible outcomes of the fault location experiment. After applying the first test pattern, we would like to narrow the possibilities down as much as possible. This can be accomplished by attempting to select the first test pattern such that the set of $n + 1$ possible outcomes is evenly divided. If the circuit passes the first test, $(n + 1)/2$ possible outcomes are identified. If the circuit fails the first test, the remaining $(n + 1)/2$ possible outcomes are identified. The test pattern applied if the first test is passed would then be selected such that the $(n + 1)/2$ possible outcomes could be evenly divided once again into two sets (one associated with passing the second test and the other associated with failing the second test), each containing $(n + 1)/4$ possible outcomes. Likewise, the test pattern applied if the circuit fails the first test would also be selected to evenly divide its $(n + 1)/2$ possible outcomes into $(n + 1)/4$ possible outcomes. This basic concept is illustrated in Fig. 7.8.

In all likelihood, it will not be possible to construct the perfect adaptive experiment of Fig. 7.8. If, however, the structure of Fig. 7.8 could be obtained, the number of test patterns that would have to be applied to locate a fault would be $\log_2(n + 1)$, where n is the total number of faults that can occur within the circuit. This can be easily shown as follows. If n faults can occur in a circuit, the adaptive experiment begins with $n + 1$ possible outcomes; the n faults plus one outcome that corresponds to the fault-free case. If, at every node of the adaptive experiment, the possible outcomes could be divided into exactly even sets, the first test pattern would divide the $n + 1$ outcomes into two sets of $(n + 1)/2$ possible outcomes. A second test pattern would then divide the $(n + 1)/2$ possible outcomes into two sets, each containing $(n + 1)/4$ possible outcomes. A third test pattern would then divide the $(n + 1)/4$ possible outcomes into two sets, each containing $(n + 1)/8$ possible outcomes. In general, the kth test pattern will produce two sets, each containing $(n + 1)/2^k$ possible outcomes. The experiment ends when

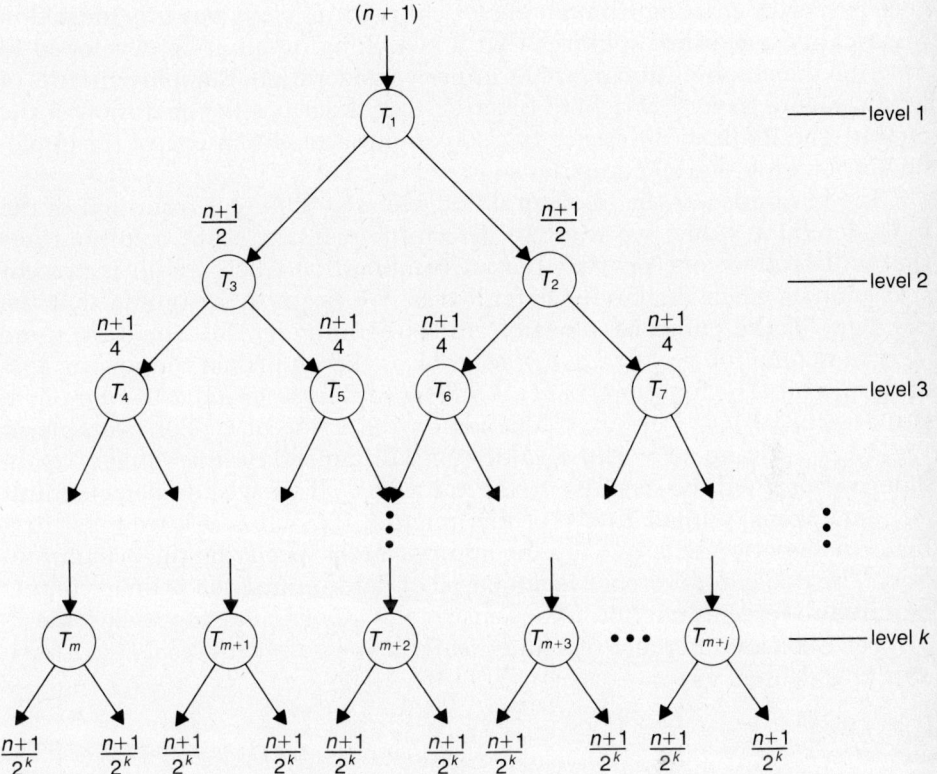

Fig. 7.8 The perfect adaptive experiment.

$$\frac{(n+1)}{2^k} = 1$$

or

$$2^k = n + 1$$

Solving the above equation for k yields

$$k = \log_2(n+1)$$

As an example computation, an ideal adaptive experiment for a circuit containing 1000 potential faults would result in the application of approximately 10 test patterns.

7.3.3 Boolean Differences

As mentioned previously, the **Boolean difference** is an algebraic technique for developing test patterns for combinational circuits. The Boolean difference is very interesting from an analytical point of view, but it is limited in practicality because it requires that a switching function be developed to describe the combinational circuit under consideration. Complex circuits of today seldom have a switching function that describes the operation of the circuit. The Boolean difference is also computationally intensive for implementation on a digital computer.

The basic idea of the Boolean difference is no different from that of the fault detection table; we want to determine primary input combinations that will produce one primary output combination when a fault is present and another when the circuit is fault free. We begin by assuming that the behavior of the combinational circuit is completely described by some switching function $f(x_1, x_2, \ldots, x_n)$, where x_i is the ith primary input variable of the circuit. The function $f(x_1, x_2, \ldots, x_n)$ is said to be *sensitive* to an input x_i if the value of $f(x_1, x_2, \ldots, x_n)$ depends on the value of x_i. For example, if $f(x_1, x_2) = x_1 x_2$ and $x_1 = 1$, the value of f will depend on the value of x_2. In other words, f will be sensitive to the value of x_2. If we wish to detect a fault on some primary input line x_i of the function $f(x_1, x_2, \ldots, x_n)$, we must first make the function sensitive to x_i by appropriately specifying the primary inputs. The Boolean difference is a method of determining the primary inputs required to force a function to be sensitive to a particular input variable.

The Boolean difference of a function $f(x_1, x_2, \ldots, x_n)$ with respect to a variable x_i is defined as

$$\frac{df}{dx_i} = f(x_1, \ldots, x_{i-1}, 0, x_{i+1}, \ldots, x_n) \oplus f(x_1, \ldots, x_{i-1}, 1, x_{i+1}, \ldots, x_n)$$

where $f(x_1, \ldots, x_{i-1}, 0, x_{i+1}, \ldots, x_n)$ is the function f evaluated with x_i being 0, $f(x_1, \ldots, x_{i-1}, 1, x_{i+1}, \ldots, x_n)$ is the function f evaluated with x_i being 1, and $\oplus$ is the EXCLUSIVE-OR operator. If df/dx_i is 0, the function f is completely independent of the input x_i. In other words, the value of f does not change when the value of x_i changes. If df/dx_i is 1, the function f depends directly on the value of x_i. The more likely result is that df/dx_i will depend on the inputs $(x_1, \ldots, x_{i-1}, x_{i+1}, \ldots, x_n)$. Appropriate specification of $(x_1, \ldots, x_{i-1}, x_{i+1}, \ldots, x_n)$ can then force df/dx_i to be 1.

A shorthand notation is typically used to simplify the representation of the Boolean difference. We define

$$f_{x_i}(0) = f(x_1, \ldots, x_{i-1}, 0, x_{i+1}, \ldots, x_n)$$
$$f_{x_i}(1) = f(x_1, \ldots, x_{i-1}, 1, x_{i+1}, \ldots, x_n)$$

Verbally, $f_{x_i}(0)$ is the function f evaluated with the input x_i being 0 and $f_{x_i}(1)$ is the function f evaluated with the input x_i being 1. The Boolean difference with respect to x_i is then given by

$$\frac{df}{dx_i} = f_{x_i}(0) \oplus f_{x_i}(1)$$

7.3 ■ Test Pattern Generation

As an example of the Boolean difference, consider the switching function given by

$$f(x_1,x_2,x_3) = x_1x_2x_3 + x_1'x_2'x_3 + x_1x_2'x_3'$$

The Boolean difference of f with respect to x_3 can be computed as follows. First, we evaluate f with x_3 being 0 and then we evaluate f with x_3 being 1 to determine

$$f_{x_3}(0) = f(x_1,x_2,0) = x_1x_2'$$
$$f_{x_3}(1) = f(x_1,x_2,1) = x_1x_2 + x_1'x_2'$$

The Boolean difference with respect to x_3 is then calculated as

$$\frac{df}{dx_3} = f_{x_3}(0) \oplus f_{x_3}(1) = x_1x_2' + x_1'x_2' + x_1x_2$$

or

$$\frac{df}{dx_3} = x_1 + x_2'$$

The Boolean difference with respect to x_3 states that if x_1 is 1 or x_2 is 0, the value of f will depend on whether x_3 is 0 or 1. This can be easily seen by examining the original function. When x_1 is 1, we find

$$f(1,x_2,x_3) = x_2x_3 + x_2'x_3'$$

As we can see from the above expression,

$$f(1,x_2,1) = x_2$$

and

$$f(1,x_2,0) = x_2'$$

which indicates that the value of f will change when x_3 changes regardless of the value of x_2, as long as x_1 is 1. Similar results can be shown when x_2 is 0.

The use of the Boolean difference to determine test patterns for faults that occur on the primary inputs consists of two fundamental steps. First, if we are trying to develop a test pattern for primary input x_i stuck-at-1, we must select x_i as 0 to attempt to force the line to deviate from its faulty value. Likewise, if we are attempting to develop a test pattern for x_i stuck-at-0, we must select x_i as 1. Second, we must select the remaining primary inputs such that the output is sensitive to the value of x_i. This second step is accomplished by forcing the Boolean difference with respect to x_i to be 1. These two steps can be placed in equation form as

$$x_i'\frac{df}{dx_i} = 1$$

for stuck-at-1 faults and

$$x_i \frac{df}{dx_i} = 1$$

for stuck-at-0 faults.

As an example of the use of the Boolean difference, consider the circuit shown in Fig. 7.9. The switching function that describes the output of this circuit in terms of the inputs can be written as

$$f(x_1, x_2, x_3, x_4) = x_1 x_2 + x_3 x_4$$

We wish to determine test patterns for both the stuck-at-0 and the stuck-at-1 faults that can appear on the primary input x_1. The Boolean difference of f with respect to x_1 can be computed by determining $f_{x_1}(0)$ and $f_{x_1}(1)$ as

$$f_{x_1}(0) = f(0, x_2, x_3, x_4) = x_3 x_4$$

and

$$f_{x_1}(1) = f(1, x_2, x_3, x_4) = x_2 + x_3 x_4$$

The Boolean difference is then given by

$$\frac{df}{dx_1} = f_{x_1}(0) \oplus f_{x_1}(1) = x_2 x_3' + x_2 x_4'$$

The test patterns for x_1 being stuck-at-0 are derived from the equation

$$x_1 \frac{df}{dx_1} = x_1(x_2 x_3' + x_2 x_4') = x_1 x_2 x_3' + x_1 x_2 x_4' = 1$$

The primary input combinations that satisfy the above equation are (x_1, x_2, x_3, x_4)=(1100, 1101, and 1110).

The test patterns for x_1 stuck-at-1 are derived from the equation

$$x_1' \frac{df}{dx_1} = x_1'(x_2 x_3' + x_2 x_4') = x_1' x_2 x_3' + x_1' x_2 x_4' = 1$$

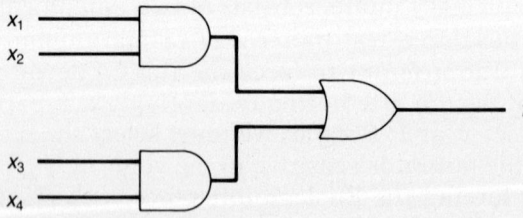

Fig. 7.9 Combinational circuit to illustrate test pattern generation using Boolean differences.

The primary input combinations that satisfy the above equation are $(x_1, x_2, x_3, x_4) = (0100, 0101,$ and $0110)$.

We have seen how the Boolean difference can be used to develop test patterns for faults that appear on the primary inputs of a circuit. Of more importance, however, is the technique for using the Boolean difference to develop test patterns for faults that appear internal to a circuit. The basic procedure is the same, but now the Boolean difference must be taken with respect to a line internal to the circuit, and the function must be made sensitive to the value that appears on that particular line. As an example, consider the circuit shown in Fig. 7.10.

Suppose that we wish to develop test patterns for faults that occur on line m in the circuit of Fig. 7.10. The switching function implemented by the circuit can be written in terms of m as

$$f(m, x_3, x_4) = mx_4' + x_3 x_4'$$

where

$$m = x_1' x_2'$$

The Boolean difference of f with respect to m can be determined from

$$f_m(0) = x_3 x_4'$$
$$f_m(1) = x_4' + x_3 x_4' = x_4'$$

as

$$\frac{df}{dm} = f_m(0) \oplus f_m(1) = x_3' x_4'$$

The test patterns for m stuck-at-1 can be determined from the equation

$$m' \frac{df}{dm} = m' x_3' x_4' = 1$$

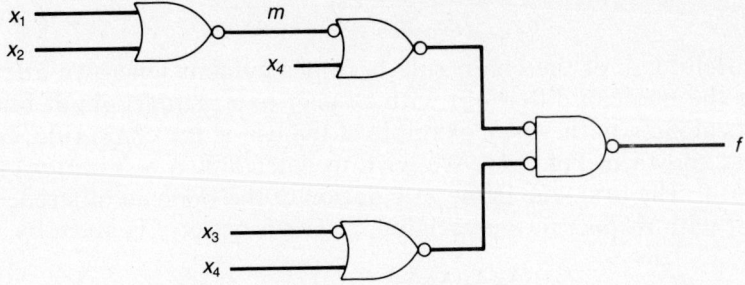

Fig. 7.10 Combinational circuit to illustrate the Boolean difference with respect to lines that are not primary input lines.

which for
$$m = x_1' x_2'$$
becomes
$$(x_1' x_2')' x_3' x_4' = x_1 x_3' x_4' + x_2 x_3' x_4' = 1$$

The test patterns for m stuck-at-1 are then given by $(x_1, x_2, x_3, x_4) = (1000, 1100, 0100)$.

In a similar manner, the test patterns for m stuck-at-0 can be derived from
$$m \frac{df}{dm} = x_1' x_2' x_3' x_4' = 1$$

The single test pattern for m stuck-at-0 is then $(x_1, x_2, x_3, x_4) = (0000)$.

A primary advantage of the Boolean difference is that it is an exact method of generating test patterns; there is no trial and error involved as will be found in some path sensitization techniques. A major disadvantage of the Boolean difference method is the effort required to calculate the Boolean difference and the resulting test patterns for each line within a circuit. One property of the Boolean difference that overcomes this disadvantage to a certain degree is the chain rule [Chiang, Reed and Banes 1972]. The chain rule states that if line k within a circuit depends on line m, and line m in turn depends on line n, then the Boolean difference of k with respect to n can be written as

$$\frac{dk}{dn} = \frac{dk}{dm} \frac{dm}{dn}$$

The chain rule can be generalized to an arbitrary number of variables. For example, if x_{i-1} depends on x_i for all i from 1 to n, the Boolean difference of x_1 with respect to x_n can be written as

$$\frac{dx_1}{dx_n} = \frac{dx_1}{dx_2} \frac{dx_2}{dx_3} \frac{dx_3}{dx_4} \cdots \frac{dx_{n-1}}{dx_n}$$

The usefulness of the chain rule becomes evident when we attempt to calculate the Boolean difference with respect to a primary input line of all the points along a path. As an example of the use of the chain rule, consider the circuit shown in Fig. 7.11. We wish to determine a test pattern for line x_1 stuck-at-0. The first step is the calculation of the Boolean difference of the function f with respect to the variable x_1. The function f is given by

$$f(x_1, x_2, x_3, x_4, x_5) = x_1 x_2 x_3 + x_4 x_5$$

The Boolean difference of f with respect to x_1 can be calculated in one of two ways: directly from the function f or using the chain rule. In the present ex-

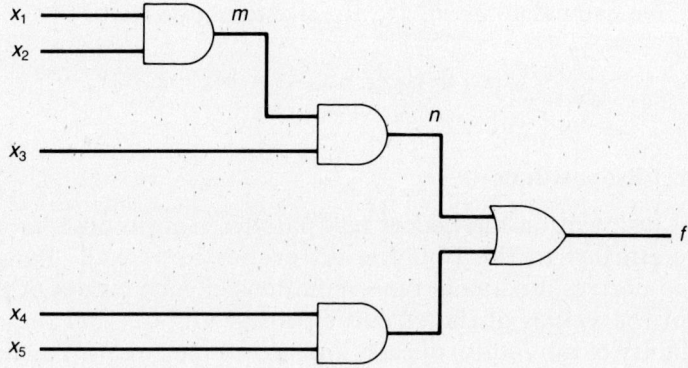

Fig. 7.11 Combinational circuit to illustrate the chain rule in the calculation of the Boolean difference.

ample, it is relatively easy to calculate the Boolean difference directly from the function f. In more complex circuits, the calculations can be more lengthy, and, in many cases, the functional form of the circuit can be unknown. Therefore, the chain rule must be used.

The chain rule, as applied to the circuit of Fig. 7.11, states that the Boolean difference of f with respect to x_1 can be found from

$$\frac{df}{dx_1} = \frac{df}{dn}\frac{dn}{dm}\frac{dm}{dx_1}$$

where

$$m = x_1 x_2$$
$$n = m x_3$$
$$f = n + x_4 x_5$$

The Boolean differences required for the chain rule evaluation are

$$\frac{df}{dn} = x_4 x_5 \oplus 1 = x_4' + x_5'$$

$$\frac{dn}{dm} = 0 \oplus x_3 = x_3$$

$$\frac{dm}{dx_1} = 0 \oplus x_2 = x_2$$

Therefore, df/dx_1 can be calculated as

$$\frac{df}{dx_1} = \frac{df}{dn}\frac{dn}{dm}\frac{dm}{dx_1} = (x_4' + x_5')x_2 x_3$$

As a check, we can calculate df/dx_1 directly to obtain

$$\frac{df}{dx_1} = (x_4 x_5) \oplus (x_2 x_3 + x_4 x_5) = (x_4' + x_5') x_2 x_3$$

7.3.4 Literal Propositions

The literal proposition method of test pattern generation is an algebraic technique similar to the Boolean difference approach [Poage 1963]. The method derives functional representations of each output of the circuit in terms of the values of the circuit's primary inputs *and* the condition (whether faulty or nonfaulty) of each line within the circuit. To allow the effects of faults to be incorporated into the functional representation of the circuit, each line p is characterized by three binary variables: p_n, p_1, and p_0. p_n is 1 if line p is fault free and 0 if line p contains a fault. p_1 is 1 if line p is stuck-at-1 and is 0, otherwise. Finally, p_0 is 1 if line p is stuck-at-0 and is 0, otherwise.

The value that appears on each line of a circuit is represented by a **proposition**. A proposition is a declarative statement that can be either true or false, but never both. For example, the statement "it is raining" is true if rain is indeed falling but is false, otherwise. The statement that line m in a circuit is 1 is true if the line is in fact 1 but is false, otherwise. Whether a proposition is true or false is represented by the value of a binary variable that is usually denoted by the letter P. P assumes a value of 1 if the proposition associated with P is true, and P is 0 if the proposition is false.

New propositions can be derived from old ones by combining the old propositions using traditional logic operations such as AND and OR. Also, a proposition is said to be the *negation* of another if and only if when one proposition is true, the other is false. The negation of the proposition P is denoted as P'.

As an example, consider the simple AND gate shown in Fig. 7.12. Let P_a be the proposition that line a will be 1. P_a assumes a value of 1 if line a is 1 and a value of 0, otherwise. P_a' is the proposition that line a will be 0. Likewise, P_b and P_b' are the propositions that line b will be 1 and 0, respectively.

Each line of the AND gate can become 1 in two ways: (1) the signal applied to the line is 1 and the line is functioning normally, or (2) the line is stuck-at-1. The proposition that line a in the AND gate of Fig. 7.12 will be 1 can be written as

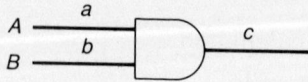

Fig. 7.12 A simple two-input AND gate to illustrate basic propositions.

$$P_a = Aa_n + a_1$$

where A is the signal applied to line a. The proposition P_a (that is, the proposition that line a will be 1) is true ($P_a = 1$) if $A = 1$ and $a_n = 1$, or $a_1 = 1$. In other words, line a is 1 if the line is fault free and the signal applied to line a is 1, or line a is stuck-at-1. The proposition that line a will be 0 can be written as

$$P_a' = A'a_n + a_0$$

The proposition P_a' is true if $A = 0$ and $a_n = 1$, or $a_0 = 1$. In other words, line a is 0 if the line is fault free and the signal applied to the line is 0, or line a is stuck-at-0.

Similar propositions can be developed for line b of the AND gate. More specifically, the propositions for line b are given by

$$P_b = Bb_n + b_1$$

and

$$P_b' = B'b_n + b_0$$

where B is the signal applied to line b.

The propositions for the output of the AND gate of Fig. 7.12 can be developed in terms of the propositions of the inputs of the gate. For example, the proposition that the output is 1 will be true if both the propositions of the input lines being 1 are also true and the output line is functioning correctly. If either input line is 0, the output line will also be 0. The output line can also be 0 if the line is stuck-at-0. Therefore, the propositions for the output line c of the AND gate of Fig. 7.12 can be written as

$$P_c = (P_a P_b)c_n + c_1$$

and

$$P_c' = (P_a' + P_b')c_n + c_0$$

The equation describing P_c states that line c will be 1 if either both lines a and b are 1 *and* c is fault free, or line c is stuck-at-1. P_c' states that two events lead to the proposition that the output c is 0 being true ($P_c' = 1$): (1) either P_a' or P_b' is true and c is fault free, or (2) line c is stuck-at-0.

Propositions can be developed for all the basic gates. Table 7.9 shows the propositions for each of the traditional gates.

Test patterns can be generated using the literal proposition approach by first developing the propositions for the output of a circuit when all lines within the circuit are fault free. Next, the propositions for the output are developed when certain faulty conditions exist. An input combination is then selected that makes the proposition that the output is 1 (or 0) under fault-free conditions true, while at the same time making the proposition that the output is 0 (or 1) under faulty conditions true. For example, sup-

TABLE 7.9 Propositions for the AND, OR, NOT, NAND, NOR, and XOR gates

Logic gate	Propositions
AND (A, B → c)	$P_a = Aa_n + a_1$ $P'_a = A'a_n + a_0$ $P_b = Bb_n + b_1$ $P'_b = B'b_n + b_0$ $P_c = P_a P_b c_n + c_1$ $P'_c = (P'_a + P'_b)c_n + c_0$
OR (A, B → c)	$P_a = Aa_n + a_1$ $P'_a = A'a_n + a_0$ $P_b = Bb_n + b_1$ $P'_b = B'b_n + b_0$ $P_c = (P_a + P_b)c_n + c_1$ $P'_c = (P'_a P'_b)c_n + c_0$
NOT (A → b)	$P_a = Aa_n + a_1$ $P'_a = A'a_n + a_0$ $P_b = P'_a b_n + b_1$ $P'_b = P_a b_n + b_0$
NAND (A, B → c)	$P_a = Aa_n + a_1$ $P'_a = A'a_n + a_0$ $P_b = Bb_n + b_1$ $P'_b = B'b_n + b_0$ $P_c = (P'_a + P'_b)c_n + c_1$ $P'_c = (P_a P_b)c_n + c_0$
NOR (A, B → c)	$P_a = Aa_n + a_1$ $P'_a = A'a_n + a_0$ $P_b = Bb_n + b_1$ $P'_b = B'b_n + b_0$ $P_c = (P'_a P'_b)c_n + c_1$ $P'_c = (P_a + P_b)c_n + c_0$
XOR (A, B → c)	$P_a = Aa_n + a_1$ $P'_a = A'a_n + a_0$ $P_b = Bb_n + b_1$ $P'_b = B'b_n + b_0$ $P_c = (P_a P'_b + P'_a P_b)c_n + c_1$ $P'_c = (P_a P_b + P'_a P'_b) + c_n + c_0$

pose that $P_f(n)$, $P_f'(n)$, $P_f(f)$, and $P_f'(f)$ are the output propositions for f being 1 in the normal case, f being 0 in the normal case, f being 1 in the faulty case, and f being 0 in the faulty case, respectively. The objective, as far as test pattern generation is concerned, is to force one of two events to occur: (1) $P_f(n)$ and $P_f'(f)$ are both true, or (2) $P_f'(n)$ and $P_f(f)$ are both true. In equation form, we have

$$P_f(n)P_f'(f) + P_f'(n)P_f(f) = 1$$

As an example of the literal proposition technique, consider the circuit shown in Fig. 7.13. Table 7.10(a) shows the propositions for each line of the circuit of Fig. 7.13. Suppose that we wish to develop a test pattern for a stuck-at-1 fault on line c. Table 7.10(b) shows the propositions for each line under both faulty and normal circumstances. The equation for determining the test patterns is

$$P_e(n)P_e'(f) + P_e'(n)P_e(f) = (ABC')(1) + (A' + B' + C)(0) = ABC' = 1$$

Inspection of the above equation yields the test pattern $(A,B,C) = (110)$.

Perhaps one of the major benefits of the literal proposition technique is that it allows test patterns for multiple fault conditions to be developed. For example, suppose that we wish to develop a test pattern for the multiple fault condition c stuck-at-1 and a stuck-at-1 in the circuit of Fig. 7.13. The propositions under the faulty conditions can now be derived by setting both c_n and a_n to 0 and both c_1 and a_1 to 1. Table 7.11 shows the propositions for the multiple fault condition of both c stuck-at-1 and a stuck-at-1. The appropriate test pattern is derived from

$$P_e(n)P_e'(f) + P_e'(n)P_e(f) = (ABC')(1) + (A' + B' + C)(0) = ABC' = 1$$

which states that the test pattern (110) is also a test for the multiple fault, a stuck-at-1 and c stuck-at-1.

7.3.5 Path Sensitization

The path sensitization approach to test pattern generation requires that one or more paths from the site of the potential fault to a primary output be selected, and the value on the faulty line propagated along those paths. If a single path is chosen, the approach is called *single-path* or *one-dimensional*

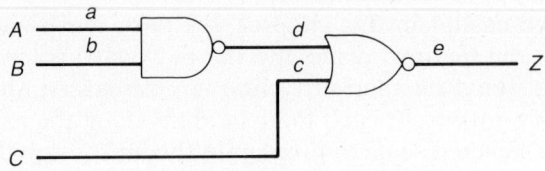

Fig. 7.13 Circuit used to illustrate the literal proposition technique.

TABLE 7.10(a) Propositions for each line of the circuit of Fig. 7.13

Line	Propositions
a	$P_a = Aa_n + a_1$ $P_a' = A'a_n + a_0$
b	$P_b = Bb_n + b_1$ $P_b' = B'b_n + b_0$
c	$P_c = Cc_n + c_1$ $P_c' = C'c_n + c_0$
d	$P_d = (P_a' + P_b')d_n + d_1$ $P_d' = (P_a P_b)d_n + d_0$
e	$P_e = (P_d' P_c')e_n + e_1$ $P_e' = (P_d + P_c)e_n + e_0$

TABLE 7.10(b) Propositions for each line of the circuit of Fig. 7.13

Line	Fault-free propositions	c stuck-at-1 propositions
a	$P_a = A$ $P_a' = A'$	$P_a = A$ $P_a' = A'$
b	$P_b = B$ $P_b' = B'$	$P_b = B$ $P_b' = B'$
c	$P_c = C$ $P_c' = C'$	$P_c = 1$ $P_c' = 0$
d	$P_d = A' + B'$ $P_d' = AB$	$P_d = A' + B'$ $P_d' = AB$
e	$P_e = ABC'$ $P_e' = A' + B' + C$	$P_e = 0$ $P_e' = 1$

path sensitization. If more than one path is chosen, the approach is called *multi-dimensional* path sensitization.

As mentioned earlier in this chapter, the path sensitization process has two basic steps: the forward drive and the backward trace. During the forward drive, the signal on the faulty line is propagated along the selected path to a primary output. At each logic module along the path, the inputs to that module are selected so as to propagate the faulty signal from one of the

TABLE 7.11 Propositions for each line of the circuit of Fig. 7.13

Line	c stuck-at-1 and a stuck-at-1 propositions
a	$P_a = 1$
	$P'_a = 0$
b	$P_b = B$
	$P'_b = B'$
c	$P_c = 1$
	$P'_c = 0$
d	$P_d = B'$
	$P'_d = B$
e	$P_e = 0$
	$P'_e = 1$

inputs of the logic module to one of the outputs of that module. The forward drive continues until the faulty signal has been propagated to a primary output. The backward trace is the process whereby the primary inputs necessary to achieve the signal values specified during the forward drive are determined.

As an illustration of path sensitization, consider the circuit shown in Fig. 7.14. Suppose that we wish to determine a test pattern for line 7 stuck-at-0. We will denote the value on line 7, whatever it may be, as D_7. The first step in the path sensitization process is to select a path from line 7 to the primary output of the circuit. Because the circuit does not contain fanout, there is only one path from line 7 to the primary output. The path starts at line 7 and passes through Gates 3 and 4. To propagate D_7 through Gate 3 requires that line 6 be a logic 0. If line 6 were a logic 1, the value on line 8

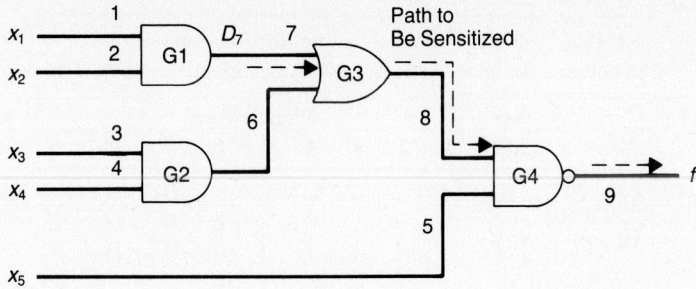

Fig. 7.14 Combinational circuit to illustrate the concept of path sensitization.

would also be 1, independent of D_7. Finally, to propagate D_7 through Gate 4 requires that line 5 be a logic 1; therefore, the output of the circuit will be D_7'. Once the primary output is reached, the forward drive is completed.

During the backward trace, we attempt to determine the primary inputs necessary to create the values specified in the forward drive. For example, line 5 being a logic 1 requires that primary input x_5 be 1. In a similar manner, line 6 being a logic 0 requires that either primary input x_3 or x_4 be 0. Finally, we know that to detect the stuck-at-0 fault on line 7, we must attempt to force line 7 to be a logic 1; therefore, we must set both x_1 and x_2 to 1. A complete test pattern for the stuck-at-0 fault on line 7 is $(x_1, x_2, x_3, x_4, x_5) = (110X1)$, where x_4 may be either 0 or 1. We could also select the test pattern $(11X01)$ since only one of x_3 and x_4 must be 0.

The implementation of the path sensitization process can be simplified through the use of a tabular method for recording the step-by-step procedure. First, a table is constructed with one column for each line in the circuit. The rows of the table are the values specified for the lines at each step of the forward drive and the backward trace processes. Each row is referred to as a *test cube*. For the example just presented using the circuit of Fig. 7.14, Table 7.12 shows the step-by-step procedure represented in a tabular form.

The path sensitization method of test pattern generation is intuitively appealing and is also very simple in fanout-free circuits. Unfortunately, most practical circuits contain numerous fanout points. If fanout is present, two or more paths will exist from the fanout point to a primary output. A faulty signal that passes through the fanout point will have multiple paths upon which it can propagate to a primary output. In one-dimensional path sensitization, one, and only one, path must be selected from the site of the fault to a primary output. Therefore, one path must be sensitized, and the remaining paths must be *desensitized*. Desensitization implies that at some point on the path, the faulty signal ceases to propagate. In other words, the output of some logic module along the path becomes independent of the faulty signal.

TABLE 7.12 The forward drive and backward trace processes for line 7 stuck-at-0 in the circuit of Fig. 7.14

	Test cube	\multicolumn{9}{c}{Circuit line}
		1	2	3	4	5	6	7	8	9
Forward drive	tc_0							D_7		
	tc_1						0	D_7	D_7	
	tc_2					1	0	D_7	D_7	D_7'
Backward trace	tc_3			0	X	1	0	D_7	D_7	D_7'
	tc_4	1	1	0	X	1	0	D_7	D_7	D_7'

As an illustration of path sensitization in a circuit with fanout, consider the circuit shown in Fig. 7.15. The signal on line A branches to form lines D and F, whereas the signal on line B creates lines H and I. Likewise, the signal on line J branches to form lines K and G.

Suppose that we wish to determine a test pattern for line J stuck-at-0. For the purposes of the forward drive process, we will let D_J represent the value of the signal that appears on line J. The signal D_J can be propagated to the output in one of several ways. For D_J to propagate along the single path through Gates 5 and 6, line H must assume a value of 1, whereas lines M and L must be 0. For line H to be 1, primary input B must also be 1. If primary input B is 1, line L will be 0, and primary input A can be selected as 0 to force line M to be 0. Finally, to test for J stuck-at-0 requires primary input C to be 0 so as to attempt to force line J to assume a value of 1. The test pattern for J stuck-at-0 is $(ABC) = (010)$. The tabular representation of the forward drive and backward trace used to sensitize the path through Gates 5 and 6 is shown in Table 7.13.

Suppose, however, that we had selected the path from J through Gates 4 and 6 to the primary output. The forward drive specifies that line F must be 1, and lines L and N must be 0 to sensitize the single path through Gates 4 and 6. The backward trace specifies that primary input A must be 1 to force F to be 1. Likewise, for line N to be 0 requires that either line H or line K must be zero. However, if we want to detect a stuck-at-0 fault on line J, we must apply a 1 to line J, so we cannot guarantee that line K will be 0. Therefore, line H must be made 0, mandating that primary input B be 0. If B is 0 and A is 1, line L will be 1, and the requirement from the forward drive that L be 0 is violated. Therefore, an *inconsistency* between the values specified during the forward drive and those found during the backward trace has been found. It must be concluded that the path from J through Gates 4 and 6 cannot be a single, sensitized path. The tabular representation of the for-

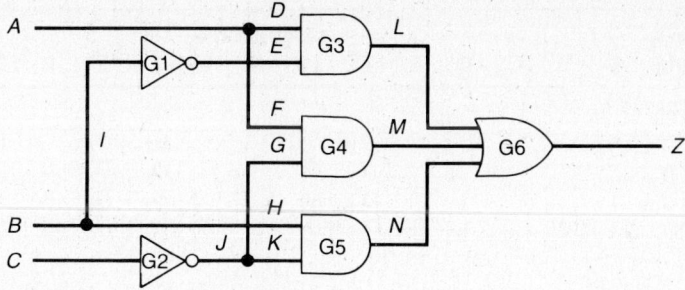

Fig. 7.15 Example of a combinational circuit with fanout to illustrate the concept and problems of path sensitization.

TABLE 7.13 The development of a test pattern for line J stuck-at-0 in the circuit of Fig. 7.15

	Test cube	A	B	C	D	E	F	G	H	I	J	K	L	M	N	Z
Forward drive	tc_0										D_J					
	tc_1							D_J	1		D_J	D_J			D_J	
	tc_2							D_J	1		D_J	D_J	0	0	D_J	D_J
Backward trace	tc_3		1			0		D_J	1	1	D_J	D_J	0	0	D_J	D_J
	tc_4	0	1		0	0	0	D_J	1	1	D_J	D_J	0	0	D_J	D_J
	tc_5	0	1	0	0	0	0	D_J	1	1	D_J	D_J	0	0	D_J	D_J

ward drive and the backward trace used to attempt to sensitize the path through Gates 4 and 6 is shown in Table 7.14.

Even though the path through Gates 4 and 6 cannot be sensitized by itself, it might be possible to simultaneously sensitize both the path through Gates 4 and 6 and the path through Gates 5 and 6. To sensitize both paths requires that both H and F be 1, therefore resulting in lines M and N being D_J. Line L must be 0 to produce a D_J on the primary output of the circuit. For line H to be 1 requires that primary input B be 1, which immediately satisfies the requirement that L be 0. Also, primary input A being 1, results in line F being 1. Finally, primary input C must be 0 to detect a stuck-at-0 fault on line J. The test pattern is, therefore, $(ABC) = (110)$. The tabular representation of the forward drive and the backward trace used to sensitize both the path through Gates 4 and 6 and the path through Gates 5 and 6 is shown in Table 7.15.

TABLE 7.14 The forward drive and backward trace processes for generating a test pattern for line J stuck-at-0 in the circuit of Fig. 7.15.

	Test cube	A	B	C	D	E	F	G	H	I	J	K	L	M	N	Z
Forward drive	tc_0										D_J					
	tc_1						1	D_J			D_J	D_J		D_J		
	tc_2						1	D_J			D_J	D_J	0	D_J	0	D_J
Backward trace	tc_3	1			1		1	D_J			D_J	D_J	0	D_J	0	D_J
	tc_4	1	0		1	1	1	D_J	0	0	D_J	D_J	0	D_J	0	D_J

Inconsistency! If A and E are both 1, L will also be 1.

7.3 ■ Test Pattern Generation

TABLE 7.15 Sensitization of two paths in the circuit of Fig. 7.15 to derive a test pattern for line J stuck-at-0

	Test cube	A	B	C	D	E	F	G	H	I	J	K	L	M	N	Z
Forward drive	tc_0										D_J					
	tc_1							D_J			D_J	D_J				
	tc_2					1		D_J	1		D_J	D_J		D_J	D_J	
	tc_3					1		D_J	1		D_J	D_J	0	D_J	D_J	D_J
Backward trace	tc_4		1			0	1	D_J	1	1	D_J	D_J	0	D_J	D_J	D_J
	tc_5	1	1		1	0	1	D_J	1	1	D_J	D_J	0	D_J	D_J	D_J
	tc_6	1	1	0	1	0	1	D_J	1	1	D_J	D_J	0	D_J	D_J	D_J

Consistency!

The preceding example illustrates several interesting points about path sensitization. First, the approach is somewhat trial and error. If the path through Gates 4 and 6 had been selected first, the forward drive and backward trace would not have yielded a test pattern, and the process would have to be repeated. Therefore, the computation time required to determine a test pattern for line J stuck-at-0 depends on the selection of the path. Second, it is possible, in many cases, to obtain a test pattern by sensitizing a single path or by sensitizing multiple paths. In the above example, two distinct test patterns, 010 and 110, were found by sensitizing one and two paths, respectively. By examination of the circuit, we can see that the test pattern 110 detects two faults (C stuck-at-1 and Z stuck-at-0) in addition to detecting J stuck-at-0. The test pattern 010, however, detects six faults (B stuck-at-0, H stuck-at-0, K stuck-at-0, N stuck-at-0, Z stuck-at-0, and C stuck-at-1) in addition to detecting J stuck-at-0. We would normally like to select the test patterns that cover the largest number of faults so that the total number of test patterns is minimized.

In many circuits, it is impossible to determine a test pattern by sensitizing only one path, consequently, multipath sensitization is required. A very simple example where multipath sensitization is required is found in the circuit of Fig. 7.16. Suppose that we wish to determine a test pattern for line A stuck-at-0. We will represent the value that appears on line A as D_A. If we select the path through Gates 2 and 4, the forward drive will specify that x_3 must be 0 and line C must be 1. However, the backward trace will indicate that C cannot be 1 unless both of the inputs to Gate 3 are 0. Because D_A is one of the inputs to Gate 3, we cannot guarantee that the output of Gate 3 will be 1. A similar situation is encountered if we attempt to propagate D_A through Gates 3 and 4. To sensitize the path through Gates 3 and 4, we must

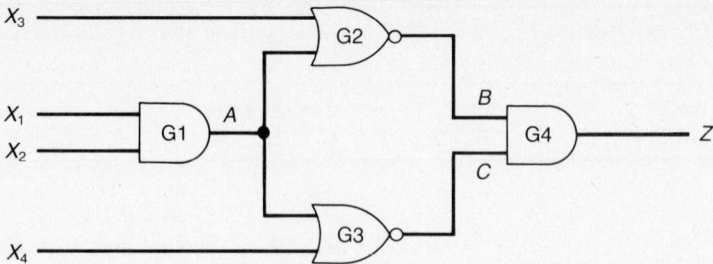

Fig. 7.16 A simple combinational circuit with fanout that requires sensitizing two paths to detect a fault on one A.

force line B to be 1. Because one of the inputs to Gate 2 is D_A, we cannot guarantee that line B will be 1. The conclusion is that we cannot sensitize a single path. We can, however, sensitize both paths by making both x_3 and x_4 0 so that lines B and C both become D'_A. The output of the circuit is then D'_A. To detect the stuck-at-0 fault on line A, we select x_1 and x_2 as 1 to attempt to force line A to become 1. The test pattern is $(x_1 x_2 x_3 x_4) = (1100)$.

7.3.6 The *D*-Algorithm

The *D*-algorithm was developed as an algorithmic approach to implementing the process of path sensitization. The algorithm is guaranteed to find a test pattern for any single, stuck-type fault in any nonredundant, combinational circuit, provided that the algorithm is allowed to execute for a sufficient length of time. The *D*-algorithm is a method, or technique, of implementing the forward drive and backward trace processes of path sensitization. In the *D*-algorithm, the forward drive is called the *D-drive*, and the backward trace is called the *consistency operation* [Roth 1966].

Before examining the theory of the *D*-algorithm, it is beneficial to intuitively investigate the basic concepts. The fundamental idea of the *D*-algorithm can be illustrated using Fig. 7.17. In Fig. 7.17, each box represents a logic module that can be a collection of logic gates or a single gate. The logic modules are interconnected to create the overall function of the circuit. Suppose that we wish to develop a test pattern for a single, stuck-type fault that occurs on a line within Module 4. The first step of the *D*-algorithm is to select inputs for Module 4 that will force the effect of the fault to appear on the output of Module 4. In other words, the inputs to Module 4 are selected such that the output of Module 4 will depend on whether or not the suspected fault is present. If the fault is present, the output of Module 4 will be one value (either 1 or 0), and if the fault is not present, the output of Module 4 will be the opposite value (either 0 or 1). The input combination and the associated output selected for Module 4 is called a **primitive *D*-cube of fault**

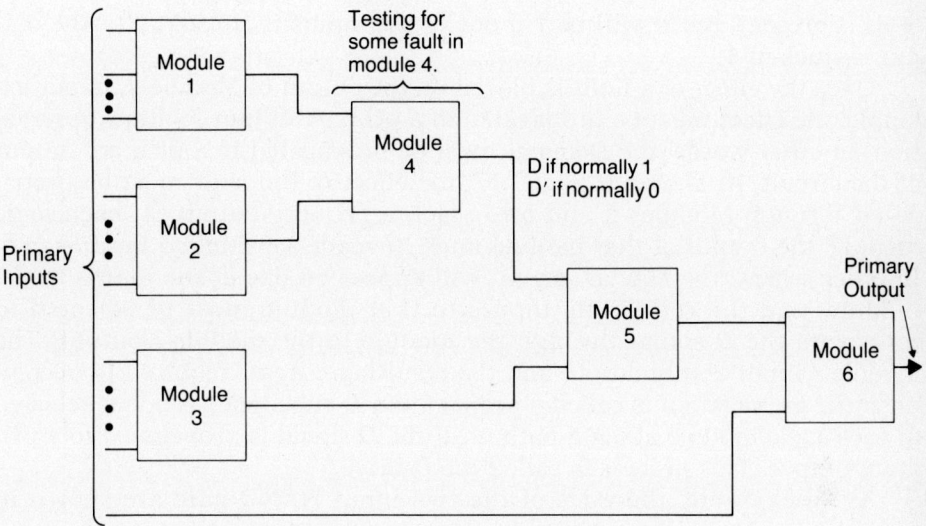

Fig. 7.17 Illustration of the basic concept of the D-algorithm.

(pdcf). The value assigned to the output of Module 4 is D if the output should normally be 1 but will be 0 if the fault is present and is D' if the output should normally be 0 but will be 1 if the fault is present.

As an example, Fig. 7.18 shows a simple two-input NAND gate and the pdcf for each single, stuck-type fault that can occur on the lines of that NAND gate. For example, the pdcf (11D') brings the effect of either input being stuck-at-0 or the output being stuck-at-1 to the output of the gate. If the input combination 11 is applied to the NAND gate, the output will be 0 if no

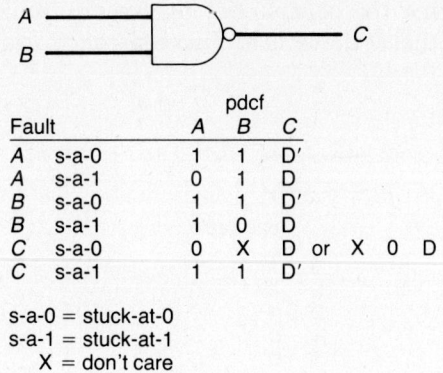

		pdcf		
Fault		A	B	C
A	s-a-0	1	1	D'
A	s-a-1	0	1	D
B	s-a-0	1	1	D'
B	s-a-1	1	0	D
C	s-a-0	0	X	D or X 0 D
C	s-a-1	1	1	D'

s-a-0 = stuck-at-0
s-a-1 = stuck-at-1
X = don't care

Fig. 7.18 pdcf for each possible fault in a two-input NAND gate.

fault is present, but it will be 1 if one of the inputs is stuck-at-0 or the output is stuck-at-1.

Once the effect of a fault is placed on the output of Module 4, in our example, the effect must be propagated to a primary output to allow observation. In other words, the D signal must be propagated to a primary output of the circuit. As shown in Fig. 7.17, the effect of the fault must be propagated through Modules 5 and 6 to reach a primary output. At each logic module, the output of that module must be made sensitive to the D signal. In other words, the D or D' signal will appear on one of the inputs to the module, and the remaining inputs to that module must be selected to propagate the D signal through the module to the module's output. The module's input combination, and the resulting output, required to accomplish the propagation is called a **propagation D cube (pdc)**. A pdc is selected at each logic module along a path until the D signal is propagated to a primary output. This process is called the D-drive.

As an example, the pdcs of the two-input NAND gate are shown in Fig. 7.19. If a D signal is applied to either input of the NAND gate, the other input must be a logic 1 to propagate the information contained in that D signal to the output of the gate. If one input of the NAND gate is D and the other input is a logic 1, the output will be D'. If one input of the NAND gate is D' and the other input is a logic 1, the output will be D. The pdcs $(D1D')$ and $(D'1D)$ are called *polarized* pdcs because the only difference between the two is the polarity of the D signal on the input and the output.

During the D-drive, numerous values are specified for lines throughout the circuit. The *consistency operation* determines the primary inputs required to create each of the signal values defined during the D-drive. The consistency operation is performed by working from the primary outputs to the primary inputs. At each step where the output of a logic module has been specified, the inputs to the module that are necessary to create the specified output are determined. At each step, it is verified that the signal values specified during the consistency operation are consistent with those determined during the D-drive. Any inconsistency requires that the algo-

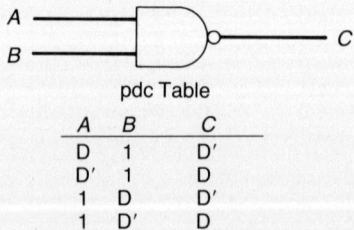

pdc Table

A	B	C
D	1	D'
D'	1	D
1	D	D'
1	D'	D

Fig. 7.19 pdc table for a two-input NAND gate.

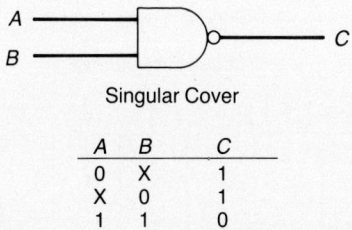

Singular Cover

A	B	C
0	X	1
X	0	1
1	1	0

Fig. 7.20 Singular cover for the two-input NAND gate.

rithm back up and *try again*. The algorithm stops once a primary input combination that can achieve the signal values defined by the pdcf and the pdcs is determined.

The consistency operation is performed using the **singular covers** of each logic module. The singular cover is a reduced truth table that shows the output produced by a logic module for each possible input combination. The reduction in the truth table takes the form of *don't cares* that can occur in the truth table. For example, if one input to an AND gate is 0, the remaining inputs are don't cares because the output will be independent of their values. As an example, the singular cover of the two-input NAND gate is shown in Fig. 7.20. For any possible output, the singular cover provides the input combinations that can be used to produce that output.

As a complete illustration of the *D*-algorithm, consider the circuit shown in Fig. 7.21. The circuit is a very simple one constructed using only two-input NAND gates. The first step in the *D*-algorithm is to number the lines and gates of the circuit, as shown in Fig. 7.21. The numbering begins at the

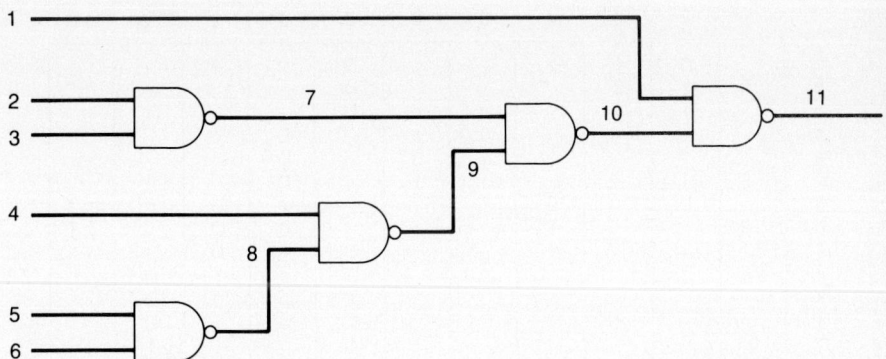

Fig. 7.21 Combinational circuit designed with two-input NAND gates and used to illustrate the D-algorithm.

Singular Cover Table

Circuit Line Number

	1	2	3	4	5	6	7	8	9	10	11
a		0	X				1				
b		X	0				1				
c		1	1				0				
d					0	X		1			
e					X	0		1			
f					1	1		0			
g				0				X	1		
h				X				0	1		
i				1				1	0		
j							0		X	1	
k							X		0	1	
l							1		1	0	
m	0									X	1
n	X									0	1
o	1									1	0

Fig. 7.22 Singular cover table for the circuit of Fig. 7.21.

primary inputs of the circuit and proceeds to the primary outputs. In a circuit that has only one primary output, that output should have the highest number of any line. Next, the singular covers, pdcfs, and pdcs of each element within the circuit are organized in a tabular format, as shown in Figs. 7.22, 7.23, and 7.24. The tables containing the singular covers, pdcfs, and

pdcf Table

Circuit Line Number

	1	2	3	4	5	6	7	8	9	10	11
a		0	1				D				
b		1	0				D				
c		1	1				D'				
d					0	1		D			
e					1	0		D			
f					1	1		D'			
g				0				1	D		
h				1				0	D		
i				1				1	D'		
j							0		1	D	
k							1		0	D	
l							1		1	D'	
m	0									1	D
n	1									0	D
o	1									1	D'

Fig. 7.23 pdcf table for the circuit of Fig. 7.21.

pdc Table

Circuit Line Number

	1	2	3	4	5	6	7	8	9	10	11
a		D	1				D'				
b		1	D				D'				
c		D'	1				D				
d		1	D'				D				
e					D	1		D'			
f					1	D		D'			
g					D'	1		D			
h					1	D'		D			
i				D				1	D'		
j				1				D	D'		
k				D'				1	D		
l				1				D'	D		
m							D		1	D'	
n							1		D	D'	
o							D'		1	D	
p							1		D'	D	
q	D									1	D'
r	1									D	D'
s	D'									1	D
t	1									D'	D

Fig. 7.24 pdc table for the circuit of Fig. 7.21.

pdcs each have one column for each line of the circuit and one row for each cube associated with each logic module. The columns are labeled according to the line numbers, and the rows are labeled in alphabetical order. For any primary input combination, the singular cover table can be used to trace through the circuit to determine the primary output that results from that input. Likewise, the pdc table can be used to determine the values applied to individual lines to propagate a D signal from any point in the circuit to a primary output. Associated with every pdcf is a list of faults in that logic module that the pdcf will detect.

Suppose that we wish to determine a test pattern for line 9 stuck-at-0. The first step in the D-algorithm is to select a pdcf that exposes a stuck-at-0 fault on line 9. We go to the table of pdcfs and select one that forces line 9 to be a logic 1 if the fault is not present. Clearly, line 9 will be 0 if the fault is present. As we can see, there are two pdcfs that will expose the stuck-at-0 fault, so the selection is arbitrary and is typically performed by selecting the pdcf that occurs first, in an alphabetical sense, in the pdcf table. The selected pdcf is (0XD), which specifies that line 4 is a 0 and line 8 is a don't care. The pdcf represents the first step in the construction of the test pattern. The process of performing the D-drive and the consistency operation is represented in tabular form just as we did in the path sensitization process. Table 7.16 shows the complete step-by-step process.

Once the appropriate pdcf is selected, the D signal must be propagated to the primary output using the pdcs. The *activity vector* shown in Table 7.16 is the number of the line to which the D signal has been propagated. Once the activity vector becomes equal to the number of a primary output line, the D-drive can be stopped. By looking at the pdcs that have a D value on line 9, we can determine that line 7 must be a logic 1 to propagate the D signal to line 10. Line 10 becomes D', line 7 becomes a logic 1, and the activity vector becomes 10, as shown in Table 7.16. By looking at the pdcs that have a D' on line 10, we see that line 1 must be a logic 1 to force line 11 to become D. Line 1 becomes a logic 1, line 11 becomes D, the activity vector becomes 11, and the D-drive is completed.

The consistency operation determines the primary inputs that are necessary to achieve the signal values specified in the D-drive. During the D-drive, we specified that lines 1, 7 and 8 must each be logic 1, and line 4 must be logic 0. We begin the consistency operation with the line that has the highest number, so we first consider line 8. We search the singular cover table from bottom to top for a gate output that has line 8 being a logic 1, and we find that the first entry encountered specifies that line 6 must be 0 and line 5 can be a don't care to force line 8 to a 1. This specification does not contradict previous specifications. Lines 5 and 6 are primary inputs and need not be considered in the consistency operation. All the values specified during the D-drive have now been obtained without any conflicts, so the consistency operation is completed. The final test pattern is (1X00X0). Table 7.16 shows each step of the consistency operation.

Now that we have a basic understanding of the concept of the D-algorithm, we can address, in more detail, the actual implementation of the algorithm. The real purpose of the D-algorithm is to allow test pattern generation to be automated on a computer; therefore, we must develop mechanisms that allow the automation to be accomplished in an easy manner. More specifically, we need a procedure for determining, in a systematic way, the pdcs and pdcfs of any arbitrary logic module. Recall that a logic module can be a single gate or a function requiring several gates. Finally, a procedure for performing the D-drive and the consistency operations must be devised.

TABLE 7.16 Forward drive and backward trace operations for developing a test vector for line 9 stuck-at-0 in the circuit of Fig. 7.21

	1	2	3	4	5	6	7	8	9	10	11	Activity vector
Forward drive				0				1	D			9
				0			1	1	D	D'		10
	1			0			1	1	D	D'	D	11
Backward trace	1	1	0	0			1	1	D	D'	D	
	1	X	0	0	X	0	1	1	D	D'	D	

7.3 ■ Test Pattern Generation

The calculation of either the pdcs or the pdcfs begins with the singular cover of the logic module. As stated earlier, the singular cover of a logic module is simply a reduced truth table. Table 7.17 shows the singular covers of several basic gates. Each entry, or row, in the singular cover is called a *cube*. A set of cubes that completely defines the output of a module for any

TABLE 7.17 Singular covers of several basic logic gates

Logic gate	Singular cover		
	A	B	C
AND (A,B → C)	0	X	0
	X	0	0
	1	1	1
	A	B	C
OR (A,B → C)	1	X	1
	X	1	1
	0	0	0
	A	B	
NOT (A → B)	1	0	
	0	1	
	A	B	C
NAND (A,B → C)	0	X	1
	X	0	1
	1	1	0
	A	B	C
NOR (A,B → C)	1	X	0
	X	1	0
	0	0	1
	A	B	C
XOR (A,B → C)	1	1	0
	0	1	1
	1	0	1
	0	0	0

possible input combination is the singular cover. The notation SC_N represents the singular cover of a normal, or fault-free, logic module. The singular cover of a module containing several gates can be created by combining the singular covers of the individual gates, as shown in Fig. 7.25. The notation N_0 represents the set of cubes from SC_N that have 0 as an output, whereas N_1 represents the set of cubes from SC_N that have 1 as an output.

To determine the pdcfs of a logic module, we must first find the singular cover SC_N of the fault-free module and then find the singular covers SC_{f_i} of the module when each possible single, stuck-type fault f_i is present in the module. The composite singular cover of the faulty module, denoted as SC_F, is the collection of all unique cubes from all the singular covers SC_{f_i}. The set

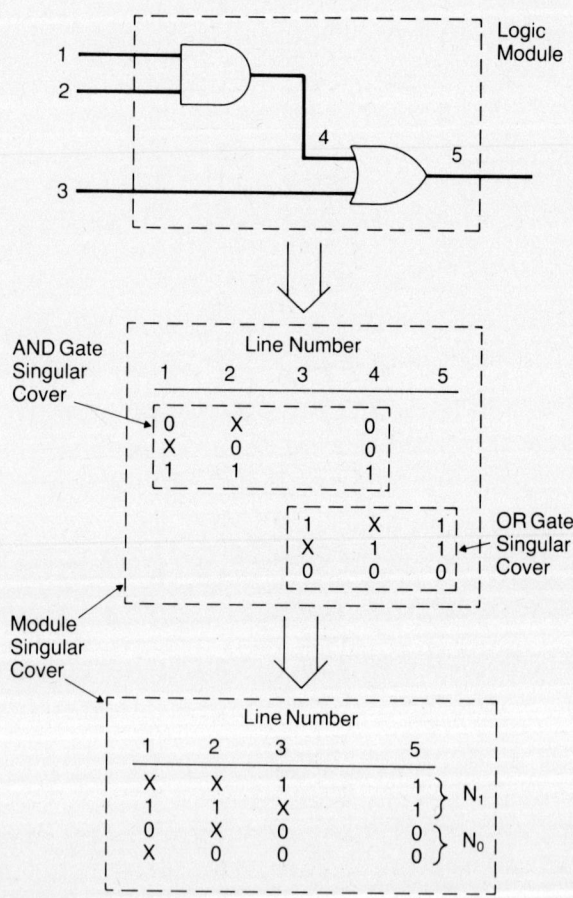

Fig. 7.25 The singular cover of a logic module can be constructed from the singular covers of the logic gates that make up that module.

of cubes in SC_F that have 0 as an output is called F_0, whereas the set of cubes that have 1 as an output is denoted F_1.

A pdcf contains an input combination that gives one output when the module is normal and the opposite output when a fault is present. Consequently, the pdcfs can be found by comparing N_0 with F_1 to find input combinations that normally produce a 0 output but produce a 1 when some fault is present, and by comparing N_1 with F_0 to find input combinations that normally produce a 1 output but produce a 0 when some fault is present. Once the input combinations are found, the output is assigned a value of D in each case where the output is normally a 1 but is 0 when a fault is present; that is, cubes found from comparison of N_1 and F_0. The output is assigned a value of D' when the output is normally a 0 but is 1 when a fault is present; that is, cubes formed by comparing N_0 and F_1.

The comparison process is facilitated by an intersection operation. The intersection of cube A with cube B is denoted $A \cap B$ and is formed by intersecting each element, or coordinate, of A with each corresponding element of B using certain rules of intersection. For example,

$$A \cap B = (a_1 \cap b_1, a_2 \cap b_2, \ldots, a_n \cap b_n)$$

where

$$A = (a_1, a_2, \ldots, a_n)$$
$$B = (b_1, b_2, \ldots, b_n)$$

with the rules of intersection being

$$0 \cap 0 = 0$$
$$1 \cap 1 = 1$$
$$0 \cap 1 = \phi$$
$$1 \cap 0 = \phi$$
$$0 \cap X = 0$$
$$X \cap 0 = 0$$
$$1 \cap X = 1$$
$$X \cap 1 = 1$$

where ϕ represents the empty set, or the null set. As an example, (0100) $\cap$ (00X0) = (0ϕ00). When using the intersection operation to form pdcfs, the output coordinates of each cube are ignored. The outputs of each pdcf are assigned a D or D' based on the value of the normal and faulty outputs.

The construction of the pdcfs for the two-input AND gate is illustrated in Fig. 7.26, where the normal singular cover (SC_N), each possible faulty sin-

Normal Singular Cover	Faulty Singular Covers		Composite Faulty Singular Cover	Intersections	pdcfs		
SC_N	SC_{A0}	SC_{A1}	SC_F	$N_1 \cap F_0$	A	B	C
A B C	A B C	A B C	A B C		1	1	D
0 X 0 ⎫	X X 0	X 1 1	X X 0 ⎫	$111 \cap XX0 = 11D$	0	1	D'
X 0 0 ⎬ N_0		X 0 0	X 0 0 ⎬ F_0	$111 \cap X00 = \phi$	1	0	D'
1 1 1 ⎭ N_1			0 X 0 ⎭	$111 \cap 0X0 = \phi$	0	X	D'
	SC_{B0}	SC_{B1}	X 1 1 ⎫		X	0	D'
	A B C	A B C	1 X 1 ⎬ F_1	$N_0 \cap F_1$			
	X X 0	1 X 1	X X 1 ⎭				
		0 X 0		$0X0 \cap X11 = 01D'$			
				$0X0 \cap 1X1 = \phi$			
	SC_{C0}	SC_{C1}		$0X0 \cap XX1 = 0XD'$			
	A B C	A B C		$X00 \cap X11 = \phi$			
	X X 0	X X 1		$X00 \cap 1X1 = 10D'$			
				$X00 \cap XX1 = X0D'$			

Fig. 7.26 Procedure for finding the pdcfs for the two-input AND gate.

gular cover (SC_{f_i}), the composite faulty singular cover (SC_F), the intersection operations, and the resulting pdcfs are shown.

The construction of the pdcs of a logic module is somewhat simpler than the construction of the pdcfs. Recall that a pdc is an input specification that propagates a D signal from one input of a module to the output of that module. The pdcs of a logic module can be found from the singular cover SC_N of that module. A pdc can be found by comparing cubes from N_0 with cubes from N_1 while ignoring the output coordinates. The objective is to find a cube in N_0 and one in N_1 that differ in only one input coordinate. By virtue of the two cubes being in N_0 and N_1, we know that their output coordinates differ, so if only one input differs, the change in output must have been caused by the one change in input. In other words, the output is sensitive to changes on that one input line when the remaining inputs are specified accordingly. The comparison can be performed by first intersecting cubes from N_0 with cubes from N_1 and then intersecting cubes from N_1 with cubes from N_0 using the following rules of intersection:

$$0 \cap 0 = 0$$
$$1 \cap 1 = 1$$
$$0 \cap 1 = D'$$
$$1 \cap 0 = D$$
$$0 \cap X = 0$$

$$X \cap 0 = 0$$
$$1 \cap X = 1$$
$$X \cap 1 = 1$$

As you can see, the order in which the intersection is performed does have an affect on the result.

As an example, consider the three-input AND gate shown in Fig. 7.27 along with its singular cover. Figure 7.27 also shows the intersections that must be performed and the resulting pdcs of the three-input AND gate.

Now that we have a means of finding the pdcfs and the pdcs of any logic module, we can turn our attention to the implementation of the complete D-algorithm. The basic steps of the algorithm are:

1. Select a pdcf for the fault for which a test pattern is to be generated.
2. Use the pdcs to propagate the D signal to a primary output of the circuit (this is the D-drive).
3. Use the singular covers of each logic module to perform the consistency operation.

The selection of the pdcf forms the first attempt at the test pattern by specifying the values of certain lines within the circuit. This first attempt is called test cube tc_0. The next test cube, tc_1, is formed by intersecting tc_0 with the appropriate pdc from the pdc table. The intersection is performed using the same rules used to construct the pdcfs. The D-drive continues until the activity vector has a value corresponding to a primary output of the circuit. The consistency operation is performed by intersecting the test cubes with appropriate cubes from the singular covers. The process is completed when

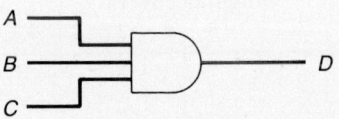

Singular Cover				Intersections															pdcs			
A	B	C	D																A	B	C	D
0	X	X	0 ⎫	0	X	X	0	∩	1	1	1	1	=	D'	1	1	D'					
X	0	X	0 ⎬ N_0	X	0	X	0	∩	1	1	1	1	=	1	D'	1	D'		D'	1	1	D' ⎫
X	X	0	0 ⎭	X	X	0	0	∩	1	1	1	1	=	1	1	D'	D'		1	D'	1	D' ⎬ Polarized
1	1	1	1 N_1	1	1	1	1	∩	0	X	X	0	=	D	1	1	D		1	1	D'	D' ⎭ pdcs
				1	1	1	1	∩	X	0	X	0	=	1	D	1	D		D	1	1	D ⎫
				1	1	1	1	∩	X	X	0	0	=	1	1	D	D		1	D	1	D ⎬
																			1	1	D	D ⎭

Fig. 7.27 Example construction of the pdcs for the three-input AND gate.

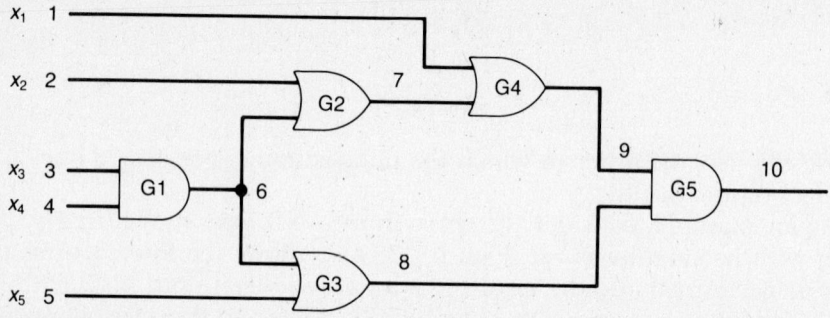

Fig. 7.28 Combinational circuit used to illustrate the D-algorithm.

a consistent set of primary inputs that achieves all signal values specified by the pdcf and the *D*-drive is obtained.

As an illustration of the complete *D*-algorithm, consider the process of determining a test pattern for a stuck-at-1 fault on line 6 of the circuit in Fig. 7.28. Table 7.18 illustrates the complete process.

7.3.7 Fault Simulation for Test Pattern Generation

The process of test pattern generation is clearly a complicated and time-consuming procedure for large digital circuits when using the techniques that have been presented thus far. All of the approaches that we have con-

TABLE 7.18(a) Singular cover table for the circuit of Fig. 7.28

	Singular covers									
	1	2	3	4	5	6	7	8	9	10
a			0	X			0			
b			X	0			0			
c			1	1			1			
d	1					X	1			
e	X					1	1			
f	0					0	0			
g					1	X		1		
h					X	1		1		
i					0	0		0		
j	1						X		1	
k	X						1		1	
l	0						0		0	
m								0	X	0
n								X	0	0
o								1	1	1

TABLE 7.18(b) pdcf table for the circuit of Fig. 7.28

| | \multicolumn{10}{c}{pdcfs} |
	1	2	3	4	5	6	7	8	9	10
a			0	1		D'				
b			1	0		D'				
c			1	1		D				
d		0				0	D'			
e		1				0	D			
f		0				1	D			
g					0	0		D'		
h					1	0		D		
i					0	1		D		
j	0						0		D'	
k	1						0		D	
l	0						1		D	
m								0	1	D'
n								1	0	D'
o								1	1	D

TABLE 7.18(c) pdc table for the circuit of Fig. 7.28

| | \multicolumn{10}{c}{pdcs} |
	1	2	3	4	5	6	7	8	9	10
a			D	1		D				
b			1	D		D				
c			D'	1		D'				
d			1	D'		D'				
e		D				0	D			
f		0				D	D			
g		D'				0	D'			
h		0				D'	D'			
i					D	0		D		
j					0	D		D		
k					D'	0		D'		
l					0	D'		D'		
m	D						0	D		
n	0						D	D		
o	D'						0	D'		
p	0						D'	D'		
q								D	1	D
r								1	D	D
s								D'	1	D'
t								1	D'	D'

TABLE 7.18 (d) Forward drive and backward trace processes for a stuck-at-1 fault on line 6 in the circuit of Fig. 7.28

Test cube	1	2	3	4	5	6	7	8	9	10	Activity vector
pdcf a = tc_0			0	1		D'					6
$tc_1 = tc_0 \cap$ pdc h		0	0	1		D'	D'				7
$tc_2 = tc_1 \cap$ pdc p	0	0	0	1		D'	D'		D'		9
$tc_3 = tc_2 \cap$ pdc t	0	0	0	1		D'	D'	1	D'	D'	10
Forward drive complete. Begin backward trace. ↓											
$tc_4 = tc_3 \cap$ sc g Backward trace complete.	0	0	0	1	1	D'	D'	1	D'	D'	

Test vector = $(x_1 x_2 x_3 x_4 x_5)$ = (00011)

sidered begin by first selecting a fault and then using some deterministic algorithm, such as the Boolean difference, to determine a test pattern for that fault. The basic concept of the *fault simulation* approach to test pattern generation is to select a primary input combination and determine its fault detection capabilities by simulation. The process is repeated until: (1) all faults are covered, (2) at least an acceptable number of faults are covered, or (3) some predefined stopping point is reached.

The process of determining the fault detection capabilities of a primary input combination through simulation is illustrated in Fig. 7.29. The input combination is applied to a simulation of the *good*, or fault-free circuit, and also applied to a simulation of the *faulty* circuit. The faulty circuit contains the fault, or faults, for which a test pattern is needed. If the outputs of the good and faulty circuits disagree, the primary input combination is retained as a test pattern for the faults contained in the faulty circuit.

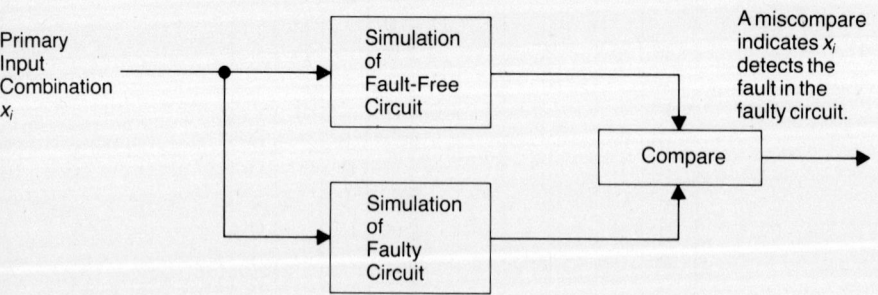

Fig. 7.29 The use of fault simulation to determine test patterns.

There are two fundamental methods of generating test patterns using the fault simulation approach: the **row method** and the **column method.** The row method derives its name because the objective is to find, at least partially, the rows of the fault detection table. In the column method, the goal is to create, again only partially, the columns of the fault detection table.

The Row Method

The basic concept of the row method is to select a primary input combination TP_1, and then attempt to determine which faults are detected by TP_1. Suppose that a circuit can contain any one of n stuck-type faults, $(f_1, f_2, \ldots, f_n)$. The simulation of the good circuit is conducted with TP_1 as the primary input combination, and the *good* output z_{good} is recorded. A simulation is then performed of the faulty circuit containing f_1 and having TP_1 as its primary input, and the *faulty* output z_{f_1} is recorded. If z_{good} and z_{f_1} disagree, TP_1 is a test pattern for f_1. The process is repeated with the faulty circuit containing the single fault f_2, then with the faulty circuit containing fault f_3, and so on until all n faults have been considered. In each simulation, the input combination is TP_1. The result is the row of the fault detection table for the input combination TP_1.

The next step in the row method is to select a second primary input combination, TP_2. The preceding process is repeated but only those faults not covered by TP_1 are considered. Therefore, if we were lucky enough that TP_1 covered half of the n faults, only $n/2$ faults need be considered for TP_2. This process continues until: (1) all faults are covered, (2) a specified percentage of the faults are covered, or (3) a user-defined stopping point is reached. Normally, the stopping point is defined in terms of the number of primary input combinations selected or the amount of processing time consumed.

The row method is illustrated in Fig. 7.30. A simple combinational circuit is shown in Fig. 7.30, as well as the portions of the fault detection table that are created during the process.

The Column Method

The column method is only slightly different from the row method. Suppose that a circuit can contain any one of n faults, $(f_1, f_2, \ldots, f_n)$. The fault detection table, therefore, contains n columns, one for each of the n faults. The column method first selects fault f_1, from the set $(f_1, f_2, \ldots, f_n)$. The faulty circuit containing f_1 is simulated with a primary input combination TP_r. If the output of the fault-free circuit with input TP_r differs from the output of the faulty circuit containing f_1 with input TP_r, TP_r is saved as a test pattern TP_1 for the fault f_1. If TP_r is not a test pattern for f_1, another primary input combination is selected and the process is repeated. This continues until a test pattern is found for f_1. The fault f_2 is then selected, and the complete proce-

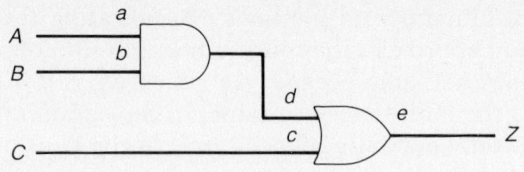

Input Pattern A B C	Fault-Free Output Z_{good}	Simulation Results Faulty Outputs										Faults Detected
		Z_{a_0}	Z_{b_0}	Z_{c_0}	Z_{d_0}	Z_{e_0}	Z_{a_1}	Z_{b_1}	Z_{c_1}	Z_{d_1}	Z_{e_1}	
0 0 0	0	0	0	0	0	0	0	0	1	1	1	c_1, d_1, e_1
0 0 1	1	1	1	0	1	0	1	1	*	*	*	c_0, e_0
0 1 0	0	0	0	*	0	*	1	0	*	*	*	a_1
0 1 1	1	1	1	*	1	*	*	1	*	*	*	none
1 0 0	0	0	0	*	0	*	*	1	*	*	*	b_1
1 0 1	1	1	1	*	1	*	*	*	*	*	*	none
1 1 0	1	0	0	*	0	*	*	*	*	*	*	a_0, b_0, d_0

Patterns simulated sequentially to generate rows of table

Simulation Stops

Test set = {000, 001, 010, 100, 110}

* – not simulated for this input pattern because the fault was detected by a previous pattern.

Fig. 7.30 The row method as used to determine a test set for a simple combinational circuit. Simulations are performed in a fashion that fills in the rows of the table.

dure is repeated until a test pattern is found for f_2. The program may place limitations on the number of primary input patterns that will be tried for a given fault in an effort to reduce the time required for test pattern generation.

Once all n faults have been considered, the test patterns that have been determined will form a test experiment, $T = (TP_1, TP_2, \ldots, TP_n)$, consisting of n test patterns. Several of the patterns may be identical and should be eliminated from the test set to reduce the number of test patterns to some number k that is less than n.

The major problem with the column method is that n unique test patterns can be found for the n faults; therefore, a very large test experiment can be produced. One approach that attempts to reduce this problem is as follows. As faults are considered, the set of test vectors is gradually created. For example, after considering f_1, the test set contains TP_1, the test pattern for f_1. To reduce the number of test patterns found, TP_1 should be the first input pattern tried for the fault f_2. If TP_1 is a test pattern for f_2, no further simulations must be run for f_2, and no additional test patterns must be added to the test set. If TP_1 is not a test pattern for f_2, a new potential test pattern must be selected from the set of primary input combinations.

As an example of the column method, Fig. 7.31 shows the complete process and the resulting test patterns for a simple combinational circuit. The column method starts with the first fault, a_0, and simulates each input combination until a test pattern is found. As shown in Fig. 7.31, the first test pattern found for a_0 is 110. The technique then uses 110 as the first input combination for the next fault, b_0. Because 110 is also a test pattern for b_0, only one simulation is required for b_0. The process continues in this manner until a test set is completed. The circuit in Fig. 7.31 is the same one used to illustrate the row method, so the two procedures and the two test sets can be compared. Note that the row method requires 36 simulations of faulty circuits compared to 30 simulations for the column method.

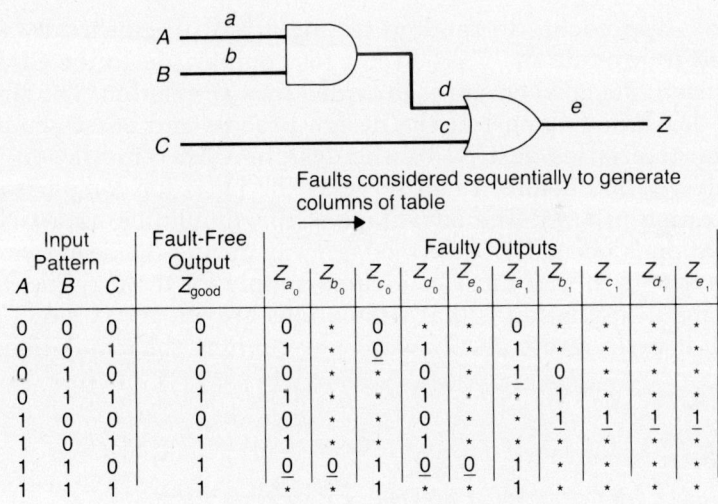

Faults considered sequentially to generate columns of table

Input Pattern A B C	Fault-Free Output Z_{good}	Z_{a_0}	Z_{b_0}	Z_{c_0}	Z_{d_0}	Z_{e_0}	Z_{a_1}	Z_{b_1}	Z_{c_1}	Z_{d_1}	Z_{e_1}
0 0 0	0	0	*	0	*	*	0	*	*	*	*
0 0 1	1	1	*	<u>0</u>	*	*	1	*	*	*	*
0 1 0	0	0	*	*	*	0	1	0	*	*	*
0 1 1	1	1	*	*	1	*	1	*	*	*	*
1 0 0	0	0	*	*	0	*	*	1	1	1	1
1 0 1	1	1	*	*	1	*	*	*	*	*	*
1 1 0	1	<u>0</u>	<u>0</u>	1	0	<u>0</u>	1	*	*	*	*
1 1 1	1	*	*	1	*	*	1	*	*	*	*

Test set = {001, 010, 100, 110}

* = input pattern not simulated for this fault because a test pattern was found in a previous simulation.

<u> </u> = underlined entries denote simulation result that identified the test pattern.

Fig. 7.31 The column method used to determine a test set for a simple combinational circuit. Simulations are performed in a fashion that fills in the columns of the table.

7.4 Random Testing

Another test procedure that often uses simulation is called **random testing** [David and Blanchet 1976]. The basic idea of random testing is that a test experiment is not determined prior to the performance of the test process. Instead, test patterns are selected at random *during* the test procedure. The major advantage of random testing is that the time-consuming process of test pattern generation is eliminated. The disadvantage of random testing is that it is very difficult to ascertain the fault coverage that is provided by the test procedure.

Random testing is normally conducted in one of two ways. Both techniques use the same concept illustrated in Fig. 7.32. The responses of a *good* circuit and the *circuit under test* (CUT) are compared as a means of detecting faults. The primary input combinations that are applied to the circuits are selected at random, thus the name *random testing*. If the CUT contains a fault, the hope is that one of the primary inputs selected at random will sensitize the fault and cause an output from the CUT that differs from the output of the good circuit.

The two approaches to random testing are distinguished by the technique used to provide the *good* circuit for comparison to the CUT. In the first approach, the good circuit is provided by a simulation. The simulation has been developed as part of the design process and possesses the functional characteristics desired by the designer. The simulation must be completely verified before it can serve as the basis for comparison in the random testing process. The advantage of the simulation approach is that the simulation is normally performed anyway during the design process, so it is not a new item that must be created simply for testing. The disadvantage of using a simulation is that simulations are often very slow. This disadvantage is easily overcome, however, by running the simulation off-line.

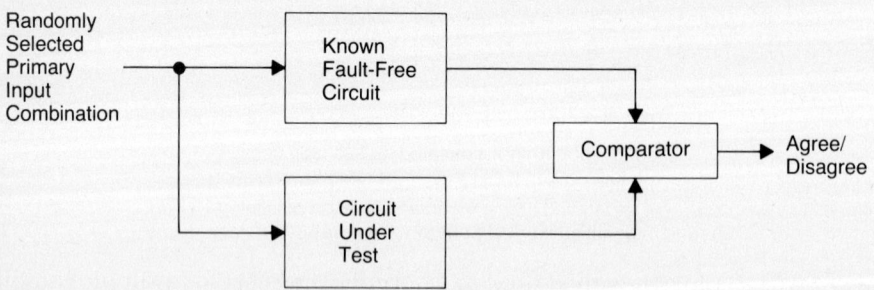

Fig. 7.32 Random testing involves comparing a known "good" circuit to a circuit under test. The input combinations are selected at random.

In other words, the random patterns are selected, the simulations run, and the responses recorded before placing a circuit in the test process. The responses of the CUT to the same patterns applied to the simulation are determined and compared to those previously derived from the simulation.

The second approach uses a second physical circuit as the *good* unit. The manufacturer of a circuit may invest the time required to guarantee that one of a type of integrated circuit is completely correct by performing an exhaustive test of one device. This device then becomes the *perfect* chip to which all other chips must be compared. The response of the CUT to certain random input patterns is then compared to the response of the *perfect* circuit to those same patterns. The perfect circuit is often called the *gold* circuit. The disadvantage of this second approach is that the gold circuit can fail sometime during its operation, and you must be certain that disagreements between the gold circuit and the CUT are due to faults in the CUT and not to faults in the gold circuit. Again, this disadvantage can be overcome by simply recording the responses of the gold device to the set of random patterns and comparing the response of each CUT to those recorded responses. Therefore, the CUT only needs to perform correctly the one time. Or, conversely, the responses of the gold device can simply be obtained from the specifications.

7.5 Signature Analysis

The basic concept of **signature analysis** is very simple [Nadig 1977]. If a sequence of primary input combinations (TP_1 followed by TP_2 and so on until TP_k) is applied to a circuit under test (CUT), every point in the circuit responds with a certain sequence of values. If certain critical points are selected as test points and the response of each test point monitored, a decision can be made concerning whether or not the circuit is faulty. If the circuit becomes faulty, the sequence of values found on at least one test point within the circuit should change, provided that the test points and the inputs are selected appropriately. These changes can identify the circuit as faulty.

The problem with using the information provided by data streams at test points within the circuit is that the amount of information that must be handled can become overwhelming. For example, k data bits at each of n test points within a circuit represent kn total data bits. The approach used in signature analysis is to compress, or encode, the data at each test point into quantities called *signatures*. The signatures of a circuit characterize the response of the circuit.

To use signature analysis for testing purposes the following procedure must be followed. First, a sequence of primary input combinations is selected to sufficiently exercise the circuit. The primary inputs are selected at random or using some other approach. Second, several test points within the circuit are selected, and each test point is connected to an encoding circuit. Third, the selected primary input combinations are applied to a *good* (fault free) circuit, and the encoding circuits create the *good* signatures of the circuit. Finally, during the testing of another circuit of the same type, the good signatures are compared to those produced by the CUT in response to the same set of input patterns. If the signatures of the good circuit disagree with those of the CUT, the CUT is labeled as faulty; otherwise, the CUT is labeled as fault-free. The basic concept of signature analysis is illustrated in Fig. 7.33.

The encoding circuits used in the signature analysis approach are called linear feedback shift registers (LFSR). The LFSR is simply a shift register formed from standard flip flops, with the outputs of selected flip flops being added (modulo-2 addition) to the incoming data stream, as illustrated in Fig. 7.34. The modulo-2 addition can be performed using a simple EXCLUSIVE-OR function. Figure 7.34 shows a register of some arbitrary length with the taps being controlled by switches. By appropriately setting each switch, the output of any flip flop can be provided as feedback.

As we saw in Chapter 3 when we discussed cyclic codes, the LFSR is simply a circuit that can be used to provide multiplication or division. Re-

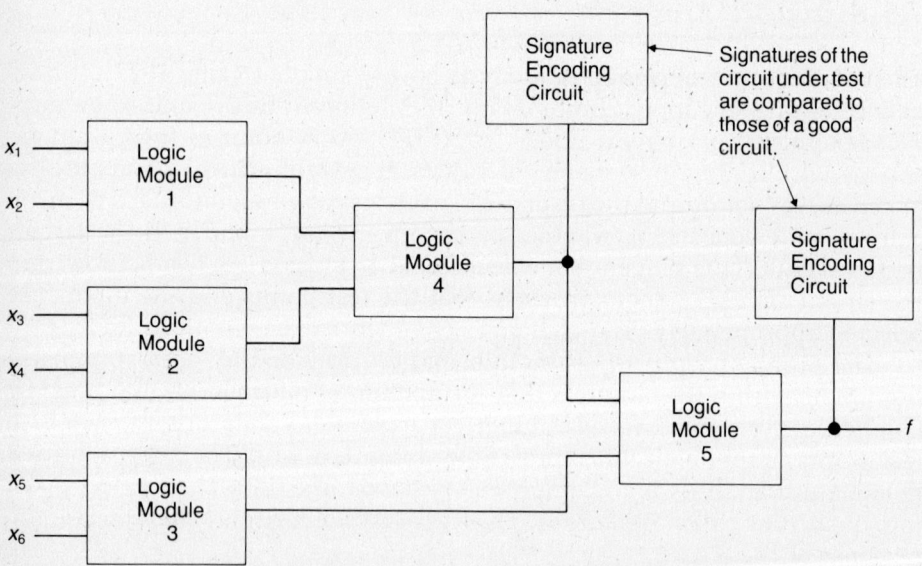

Fig. 7.33 In signature analysis special signature encoding circuits are placed at specified points within the circuit. Signatures are recorded for a given input sequence and compared to known signatures of a fault-free circuit.

7.5 ■ Signature Analysis

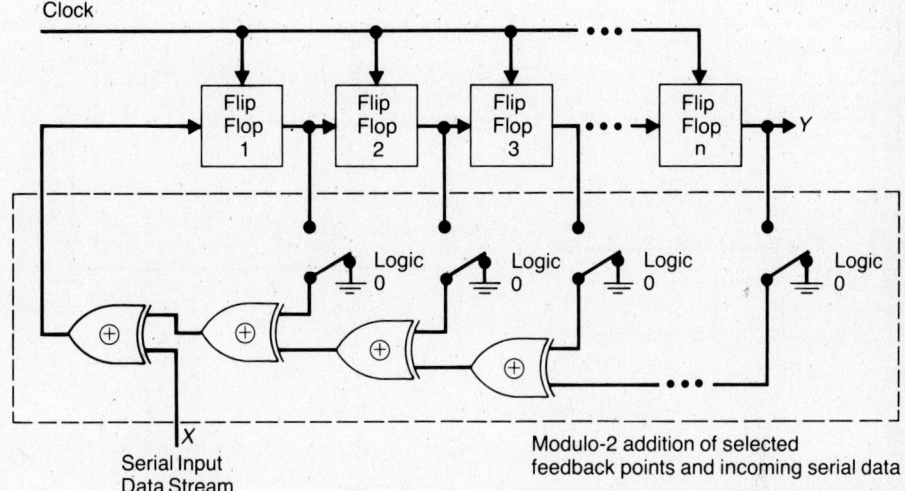

Fig. 7.34 The general linear feedback shift register allows the feedback to be selected from any (or all) combination of flip flop outputs. The feedback points are summed (modulo-2) with the incoming serial data.

ferring to Fig. 7.34, we see that the output Y of the LFSR can be written in terms of the input X as

$$Y = D^n(X + Ya_n + D^{-(n-1)}Ya_1 + D^{-(n-2)}Ya_2 + \cdots + D^{-1}Ya_{n-1})$$

where each flip flop provides a delay of D and all additions are modulo-2 additions. The variables a_i represent the position of the ith switch; a_i is 0 if the switch is connected to logic 0 and a_i is 1, otherwise. Rearranging the above equation yields

$$Y = \frac{X}{(D^{-n} + D^{-(n-1)}a_1 + D^{-(n-2)}a_2 + D^{-(n-3)}a_3 + \cdots + D^{-1}a_{n-1} + a_n)}$$

One way to visualize the operation of the LFSR is, as we did for cyclic codes, to represent the incoming data stream as a polynomial with coefficients equal to the bits of the data stream. The output data stream is the coefficients of a polynomial formed by dividing the incoming polynomial by the characteristic (often called the generator polynomial) polynomial of the circuit. After the incoming data stream is passed through the LFSR, the contents of the LFSR will be the remainder (usually called the residue or the syndrome) of the division process. The coefficients a_i are used to appropriately specify the desired generator polynomial. Figure 7.35 shows a simple 4-bit LFSR and the syndrome produced for a specific input stream. The syndrome is the signature used in signature analysis.

520 Testing

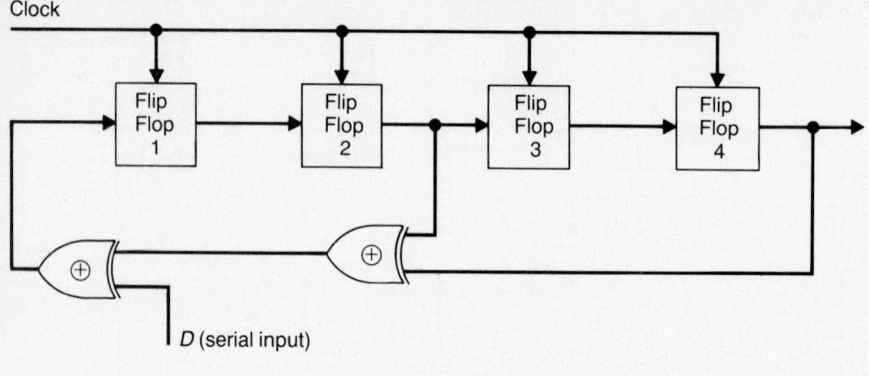

Fig. 7.35 The contents of the LFSR after the occurrence of a clock pulse depends on the contents prior to the occurrence of the clock pulse and the value of the serial input line.

As an example application of signature analysis, consider the small system shown in Fig. 7.36. The system contains four modules that are simple combinational circuits, and the outputs of two of those modules are monitored to obtain signatures using the simple LFSR also shown in Fig. 7.36. Table 7.19 shows the *good* signatures obtained from three randomly selected primary input combinations as well as the faulty signatures obtained when line α is stuck-at-1. Clearly, the signatures are different, allowing the faulty condition to be detected.

7.6 Design for Testability

We have examined in the previous sections of this chapter several techniques for generating test patterns for combinational circuits. We have not, however, addressed the problem of test pattern generation for sequential circuits. The importance of testing sequential devices is clear because of the frequency of their occurrence in practical designs. Very few complex de-

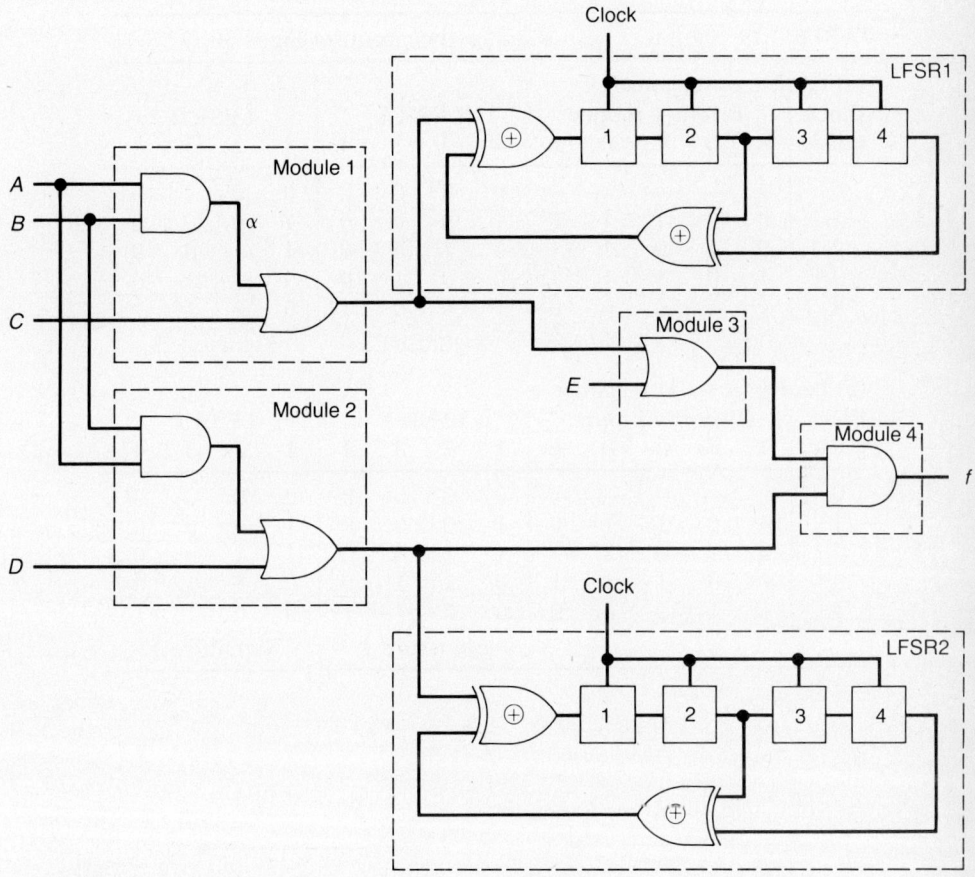

Fig. 7.36 Signature registers (LFSR 1 and LFSR 2) are placed at strategic points within the circuit. In this example, two signature registers are used.

signs can be achieved using only combinational logic; therefore, it is crucial that test techniques be available for sequential circuits.

As a quick review, the basic structure of a sequential circuit is shown in Fig. 7.37. The circuit is composed of combinational logic and memory elements; the memory elements usually being some type of flip flop. The circuit has sets of n *primary inputs* $(x_1, \ldots, x_n)$, m *primary outputs* $(z_1, \ldots, z_m)$, k *state variables* $(y_1, \ldots, y_k)$, and k *excitation variables* $(Y_1, \ldots, Y_k)$. The primary outputs are functions of either, or both, the primary inputs and the state variables. If the primary outputs depend only on the state variables, the circuit is called a *Moore sequential machine*. If the primary outputs depend on both the state variables and the primary inputs, the circuit is

522   Testing

TABLE 7.19 Signature responses for the circuit of Fig. 7.36

(a) Fault-free response

Clock pulse	Primary inputs					LFSR 1				LFSR 2			
	A	B	C	D	E	1	2	3	4	1	2	3	4
0	—	—	—	—	—	0	0	0	0	0	0	0	0
1	0	0	1	1	0	1	0	0	0	1	0	0	0
2	0	1	0	1	1	0	1	0	0	1	1	0	0
3	1	0	1	1	1	0	0	1	0	0	1	1	0
4	1	1	0	0	0	1	0	0	1	0	0	1	1
						Signature 1				Signature 2			

(b) Faulty (α s-a-1) response

Clock pulse	Primary inputs					LFSR 1				LFSR 2			
	A	B	C	D	E	1	2	3	4	1	2	3	4
0	—	—	—	—	—	0	0	0	0	0	0	0	0
1	0	0	1	1	0	1	0	0	0	1	0	0	0
2	0	1	0	1	1	1	1	0	0	1	1	0	0
3	1	0	1	1	1	0	1	1	0	0	1	1	0
4	1	1	0	0	0	0	0	1	1	0	0	1	1
						Signature 1				Signature 2			

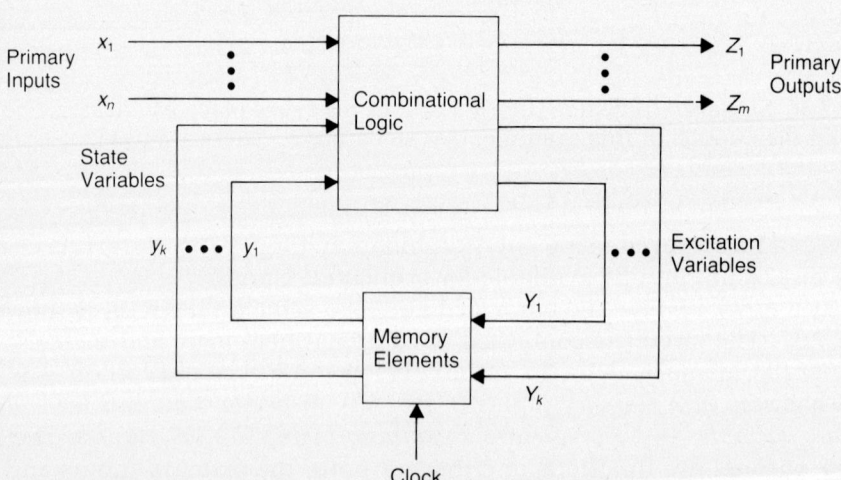

Fig. 7.37 The basic structure of a sequential circuit includes both combinational logic and memory elements (usually flip flops).

called a *Mealy sequential machine*. The excitation variables are functions of the *present* state variables and the primary inputs, and specify the *next state* of the memory elements. The memory elements are all clocked by a single clock signal.

Sequential circuits are extremely difficult to test. Not only must we verify that the circuit provides the correct primary outputs for a given set of primary inputs, but we must also verify that the correct state transition occurs. To completely test a sequential machine requires that the primary outputs and the state transitions for all possible primary inputs *and* all possible initial states be verified. As k and n become large, the testing of sequential machines becomes prohibitively complex. In fact it is impossible from a practical standpoint to completely, functionally test present-day sequential devices unless some constraints are placed on the design of such devices. Constraints on the design of digital circuits are one form of design for testability (DFT) [Williams and Parker 1983]. The objective of DFT is to create a design that is easy to test in an economical fashion.

Design for testability (DFT) techniques can be divided into two categories: *ad hoc* methods and *structured* approaches. Ad hoc techniques consist of heuristic methods such as circuit partitioning and adding extra test points. Circuit partitioning is a method by which a circuit is divided into small modules with each module being independently tested. The idea is that several small circuits are much easier to test than one large circuit. Test points are lines within a circuit that are normally not accessible but that are brought out to an accessible point to improve the testability of the circuit. For example, several internal lines can be connected to package pins to allow an external test device to control and observe those points. Although ad hoc approaches can be adequate for a specific design, they are not generally applicable to all designs. Also, ad hoc approaches are more of an art than a science. Two designers working on the same circuit might create two completely different designs using ad hoc methods. In other words, there is little, if any, standardization when using ad hoc methods. The quality of an ad hoc design will certainly depend on the experience and cleverness of the designer.

Structured techniques, on the other hand, involve a set of general design rules by which a design is implemented. All designs are typically required to follow the same set of design rules. The advantages of structured design for testability are that it improves the ability to test a sequential machine, and it provides standardization of designs.

Most structured DFT techniques adhere to the principle of converting sequential machines into combinational circuits for test purposes. The conversion process is accomplished by providing a means of breaking the feedback loop in the sequential machine such that the state of the machine can be controlled and observed. The correct operation of the machine can then be more easily verified. This basic concept is illustrated in Fig. 7.38. The design technique of Fig. 7.38 provides a method of controlling the state variables of the circuit and observing the excitation variables. The test process is then reduced to one of testing the combinational logic that has as its inputs the primary inputs and the state variables and that has as its outputs the primary outputs and the excitation variables. With the feedback loop broken and the state variables and excitation variables accessible, all inputs to the combinational logic become completely controllable and all outputs become completely observable.

In this chapter, we examine several structured design for testability techniques including Level Sensitive Scan Design (LSSD), Scan Path, BILBO, Scan/Set Logic, and the Random Access Scan approach. An example design that compares LSSD and Scan Path is presented to illustrate the cost of using structured DFT techniques.

7.6.1 Scanning as a Method of Design for Testability

The object of DFT is to gain observability and controllability of the circuit under consideration at a reasonable cost, both in terms of time and economics. **Scan design,** which is the method used by both Level Sensitive Scan Design (LSSD) and Scan Path, requires the use of specially designed, clocked flip flops that can be placed in either the *operate* or *test* mode. The flip flops

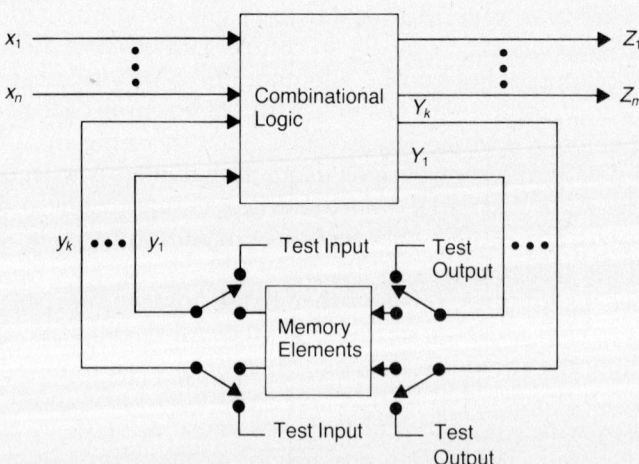

Fig. 7.38 Switches allow the feedback loop to be broken. Test inputs can then provide specified values for the state variables $(y_1, \ldots, y_k)$, and test outputs allow observation of the excitation variables $(Y_1, \ldots, Y_k)$. Consequently, all inputs to the combinational logic become controllable and all outputs become observable.

possess the capability to accept test vectors that control the present state of the circuit, and the ability to clock (or scan) out the current excitation variables of the circuit. Scanning techniques accomplish this capability by interconnecting all the flip flops in the circuit into a single shift register. An appropriate test vector can then be clocked into this shift register, a normal operation performed, and the excitation variables of the circuit may be similarly clocked out.

Figure 7.39 shows a generalized block diagram of a system using a scanning method of DFT. During normal operation, the individual flip flops of the shift register (SR) operate as totally independent flip flops and perform normal system functions. During the test process, however, each flip flop can be loaded with a specified value by shifting in a serial data stream via the scan-in line of the SR. In other words, the SR can accept a serial input via the scan-in line and provide a parallel output (the state variables). Likewise, the SR can accept a parallel input (the excitation variables) and provide a serial output via the scan-out line.

Once the flip flops are appropriately loaded for a test process, the combinational circuit can perform some normal operation. The primary outputs of the circuit can then be observed in a normal fashion, and the excitation variables observed by loading their values into the SR and shifting out the result via the scan-out line.

Both LSSD and Scan Path adhere to the basic structure shown in Fig. 7.39. The fundamental difference between LSSD and Scan Path is the structure of the flip flops used to construct the shift register. Each flip flop will be described in the subsequent sections.

Level Sensitive Scan Design (LSSD)

Level Sensitive Scan Design (LSSD) is a technique developed by IBM to deal with the testability issue [Williams and Parker 1983]. The term *level sensitive* refers to the fact that the steady-state response of a circuit to any allowable input change is independent of delays within the circuit. Also, the steady-state response of the circuit must be independent of the order in which signals within the circuit change. The term *scan design* implies that the technique uses the scanning approach illustrated in Fig. 7.39.

The key element in the LSSD methodology is the master-slave flip flop shown in Fig. 7.40. The flip flop shown in Fig. 7.40 is simply a master-slave, D-type flip flop that is provided with an extra input stage. The extra input stage allows the input to the master flip flop to come from either the normal data line or a special test line. During normal operation, the test input stage of the flip flop is disabled and the circuit performs as a normal master-slave, D-type flip flop. During testing, the normal input stage can be disabled to allow the flip flop to be loaded with the test input.

Race conditions are avoided in the flip flop by using separate, non-overlapping clocks for the master and slave flip flops, thereby providing the

526   Testing

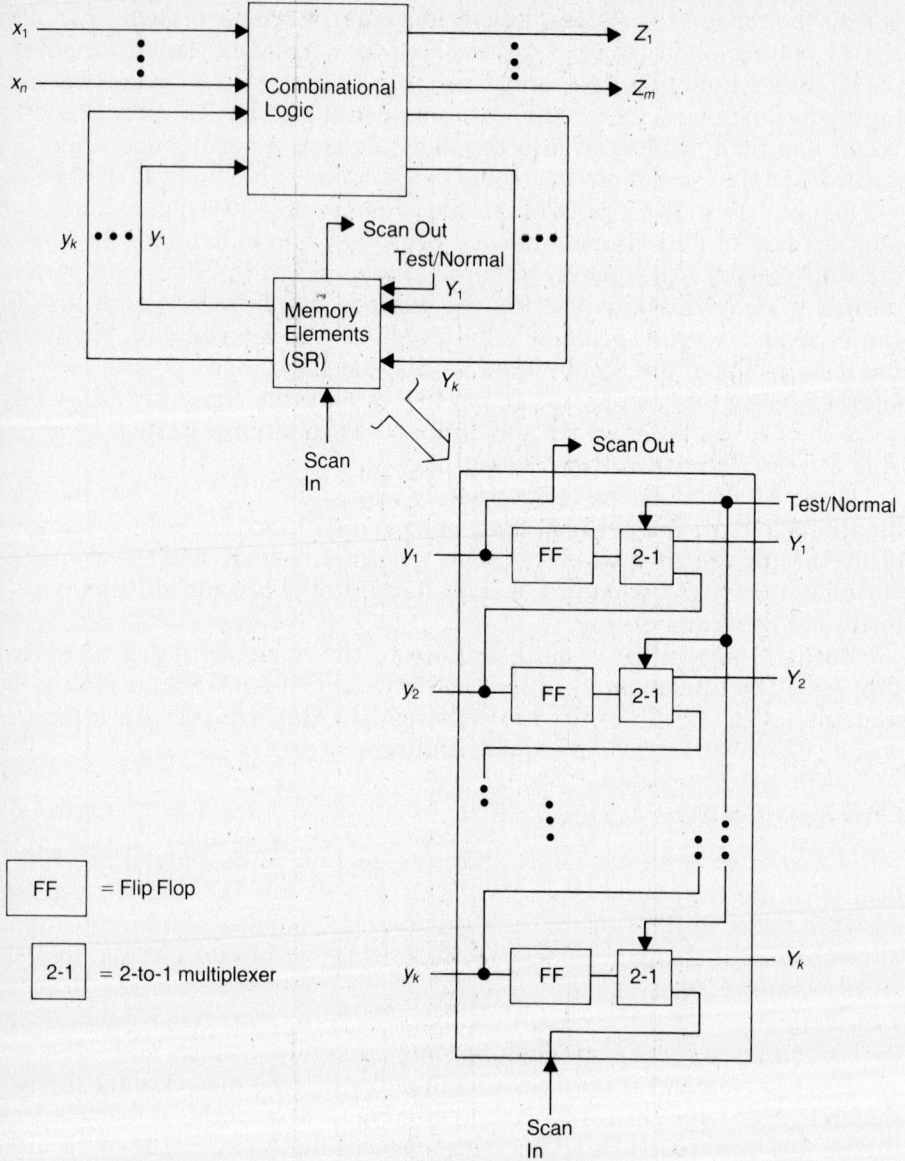

Fig. 7.39 Scan design uses a special shift register that can operate as a parallel-in parallel-out or serial-in/serial-out register.

level-sensitive operation. A third clock input is provided to be used for the scan operation. The scan process is also accomplished using nonoverlapping, two-phase clocking. The flip flops are threaded throughout the system,

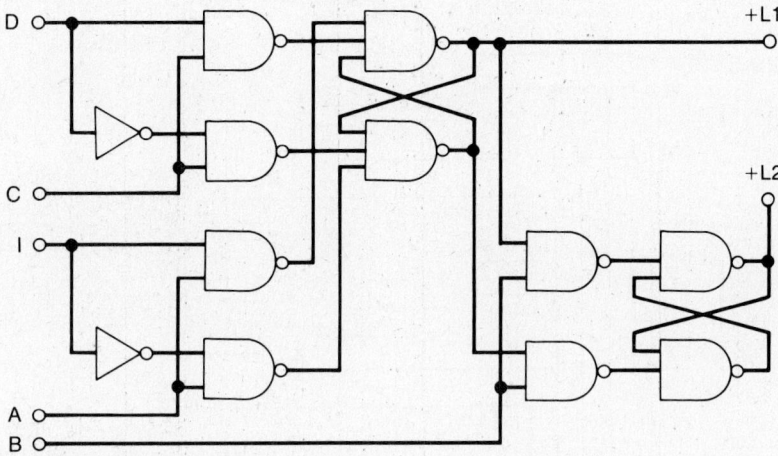

Fig. 7.40 The LSSD flip flop is a master-slave D flip flop, which uses two nonoverlapping clocks. During normal operation, clocks C and B are used in the master (L1) and slave (L2), respectively. During testing clocks A and B are used. Line I is the test input. (From [Williams and Parker 1983] © 1983 IEEE)

as shown in Fig. 7.41, to provide the scan feature by connecting the outputs of the slave portion of the flip flop to the scan input of the master section of the next flip flop. Normal operation is performed by operating clocks C and B, whereas the scan operation uses clocks A and B.

While the LSSD flip flop shown in Fig. 7.40 is the key aspect of LSSD, it is important to understand that LSSD is a *methodology* through which designs are performed. The LSSD methodology requires that the designer adhere to a set of basic rules. The purpose of the LSSD rules is to improve the testability of the resulting design. The LSSD design methodology consists of the six rules presented below [Eichelberger and Williams 1977].

1. All storage elements within a circuit must be implemented using the flip flop shown in Fig. 7.40. Without the use of the LSSD flip flops, the required scan paths could not be developed within the circuit.

2. The flip flops within the circuit must be controlled by two or more nonoverlapping clocks to provide for the level-sensitive operation of the circuit. As part of the requirement for nonoverlapping clocks, certain constraints are placed on the interconnection of flip flops within the circuit. First, the output of any arbitrary flip flop A can be the input to any other flip flop B if and only if the clock applied to B is not the clock applied to A. If the output of A is the input to B and both flip flops have the same clock, the clock width and the delay through A will determine whether or not B captures its input signal. Second, if the output of some

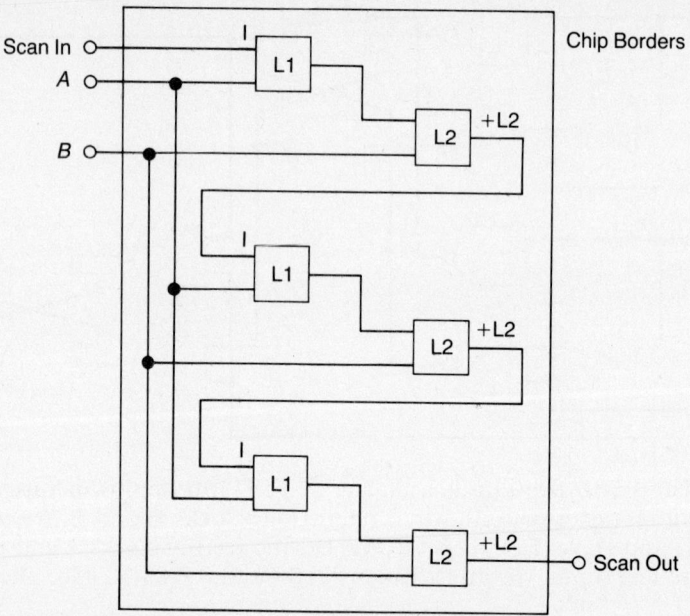

Fig. 7.41 The shift register is formed by connecting the output of L2 to the test input (I) of the next flip flop, continuing until all flip flops are connected together. The test input of the first flip flop in the string becomes the scan-in line, and the L2 output of the last flip flop in the string becomes the scan-out line. (From [Williams and Parker 1983] © 1983 IEEE)

flip flop A is used to enable some clock C_1, then C_1 can drive another flip flop if and only if the clock for A is not derived, in any way, from C_1.

3. There must exist a set of primary inputs, called the *primary clock inputs*, that control all the clocks applied to the flip flops. In other words, the clocks applied to the flip flops must be controllable from a single set of accessible inputs. The primary clock inputs can be the actual clocks or simply control the clocks. Whenever all the primary clock inputs are in the *off* state, all the clocks applied to the flip flops must also be in the *off* state. Likewise, the clocks applied to the flip flops can be placed in the *on* state by setting one or more primary clock inputs to the *on* state. Finally, clocks cannot be generated by performing the AND operation of any other combination of clocks.

4. The clock primary inputs cannot be used to alter, either directly or through combinational logic, the value of the data applied to any flip flop. The clock primary inputs must only be used to provide the clock signals to the flip flops.

5. All the flip flops in a circuit must be interconnected into unique shift registers that each has an input (scan-in line), an output (scan-out line),

and clocks available as primary inputs. This allows each flip flop to be accessed for controllability and observability purposes.

6. The circuit must be capable of being placed in a *scan state* by exercising the primary inputs (including the primary clock inputs) of the circuit. The scan state is characterized by four basic features: (a) the value contained in each flip flop is either a function of the preceding flip flop in the scan string or a function of some primary input, (b) the value found on each primary output that is a scan-out line is either a function of the output of the flip flop that drives the scan-out line or a function of a primary input, (c) all clocks except the shift clocks are disabled, and (d) the shift clocks are controllable from the primary clock inputs.

Rules 1 through 4 are required to guarantee that the design is level-sensitive, whereas Rules 5 and 6 are necessary to ensure the testability of the circuit.

Although not strictly an aspect of the LSSD design methodology, it is not unusual to see additional restrictions placed on designs that adhere to the basic LSSD rules. For example, a designer is usually limited in the number of flip flops that can be placed in a single shift register; 100 is a commonly used number. Because of such limitations, complex designs often require several scan-in and scan-out lines.

As an illustration of LSSD, Fig. 7.42 shows the LSSD implementation of a simple circuit containing two flip flops. Each of the two flip flops in the circuit is implemented using the LSSD master-slave flip flop of Fig. 7.40. During normal operation, clock A is kept in the *off* state while the nonoverlapping clocks C and B operate the master and slave portions of each flip flop, respectively. When it is desired to test the circuit clock C is held in the *off* state, and the nonoverlapping clocks A and B are used to load the flip flops with the desired values for y_1 and y_2. For example, suppose that we wish to make $y_1 = 1$ and $y_2 = 0$. Figure 7.43 shows a timing diagram that illustrates the procedure for making $y_1 = 1$ and $y_2 = 0$. With y_1 and y_2 specified, the circuit's primary input x can be specified to provide a complete test pattern for the combinational logic. Once the combinational logic has had time to respond to the test pattern, clocks C and B may be used to capture the values of the excitation variables Y_1 and Y_2. Clocks A and B may then be used to *scan out* Y_1 and Y_2 via the scan-out line. The LSSD approach provides a means of controlling all the inputs to the combinational logic and observing all the outputs of the combinational logic.

Scan Path

Scan Path is a name of a design for testability method, similar to LSSD, developed by Nippon Electric Co., Ltd. [Funatsu, Wakatsuki, and Arima 1975]. Like LSSD, the internal flip flops in the data path are replaced with flip flops that can be connected into a shift register. The flip flop is a *D*-type

530 Testing

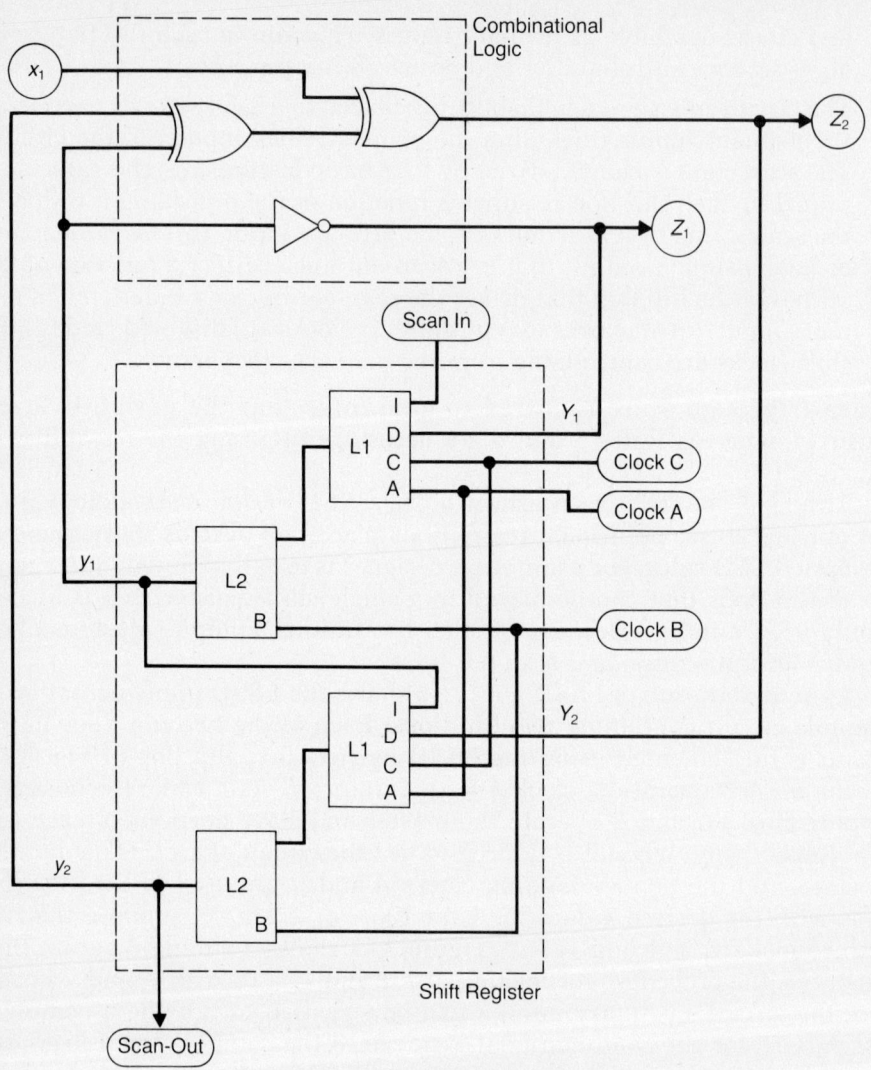

Fig. 7.42 A simple combinational circuit designed using the LSSD approach.

flip flop designed in a raceless, master-slave configuration, as shown in Fig. 7.44. The design uses two single-input clocks; one for normal operation and the other for scan operation.

As shown in Fig. 7.44, Clock 1 is the clock for normal operation and Clock 2 is the scan clock. As long as Clock 2 is at a logic 1, the test input is blocked. When Clock 1 is 0, the master portion of the flip flop is enabled and

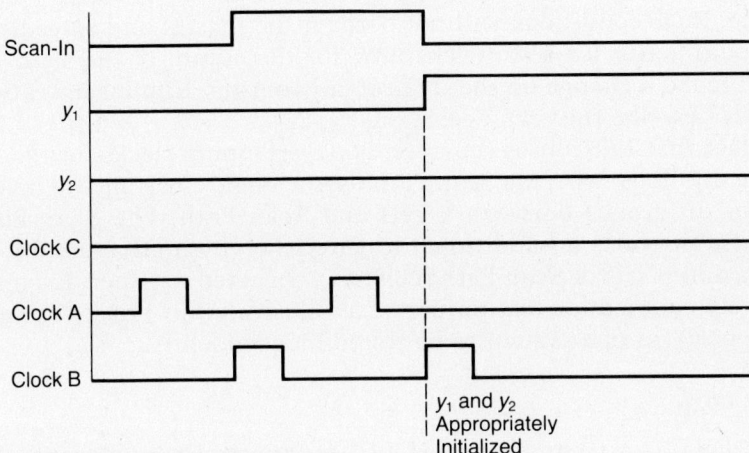

Fig. 7.43 y_1 and y_2 in Fig. 7.42 can both be initialized via a sequence of clock A and clock B pulses provided clock C is low and the scan-in line is placed at the correct value at the appropriate times.

data can be stored as long as the clock remains at 0 for a sufficient time. As the clock changes to a logic 1, the slave portion of the flip flop is enabled. The clock must remain in the logic 1 state sufficiently long to store the in-

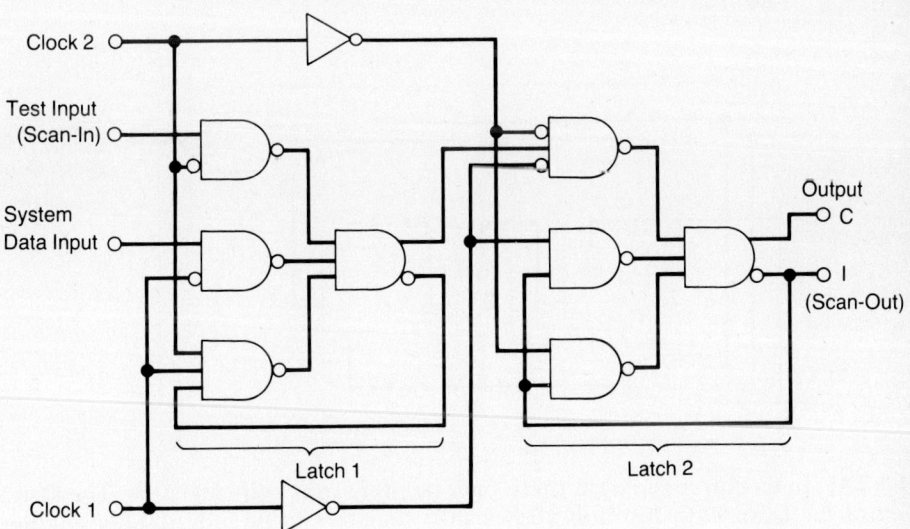

Fig. 7.44 The logic diagram for a Scan Path flip flop. (From [Parker 1983] © 1983 IEEE)

532 Testing

formation. Race conditions will not occur if the master and slave flip flops are not both active for a sufficient time for the output of the slave to feed back and make a change on the input to any master. Similar operation and conditions describe the scan operation.

The fact that LSSD uses separate, nonoverlapping clocks for the master and slave flip flops, whereas Scan Path uses a single clock and an inverter is the major difference between LSSD and Scan Path. The Scan Path approach also provides an additional feature whereby a large circuit can be partitioned into sets of Scan Path registers connected together. Each set can be uniquely enabled for test purposes, as illustrated in Fig. 7.45, therefore allowing portions of a system to be tested independently.

Scan/Set Logic

Scan/Set logic is a technique developed by Sperry-Univac to enhance the testability of a design [Stewart 1977]. Scan/Set logic is similar in nature to LSSD and Scan Path with the exception of one key difference; the shift register (SR) in Scan/Set logic is not an integral part of the data path of the circuit. Instead, Scan/Set logic provides the SR as an additional component that is completely independent of the normal, system flip flops.

As illustrated in Fig. 7.46, the SR in Scan/Set logic can be used to sample, or observe, the values of various points within the circuit and control the values of certain lines. The SR has a scan-in line and a scan-out line just as the LSSD and Scan Path approaches do. A serial data stream can be clocked into the SR via the scan-in line and then applied to the system's

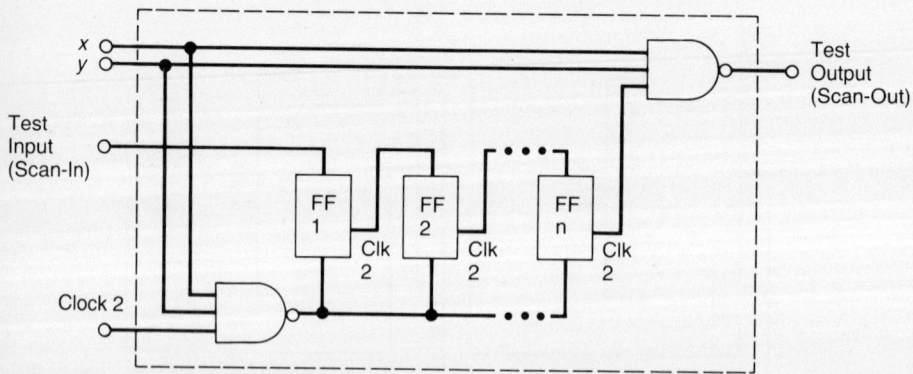

Fig. 7.45 In a complex system there may be numerous shift registers. The structure of the Scan Path flip flop allows shift registers to be individually enabled using control lines (x and y). In this manner one scan-in and one scan-out line may be used for several shift registers. (From [Williams and Parker 1983] © 1983 IEEE)

7.6 ■ Design for Testability 533

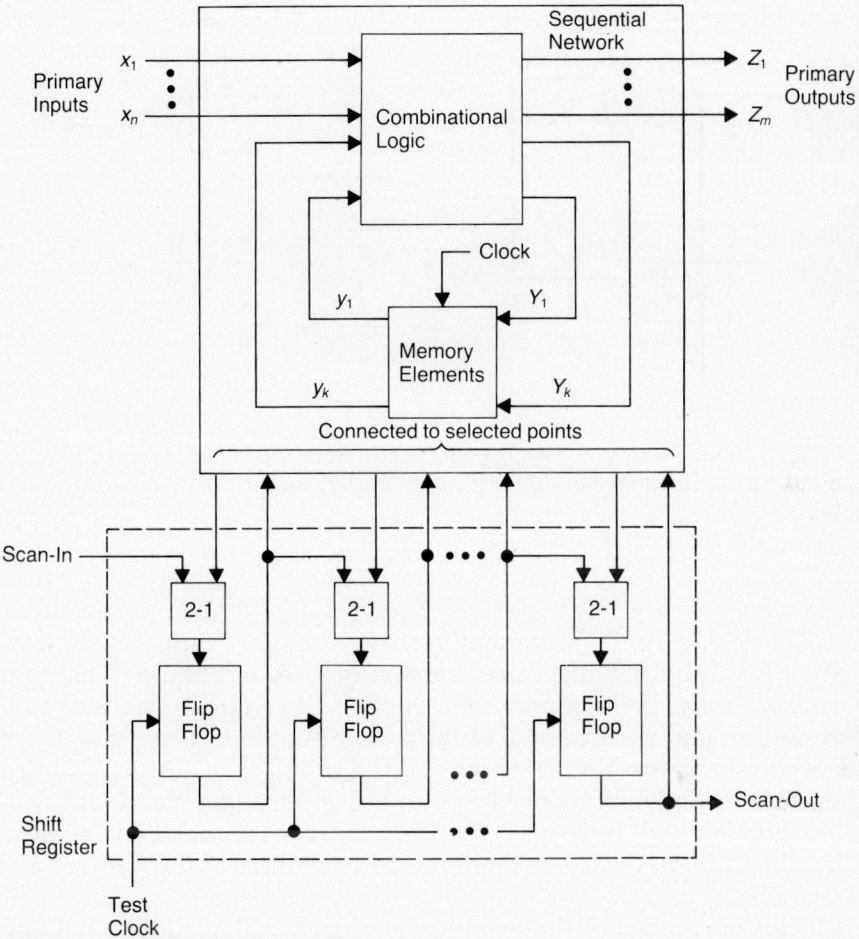

Fig. 7.46 In Scan/Set logic an external shift register allows controlling and observing various points within the sequential circuit.

logic. Once the circuit has responded to the input conditions, the values on certain lines can be sampled by the SR and clocked out on the scan-out line. The flip flops used to provide the normal system operation must be modified to accept inputs from the SR at the appropriate time. In other words, the normal inputs to a flip flop must be multiplexed with the test inputs provided by the SR, as illustrated in Fig. 7.47.

In general, the SR of Scan/Set logic is provided to the designer as a module that can be used, or not used, at the discretion of the designer. For example, some commercially available gate arrays come with the Scan/Set

534 Testing

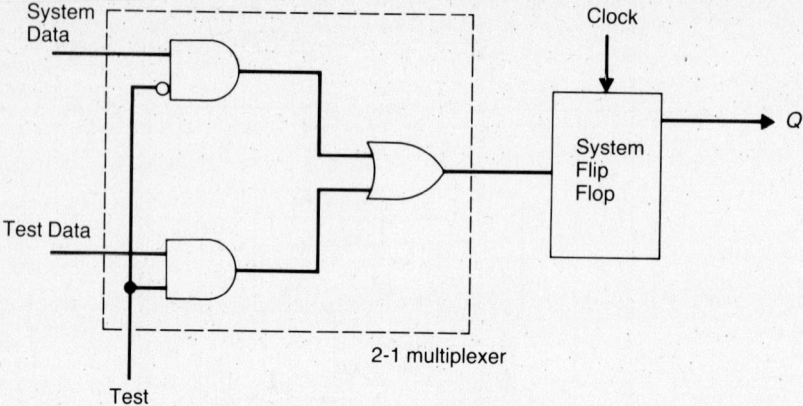

Fig. 7.47 In Scan/Set logic, any flip flop that must be set via the shift register has to have a multiplexed input so that normal system data or test data may be selected.

SR provided on-chip, regardless of whether the user wants to employ the SR. The SR is usually a fixed size (for example, 64 or 128 flip flops) so that the designer may, or may not, have enough flip flops in the SR to completely control and observe each of the system flip flops. If some system flip flops cannot be controlled or observed, the problem of testing the circuit has not been completely reduced to that of testing a combinational circuit. The test problem will have been significantly reduced, however, by providing the capability to control and observe at least some of the internal points of the circuit.

A major advantage of the Scan/Set approach is that the normal operation of the circuit need not be stopped to observe the values of the monitored points. Because the scan SR is not an integral part of the system, it can sample points within the circuit independently of the system operation. Consequently, points within the circuit can be observed during normal operation. For example, the monitored points can be periodically sampled and clocked out to some monitoring device. A disadvantage of the Scan/Set logic is that the redundancy can be more significant than that of LSSD or Scan Path because none of the extra logic required for the SR is also used to perform some function of the system.

As an example of the Scan/Set design technique, Fig. 7.48 shows a Scan/Set implementation of the circuit used to demonstrate the LSSD concept in Fig. 7.42. In Fig. 7.48 both state variables and excitation variables are accessible through the SR. A test pattern can now be scanned into the SR, applied to the flip flops, the circuit's response loaded into the SR and scanned out.

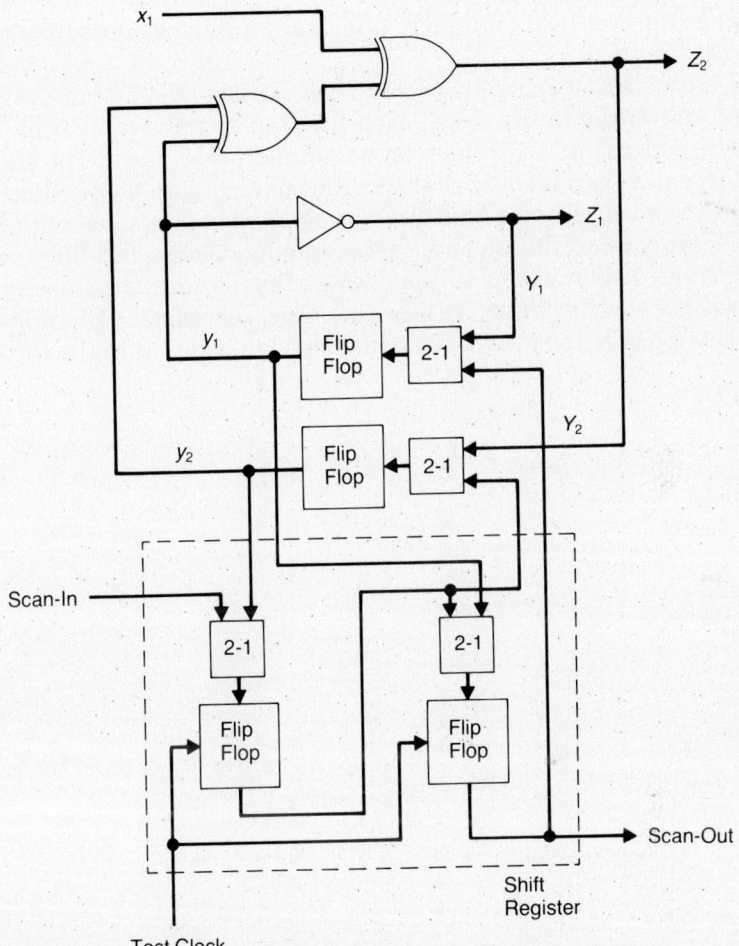

Fig. 7.48 The Scan/Set approach allows state variables y_1 and y_2 to be observed while providing a means of controlling excitation variables Y_1 and Y_2.

Random-Access Scan

The Random-Access Scan technique of design for testability has the same goals as LSSD, Scan Path, and Scan/Set; that is, providing the ability to completely control and observe the values stored in flip flops within a circuit [Ando 1980]. While LSSD, Scan Path, and Scan/Set accomplish controllability and observability by using shift registers that create scan strings within the circuit, Random-Access Scan uses an addressing scheme that allows each flip flop to be uniquely addressed and subsequently either controlled or observed.

The basic idea of Random-Access Scan is illustrated in Fig. 7.49. Rather than use conventional flip flops, each flip flop in the circuit is an addressable storage element (ASE) that can be uniquely addressed. The address decoders accept a *scan address* that specifies one, and only one, flip flop. The flip flops in the circuit are addressed as elements of an array with each flip flop having an x-coordinate and a y-coordinate. Once a flip flop is selected, its data input is connected to the *scan-in* line so that the flip flop can be loaded with a specific value. At the same time, the selected flip flop's output is connected to the *scan-out* line so that the value stored in the flip flop can

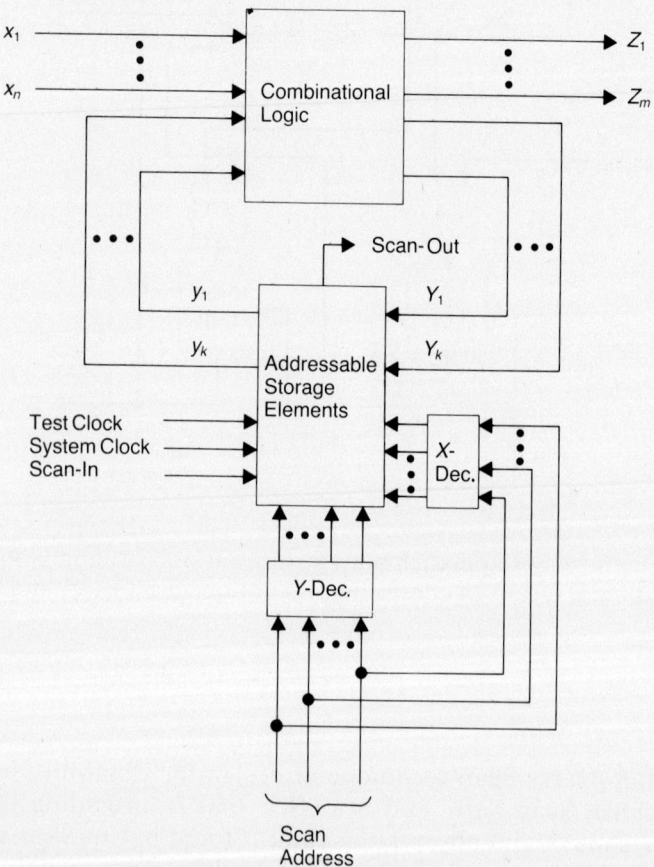

Fig. 7.49 In Random-Access Scan each flip flop is uniquely addressed and can be directly controlled or observed via the scan-in or scan-out lines, respectively.

be observed. The scan-in line is a primary input while the scan-out line is a primary output. In many cases, several scan-in and scan-out lines will be provided; one for each independent section of a circuit.

The addressable flip flop, called the Polarity-Hold Addressable Storage Element (PH-ASE), used in a Random-Access Scan design is shown in Fig. 7.50. During normal operation, with the x-address (ENX) and y-address (ENY) lines at logic 0, the output Q will become equal to the input D when the clock line CLK, becomes a logic 0. When both the x-address and the y-address lines are at logic 1, however, the value on the scan-in line will be loaded into the flip flop as soon as the *test clock* (SCK) line becomes a logic 1. Therefore, the flip flop can be loaded with whatever value is placed on the scan-in line. The designer must ensure that the CLK and the SCK lines are not active at the same time to prevent conflicts between the normal inputs and the test inputs. In a similar fashion, the output of the flip flop is placed on the scan-out line whenever both the x-address and the y-address lines are at logic 1 values. Because only one flip flop is addressed at any given time, the scan-out lines of all flip flops can be wire-ORed to create a single scan output from the circuit.

A second type of flip flop, called the Set/Reset Addressable Storage Element (SR-ASE), often used in Random-Access Scan designs is shown in Fig. 7.51. The SR-ASE does not have a separate scan-in line as was found in the PH-ASE. Instead, the SR-ASE provides for each flip flop to be uniquely

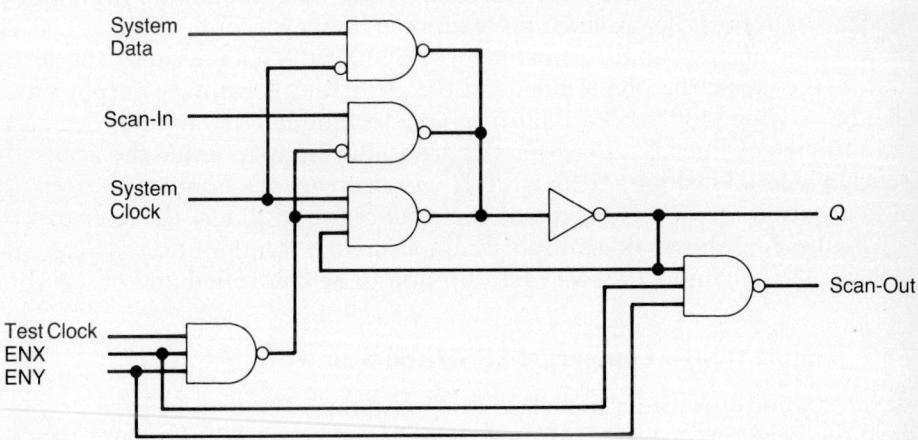

Fig. 7.50 The Polarity-Hold Addressable Storage Element (PH-ASE) can be set to the value on the scan-in line when ENX (enable X) and ENY are high, the test clock is high, and the system clock is high. Likewise, the value held by the flip flop is placed on the scan-out line when ENX and ENY are both high. ENX and ENY must be generated by the x and y address decoders, respectively. (From [Williams and Parker 1983] © 1983 IEEE)

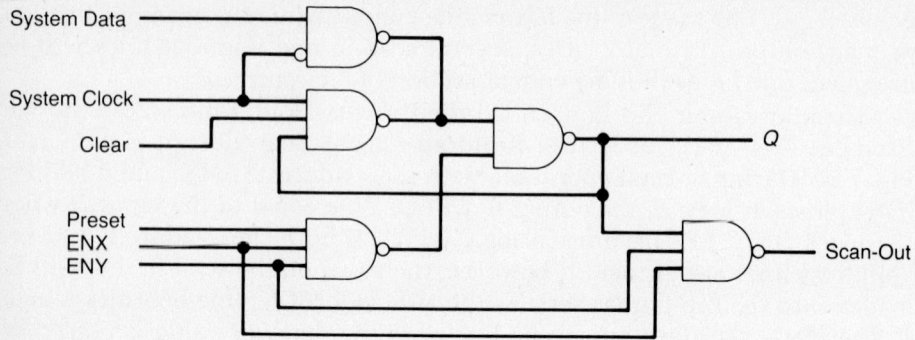

Fig. 7.51 The Set/Reset Addressable Storage Element (SR-ASE) can be set to 0 using the clear line or set to 1 using the preset line. The contents of the flip flop can be placed on the scan-out line by raising ENX and ENY to 1. (From [Williams and Parker 1983] © 1983 IEEE)

addressed and either cleared to a logic 0 value or preset to a logic 1 value. When the SR-ASE is addressed, the *clear* line can be used to set the output of the flip flop to 0 or the *preset* line can be used to set the output to 1. The value stored in an SR-ASE can be observed in the same manner as the PH-ASE. The advantage of the SR-ASE over the PH-ASE is that fewer gates are required for its implementation.

The major cost of using the Random-Access Scan method is the x and y decoders that must be provided. In addition to having to provide the decoders themselves, primary inputs must be provided for the scan address, the scan-in line, the clock, the preset line, and the clear line. A primary output must also be provided for the scan-out line. One technique often used to decrease the number of lines required for the scan address is to enter the scan address in a serial fashion. Although this will decrease the number of extra inputs required, it will also increase the time required to test the device.

Figure 7.52 shows an example design using the Random-Access Scan approach. The technique allows each flip flop to be controlled and observed.

7.6.2 Sample Design Comparing LSSD and Scan Path

We now want to consider an actual design example that can be used to compare the performance of LSSD and Scan Path [Aylor, Johnson, and Rector 1986]. Specifically, we will examine the design of a simple memory address translation unit implemented in a gate array technology. We will compare the chip area required to implement the circuit using LSSD with that required using Scan Path. Finally, we will examine the propagation delays of the LSSD master-slave flip flop and the Scan Path master-slave flip flop.

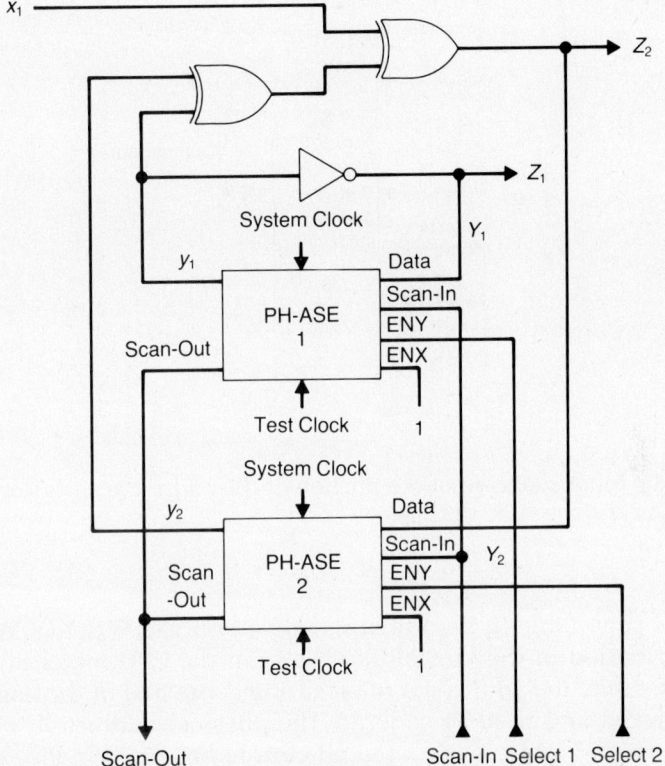

Fig. 7.52 The Random-Access Scan approach allows y_1 to be set by placing the desired value on the scan-in line, applying a logic 1 to the "select 1" line, and making the test clock high. A similar procedure sets y_2. y_1 or y_2 may be placed on the scan-out line by enabling the appropriate PH-ASE.

The system under consideration is a virtual memory Address Translation Unit (ATU) designed to be used as a component of a small microcomputer system. The design uses the classical scheme of dynamic address translation using fixed-sized pages and table look-up. Page replacement is performed by the central processing unit (CPU) in response to a *page fault* generated on-chip. The specific operation of the ATU is not important in this illustration but will be briefly described.

A block diagram of that portion of the system to be considered is shown in Fig. 7.53. The paging table and the page replacement control mechanism are not included on-chip.

The physical, or *real* address is obtained by comparing certain fields of the virtual address that is sent out by the CPU to a list of addresses con-

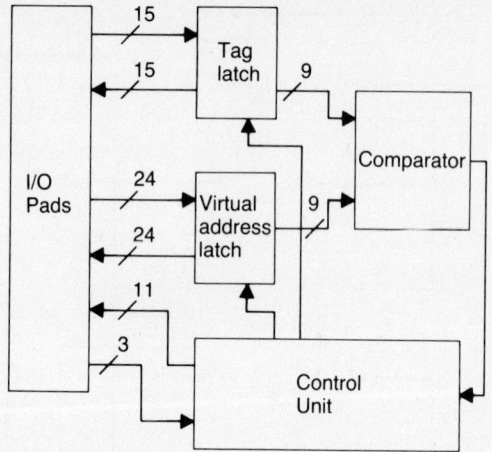

Fig. 7.53 The block diagram of the portion of the address translation unit (ATU) considered in the example design.

tained in a page table in an off-chip, dedicated Read/Write (R/W) memory. If the information in the virtual address from the CPU matches an address in the page table, the additional information contained in the page-table entry is retrieved and used to generate the physical address. If no match is found, the virtual address is routed off-chip to be stored in an external register where an additional intelligent device will use it to transfer to main memory the required page of data from the paging device. Once memory is loaded, a new *tag* or identifier is placed in the page table at an address generated on-chip. This page-table address is generated in a semi-random fashion using an internal page-fault counter to contain the new page location. After the new page has been loaded into memory, the process of looking for a match will be repeated, and the appropriate physical address for the page just loaded will be generated. Addressing of the page table is derived on-chip and routed off-chip. Both the page-fault counter and the page-table address are contained in the *control* section shown in Fig. 7.53.

The design of that portion of the virtual memory address translation unit shown in Fig. 7.53 was implemented in gate array technology using first LSSD and then Scan Path. The technology used was a two micron Complementary Metal-Oxide Semiconductor (CMOS) process. The particular target part was a 1984 gate array that actually contains 1323 array cells. Each cell contains six Metal-Oxide Semiconductor (MOS) transistors that can be used to construct various types of gates and flip flops.

The number of cells required to implement a particular design is highly dependent on the layout process. For this reason, the area required was calculated based on worst-case and best-case estimates. For best-case (minimum) cell counts, the total for the design was the sum of the exact cell count of each primitive contained in the design. For worst-case (maximum), the cell count for the design was the sum of the resulting cell counts of each macro after rounding those cell counts to the next whole number. For example, consider a design containing three macros with 2 1/3, 8 1/3, and 3 2/3 cells, respectively. The best-case cell count for the total design is 14 1/3 (simply, the sum of the individual cell counts) with the worst-case total being 16 cells (the sum of the rounded cell counts). This approach appeared to be the safest method for comparison of various techniques while accounting for both perfect routing and layout with no waste (best-case), and gross wastage of cells because of routing inefficiencies and hierarchical design techniques (worst-case).

The macros used throughout the design are found in Table 7.20. Associated with each macro is the number of occurrences and the best-case and worst-case cell counts. It should be noted that this list contains both primitive cells, such as a two-input NAND gate, as well as higher level macros, such as the implementation of the LSSD flip flop. Both the LSSD and the Scan Path flip flops are constructed using the more basic primitives such as NAND gates and inverters. As noted in Table 7.20, the number of cells re-

TABLE 7.20 Macros used in the ATU design

Macro name	Number of cells		Number of occurrences
	Best case	Worst case	
Two-input AND	1	1	20
Three-input AND	1 1/3	2	15
Four-input AND	1 2/3	2	8
Two-input OR	1	1	13
Three-input OR	1 1/3	2	2
Four-input OR	1 2/3	2	2
Two-input NAND	2/3	1	8
Three-input NAND	1	1	2
Four-input NAND	1 1/3	2	1
EXCLUSIVE-NOR	1 2/3	2	1
Inverter (I1X)	1/3	1	25
Inverter (I3X)*	1	1	3
Tristate inverter	1 2/3	2	6
EXCLUSIVE-OR	1 2/3	2	9
D-type latch	4	4	52
LSSD latch	8	10	13
Scan-path latch	9 1/3	10	13

*I3X has three times the drive capability of I1X.

quired for an LSSD flip flop as implemented, is 8 (best case) or 10 cells (worst case). For the Scan Path flip flop, the number is 9 1/3 and 10 cells for best-case and worst-case counts, respectively. For comparison, a simple *D*-type, master-slave flip flop required 4 cells.

The design of the ATU was first implemented by adhering strictly to the LSSD rules. In other words, all flip flops were made to be the special LSSD, or Scan Path, flip flops. The results of forcing all flip flops to adhere to the structured design approach are shown in Fig. 7.54. The most important thing to note from these results is that the penalty in cell count, and consequently area, for using DFT techniques throughout ranges from 69% to 93%. The reason for these large penalty figures is due to the fact that the partitioning of the design is such that the chip implemented contains a relatively large number of sequential elements compared to the total number of devices.

The results of Fig. 7.54 led to the consideration of an alternative approach. Upon further examination of the circuit, it was determined that all flip flops that were not part of the *control* circuitry of the chip were both controllable and observable because of their direct interconnection to the external world. The *comparison* circuitry was also testable because of its in-

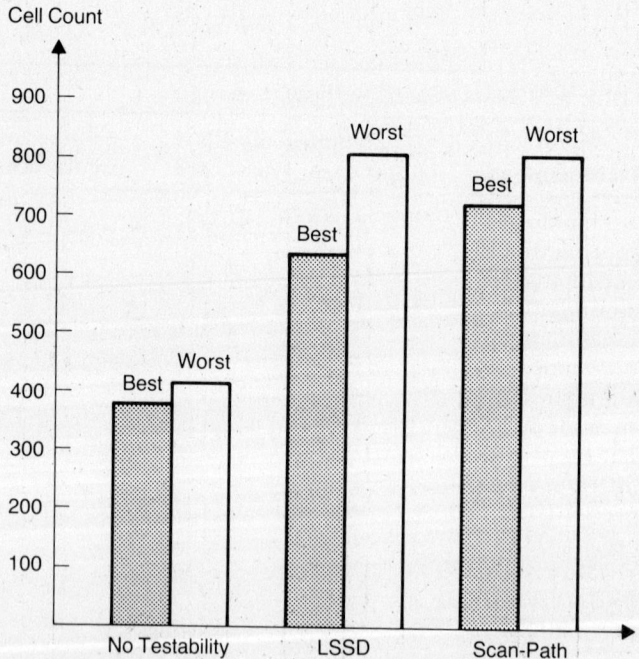

Fig. 7.54 Best- and worst-case cell counts for the case where design for testability was used for all flip flops.

terconnection to controllable and observable flip flops and the interconnection of its output to the control circuitry, which remained implemented in a structured fashion.

If only the flip flops in the control circuitry are implemented using the structured approaches, the results shown in Fig. 7.55 are obtained. As is obvious, a large reduction is noted in the area penalty. Penalty figures now range from approximately 14% to 18% which are reasonable relative to other designs analyzed. The results of Figs. 7.54 and 7.55 illustrate the importance of combining structured design approaches with *ad hoc* techniques. Certainly in the design of the ATU, the blind application of the structured approaches would have led to a much more costly design.

The percentage increase in cell count experienced when using structured DFT techniques, such as LSSD and Scan Path, is going to be determined by the ratio of sequential to total logic since the additional logic associated with the DFT approach is implemented in only the sequential devices. As the ratio increases in a circuit, the cost increases in terms of chip area increases. In this particular design, the ratio of cells used for sequential to total cells is approximately 0.20 to 0.25, depending on wiring in-

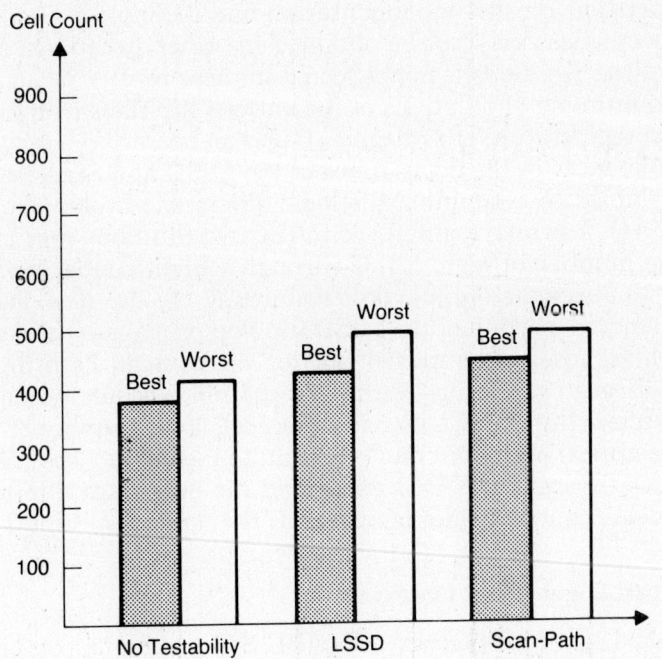

Fig. 7.55 Best- and worst-case cell counts for the case where design for testability was only used for the flip flops in the control unit.

efficiencies. With no more than approximately an 18% overhead in chip area (worst case, using LSSD), the testability problem has been greatly reduced.

Another consideration, in addition to the chip area required for a given method of designing for testability, is the relative performance of the chosen method. One performance measure is the propagation delay through the master-slave flip flop during normal operation. The actual time between data arriving at the input of the master flip flop and that same data appearing at the output of the slave flip flop depends on clock frequency, as well as other variables. To obtain propagation times that are independent of clock frequency, the delay time of the scan flip flop is defined as the sum of the times for the master flip flop to respond to an input change, assuming the master is activated when the input changes, and the time for the slave to respond to a change in the master's output, again assuming that the slave is activated when the change occurs. The resulting delay time represents the fastest possible operation of the flip flop.

Figure 7.56 shows the delay time of the LSSD and the Scan Path flip flops. The delay time of a standard, D-type master-slave flip flop is also shown. The performance degradation experienced when using structured techniques can be noted by comparing the scan flip flop delays to the simple D-type master-slave flip flop. Note, however, that delay figures are highly dependent on the implementation and technology used. The same performance figures may not be obtained for other situations. Also, output loading of these flip flops is not reflected in these results. Delay times were derived from timing simulations of the various flip flops using typical gate delays for a temperature of 27 degrees C and an operating voltage of 5 volts.

The differences in the delay times of the LSSD and Scan Path flip flops can be explained by examining the logic diagrams of each (shown in Figs. 7.40 and 7.44). A primary difference in the two flip flops that affects delay time is the number of logic levels through which feedback signals must propagate to affect the output, thus stabilizing the flip flop. Feedback signals in the master portion of the LSSD flip flop propagate through one level of logic while those in the master portion of the Scan Path flip flop must propagate through two levels. A similar situation exists in the slave portions of the respective flip flops. A second difference is the number of three-input gates in the critical path from the data input to the output. The LSSD flip flop contains two three-input NAND gates, and the Scan Path flip flop contains four. The delay time of a gate increases as the number of inputs increases.

7.6.3 Built-in Logic Block Observation

The Built-In Logic Block Observation (BILBO) technique combines the basic features of scan designs with those of the signature analysis method [Koenemann, Mucha, and Zwiehoff 1979]. BILBO uses a controllable shift

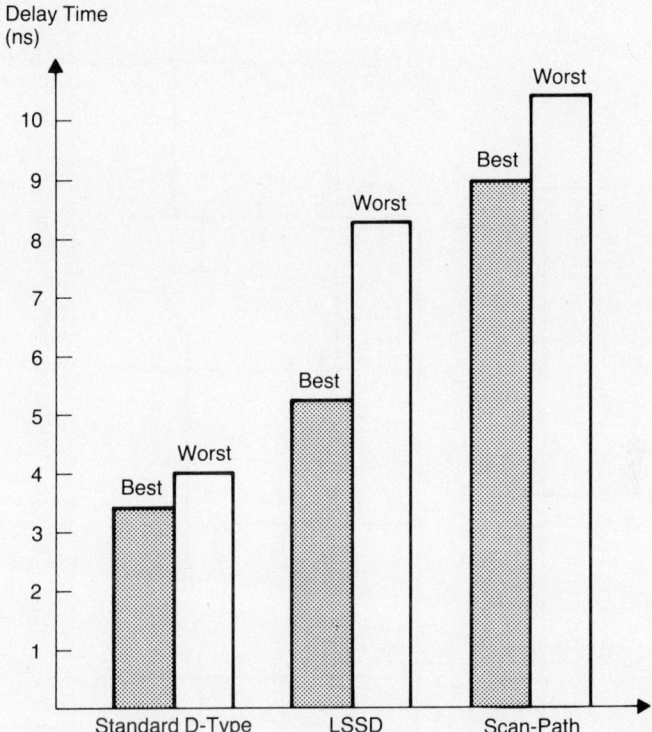

Fig. 7.56 Best- and worst-case delay times for the LSSD, Scan-Path, and standard D-type flip flops.

register (CSR) that can be placed in any one of four modes. An 8-bit CSR is shown in Fig. 7.57, and each possible mode of the CSR is shown in Figs. 7.58 through 7.61.

In the first mode shown in Fig. 7.58, with control lines C_1 and C_2 both at logic 1, each flip flop of the CSR operates independently and can be used in the normal operation of the circuit. The inputs to the flip flops are $(D_1, \ldots, D_8)$ and the outputs are $(Q_1, \ldots, Q_8)$.

In the second mode shown in Fig. 7.59, with control lines C_1 and C_2 both at logic 0, the flip flops of the CSR are connected into a single shift register that can be used to form a scan string from the scan-in line to the scan-out line. The second mode allows the contents of each flip flop to be clocked out as a serial data stream or the flip flops can be loaded by clocking in a serial data stream. The multiplexer at the input of flip flop FF1 also allows the second mode of operation to be that of a linear feedback shift register that can only be loaded serially. When the multiplexer selects the scan-in line, the

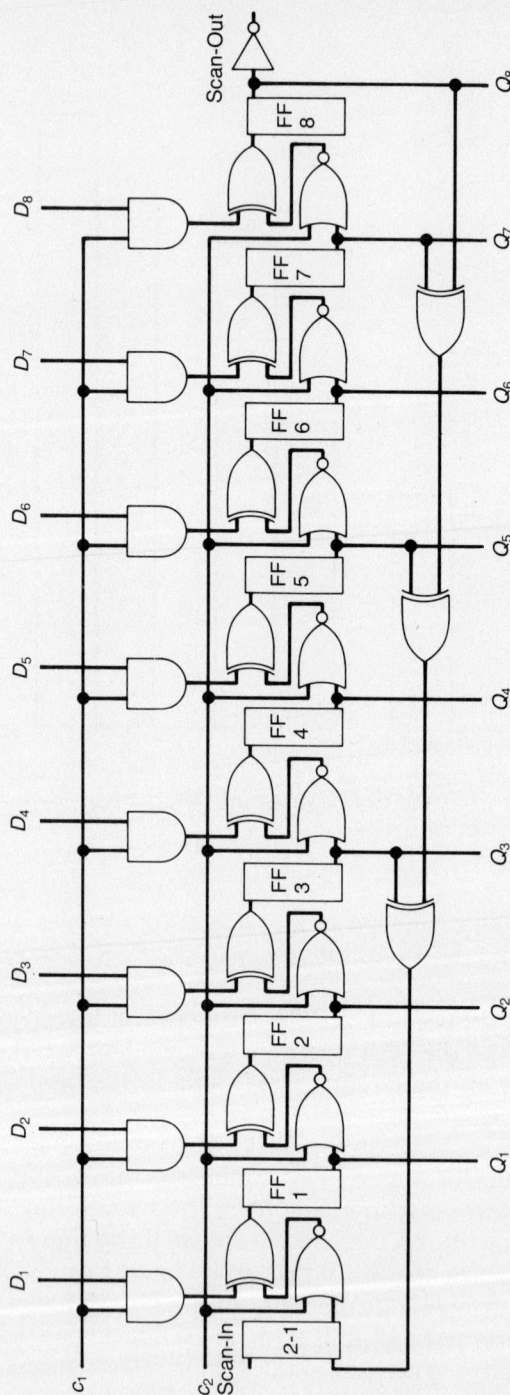

Fig. 7.57 The 8-bit controllable shift register can function in any one of four modes specified by the control signals C_1 and C_2. (From [Williams and Parker 1983] © 1983 IEEE)

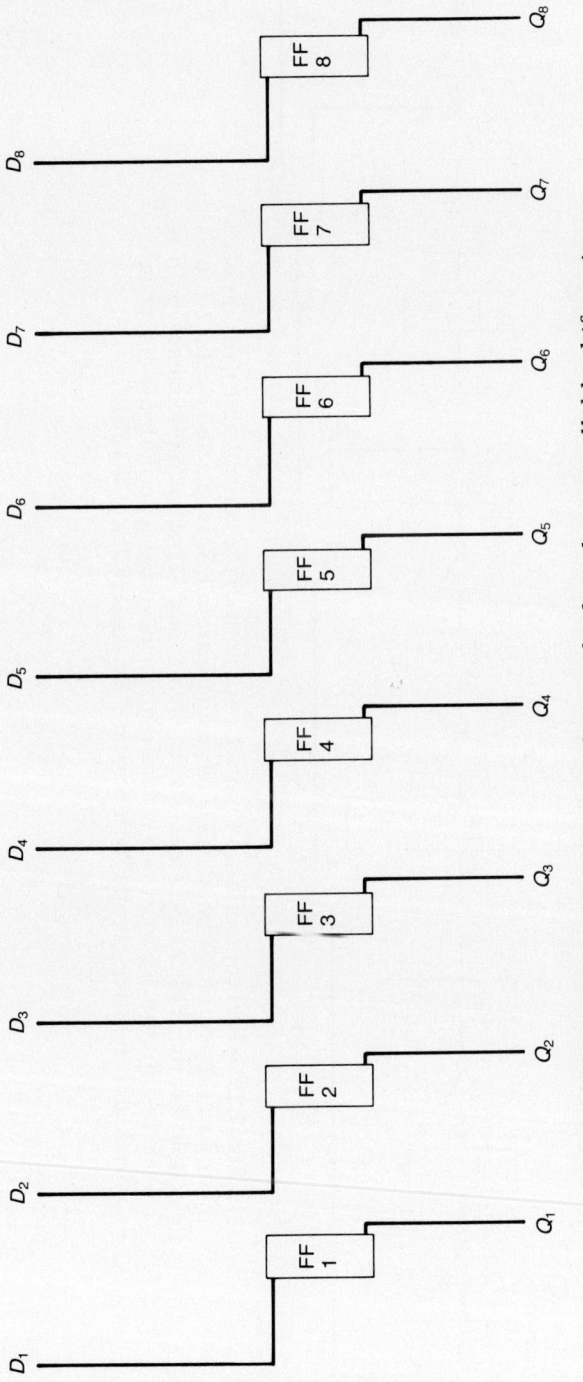

Fig. 7.58 When control lines C_1 and C_2 are both 1, the controllable shift register reduces to eight independent flip flops that may be used in the normal operation of the system. (From [Williams and Parker 1983] © 1983 IEEE)

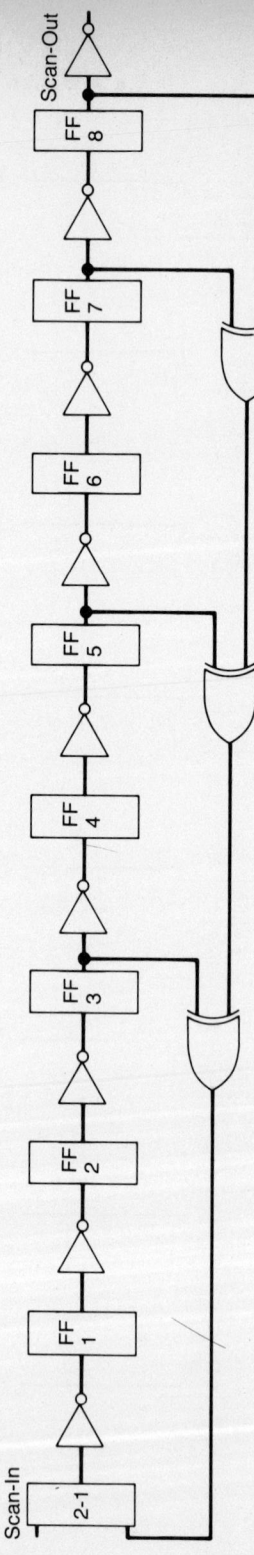

Fig. 7.59 When control lines C_1 and C_2 are both 0, the controllable shift register reduces to one of two modes: (1) when the scan-in line is selected by the 2-1 multiplexer, the register becomes a scan string, and (2) when the feedback path is selected by the 2-1 multiplexer, the register becomes a serial-input linear feedback shift register. (From [Williams and Parker 1983] © 1983 IEEE)

CSR is simply a single shift register. However, when the multiplexer selects the feedback line, the CSR becomes a linear feedback shift register. The linear feedback shift register can be used to generate pseudo-random test patterns on the outputs of the flip flops. The pseudo-random patterns can be applied to combinational logic to provide a random test of that logic.

In the third mode shown in Fig. 7.60, with C_1 at a logic 1 and C_2 at a logic 0, the CSR converts into a linear feedback shift register with parallel inputs $(D_1, \ldots, D_8)$. This mode is called the signature analysis mode. By capturing the parallel inputs in the feedback shift register, a signature of the performance of the circuit is formed and can be used to classify the circuit as operating correctly or incorrectly.

The fourth, and final, mode shown in Fig. 7.61, with C_1 at a logic 0 and C_2 at a logic 1, simply resets each flip flop. The fourth mode can be used to initialize all flip flops to a logic 0 value if desired.

Let us consider in more detail how to use the BILBO technique in the design of a sequential circuit. One approach, though certainly not the only approach, is illustrated in Fig. 7.62. The normal flip flops of the sequential circuit are implemented using two BILBO CSRs. The outputs of CSR B are the inputs to CSR A. The inputs to CSR B are the excitation variables and the outputs of CSR A are the state variables. Each CSR has scan-in and scan-out lines. When both CSRs are in the normal operating mode, the non-overlapping clocks, CLK A and CLK B, can be used to create a master-slave flip flop operation.

The testing of the combinational logic proceeds as follows. A fixed pattern is scanned into CSR A by placing A in the scan mode and forcing the multiplexer to select the scan-in line. During testing, A is held in the scan mode such that it ignores its data inputs from B, but the multiplexer of A selects the feedback signal such that the outputs of A become pseudo-random test patterns for the combinational logic. CSR B is placed in the signature analysis mode such that a signature of the response of the combinational logic to the pseudo-random inputs is recorded. Once a sufficient number of test patterns has been applied to the combinational logic, CSR B is converted to the scan mode, and the signature of the circuit is clocked out.

The real advantage of the BILBO technique over normal scanning methods such as LSSD and Scan Path is the speed at which the test process can be performed. In normal scan methods, each test pattern must be scanned in serially and the results scanned out serially. These serial operations are very time consuming. In BILBO, however, only the initial *seed* for the pseudo-random test pattern generation and the resulting signature must be scanned in or out. The test patterns are generated by the BILBO registers.

The disadvantage of the BILBO approach is that some combinational circuits cannot be sufficiently tested using a random test pattern approach. A second disadvantage of BILBO is that there is more delay between the ex-

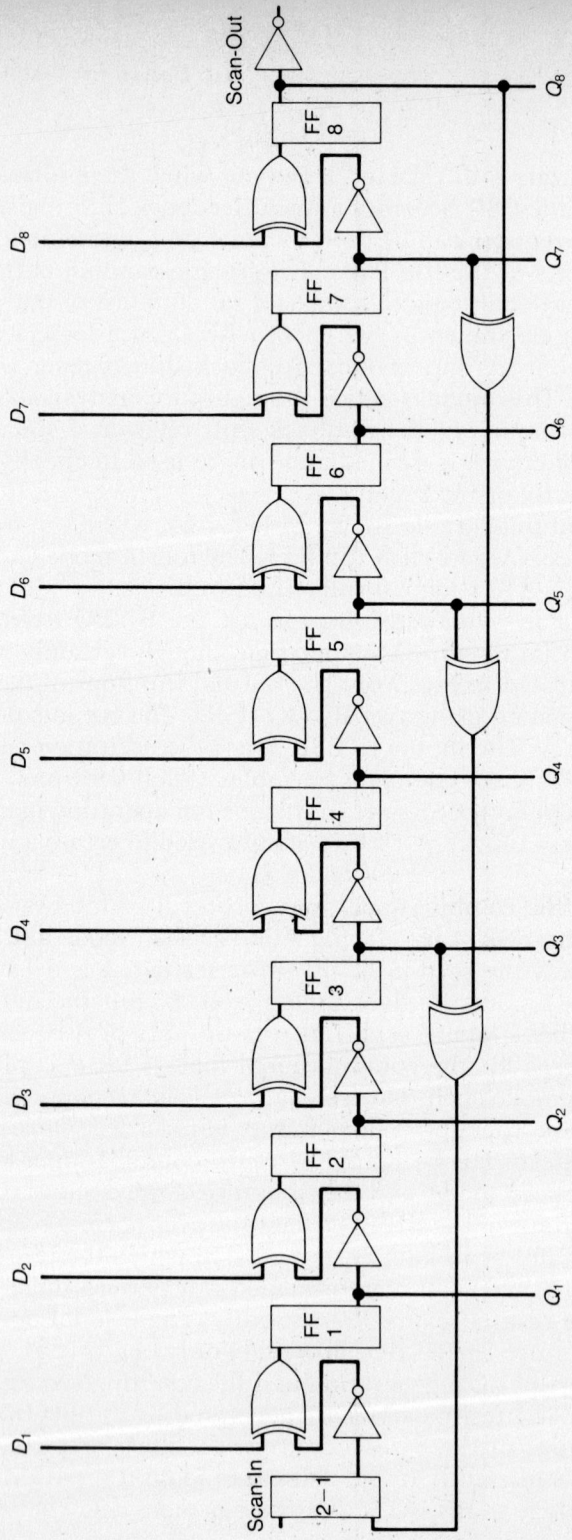

Fig. 7.60 When C_1 is 1 and C_2 is 0 the controllable shift register becomes a parallel-input, parallel-output, linear feedback shift register.

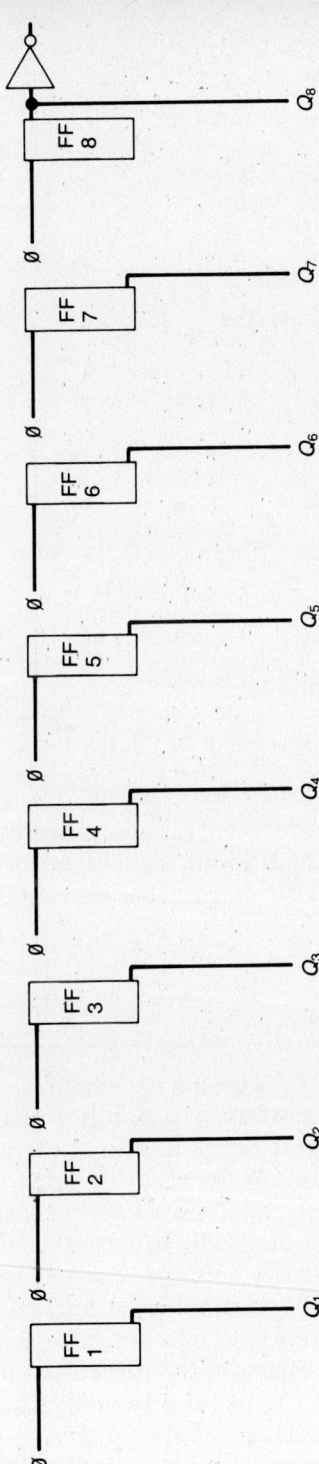

Fig. 7.61 When C_1 is 0 and C_2 is 1, each flip flop is loaded with a value of 0.

552 Testing

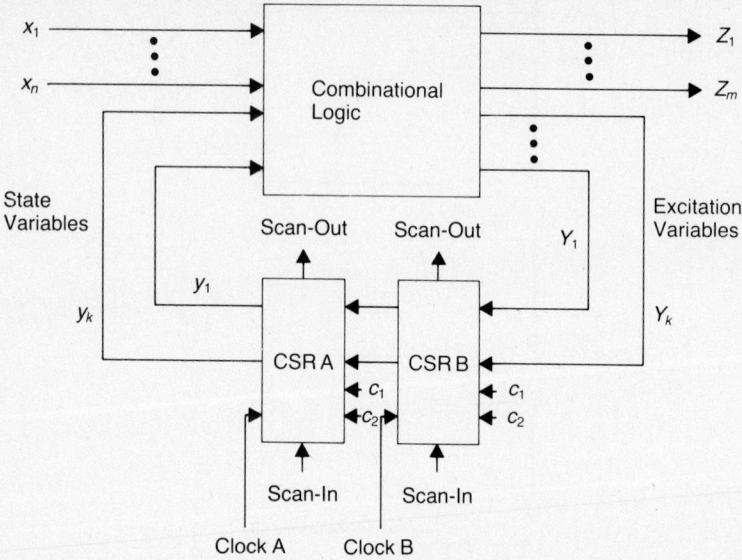

Fig. 7.62 Example structure of a sequential machine designed using the BILBO approach.

citation variables and the state variables because of the AND gate and EXCLUSIVE-OR gate that each input must pass through before being applied to the flip flop input. This extra delay decreases the speed at which a circuit can be operated.

7.7 Testability Analysis

The process of characterizing the ease with which a circuit may be tested is typically called **testability analysis** or **testability measurement**. Testability analysis can be an integral part of the design of digital circuits by identifying specific points within a circuit that are difficult to test. The designer can use the results of a testability analysis to define test points or to identify areas of the design that must be modified to improve the testability.

Testability analysis is usually coupled with techniques such as LSSD to significantly improve a design's testability. LSSD can convert the problem of testing a sequential machine into one of testing a combinational circuit, and testability analysis can estimate the difficulty of testing that combinational circuit. The results of the testability analysis can be used to modify the combinational logic so that it is easier to test.

Testability analyzers presently exist, including COPTR (Controllability, Observability, Predictability, Testability Report) [Calma 1984], SCOAP

(Sandia Controllability/Observability Analysis Program) [Goldstein 1979], TMEAS (Testability Measurement) [Grason 1979], CAMELOT (Computer-Aided Measure of Logic Testability) [Bennetts 1984], PREDICT (Probabilistic Estimation of Digital Circuit Testability) [Seth, Pan, and Agrawal 1985], and VICTOR (VLSI Identifier of Controllability, Testability, Observability, and Redundancy) [Ratiu, Sangiovanni-Vincentelli, and Pederson 1982]. Of these, COPTR, SCOAP, and CAMELOT seem to have been the most widely used thus far.

The purpose of this section is to introduce the concept of testability analysis and to describe, and illustrate with examples, a couple of existing testability analysis techniques. In addition, the shortcomings of existing analyzers will be examined. Unfortunately, we will not discuss all the existing analyzers due to time and space limitations. However, the analyzers that are discussed will give you a good feel for the fundamental idea behind many existing approaches.

We will begin the section with several fundamental definitions that are important to the understanding of testability analysis. In some cases, the definitions will be reviews of those presented earlier in this text, but it is important to refresh our memories at this point.

7.7.1 Important Definitions

The concepts of controllability, observability, and testability are fundamental to all testability analysis techniques, but different definitions are present throughout the literature. One definition, for example, presents the controllability of a line as a measure of the difficulty associated with setting that line to a particular logic value. A second definition presents a more intuitive meaning of controllability as simply the ability to set a line in a circuit by specifying the values of the circuit's primary inputs. Although the differences between the above definitions may be slight, they are extremely important and serve to illustrate the varying interpretations held by people working in the testability field.

The following definitions appear to best describe the meanings of several important terms and will be used throughout this text:

> The **controllability** of a line in a digital circuit is the *ability* to set that line to a specific logic value by specifying the primary inputs of the circuit. A line that can be set to either logic state is said to be *completely controllable*. To distinguish between logic states, a line that can be set to a logic 1 is called *1-controllable* while a line that can be set to a logic 0 is called *0-controllable*.
>
> The **cost of 0-controllability** for a particular line is a *measure* of the ease with which the line may be set to a logic 0 by specifying the primary inputs of the circuit. The notation $CC0(i)$ represents the cost of controlling line i to a 0.

The **cost of 1-controllability** for a particular line is a *measure* of the ease with which the line may be set to a logic 1 by specifying the primary inputs of the circuit. The notation $CC1(i)$ represents the cost of controlling line i to 1.

The **observability** of a line in a digital circuit is the *ability* to propagate the value on that line, regardless of whether that value is a 1 or a 0, to a primary output or some other measurable point by specifying the primary inputs of the circuit. In other terms, observability is the ability to sensitize a path from the line in question to a measurable point.

The **cost of observing** a line is a *measure* of the ease with which a sensitized path may be established from that line to a measurable output by specifying the primary inputs of the circuit. The notation $CO(i)$ represents the cost of observing line i.

The **testability** of a line is the *ability* to find test vectors for faults on that line. Recall that a test vector is a primary input combination that yields a different value on at least one primary output when the line is faulty than when the line is not faulty. Considering only stuck-type faults, a line for which test vectors exist for both stuck-at-1 and stuck-at-0 faults is said to be *completely testable*. A line for which test vectors exist for the stuck-at-1 case is said to be *1-testable*, whereas a line for which test vectors exist for the stuck-at-0 case is said to be *0-testable*.

The **cost of testing a line for a stuck-at-0** fault is a *measure* of the ease with which a test vector can be found for that stuck-at-0 fault. The notation $CT0(i)$ represents the cost of testing line i for a stuck-at-0 fault.

The **cost of testing a line for a stuck-at-1 fault** is a *measure* of the ease with which a test vector can be found for that stuck-at-1 fault. The notation $CT1(i)$ represents the cost of testing line i for a stuck-at-1 fault.

The primary purpose of a testability analyzer is to estimate $CC0(i)$, $CC1(i)$, $CO(i)$, $CT0(i)$, and $CT1(i)$ for all lines i in a given circuit. The basic differences between existing analyzers are the techniques used to estimate these quantities. In addition, we would hope that an analyzer would identify any lines within a circuit that are not controllable, observable, or testable. Unfortunately, few existing analyzers have this added feature.

7.7.2 Testability Analyzers

This section describes two testability analyzers and illustrates the use of each on some simple combinational circuits. The two analyzers are SCOAP and CAMELOT.

Note that most of the analyzers are capable of considering both sequential and combinational networks. For our purposes, however, only combinational networks will be considered. It is assumed that sequential networks are designed using an approach such as Level Sensitive Scan Design (LSSD), or the like, such that the testing of sequential circuits is reduced to that of testing combinational circuits. Therefore, the analyzer is only used on the combinational portion of the circuit.

SCOAP

Both SCOAP and COPTR are based on the same concepts, therefore, only SCOAP will be discussed. SCOAP relates the cost of controlling an output of a logic module to the costs of controlling the inputs to that module. For example, the cost of controlling the output of an AND gate to 0 is a function of the cost of controlling *one* of the inputs to 0 because only one input must be 0 to force the output to be 0. On the other hand, the cost of controlling the output of an AND gate to 1 is a function of the sum of the costs of controlling each input to 1 because all the inputs must be 1 to make the output 1.

To generalize the concepts presented above, consider a logic module that provides some arbitrary function, as shown in Fig. 7.63. The logic module has n inputs $(x_1, \ldots, x_n)$, and m outputs $(z_1, \ldots, z_m)$. SCOAP determines the controllability cost $CC1$ for each output z_i as

$$CC1(z_i) = \begin{cases} \sum_{\text{all } j \in J} CC1(x_j) + \sum_{\text{all } k \in K} CC0(x_k) + CD & \text{if multiple inputs must be set} \\ \min[CC1(x_j), CC0(x_k)] + CD & \text{if only one input must be set} \end{cases}$$

where J is the set of all $x_j, j \in \{1, 2, \ldots, n\}$, that must be controlled to 1 to control z_i to 1, K is the set of all $x_k, k \in \{1, 2, \ldots, n\}$, that must be controlled to 0 to control z_i to 1, CD is the cell depth, and min is the minimum function. The cell depth is the number of logic levels in the module.

The preceding equation can be explained as follows. If several inputs to the logic module must be controlled to achieve a 1 on the output (for example, in the case of an AND gate), the cost of controlling the output to 1 will be the sum of the costs of controlling the inputs to their required values, plus the cell depth. On the other hand, if only one input must be controlled to control the output to 1 (for example, in the case of an OR gate), the controllability cost will be the minimum of the controllability costs of those lines that can be controlled to achieve the desired output, plus the cell depth.

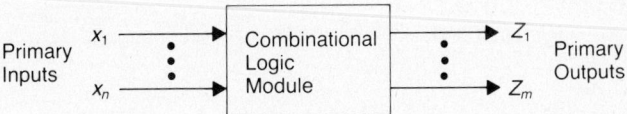

Fig. 7.63 A generalized combinational logic circuit with primary inputs $(x_1, \ldots, x_n)$ and primary outputs $(Z_1, \ldots, Z_m)$.

Similarly, the cost of controlling an output of the logic module in Fig. 7.63 to 0 may be written as

$$CC0(z_i) = \begin{cases} \sum_{\text{all } j \in J} CC1(x_j) + \sum_{\text{all } k \in K} CC0(x_k) + CD & \text{if multiple inputs must be set} \\ \min[CC1(x_j), CC0(x_j)] + CD & \text{if only one input must be set} \end{cases}$$

where J is the set of all $x_j, j \in \{1, 2, \ldots, n\}$, that must be controlled to 1 to control z_i to 0, K is the set of all $x_k, k \in \{1, 2, \ldots, n\}$, that must be controlled to 0 to control z_i to 0, CD is the cell depth, and min is the minimum function.

The preceding equation can be explained as follows. If several inputs to the logic module must be controlled to achieve a 0 on the output (for example, in the case of an OR gate), the cost of controlling the output to 0 is the sum of the costs of controlling the inputs to their required values, plus the cell depth. On the other hand, if only one input must be controlled to achieve a 0 on the output (for example, in the case of an AND gate), the controllability cost is the minimum of the controllability costs of those lines that can be set to achieve the desired output, plus the cell depth.

As an example, Table 7.21 shows the controllability cost equations for the outputs of several basic gates in terms of the controllability costs of the input lines. In Table 7.21, the cell depth is 1 for all gates except the inverter. The cell depth of the inverter is 0 because the output of the inverter is no more difficult to control than the input. Consequently, it makes sense to

TABLE 7.21 Controllability equations for several basic gates

Equations	Gate
$CC1(Z) = CC1(x_1) + CC1(x_2) + 1$ $CC0(Z) = \min\{CC0(x_1), CC0(x_2)\} + 1$	AND
$CC1(Z) = \min\{CC1(x_1), CC1(x_2)\} + 1$ $CC0(Z) = CC0(x_1) + CC0(x_2) + 1$	OR
$CC1(Z) = CC0(x_1)$ $CC0(Z) = CC1(x_1)$	INVERTER
$CC1(Z) = \min\{CC0(x_1), CC0(x_2)\} + 1$ $CC0(Z) = CC1(x_1) + CC1(x_2) + 1$	NAND
$CC1(Z) = CC0(x_1) + CC0(x_2) + 1$ $CC0(Z) = \min\{CC1(x_1), CC1(x_2)\} + 1$	NOR

have the controllability costs of the output of the inverter the same as the costs of the input.

When calculating the controllability costs in a logic circuit, you proceed from the primary inputs to the primary outputs, calculating costs in an iterative fashion. The controllability costs, both $CC1$ and $CC0$, of each primary input are assumed to be 1. Figure 7.64 shows a simple combinational circuit and the controllability costs of each line.

In SCOAP, the cost of observing an input of a logic module is related to the cost of observing the output of that module *and* the cost of sensitizing a path from the input to the output. The cost of observing some input x_i at some output z_j, of the logic module of Fig. 7.63 may be written as

$$CO(x_i) = CO(z_j) + \sum_{\text{all } k \in K} CC1(x_k) + \sum_{\text{all } m \in M} CC0(x_m) + CD$$

where $CO(z_j)$ is the observability cost of the output, K is the set of all $x_k, k \in \{1, 2, \ldots, n\}$, that must be controlled to 1 to propagate the signal on input x_i to output z_j, M is the set of all $x_m, m \in \{1, 2, \ldots, n\}$, that must be controlled to 0 to propagate the signal on input x_i to output z_j, and CD is the cell depth.

As an example, Table 7.22 shows the observability cost equations for several basic logic gates using a cell depth of 1 for all gates except the inverter which has a cell depth of 0. The observability costs in a circuit are computed by working from the primary outputs to the primary inputs, assuming that all primary outputs have observability costs of 0. Figure 7.65 shows the simple combinational circuit of Fig. 7.64 and the observability cost of each line.

SCOAP generates values for $CC0(i)$, $CC1(i)$, and $CO(i)$ for every line i in the circuit. These values may be used to compute $CT0(i)$ and $CT1(i)$ as

$$CT0(i) = CC1(i) + CO(i)$$
$$CT1(i) = CC0(i) + CO(i)$$

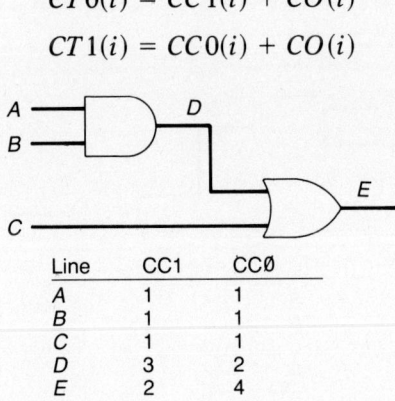

Line	CC1	CCØ
A	1	1
B	1	1
C	1	1
D	3	2
E	2	4

Fig. 7.64 Controllability costs calculated using SCOAP for a simple combinational circuit.

558 Testing

TABLE 7.22 Observability equations for several basic gates

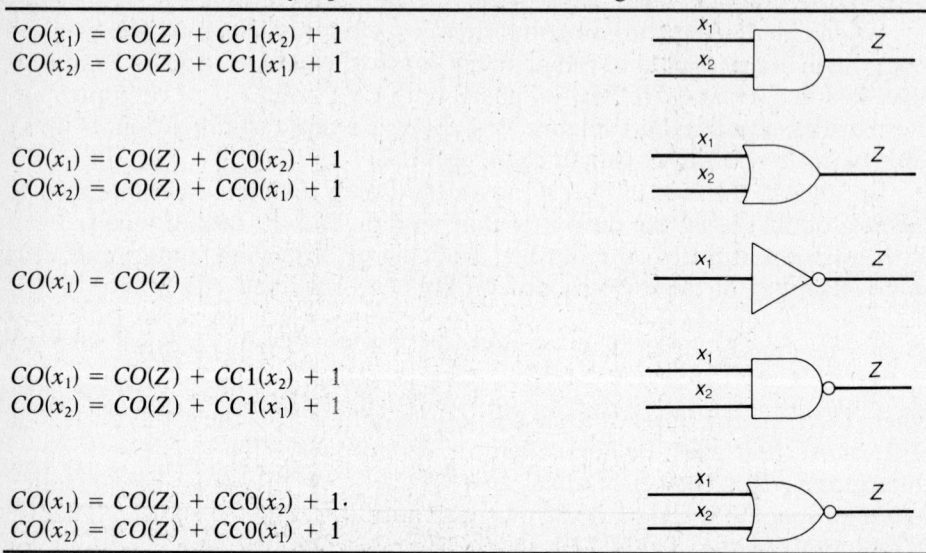

$CO(x_1) = CO(Z) + CC1(x_2) + 1$
$CO(x_2) = CO(Z) + CC1(x_1) + 1$

$CO(x_1) = CO(Z) + CC0(x_2) + 1$
$CO(x_2) = CO(Z) + CC0(x_1) + 1$

$CO(x_1) = CO(Z)$

$CO(x_1) = CO(Z) + CC1(x_2) + 1$
$CO(x_2) = CO(Z) + CC1(x_1) + 1$

$CO(x_1) = CO(Z) + CC0(x_2) + 1.$
$CO(x_2) = CO(Z) + CC0(x_1) + 1$

In other words, if we wish to test a line for a stuck-at-1 fault, we must first attempt to control the line to 0 and then observe the value of the line. Likewise, if we wish to test a line for a stuck-at-0 fault, we must first attempt to control the line to 1 and then observe the value of the line. The testability cost is the sum of the appropriate controllability cost and the observability cost. Figure 7.66 shows the simple combinational circuit of Figs. 7.64 and 7.65 and the resulting testability costs for each line.

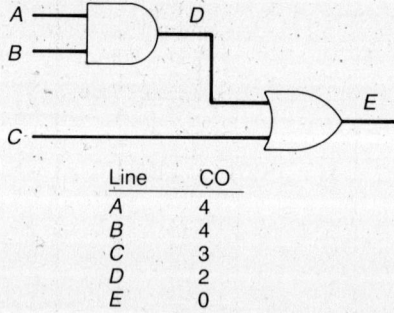

Line	CO
A	4
B	4
C	3
D	2
E	0

Fig. 7.65 Observability costs calculated using SCOAP for a simple combinational circuit.

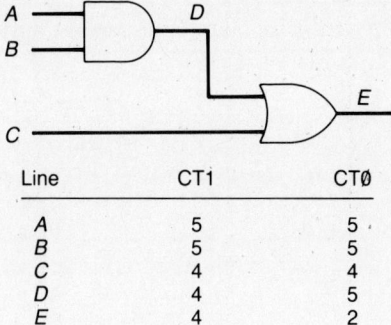

Line	CT1	CT0
A	5	5
B	5	5
C	4	4
D	4	5
E	4	2

Fig. 7.66 Testability costs calculated using SCOAP for a simple combinational circuit.

CAMELOT

CAMELOT defines the costs of controllability and observability in a manner such that these costs lie between 0 and 1, with 1 representing low cost and 0 high cost. In addition, the cost of 0-controllability and 1-controllability are not distinguished but rather combined to yield a single controllability cost. Finally, the cost of testability is defined as the product of the cost of controllability and the cost of observability.

CAMELOT calculates the cost of controllability by progressing through a circuit from its primary inputs to its primary outputs. In a similar manner, the cost of observability is computed by working backwards from the primary outputs to the primary inputs. As with SCOAP, the analysis is complete after two passes through the circuit.

Once again, consider the logic module shown in Fig. 7.63 to illustrate the manner in which controllability and observability are calculated. The controllability of an output line z_i is quantified by CAMELOT as

$$CC1(z_i) = CC0(z_i) = CC(z_i) = \frac{CTF(z_i)}{n} \sum_{j=1}^{n} CC(x_j)$$

where $CTF(z_i)$ is a controllability transfer factor for the output z_i, $CC(x_j)$ is the controllability cost for the x_j input, and $CC(z_i)$ is the value of the controllability cost at the z_i output node. $CTF(z_i)$ is computed as

$$CTF(z_i) = 1.0 - \frac{|N(0) - N(1)|}{N(0) + N(1)}$$

where $N(0)$ is the number of input combinations that produce a 0 on the output node z_i, and $N(1)$ is the number of input combinations that produce a 1 on the output node. Multiple-output devices will have multiple CTFs; one for each output. Table 7.23 shows the CTF factors for several basic gates.

TABLE 7.23 Controllability transfer factors for several basic logic gates

$CTF(Z) = 0.5$	AND gate: $x_1, x_2 \to Z$
$CTF(Z) = 0.5$	OR gate: $x_1, x_2 \to Z$
$CTF(Z) = 1$	NOT gate: $x_1 \to Z$
$CTF(Z) = 0.5$	NAND gate: $x_1, x_2 \to Z$
$CTF(Z) = 0.5$	NOR gate: $x_1, x_2 \to Z$
$CTF(Z) = 0.25$	3-input AND gate: $x_1, x_2, x_3 \to Z$
$CTF(Z) = 1$	XOR gate: $x_1, x_2 \to Z$

As mentioned earlier, the controllability costs are calculated by passing through the circuit from the primary inputs to the primary outputs, calculating the controllability cost of the output of each gate in terms of the controllability cost of each input to each gate. The controllability costs of primary inputs are assumed to be 1. Figure 7.67 shows a simple combinational circuit and the controllability costs of each line within that circuit.

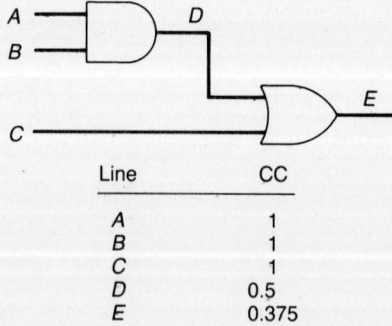

Line	CC
A	1
B	1
C	1
D	0.5
E	0.375

Fig. 7.67 Controllability costs calculated by CAMELOT for a simple combinational circuit.

7.7 ■ Testability Analysis

The observability cost is computed by working from the primary outputs to the primary inputs. The observability cost of a primary output is assumed to be 1. The cost of observability of an input x_j, of the logic module of Fig. 7.63 is computed as a function of the observability cost, $CO(z_i)$, of the output as

$$CO(x_j) = CO(z_i) \frac{OTF(j-i)}{n-1} \sum_{k=1}^{n} CC(x_k) \quad k \neq j$$

where $OTF(j-i)$ is an observability transfer factor from input x_j to output z_i, $CC(x_k)$ is the controllability cost of the x_k input, and $CO(x_j)$ is the resulting observability cost of the x_j input node. The $OTF(j-i)$ is computed as

$$OTF(j-i) = \frac{N(SP:j-i)}{N(SP:j-i) + N(IP:j-i)}$$

where $N(SP:j-i)$ is the total number of distinct input combinations that sensitize a path from input x_j to output z_i and $N(IP:j-i)$ is the total number of distinct input combinations that do not sensitize a path from x_j to z_i.

Table 7.24 shows the OTF factors for several basic gates. Figure 7.68 shows a simple combinational circuit and the observability cost of each line

TABLE 7.24 Observability transfer factors for several basic logic gates

OTF values	Gate
$OTF(x_1 - Z) = 0.5$ $OTF(x_2 - Z) = 0.5$	AND (2-input)
$OTF(x_1 - Z) = 0.5$ $OTF(x_2 - Z) = 0.5$	OR (2-input)
$OTF(x_1 - Z) = 1.0$	NOT
$OTF(x_1 - Z) = 0.5$ $OTF(x_2 - Z) = 0.5$	NAND (2-input)
$OTF(x_1 - Z) = 0.5$ $OTF(x_2 - Z) = 0.5$	NOR (2-input)
$OTF(x_1 - Z) = 0.25$ $OTF(x_2 - Z) = 0.25$ $OTF(x_3 - Z) = 0.25$	AND (3-input)

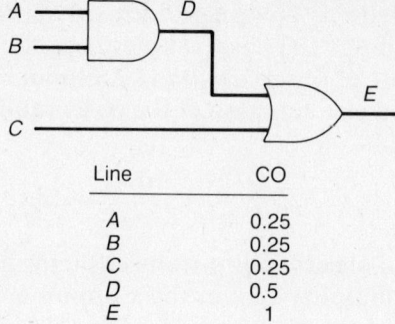

Line	CO
A	0.25
B	0.25
C	0.25
D	0.5
E	1

Fig. 7.68 Observability costs calculated using CAMELOT for a simple combinational network.

within that circuit. Figure 7.69 shows the simple circuit of Figs. 7.67 and 7.68 and the testability cost (the product of the controllability cost and the observability cost) produced by CAMELOT for each line.

7.7.3 Comparison of Testability Analyzers

Now that we have examined several testability analyzers, we need some way to compare one analyzer with another. The comparison approach that we will use is fairly intuitive. It seems reasonable that the more test patterns that exist for a given fault, the easier it will be to find one of those test patterns with a test pattern generation technique. For example, if a circuit has eight primary inputs, there will be 2^8 different primary input combinations. If Fault A is detected by only one of those input combinations and

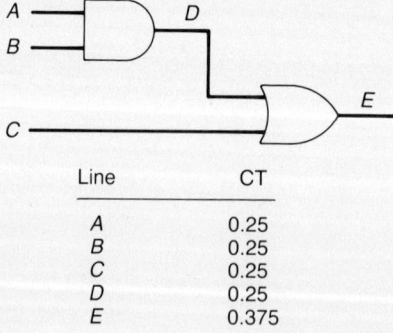

Line	CT
A	0.25
B	0.25
C	0.25
D	0.25
E	0.375

Fig. 7.69 Testability costs calculated using CAMELOT for a simple combinational circuit.

Fault B is detected by half of the input combinations, it will be much easier to find a test pattern for Fault B than for Fault A.

To provide some means of evaluating and comparing the testability analyzers, we will first determine the test patterns, if they exist, for each fault that can occur within a circuit. The number of test patterns for a particular fault divided by the total number of primary input combinations will be called the *exact testability cost* [Savir 1983]. The results obtained by both SCOAP and CAMELOT will be compared to the exact testability cost. In a similar manner, the *exact 1-controllability cost* of a line will be the number of primary input patterns that produce a 1 on that line divided by the total number of primary input patterns. The *exact 0-controllability cost* of a line will be the number of primary input patterns that produce a 0 on that line divided by the total number of primary input patterns. Finally, the *exact observability cost* of a line will be the number of primary input patterns that sensitize a path from that line to a primary output divided by the total number of primary input patterns.

The circuit that will be used for comparison is shown in Fig. 7.70. Table 7.25 shows the comparison between the testability calculated by SCOAP and CAMELOT and the exact testability.

The numbers shown in Table 7.25 cannot be compared directly because of differences in the way that they are calculated; for example, the test-

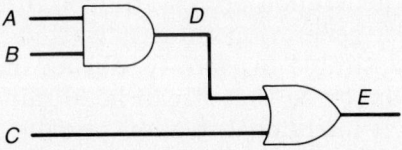

Fig. 7.70 Combinational circuit for SCOAP and CAMELOT.

TABLE 7.25 Comparison of testability costs for the circuit of Fig. 7.70.

Line	SCOAP		CAMELOT	Exact	
	CT0	CT1	CT	CT0	CT1
A	5	5	0.250	0.125	0.125
B	5	5	0.250	0.125	0.125
C	4	4	0.250	0.375	0.375
D	5	4	0.250	0.125	0.375
E	2	4	0.375	0.625	0.375

ability cost calculated by CAMELOT is between 0 and 1, whereas that produced by SCOAP is some number greater than 1. However, the results drawn from the numbers may be compared.

The results of SCOAP are somewhat encouraging when compared to the exact values. The fault E_0 is certainly the easiest fault for which a test can be derived, and this fact is indicated by $CT0(E)$ having the smallest test cost. Likewise, faults A_0, A_1, B_0, B_1 and D_0 are equally difficult to develop tests for, and the testability costs reflect this fact. Finally, C_0, C_1, D_1, and E_1 are also equally difficult to test, and the analyzer indicates this condition. Based on the results shown in Table 7.25, one must conclude that SCOAP has performed well on this circuit.

The results of CAMELOT are somewhat less appealing than those of SCOAP. The CAMELOT results in Table 7.25 indicate that nodes A, B, C, and D are equally difficult to test. As shown by the exact values, however, it should be substantially easier to find a test for line C than either lines A or B simply because of the existence of more test vectors. One difficulty with CAMELOT is that using the product of CC and CO to form CT tends to obscure true testability cost information.

7.7.4 Analysis of Circuits Containing Redundancy

The existence of redundancy in combinational circuits significantly complicates the test pattern generation process. In fact, it has been reported that almost 90% of the test generation effort is wasted because of redundant faults. Clearly, it is desirable to identify, during the design process, any unintentional redundancy so that it can be eliminated. In many cases, however, redundancy is intentional to allow circuits to tolerate faults or to remove hazards from a design. In these cases, we do not want to remove the redundancy but instead wish to modify the design to make the redundant circuit testable. In any event, we want a testability analyzer that is capable of pinpointing potential problems so that design actions may be taken.

In light of the preceding discussion, it is interesting to explore the performance of existing analyzers on a circuit that contains redundancy. The redundant circuit is shown in Fig. 7.71. This circuit is a common example of a design that uses redundant gates to eliminate static hazards. The redundant gate is the AND gate with line M as its output. As we will see momentarily, the redundancy results in a line of the circuit being untestable.

The results of applying SCOAP and CAMELOT to the circuit of Fig. 7.71 are shown in Tables 7.26 and 7.27. As can be seen, SCOAP and CAMELOT give no indication about the location of a redundancy in the circuit even though a test vector for a stuck-at-0 fault on line M does not exist. Table 7.28 shows the exact testability cost of each line in the circuit and compares it with those values generated by SCOAP and CAMELOT.

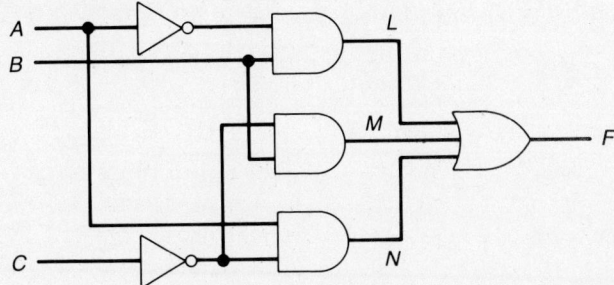

Fig. 7.71 A simple combinational circuit containing redundancy.

Summary

This chapter has presented the fundamental definitions and basic techniques necessary to perform fault testing, design for testability, and testability analysis. The three topics are intimately related. For example, testability analysis is often a means of design for testability. Also, both testability analysis and design for testability can significantly decrease the difficulty of test pattern generation. The specific topics covered include: fault tables, Boolean differences, literal propositions, the D-algorithm, test pattern generation through fault simulation, random testing, signature analysis, LSSD, Scan Path, Scan/Set, Random-Access Scan, BILBO, SCOAP, and CAMELOT. The following list highlights the important concepts developed in this chapter.

Adaptive Experiment—a test experiment where the next test pattern applied to the circuit is selected based on the circuit's response to the present test pattern.

TABLE 7.26 Testability costs from SCOAP for the circuit of Fig. 7.71

Line	CT0	CT1
A	8	8
B	8	8
C	8	8
L	8	7
M	8	7
N	8	7
F	4	7

TABLE 7.27 Testability costs from CAMELOT for the circuit of Fig. 7.71

Line	CT
A	0.0625
B	0.0625
C	0.0625
L	0.0625
M	0.0625
N	0.0625
F	0.1250

At-Speed Test—a test performed while running a device at its normal operational speed.

Boolean Difference—an algebraic expression used to determine under what conditions a function is dependent on a particular variable.

Built-in Test Techniques—test techniques incorporated into the design of a system.

Column Method—a simulation-based test pattern generation technique that partially creates the columns of the fault detection table.

Concurrent Test Techniques—test techniques that can be performed while the system is operating normally.

Controllability—the ability to set a line to a specific logic value.

TABLE 7.28 Comparison of testability costs for the circuit of Fig. 7.71

Line	SCOAP		CAMELOT	Exact	
	CT0	CT1	CT	CT0	CT1
A	8	8	0.0625	0.250	0.25
B	8	8	0.0625	0.250	0.25
C	8	8	0.0625	0.250	0.25
L	8	7	0.0625	0.125	0.50
M	8	7	0.0625	0	0.50
N	8	7	0.0625	0.125	0.50
F	4	7	0.1250	0.500	0.50

Controllability Cost — a measure of the difficulty of controlling a given line.

Design for Testability — the process of including special features to make a device easily testable.

Dynamic Test — a test of the timing characteristics of a device.

Essential Test Pattern — a test pattern that detects at least one fault that no other test pattern detects.

Experiment, Fault Detection Experiment, or Test Experiment — a collection of test patterns used to determine if a device is faulty or fault free.

External Test Techniques — test techniques that normally require test equipment outside of the device under test.

Fault Detection Table — a table containing one row for each input combination and one column for each fault. The entries in the table specify which input combinations detect which faults.

Fault Dictionary — a table that shows the response of a circuit to each test pattern under fault-free conditions and when each possible fault is present.

Fault Location Experiment — a set of test vectors used for detecting and locating faults in a circuit.

Fault Location Table — a table containing one row for each input combination, one column for each fault, and one column for each pair of faults. The entries in the table identify which faults are detected by which input combinations and which pairs of faults are distinguished by the input combinations.

Fault Testing — the process of checking for physical hardware faults.

Functional Testing — verification that a device performs its design-specified operations.

Nonconcurrent Test Techniques — test techniques that require a device to discontinue normal operations while being tested.

Observability — the ability to propagate the value on a line to an output where the value can be viewed.

Observability Cost — a measure of the difficulty of observing a given line.

Parametric Testing — verification of characteristics such as timing parameters.

Path Sensitization — the process of propagating a signal through a logic circuit to a primary output.

Preset Experiment — a collection of test patterns, all of which must be applied to complete the test experiment.

Primitive D-Cube of Fault (pdcf) — a test pattern for a logic module that brings the effect of a fault within the module to the module's output.

Propagation D-Cube (pdc) — an input combination for a module that forces the module's output to be dependent upon one of the module's inputs.

Proposition—a declarative statement that is either true or false, but never both.

Random Testing—a test experiment where test patterns are selected at random.

Row Method—a simulation-based test pattern generation technique that partially creates the rows of the fault detection table.

Scan Design—a design for testability method where all flip flops are connected to form one or more shift registers to allow easy controllability and observability of internal points within a circuit. LSSD, Scan Path, BILBO, Scan/Set Logic, and Random-Access Scan are examples of scan design.

Signature Analysis—a test procedure where the response of a device over a period of time is compressed into characteristic values called *signatures*. The signatures of the device under test are compared to those of a known fault-free device.

Singular Cover—a reduced truth table that completely characterizes the response of a device.

Static Test—a test of the steady-state characteristics of a device.

Test, Test Pattern, Test Vector, or Fault Detection Test—a primary input combination that causes a device to produce an erroneous output when a fault is present.

Testability—the ability to find test vectors for faults on a specific line.

Testability Analysis—the process of characterizing the ease with which a circuit can be tested.

Testability Analyzers—programs capable of estimating the cost of testability. Examples include SCOAP, COPTR, TMEAS, CAMELOT, PREDICT, and VICTOR.

Testability Cost—a measure of the difficulty of testing a given line.

Test Pattern Generation—the process of determining a set of test patterns.

References

1. Ando, H. "Testing VLSI with random access scan," *Digest of Papers of COMPCON '80*, February 1980, pp. 50–52.
2. Aylor, J.H., B.W. Johnson, and B.J. Rector. "Structured design for testability in semicustom VLSI," *IEEE Micro*, Vol. 6, No. 1, February 1986, pp. 51–58.
3. Bennetts, R.G. *Design of Testable Logic Circuits*, Addison-Wesley Publishing Company, Reading, Mass., 1984.
4. Breuer, M.A., and A.D. Friedman. *Diagnosis and Reliable Design of Digital Systems*, Computer Science Press, Woodland Hills, Calif., 1976.

5. Buehler, M.G., and M.W. Sievers. "Off-line, built-in test techniques for VLSI circuits," *Computer*, Vol. 15, No. 6, June 1982, pp. 69–82.
6. Calma Company Technical Report. "COPTR/ATG technical summary," 1984.
7. Chiang, A.C., I.S. Reed, and A.V. Banes. "Path sensitization, partial Boolean difference, and automated fault diagnosis," *IEEE Transactions on Computers*, Vol. C-21, No. 2, February 1972, pp. 189–195.
8. David, R., and G. Blanchet. "About random fault detection of combinational networks," *IEEE Transactions on Computers*, Vol. C-25, No. 6, June 1976, pp. 659–664.
9. Eichelberger, E.B., and T.W. Williams. "A logic design structure for LSI testability," *Proceedings of the 14th Design Automation Conference*, New Orleans, June 1977, pp. 462–468.
10. Funatsu, S., N. Wakatsuki, and T. Arima. "Test generation systems in Japan," *Proceedings of the 12th Design Automation Conference*, June 1975, pp. 114–122.
11. Goel, P., and P.R. Moorby. "Fault simulation techniques for VLSI circuits," *VLSI Design*, Vol. 5, No. 7, July 1984.
12. Goldstein, L.H. "Controllability/observability analysis of digital circuits," *IEEE Transactions on Circuits and Systems*, Vol. CAS-26, No. 9, September 1979, pp. 685–693.
13. Grason, J. "TMEAS—A testability measurement program," *Proceedings of the 16th IEEE Design Automation Conference*, 1979, pp. 156–161.
14. Koenemann, B., J. Mucha, and G. Zwiehoff. "Built-in logic block techniques," *Digest of Papers of the 1979 Test Conference*, Cherry Hill, N.J., October 23–25, 1979, pp. 37–41.
15. Kohavi, Z. *Switching and Finite Automata Theory*, McGraw-Hill, New York, 1978.
16. McCluskey, E.J. *Logic Design Principles with Emphasis on Testable Semicustom Circuits*, Prentice-Hall, New York, 1986.
17. Nadig, H.J. "Signature analysis: Concepts, examples, and guidelines," *Hewlett-Packard Journal*, May 1977, pp. 15–21.
18. Poage, J.F. "Derivation of optimum tests to detect faults in combinational circuits," *Proceedings of Mathematical Theory of Automata*, New York, Polytechnic Press of Polytechnic Institute of Brooklyn, New York, 1963, pp. 483–528.
19. Ratiu, I.M., A.S. Sangiovanni-Vincentelli, and D.O. Pederson. "VICTOR—A fast VLSI testability analysis program," *Proceedings of the 1982 IEEE Test Conference*, Philadelphia, November 15–18, 1982, pp. 397–401.
20. Roth, J.P. "Diagnosis of automata failures: A calculus and a method," *IBM Journal of Research and Development*, Vol. 10, No. 7, July 1966, pp. 278–291.
21. Savir, J. "Good controllability and observability do not guarantee good testability," *IEEE Transactions on Computers*, Vol. C-32, No. 12, December 1983, pp. 1198–1200.
22. Sellers, F.F., M.Y. Hsiao, and C.L. Bearnson. "Analyzing errors with the Boolean difference," *IEEE Transactions on Computers*, Vol. C-17, No. 7, July, 1968, pp. 676–683.

23. Seth, S.C., L. Pan, and V.D. Agrawal. "PREDICT—Probabilistic estimation of digital circuit testability," *Proceedings of the 15th Annual International Symposium on Fault-Tolerant Computing*, Ann Arbor, Mich., June 1985, pp. 220–225.
24. Stewart, J.H. "Future testing of large LSI circuit cards," *Digest of Papers of the 1977 Semiconductor Test Symposium*, October 1977, pp. 6–17.
25. Williams, T.W., and K.P. Parker. "Design for testability—A survey," *Proceedings of the IEEE*, Vol. 71, No. 1, January 1983, pp. 98–112.

Additional Reading

The following references provide additional reading material on the topics, as well as related ones, discussed in this chapter. These references cover a broad selection of material for the reader interested in further pursuing testing.

Agarwal, P., and V. Agarwal. "Probabilistic analysis of random test generation method for irredundant combinational logic circuits," *IEEE Transactions on Computers*, Vol. C-24, No. 7, July 1975, pp. 691–695.

Agarwal, V.D. "When to use random testing," *IEEE Transactions on Computers*, Vol. C-27, No. 11, November 1978, pp. 1054–1055.

Agarwal, V.K., and A.S. Fung. "Multiple fault testing of logic circuits by single fault test sets," *IEEE Transactions on Computers*, Vol. C-30, No. 11, November 1981, pp. 854–855.

Armstrong, D.B. "On finding a nearly minimal set of fault detection tests for combinational logic nets," *IEEE Transactions on Electronic Computers*, Vol. C-15, No. 2, February 1966, pp. 66–73.

Chin, C.K., and E.J. McCluskey, "Test length for pseudo random testing," *Proceedings of the 1985 International Test Conference*, November 1985, pp. 94–99.

Eichelberger, E.B., and T.W. Williams. "A logic design structure for LSI testability," *Journal of Design Automation and Fault Tolerant Computing*, Vol. 2, No. 2, May 1978, pp. 165–178.

Fasang, P.P., J.P. Shen, M.A. Schuette, and W.A. Gwaltney. "Automated design for testability of semicustom integrated circuits," *Proceedings of the 1985 International Test Conference*, November 1985, pp. 558–564.

Goel, P. "An implicit enumeration algorithm to generate tests for combinational logic circuits," *IEEE Transactions on Computers*, Vol. C-30, No. 3, March 1981, pp. 215–222.

Hayes, J.P. "Rapid count testing for combinational logic circuits," *IEEE Transactions on Computers*, Vol. C-25, No. 6, June 1976, pp. 613–620.

Hayes, J.P. "A NAND model for fault diagnosis in combinatorial logic circuits," *IEEE Transactions on Computers*, Vol. C-20, No. 12, December 1971, pp. 1496–1506.

Hayes, J.P. "Transition count testing of combinational logic networks," *IEEE Transactions on Computers*, Vol. C-25, No. 6, June 1976, pp. 613–620.

Hung, A.C., and F.C. Wang. "A method of test generation directly from testability analysis," *Proceedings of the 1985 International Test Conference*, November 1985, pp. 62–78.

Hewlett-Packard Ltd. "A designer's guide to signature analysis," Application Note 222, 1977.

Keiner, W.L., and R.P. West. "Testability measures," *Proceedings of AUTOTESTCON '77*, November 1977.

Krishnamurthy, B., and R.L. -C. Sheng. "A new approach to the use of testability analysis," *Proceedings of the 1985 International Test Conference*, November 1985, pp. 769–778.

Ku, C.T., and G.M. Masson. "The Boolean difference and multiple fault analysis," *IEEE Transactions on Computers*, Vol. C-24, No. 7, July 1975, pp. 691–695.

McCluskey, E.J., and F.W. Clegg. "Fault equivalence in combinational logic networks," *IEEE Transactions on Computers*, Vol. C-20, No. 11, November 1971, pp. 1286–1293.

Muehldorf, E.I., and A.D. Savkar. "LSI logic testing— An overview," *IEEE Transactions on Computers*, Vol. C-30, No. 1, January 1981, pp. 1–17.

Richman, J., and K.R. Bowden. "The modern fault dictionary," *Proceedings of the 1985 International Test Conference*, November 1985, pp. 696–702.

Savir, J., G.S. Ditlow, and P.H. Bardell. "Random pattern testability," *IEEE Transactions on Computers*, Vol. C-33, No. 1, January 1984, pp. 79–90.

Schneider, P.R. "On the necessity to examine D-chains in diagnostic test generation: An example," *IBM Journal of Research and Development*, November 1967, pp. 114.

Turner, M.E., D.G. Leet, R.J. Prilik, and D.J. McLean. "Testing CMOS VLSI: Tools, concepts, and experimental results," *Proceedings of the 1985 International Test Conference*, November 1985, pp. 322–328.

Williams, T.W., and K.P. Parker. "Testing logic networks and design for testability," *Computer*, Vol. 12, No. 10, October 1979, pp. 9–21.

Williams, T.W. "VLSI testing," *Computer*, Vol. 17, No. 10, October 1984, pp. 126–136.

Wadsack, R.L. "Fault coverage in digital integrated circuits," *The Bell System Technical Journal*, Vol. 57, No. 5, May-June 1978, pp. 1475–1488.

Wadsack, R.L. "Fault modeling and logic simulation of CMOS and MOS integrated circuits," *The Bell System Technical Journal*, Vol. 57, No. 5, May-June 1978, pp. 1449–1474.

Problems

7.1 For the circuit shown in Fig. 7.72 determine a preset fault location experiment capable of locating all distinguishable stuck-at-0 faults. Show the corresponding fault dictionary. Now design an adaptive experiment to locate all of the stuck-at-0 faults.

7.2 The perfect adaptive experiment is one in which the same number of test patterns are required to locate each fault. In other words, fault 1 is located with k test patterns, fault 2 is located with k test patterns, the fault-free condition is identified after k test patterns, and so on. Develop a relationship between k and n, where n is the number of lines within the circuit and k is the number of test patterns as mentioned above. Now develop a relationship between k and n for the best-case, preset fault location experiment and compare it to that for the perfect adaptive experiment.

7.3 Use the literal proposition method to determine the minimal set of test patterns that will detect all single stuck-at-1 faults in the circuit shown in Fig. 7.73.

7.4 Determine the primitive d-cubes of fault (pdcfs) and the propagation d-cubes (pdcs) for the logic block shown in Fig. 7.75.

7.5 Use the Boolean difference to determine all test patterns (if any exist) that detect the multiple fault condition x_4 stuck-at-1 and x_3 stuck-at-0 in the circuit shown in Fig. 7.75.

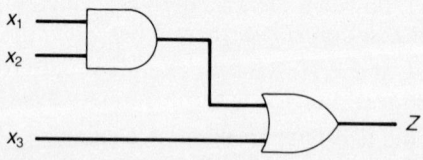

Fig. 7.72

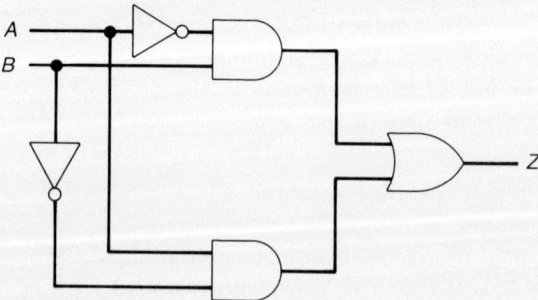

Fig. 7.73

Problems 573

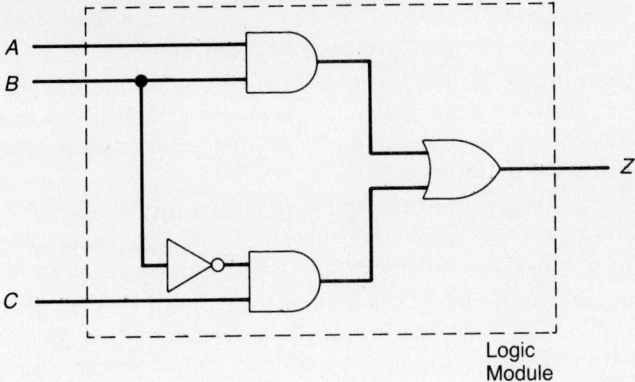

Fig. 7.74

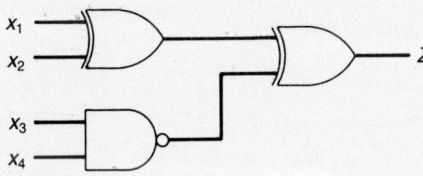

Fig. 7.75

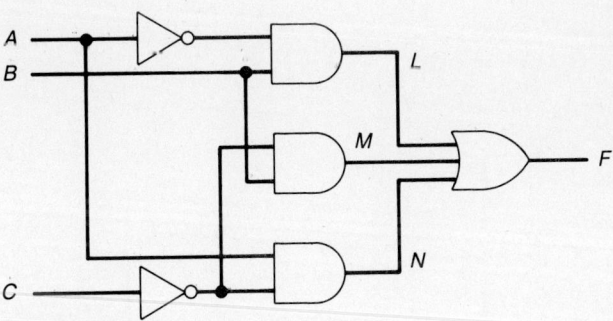

Fig. 7.76

7.6 The circuit shown in Fig. 7.76 is a hazard-free implementation of the function $Z(A,B,C) = AB' + BC'$. Use the Boolean difference to determine whether or not a test for line M stuck-at-0 exists. If a test does exist, simply show the test pattern. If, however, a test pattern does not exist, explain the feature of this circuit that prevents a test pattern from being found.

7.7 For the combinational circuit shown in Fig. 7.77, use the *D*-algorithm to determine a test for line 5 being stuck-at-0. Be sure to show all the steps of the algorithm. What other (if any) stuck-at-1 or stuck-at-0 faults will be detected by the resulting test pattern?

7.8 Use the SCOAP algorithm to estimate the testability of each line in the circuit of Fig. 7.78. Repeat the problem using the CAMELOT technique, and compare the results produced by the two approaches.

7.9 The circuit shown in Fig. 7.79 is the implementation of a simple counter. First, convert the circuit to an LSSD implementation and then show the Scan Path implementation. Using the method of your choice, develop the test patterns necessary to test the circuit for stuck-type faults. Explain the procedure for testing the resulting LSSD (and Scan Path) circuit.

7.10 The combinational circuit shown in Fig. 7.80 has been designed by an engineer within your section. Your job is to improve the testability of the circuit by selecting test points that will be connected to package pins to improve observability. Unfortunately, you are restricted by pin limitations to two test points. Show the two test points that you would select, and justify your answer.

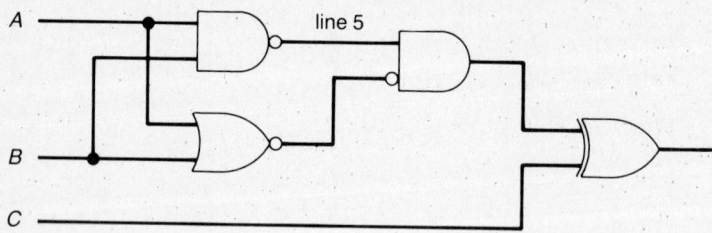

Fig. 7.77

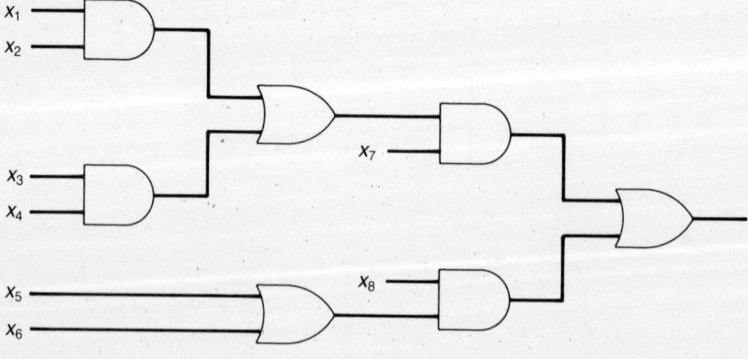

Fig. 7.78

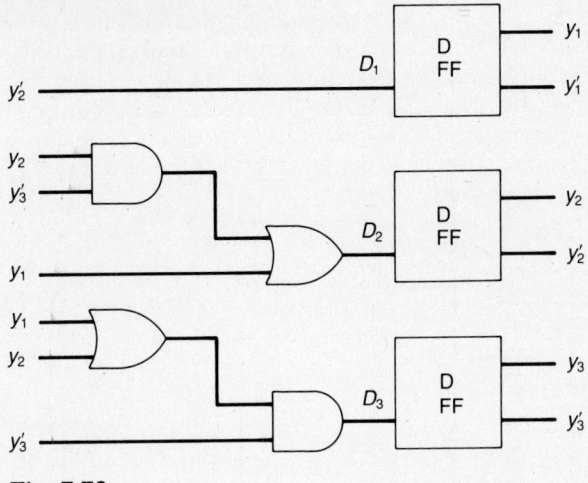

Fig. 7.79

7.11 Consider the circuit of Fig. 7.80 once again. Suppose that you wanted to use the signature analysis approach to testing, and you are limited to two signature analysis registers for the circuit. Using 4-bit signature analysis registers, illustrate the concept of signature analysis in the circuit of Fig. 7.80. Be sure to show the logical circuit for your signature analysis registers. How many test patterns should be applied to create the signatures, and what are those patterns? What is the coverage obtained for stuck-type faults?

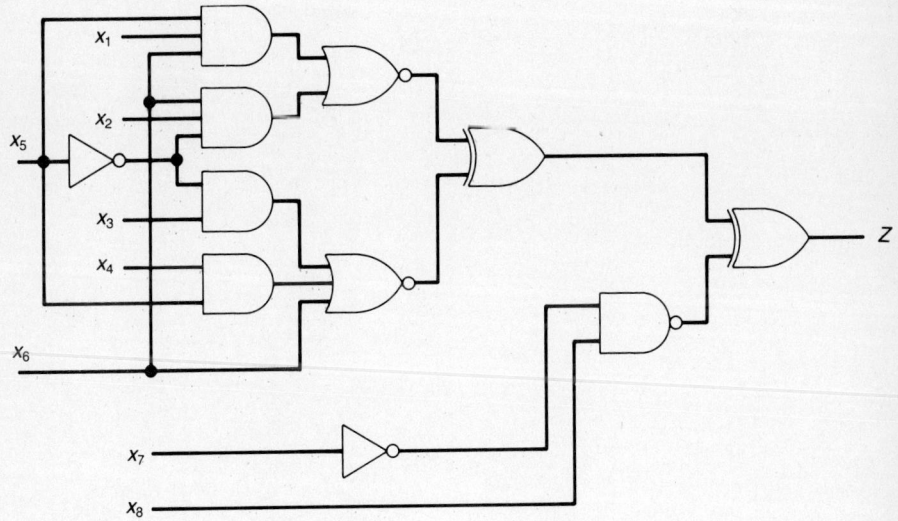

Fig. 7.80

7.12 The circuit shown in Fig. 7.81 is an implementation of a triple modular redundant half-adder. The voter used is simply a digital, majority voter. Use the SCOAP testability analysis technique to determine the testability cost of each line in the circuit. Does SCOAP indicate that there will be any problems in testing any points within the circuit? Are there any problems with testing any points within the circuit? If there are problems, how could you improve the testability of those lines?

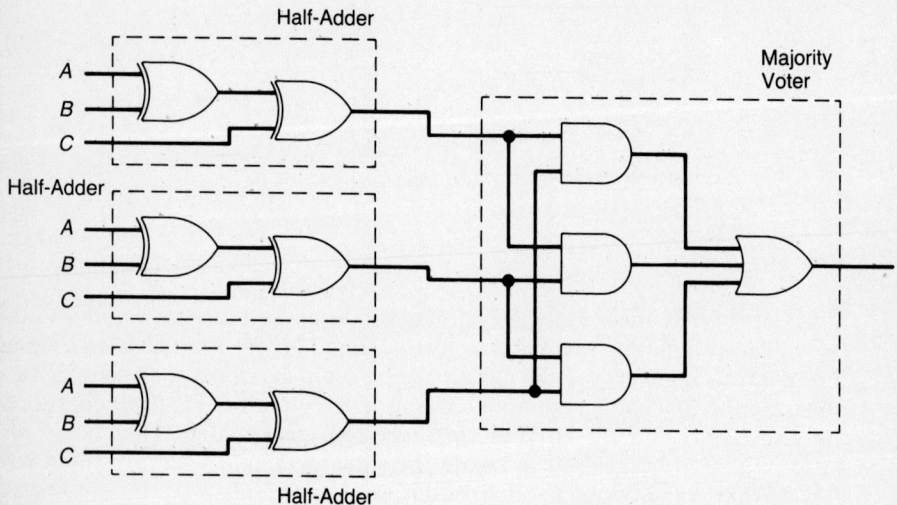

Fig. 7.81

Index

A

Active hardware redundancy, 62–69
  cold standby sparing, 66
  duplication with comparison, 63–65
  hot standby sparing, 66
  pair-and-a-spare technique, 67–68
  standby sparing, 65–67
  watchdog timer, 68–69
Adaptive experiment, 477–481
Agusta A129 Integrated Multiplex
    System, 336–345
Algorithm-based fault tolerance,
    430–438
Alternating logic, 137–139
AN codes, 112–115
Area defects, 439
Arithmetic codes, 112–123
  AN codes, 112–115
  residue codes, 115–119
  inverse-residue codes, 119
  residue number system, 119–123
At-speed test, 465
August Systems CS-3001 Control
    Computer, 332–333
Availability, 5
Availability models, 219–223

B

Bathtub curve, 172–173
Bell Electronic Switching System, 13,
    350–353

Berger codes, 123–125
Binary code, 81
Binary tree, 478
Bit-per-byte parity, 87, 88
Bit-per-chip parity, 87, 88
Bit-per-multiple-chips parity, 87, 88
Bit-per-word parity, 87
Boolean difference, 481–488
Built-in Logic Block Observation
    (BILBO), 544–552
Built-in test techniques, 464
Burn-in, 172
Burst error, 103

C

CAMELOT, 553, 559–562
Capability check, 154
Chain rule, 486–488
Checksums, 98–102
  double-precision checksum, 100
  Honeywell checksum, 100
  residue checksum, 101
  single-precision checksum, 99
Code, 81
Code disjoint, 397
Code distance, 83
Code polynomial, 104
Code selection issues, 133
Code word, 81
Cold standby sparing, 66
Column method, 513
Combinatorial models, 185–199

578   Index

parallel systems, 185, 189–193
series systems, 185, 186–189
Common-mode failure, 387–389
Compile-time reconfiguration, 409, 410–422
   fault stealing replacement strategy, 413–417
   repair-most replacement strategy, 417–422
   rippling replacement strategy, 411–413
Complemented duplication, 95–97
Component defects, 29
Conceptual design review, 275
Conceptual phase, 266
Concurrent test techniques, 464
Consistency operation, 498
Consistency check, 153
Continuous-time Markov model, 205
Controllability, 467, 553
Controllability cost, 553–554
Convenience functions, 268
COPTR, 552
Critical-computation application, 10, 319–345
   August Systems CS-3001 Control Computer, 332–333
   Agusta A129 Integrated Multiplex Ssytem, 336–345
   Fault-Tolerant Multiprocessor (FTMP), 324–329
   Multi-Microprocessor Flight Control System (MMFCS), 333–336
   Software implemented fault tolerance (SIFT), 329–332
   space shuttle, 319–324
Cyclic code, 102–112

D

D-Algorithm, 498–510
   D-drive, 498
   Primitive D-cube of fault (pdcf), 498, 505–508
   Propagation D-cube (pdc), 500, 508–509
D-drive, 498

Decoding process, 82
Dependability, 8
Design diversity, 389
Design for testability, 2, 466, 520–552
Design phase, 266
Design process, 265–273
Design rules, 277
Detailed design review, 276
Determinate fault, 31
Diffusion, 380
Discrete component, 377
Discrete-time Markov model, 205
Documentation, 277
Double-precision checksum, 100
Duplication codes, 95–98
Duplication with comparison, 63–65
Duplication with complementary logic, 390, 391–394
Dynamic test, 465

E

Easily-testable system, 1
Encoding process, 82
Error, 24
Error containment, 40
Error correcting codes, 83
   hamming error-correcting codes, 127–131
   horizontal and vertical parity, 125–126
Error correcting memory chips, 131–133
Error detecting codes, 82
   arithmetic codes, 112–123
   Berger codes, 123–125
   checksums, 98–102
   cyclic codes, 102–112
   duplication codes, 95–98
   m-of-n codes, 93–95
   parity codes, 84–93
   residue codes, 115–119
   swap-and-compare code, 97
Error detection, 40
Error latency, 27
Error location, 40
Error models, 37

Error recovery, 40
Esential test pattern, 472
Exact observability cost, 563
Exact testability cost, 563
Experiment, 467
Exponential failure law, 174
External disturbances, 28
External test, 464
External universe, 27

F

Fabrication-time reconfiguration, 409, 410–422
    fault stealing replacement strategy, 413–417
    repair-most replacement strategy, 417–422
    rippling replacement strategy, 411–413
Failure, 24
Failure density function, 172
Failure modes, 378–383
    Diffusion, 380
    Foreign material, 381
    Metal systems, 379
    misapplication, 382
    mounting, 382
    packaging and bonding, 382
    oxide, 381
Failure rate, 170–175
Failure rate calculation, 175–178
Failure rate function, 172
Fault, 24
Fault avoidance, 15, 38, 273–277
    conceptual design review, 275
    design rules, 277
    detailed design review, 276
    documentation, 277
    final review, 276
    parts selection, 276
    requirements design review, 274
    specifications design review, 275
Fault cause, 28
    component defects, 29
    external disturbances, 28
    implementation mistakes, 28
    specification mistakes, 28
Fault containment, 40
Fault containment coverage, 182
Fault coverage, 182–185, 193–196
Fault detection, 39
Fault detection coverage, 182
Fault detection experiment, 467
Fault detection table, 468–477
Fault detection test, 467
Fault dictionary, 474
Fault distributions, 383–385
Fault duration, 30
Fault extent, 31
Fault latency, 27
Fault location, 39
Fault location coverage, 182
Fault location experiment, 475–477
Fault location table, 468
Fault masking, 38
Fault model, 31
    Logical stuck-fault model, 32
    transistor stuck-fault model, 37
Fault nature, 30
Fault recovery, 40
Fault recovery coverage, 182
Fault secure, 395
Fault simulation, 510–516
    column method, 513
    row method, 513
Fault stealing replacement strategy, 413–417
Fault testing, 464, 466–468
Fault tolerance, 1, 39
Fault-tolerant building block computer (FTBBC), 315–319
Fault-tolerant computing, 1
Fault-Tolerant Multiprocessor (FTMP), 324–329
Fault-Tolerant Spaceborne Computer (FTSC), 310–314
Fault-tolerant system, 1
Fault value, 31
Final review, 276
Flexibility, 228
Flight-critical functions, 268
Flux summing, 61, 80
Foreign material, 381
Four-universe model, 26

Functional testing, 464

G

Generator polynomial, 105
Graceful degradation, 7

H

Hamming distance, 83
Hamming error-correcting codes, 127–131
Hardware redundancy, 49, 51–81
   active hardware redundancy, 62–69
   hybrid hardware redundancy, 69–80
   passive hardware redundancy, 51–62
Hazard function, 172
Hazard rate, 172
High availability application, 13, 345–355
   Bell Electronic Switching system, 13, 350–353
   Stratus/32 System, 348–350
   Synapse N+1 System, 353–355
   Tandem 16 NonStop System, 346–348
Honeywell checksum, 100
Horizontal and vertical parity, 125–126
Hot standby sparing, 66
Hybrid hardware redundancy, 69–80
   N-modular redundancy (NMR) with spares, 70–71
   Self-purging redundancy, 71–75
   Sift-out modular redundancy, 75–78
   Triple-duplex architecture, 78–80

I

Implementation mistakes, 28
Indeterminate fault, 31
Infant mortality phase, 172
Information redundancy, 49, 81–134
   Arithmetic codes, 112–123

   Berger codes, 123–125
   Checksums, 98–102
   Code selection issues, 133–134
   Cyclic codes, 102–112
   Duplication codes, 95–97
   Error correcting integrated circuits, 131–133
   Hamming error-correcting codes, 127–131
   Horizontal and vertical parity, 125–126
   Parity codes, 84–93
   m-of-n codes, 93–95
Informational universe, 27
Intel 432, 14
Integrated circuit, 377
Interlaced parity, 87, 90
Intermittent fault, 30
Inverse-residue codes, 119

L

Large-scale Integration (LSI), 376–377
Latent fault, 27
Level Sensitive Scan Design (LSSD), 525–529
Linear array, 406
Linear Feedback Shift Register (LFSR), 518–520
Literal propositions, 488–491
Logical stuck-fault model, 32
Long-life application, 9, 304–319
   Fault-Tolerant Bulding Block Computer (FTBBC), 315–319
   Fault-Tolerant Spaceborne Computer (FTSC), 310–314
   Self-Testing and Repairing (STAR) Computer, 305–310
Loose synchronization, 335

M

Maintainability, 7
Maintainability models, 223–226
Maintenance postponement application, 11

Majority voter, 56
Malfunction, 24
Markov models, 199–214
  continuous-time Markov models, 205
  discrete-time Markov models, 205
Mean time between failure (MTBF), 180–182
Mean time to failure (MTTF), 178–180
Mean time to repair (MTTR), 180
Medium-scale Integration (MSI), 376–377
Metal systems, 379
Mid-value select, 60
MIL-HDBK-217, 175
Misapplication, 382
Mission time (MT[r]), 218
Mission time improvement, 219
Mission-critical functions, 268
m-of-n codes, 93–95
M-of-N system, 197–199
  N-modular redundancy (NMR), 54
  triple-modular redundancy (TMR), 52–53
Mounting, 382
Multi-dimensional path sensitization, 492
Multi-Microprocessor Flight Control System, 333–336

N

Near-neighbor mesh, 406
N-modular redundancy (NMR), 54
N-modular redundancy (NMR) with spares, 70–71
Nonconcurrent test techniques, 464
Nonlatching design, 338
Nonseparable code, 84
Nonseparable cyclic code, 102–111
Nonredundant circuits, 400
N-version programming, 154–155

O

Observability, 467, 554
Observability cost, 554

One-dimensional path sensitization, 491
Overlapped parity, 90–93
Oxide, 381

P

Package and bonding, 382
Pair-and-a-spare technique, 67–68
Parallel system, 185, 189–193
Parametric testing, 465
Parity code, 84–93
  Bit-per-byte parity, 87, 88
  Bit-per-chip parity, 87, 88
  Bit-per-multiple-chips parity, 87, 88
  Bit-per-word parity, 87
  Single-bit parity code, 84–87
Parity checker, 86
Parity generator, 86
Partitioning, 267–269
Parts selection, 276
Passive hardware redundancy, 51–62
  fault masking, 38
  N-modular redundancy (NMR), 54
  triple-modular redundancy (TMR), 52–53
Path sensitization, 468, 491–498
Performability, 6
Permanent fault, 30
Permanent fault detection, 136–137
Physical universe, 26
Polarity-Hold Addressable Storage Element (PH-ASE), 537
PREDICT, 553
Preset experiment, 477
Primitive D-cube of fault (pdcf), 498, 505–508
Problem definition, 266
Propagation D-cube (pdc), 500, 508–509
Proposition, 488

R

Random-Access Scan, 535–538
Random testing, 516–517

Real-time reconfiguration, 409,
    422–438
  Algorithm-based fault tolerance,
    430–438
  Successive column elimination
    reconfiguration, 429
  Successive row elimination
    reconfiguration, 423–429
  Successive row and column
    elimination reconfiguration,
    429–430
Reconfigurable arrays, 390, 404–438
  Compile-time reconfiguration, 409,
    410–422
  Fabrication-time reconfiguration,
    409, 410–422
  Real-time reconfiguration, 409,
    422–438
Reconfiguration, 39
Recomputing for error correction,
  151–152
Recomputing with duplication with
  comparison (REDWC), 145–151
Recomputing with shifted operands
  (RESO), 139–144
Recomputing with swapped operands
  (RESWO), 144–145
Redundancy, 49
  hardware redundancy, 49, 51–81
  information redundancy, 49, 81–134
  software redundancy, 49, 152–155
  time redundancy, 49, 134–152
Redundancy ratio, 226
Reliability, 4, 171
Reliability block diagram, 186
Repair levels, 225
Repair-most replacement strategy,
  417–422
Repair rate, 224
Repair rate function, 224
Requirements, 267, 271
Requirements design review, 274
Requirements phase, 266
Residue checksum, 101
Residue code, 115–119
Residue number system, 119–123
Restoring organ, 53
Rippling replacement strategy,
  411–413
Row method, 513

S

Safety, 6
Safety modeling, 214–216
Scan design, 524–538
  Level Sensitive Scan Design (LSSD),
    525–529
  Random-Access Scan, 535–538
  Scan path, 529–532
  Scan/Set logic, 532–535
Scan Path, 529–532
Scan/Set logic, 532–535
SCOAP, 553, 555–559
Self-checking computer module
  (SCCM), 316–319
Self-checking logic, 390, 394–402
Self-purging redundancy, 71–75
Self-testing, 395
Self-Testing and Repairing Computer
  (STAR), 305–310
Separable code, 84
Separable cyclic code, 111–112
Series system, 185, 186–189
Set/Reset Addressable Storage
  (SR-ASE), 538
Sift-out modular redundancy, 75–78
Signature analysis, 517–520
Single-bit parity code, 84–87
Single-path sensitization, 491
Single point of failure, 52
Single-precision checksum, 99
Singular cover, 501, 505
Small-scale integration (SSI), 376–377
Software implemented fault tolerance
  (SIFT), 329–332
Software redundancy, 49, 152–155
  capability checks, 154
  consistency checks, 153
  N-version programming, 154–155
Software voting, 58
Space shuttle, 319–324
Specifications, 271
Specifications design review, 275
Specifications mistakes, 28

Specifications phase, 266
Spot defects, 439
Standby sparing, 65–67
State transition, 200
Static test, 465
Steady-state availability, 221
Stratus/32 System, 348–350
Stuck-open fault, 34
Sub-Element Redundant
    Fault-Tolerant (SERF) Computer,
    311–314
Successive column elimination
    reconfiguration, 429
Successive row and column
    elimination reconfiguration,
    429–430
Successive row elimination
    reconfiguration, 423–429
Swap-and-compare code, 97
Syndrome polynomial, 108
Synapse N+1 system, 353–355
System integration, 272
System state, 200
Systolic array, 406

T

Tandem 16 NonStop system, 346–348
Technology dependence, 228
Test, 7, 467
Testability, 8, 229, 554
Testability analysis, 466, 552–554
Testability analyzers, 552, 554–565
    CAMELOT, 553, 559–562
    COPTR, 552
    PREDICT, 553
    SCOAP, 553, 555–559
    TMEAS, 553
    VICTOR, 553
Testability cost, 554
Testability measurement, 552
Test cube, 494
Test experiment, 467
Test pattern, 467
Test pattern generation, 468–515
    adaptive experiments, 477–481
    Boolean differences, 481–488
    D-algorithm, 498–510

fault simulation, 510–515
    fault tables, 468–477
    literal propositions, 488–491
    Path sensitization, 491–498
Test phase, 266
Test vector, 467
Three-universe model, 26
Threshold gate, 72
Tight synchronization, 335
Time redundancy, 49, 134–152
    alternating logic, 137–139
    permanent fault detection, 136–151
    recomputation for error correction,
        151–152
    recomputing with duplication with
        comparison (REDWC), 145–151
    recomputing with shifted operands
        (RESO), 139–144
    recomputing with swapped
        operands (RESWO), 144–145
    Transient fault detection, 135–136
TMEAS, 553
Totally self-checking logic, 395
Totally self-checking checkers,
    402–404
Tradeoffs, 264
Transient fault, 30
Transient fault detection, 135–136
Transistor stuck-fault model, 37
Transparency to user, 229
Triple-duplex architecture, 78–80
Triple modular redundancy (TMR),
    52–53
Triplicated voters in TMR, 53
Two-rail checkers, 398–402

U

United Data Systems (UDS), 318–319
Unreliability, 4, 171
Useful life phase, 172
User's universe, 27

V

Very large-scale integration (VLSI),
    375–378
VICTOR, 553

Voting techniques, 54–59
Voyager, 10

W

Watchdog timer, 68–69
Wear-out phase, 172
Weibull distribution, 174

X

X-29 Flight Control system, 12

Y

Yield, 439
Yield enhancement, 390, 439–451

ADDISON-WESLEY ∧ THE SIGN
AND COMPUTER ENGINEERING
OF EXCELLENCE IN ELECTRICA
ADDISON-WESLEY ∧ THE SIGN
AND COMPUTER ENGINEERING
OF EXCELLENCE IN ELECTRICA
ADDISON-WESLEY ∧ THE SIGN
AND COMPUTER ENGINEERING
OF EXCELLENCE IN ELECTRICA
ADDISON-WESLEY ∧ THE SIGN
AND COMPUTER ENGINEERING
OF EXCELLENCE IN ELECTRICA
ADDISON-WESLEY ∧ THE SIGN
AND COMPUTER ENGINEERING
OF EXCELLENCE IN ELECTRICA
ADDISON-WESLEY ∧ THE SIGN
AND COMPUTER ENGINEERING
OF EXCELLENCE IN ELECTRICA
ADDISON-WESLEY ∧ THE SIGN
AND COMPUTER ENGINEERING
OF EXCELLENCE IN ELECTRICA
ADDISON-WESLEY ∧ THE SIGN
AND COMPUTER ENGINEERING
OF EXCELLENCE IN ELECTRICA
ADDISON-WESLEY ∧ THE SIGN
AND COMPUTER ENGINEERING
OF EXCELLENCE IN ELECTRICA